Fractional Order Systems with Application to Electrical Power Engineering

Fractional Order Systems with Application to Electrical Power Engineering

Editors

Arman Oshnoei
Behnam Mohammadi-Ivatloo

Basel • Beijing • Wuhan • Barcelona • Belgrade • Novi Sad • Cluj • Manchester

Editors
Arman Oshnoei
Department of Energy,
Aalborg University
Aalborg
Denmark

Behnam Mohammadi-Ivatloo
LUT School of Energy
Systems, LUT University
Lappeenranta
Finland

Editorial Office
MDPI
St. Alban-Anlage 66
4052 Basel, Switzerland

This is a reprint of articles from the Special Issue published online in the open access journal *Fractal and Fractional* (ISSN 2504-3110) (available at: https://www.mdpi.com/journal/fractalfract/special_issues/electrical_power_engineering).

For citation purposes, cite each article independently as indicated on the article page online and as indicated below:

Lastname, A.A.; Lastname, B.B. Article Title. *Journal Name* **Year**, *Volume Number*, Page Range.

ISBN 978-3-7258-0489-4 (Hbk)
ISBN 978-3-7258-0490-0 (PDF)
doi.org/10.3390/books978-3-7258-0490-0

Contents

About the Editors

Arman Oshnoei

Arman Oshnoei (member of the IEEE) received his Ph.D. in Electrical Engineering from Shahid Beheshti University, Tehran, Iran, in 2021. From August 2021 to March 2022, he was a research assistant at the Department of Energy, Aalborg University, Denmark. From May 2022 to October 2023, he was a post-doctoral research fellow at Aalborg University, where he is currently an assistant professor. His current research interests include the control and stability of power electronic-based power systems, energy storage systems, and intelligent control. He was selected and awarded by the National Elite Foundation of Iran in 2019. He was the recipient of the Outstanding Researcher Award from Shahid Beheshti University in 2022.

Behnam Mohammadi-Ivatloo

Dr. Behnam Mohammadi-Ivatloo, Ph.D., is a professor at the University of Tabriz, Tabriz, Iran, where he is the head of the Smart Energy Systems Lab. He also has research experience at Aalborg University, Aalborg, Denmark, and the Institute for Sustainable Energy, Environment and Economy, University of Calgary, Canada. He obtained his MSc degree and Ph.D. in Electrical Engineering from the Sharif University of Technology, Tehran, Iran. He has a mix of high-level experience in research, teaching, administration, and voluntary jobs at national and international levels. He is the PI or Co-PI of 20 externally funded research projects. He has been a senior member of the IEEE since 2017 and a member of the Governing Board of the Iran Energy Association since 2013, where he was elected as president in 2019. His main areas of interest are integrated energy systems, renewable energies, energy storage systems, microgrid systems, and smart grids.

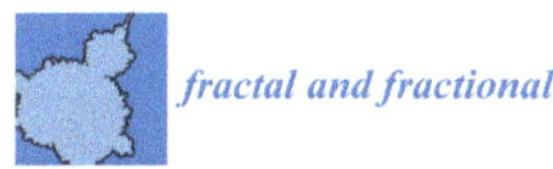 *fractal and fractional*

Review

Fractional-Order Control Techniques for Renewable Energy and Energy-Storage-Integrated Power Systems: A Review

Masoud Alilou [1], Hatef Azami [1], Arman Oshnoei [2], Behnam Mohammadi-Ivatloo [1,3,*] and Remus Teodorescu [2]

[1] Faculty of Electrical and Computer Engineering, University of Tabriz, Tabriz 51666-16471, Iran; masoud.alilou@yahoo.com (M.A.); hatef.azami1992@gmail.com (H.A.)
[2] Department of Energy, Aalborg University, 9220 Aalborg, Denmark; aros@energy.aau.dk (A.O.); ret@energy.aau.dk (R.T.)
[3] Department of Electrical Engineering, School of Energy Systems, LUT University, 53850 Lappeenranta, Finland
* Correspondence: mohammadi@ieee.org

Abstract: The worldwide energy revolution has accelerated the utilization of demand-side manageable energy systems such as wind turbines, photovoltaic panels, electric vehicles, and energy storage systems in order to deal with the growing energy crisis and greenhouse emissions. The control system of renewable energy units and energy storage systems has a high effect on their performance and absolutely on the efficiency of the total power network. Classical controllers are based on integer-order differentiation and integration, while the fractional-order controller has tremendous potential to change the order for better modeling and controlling the system. This paper presents a comprehensive review of the energy system of renewable energy units and energy storage devices. Various papers are evaluated, and their methods and results are presented. Moreover, the mathematical fundamentals of the fractional-order method are mentioned, and the various studies are categorized based on different parameters. Various definitions for fractional-order calculus are also explained using their mathematical formula. Different studies and numerical evaluations present appropriate efficiency and accuracy of the fractional-order techniques for estimating, controlling, and improving the performance of energy systems in various operational conditions so that the average error of the fractional-order methods is considerably lower than other ones.

Keywords: control methods; energy systems; renewable energy sources; energy storage systems; fractional-order system

Citation: Alilou, M.; Azami, H.; Oshnoei, A.; Mohammadi-Ivatloo, B.; Teodorescu, R. Fractional-Order Control Techniques for Renewable Energy and Energy-Storage-Integrated Power Systems: A Review. *Fractal Fract.* **2023**, *7*, 391. https://doi.org/10.3390/fractalfract7050391

Academic Editor: Norbert Herencsar

Received: 27 March 2023
Revised: 5 May 2023
Accepted: 7 May 2023
Published: 9 May 2023

1. Introduction

Population growth, climate change, and increasing electricity demand have driven governments to utilize novel energy management systems such as microgrids (MGs) and smart grids (SGs) instead of traditional power networks [1]. MGs support a flexible and efficient power network by enabling the integration of renewable energy sources (RESs) instead of conventional power plants [2,3]. Moreover, the management of green-energy-supporting technologies such as electric vehicles (EVs) and energy storage systems (ESSs) is more straightforward in small SGs [4,5]. The use of local energy units reduces energy losses in transmission and distribution systems and increases the efficiency and quality of the electricity system. Although renewable sources have lower carbon losses and their energy is available in most parts of the earth, their energy is stochastic and variable during the day and year [6,7]. Electrical storage systems are the solution for stabilizing the produced energy from renewable sources. Electric vehicles, the main system of which is a storage system, are other practical technologies for reducing air pollution and increasing social welfare [8]. In recent years, considerable development of renewable resources and energy storage systems has been achieved due to the promotion of technologies of MGs, SGs, and also EVs.

Modeling, estimating, and controlling the utilized devices such as renewable energy sources and energy storage systems in microgrids are important and complex. Fractional-order methods are a practical way to enhance energy system performance [9,10]. The FO calculus can be used in integer-order and non-integer-order models [11], and the results of the previous studies have proved that fractional-order techniques are very suitable and flexible ways to characterize the properties in various energy processes [12]. So far, researchers and operators have modeled many practical systems in energy fields such as wind turbines [13], hydro-turbine governing systems [14], chaotic systems [15], energy supply–demand systems [16], permanent magnet synchronous generator systems [17], and microgrids [18] by using fractional-order equations. The FO methods also have proper performance in the estimation of parameters of energy systems [19], especially in energy storage systems, the characteristics of which such as the state of charge [20], state of energy [21], state of health [22], and state of temperature [23] are particularly important to estimate for maintaining the safe and efficient operation of storage systems. Controlling the energy system's process, which causes the system's performance to remain unchanged, is another practical application of fractional-order methods [24,25]. The FO techniques are utilized to control various performance parameters of energy systems, such as voltage [26], current [27], power [28], temperature [29], and cost [30]. Engineers use fractional-based controllers to control the practical energy systems of different technologies such as wind turbines [31], photovoltaic panels [32], electric vehicles [33], and storage systems [34].

In recent decades, the tendency and motivation to utilize low-carbon and renewable power resources have increased considerably due to the rareness of fossil fuels and increased environmental problems. According to statistical data on world energy in 2022, more than one-third of the consumed electricity was approximately generated by low-carbon units including hydropower, nuclear, solar, and wind generations [35,36]. Renewable resources are expected to play a larger role in electricity generation in the coming years, facilitated by the use of microgrids (MGs), smart grids (SGs), and smart homes (SHs) [37]. It should be considered that due to the stochastic performance of renewable resources, their control and management systems have a high impact on their performance and the efficiency of the total power network. Renewable energy resources exhibit significant fluctuations in power generation due to the unpredictable nature of their primary energy source. Their stochastic behavior affects the power quality, frequency, and voltage of the power network [38]. The produced stochastic power of RESs and the necessity of utilization of ESSs in the power system introduce new controlling and management challenges. The utilization of the appropriate control technique has a high effect on the optimal performance of the renewable units and the power network [39]. In other words, the high penetration rate of green technologies presents critical problems for power systems. Control programs are an important way to resolve this problem in which frequency fluctuation and other dangerous issues are observable. Control methods such as fractional-order proportional integral derivative (FOPID) [40,41], fractional-order cascade controller [42], proportional integral (PI) [43], proportional derivative (PD) [44], proportional integral derivative (PID) [45], fuzzy methods [46,47], model predictive control (MPC) [48,49], tilted integral derivative (TID) [50], and a hybrid method of TID and FOPID [51] have been utilized in different studies for improving the efficiency of renewable energy systems. Moreover, these control methods have been implemented for improving the inertia of the power system when various low and high disturbances occur in the devices and systems. Fractional-order (FO) control methods are also used for power converters in renewable energy systems. This flexible control and power converter technique decreases the overshoot level and settling time and increases the rising time. Fractional-order controllers have better dynamics, steady-state error, and stability than traditional controllers such as PI, PD, and PID. Moreover, the efficiency of the fractional-order controller, which can be defined as the ratio of the controlled output parameter to the input parameter, is higher than that of the other ones [52].

ESSs are employed to enhance the stability of the overall power system by bridging the gap between energy generation and consumption. Moreover, they are used in EVs to provide the required power for the vehicle. The energy storage systems technologies can be divided into four groups: electrical ESS, electrochemical ESS, electromechanical ESS, and thermal storage system [53,54]. Lithium-ion batteries (LIBs) and supercapacitors are among the most commonly utilized ESSs in MGs and EVs. The long lifetime, low self-discharge rate, and high energy and power density are the most highlighted advantages of LIBs. On the other hand, shortcut problems and heat-runaway are one of their disadvantages. To increase their performance and prevent problems, reliable situation estimation is so essential. Supercapacitors or ultra-capacitors are powerful energy storage devices with more abilities than batteries and capacitors. They can be charged more than capacitors and discharge more power than batteries [55,56]. The control and management systems of energy storage devices are developed to control and monitor the different parameters such as the state of charge (SOC), state of power (SOP), and state of health (SOH) in order to ensure the safety and reliability of the storage system [57]. The battery's SOC presents the battery's available stored energy, while the SOP shows the rate of the battery working in extreme conditions [58,59]. The SOH of the storage system indicates its degradation and remaining capacity. In other words, this index compares the remaining charge of the storage system with its rated value [60]. The stability and robustness of the estimation value of each of these parameters are required for appropriately modeling the storage system [61]. The accurate estimation of the ESS's situation is crucial for its proper management. Although different methods such as artificial neural networks (ANNs) [62], the support vector machine [63], the learning machine [64], fuzzy logic [65], the open circuit voltage method [66], and ampere-hour integration [67] have been used for predicting the ESSs, three techniques including the electrochemical model [68], integer-order model [69], and fractional order [70] are more commonly utilized methods. The integer-order model, called the equivalent circuit model, considers resistance capacitance systems [69]. The electrochemical model presents differential algebraic formulas about the intern electrochemical reactions of the storage device [68]. The FO model, which is more complete and effective than these two methods, considers the electrochemical characteristics without considering the equivalent circuit model. This method considers the fractional impedances, including the constant phase element (CPE) and Warburg element [70]. The FO model has more stability than traditional methods and finds the proper results in a lower time. Therefore, it can be applied to online and real applications.

Figure 1 demonstrates a comprehensive structure of fractional-order methods in systems of renewable energy systems and energy storage systems. As can be shown in this figure, fractional-order methods have been utilized in renewable energy systems of WTs, PVs, geothermal systems, and some hybrid systems. In these systems, different fractional-based systems such as FOPID, FOPD, FO-TID, and FOI-TD have been investigated. In energy storage systems, the majority of fractional-order systems have been applied on LIBs, LMBs, SMESSs, and supercapacitors. The application of fractional-based systems in ESSs can be divided into three main categories including modeling, estimating, and controlling. Based on the formulation, four definitions including GLD, RLD, CD, and OSD are most utilized in fractional-order systems. Some assistive techniques are used in the fractional-order method to improve its performance. In the following, the details of different applications, definitions, objectives, and assistive techniques of fractional-order methods are presented.

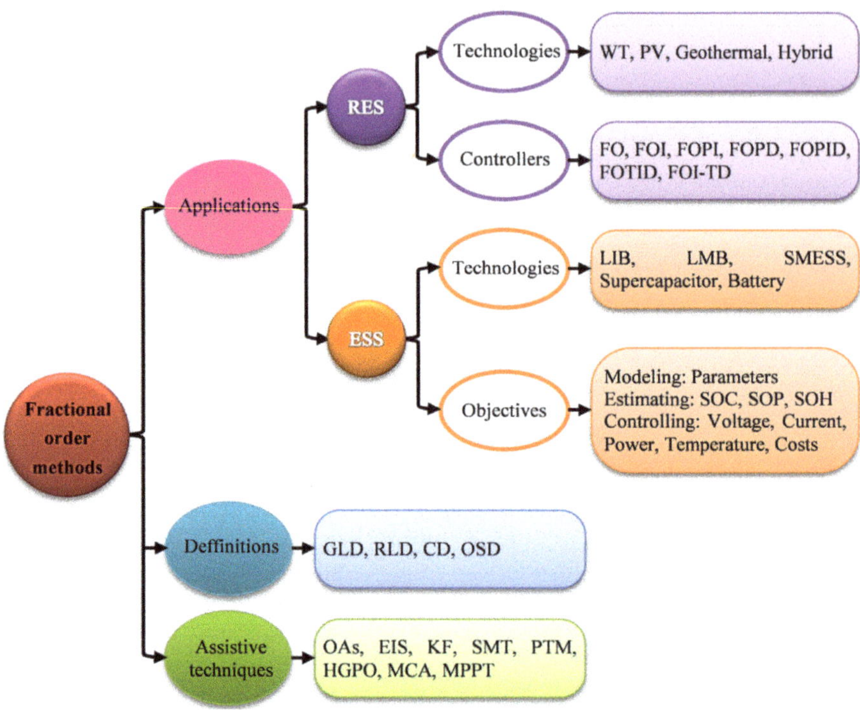

Figure 1. Comprehensive structure of FO methods in energy systems of RESs and ESSs.

1.1. Motivations

Although in recent years, some authors considered fractional-order techniques for modeling, estimating, and controlling the energy systems of RESs and ESSs and modifying their performance, there was no overall research on applications of FO methods on the energy systems of these devices. For this reason, it motivated us to prepare an overview of previous studies on fractional-order methods and the energy systems of RESs and ESSs. Moreover, we wanted to present a framework of important formulations and definitions of FO techniques that can be applied to practical and experimental energy systems for researchers and engineers. On the other hand, RESs and ESSs appropriately complement each other in the power network so that RESs inject renewable and eco-friendly energy into ESSs while ESSs increase the stability and quality of the produced energy of RESs. It was another reason to prepare a review paper about applications of FO techniques in both RESs and ESSs so that readers can easily see the advantages and disadvantages of FO-based estimating, modeling, and controlling methods of RESs and ESSs in a comprehensive article. In other words, we present important details of the latest articles about applications of fractional-order methods in RESs and ESSs, such as objectives, optimization methods, electrical circuit models, identification methods, remarkable numerical results, advantages, and disadvantages, in this review paper in order to comprehensively evaluate the performance of FO methods in energy systems of renewable units and storage devices. Thus, the main motivations of this review paper are as follows:

1. Providing a detailed explanation of fractional-order control techniques and their applications in renewable energy and energy-storage-integrated power systems.

2. Summarizing the current research findings on the use of fractional-order control techniques in these systems, including their advantages and limitations.

3. Comparing and evaluating different fractional-order control techniques used in renewable energy and energy-storage-integrated power systems based on their performance and applicability.

4. Identifying challenges and opportunities for future research in this field and suggesting possible directions for further investigation.

1.2. Methodology

To extract appropriate and adequate resources for evaluating the applications of FO methods in RESs and ESSs, published papers in the last decade that have considered FO techniques for estimating, modeling, and controlling the energy systems of RESs and ESSs were gathered from international scientific databases, such as IEEE and ScienceDirect. Some important and practical keywords, such as fractional-order controllers, estimation methods, modeling techniques, renewable sources, energy storage systems, photovoltaic panels, wind turbines, electrochemical impedance spectroscopy, lithium-ion battery, parameter identification, and state of charge were considered for selecting the proper and related papers. After reviewing these articles, the papers with high relevance to the applications of FO methods in RESs and ESSs were divided into two main categories, including renewable energy systems and energy storage systems, for more detailed evaluations. The explanations of the proposed method, the application of the fractional-order technique in the energy system, the utilized equations, objectives, constraints, assistive techniques, numerical results, and the comparison data with other methods are the extracted details from the reviewed articles. They are explained and categorized in the following sections. Moreover, the formulations and methods of some papers are considered for simulation and evaluation of the FO methods, and simulation results are presented and explained using figures and tables.

Therefore, in this review paper, after presenting the main formulations of FO methods in the energy systems, previous studies about FO methods that have been utilized in the energy systems of RESs and ESSs are considered for reviewing FO controllers and systems. These studies are divided into two sections, including (a) fractional-order techniques and renewable energy systems and (b) fractional-order techniques and energy storage systems. The papers of each section are evaluated deeply based on the methods, objectives, advantages, and disadvantages. Moreover, their details, such as the models, type of controllers, optimization methods, electrical circuits, identification methods, and objectives, are also categorized and summarized for easy access to an overview of applications of FO methods in the energy systems of RESs and ESSs. After pondering different studies, some suggestions for future research on FO methods are presented. Table 1 presents a summary of the remaining sections in this paper, with the key objectives of each section.

Table 1. Summary of the sections in this paper.

Sections	Objectives
2. Fractional-order systems	Mathematical formulation of FO systems, Definitions of FO operators, FO controller, FO-based converters and inverters
3. Fractional-order techniques and renewable energy sources	Applications of FO methods in RESs, description of different methods, Evaluation of numerical results, Discussion about the effect of FO methods on the energy system of RESs
4. Fractional-order techniques and energy storage systems	Applications of FO methods in ESSs, explanation of various techniques, Evaluation of numerical results, Discussion about the advantages and disadvantages of FO methods in the energy system of ESSs
5. Future perspectives	Future perspective on FO methods, Presenting some suggestions for future works on fractional-order systems
6. Conclusion	Conclusion of the literature review and summary of advantages and disadvantages of FO methods

2. Fractional-Order Systems

In this section, an overall description of FO techniques, which are used in energy systems for controlling, is explained. Fractional-order techniques are applied in dynamic systems to model their operation by a fractional differential equation considering a non-integer derivative. To describe the objects of the considered system as fractal properties, integrals and derivatives of fractional orders are utilized [71,72]. In recent years, the growth of engineering systems, technologies, and science has led to the implementation of FO models in many dynamic systems such as electrochemistry systems [73], physics systems [74], viscoelasticity systems [75], biological systems [76], and chaotic systems [77] for different purposes such as controlling [78], observing [79], estimating [80], and stabilizing [81].

The FO operator is mathematically defined in Equation (1). This operator describes the dynamic processes of the system with infinite dimensions.

$$_aD_t^\beta = \begin{cases} \dfrac{d^\beta}{dt^\beta} & \beta > 0 \\ 1 & \beta = 0 \\ \int_a^t (d\tau)^\beta & \beta < 0 \end{cases} \tag{1}$$

Here, β shows the order of the fractional operator, while a and t are the bounds of the operation. The value of the order of the fractional operator is variable in the real number domain $(\beta = 1, 2, 3, \dots)$.

2.1. Fractional-Order Definitions

Most definitions of the fractional integrals and derivatives, which facilitate numerical evaluations and analysis, are the Grünwald–Letnikov definition (GLD), the Riemann–Liouville definition (RLD), and the Caputo definition (CD) [82]. The mathematical descriptions of the most used definitions are presented in the following.

Grünwald–Letnikov definition: The GLD as the most utilized method presents a unique form of fractional calculus. Mathematically, this definition is mentioned in Equation (2) [83].

$$_aD_t^\beta f(t) = \lim_{h \to 0} \frac{1}{h^\beta} \sum_{j=0}^{\infty} (-1)^j \binom{\beta}{j} f(t - jh) \tag{2}$$

In this equation, the symbol h is the sampling interval, and $\binom{\beta}{j}$ is the binomial coefficients, which can be calculated by using Equation (3).

$$\binom{\beta}{j} = \begin{cases} \dfrac{\beta!}{j!(\beta - j)!} = \dfrac{\Gamma(\beta + 1)}{\Gamma(k + 1)\Gamma(\beta - j + 1)} & j > 0 \\ 1 & j = 0 \end{cases} \tag{3}$$

Here, $\Gamma(k)$ is the gamma function. Equation (4) shows this function.

$$\Gamma(k) = \int_0^\infty e^{-k} x^{k-1} dx \tag{4}$$

To calculate the coefficient $(-1)^j \binom{\beta}{j}$, a recursive method is utilized to simplify the complex process. This approach is shown in Equation (5).

$$\begin{cases} (-1)^j \binom{\beta}{j} = 1 & j = 0 \\ (-1)^j \binom{\beta}{j} = \left(1 - \dfrac{\beta + 1}{j}\right)\left[(-1)^{j-1} \binom{\beta}{j-1}\right] & j > 0 \end{cases} \tag{5}$$

Riemann–Liouville definition: The RLD, which is named after Bernhard Riemann and Joseph Liouville, is another method for presenting the possibility of fractional calculus. This definition is shown mathematically in Equation (6) [84].

$$_aD_t^\beta f(t) = \frac{1}{\Gamma(N-\beta)} \left(\frac{d}{dt}\right)^\beta \int_\beta^t \frac{f(\tau)}{(t-\tau)^{\beta-N+1}} d\tau \tag{6}$$

This equation considers $(N-1 \le \beta \le N)$. Here, N is an integer parameter.

Caputo definition: This definition is almost similar to the RLD, but the derivative element is not in the CD. Equation (7) presents the mathematical form of this definition considering the condition $(N-1 \le \beta \le N)$ [85].

$$_aD_t^\beta f(t) = \frac{1}{\Gamma(N-\beta)} \int_\beta^t \frac{f(\tau)}{(t-\tau)^{\beta-N+1}} d\tau \tag{7}$$

Oldham and Spanier definition (OSD): Oldham and Spanier introduced a definition in 1974 for presenting fractional calculus. This mathematical definition is shown in Equation (8).

$$\frac{d^q(\beta x)}{dx^q} = \beta^q \frac{d^q(\beta x)}{d(\beta x)^q} \tag{8}$$

where β function is defined as bellow:

$$\beta(m,n) = \int_0^1 (1-x)^m x^{n-1} dx \qquad m, n \in \mathbb{R} \tag{9}$$

Other definitions of fractional-order calculus can be studied in [86].

2.2. Fractional-Order Controller

The controllers have been improved in both industry and academic systems during recent years, and they have become more advanced and complex. The FOPID controllers have received great attention in previous years. However, simple tuning rules and no effectiveness still exist for FOPID controllers such as those specified for the integer PID controllers [87]. The PID controllers are mostly utilized in industrial applications due to their functional simplicity. On the other hand, the parameters of PID controllers are often adjusted using tests, experience, or error methods. Although this adjusting process can be applied easily in academic systems, it is definitely difficult to use to calculate the controller's gain in an industrial system because most industrial systems have some difficulties, such as uncertainties, nonlinearities, and structural complexity [88].

The variant of the fractional order including PD, PID, FOPID, and tilted integral derivative (TID) is presented explicitly in [89]. This paper defines the most usable commande robuste d'ordre non-entier approximation besides the FOPID controller by using Equation (10) [89].

$$S^a = C \prod_{i=1}^n \frac{1 + \left(\dfrac{s}{\omega_{z,i}}\right)}{1 + \left(\dfrac{s}{\omega_{p,i}}\right)} \tag{10}$$

The PID controller is the most popular and applicable shape of the controller, but FOPID is proposed according to the domain of variations. The FOPID has more stability than the PID controller. Moreover, the FO-based controller is more practical in complex problems with various parameters [90]. In Figure 2, the structure of the PID controller is demonstrated. According to this figure, the transfer function of fractional-order PID is presented in Equation (11) [90].

$$C(s) = K_p + K_d S^\beta + \frac{K_I}{S^\alpha} \tag{11}$$

By particular values of α and β, conventional controllers are reachable. If $\alpha = 0$ and $\beta = 0$, the values can be adjusted on the proportional controller. If $\alpha = 0$ and $\beta = 1$, the PD controller is implemented although, in the PI controller, the value of these parameters are $\alpha = 1$ and $\beta = 0$. Finally, if $\alpha \in \mathbb{R}$ and $\beta \in \mathbb{R}$, the FOPID controller can be reached with a flexible character [91]. Therefore, the FOPID controller has more flexibility. Moreover, the integral of time error is optimized by manipulating controller parameters and intercepting fuzzy logic. In [92], a novel fractional-order controller with two free degree orders is introduced by utilizing a combination of the genetic algorithm (GA) and the artificial bee colony (ABC) algorithm. The tilted integral derivative is constructed by replacing the proportional section of PID with $\dfrac{1}{S^n}$. TID is powerful enough to reduce or reject disturbance and noise [93].

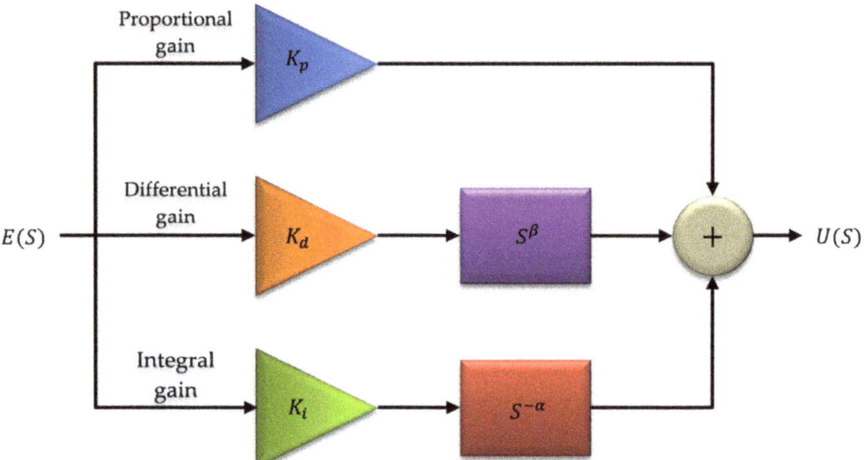

Figure 2. Structure of fractional-order PID controller.

2.3. Fractional Order in the Converters and Inverters

In the last decade, researchers have published papers about the application of FOPID and its variants in converters and inverters. As noticed, power electronics is an inevitable part of the power system and is used in a wide range of power systems such as renewable energy, high voltage DC, the drive of motors, etc.

2.3.1. DC-DC Converters

The author of [94] concluded positive results from the effects of FOPID and conventional controllers in improving buck–boost converter transient efficiency. In that paper, the results represented rise time, steady-state situation, and overshoot improvement in the FOPID case study. In [95], the authors investigated the impacts of FOPID in the DC-DC converter. In that study, by using biquadratic approximation, it was concluded that non-integer controllers are flexible and offer faster responses and more stable situations than other controllers.

The authors of [96] presented a novel cascade controller for buck–boost DC-DC converters. The cascade controller contains two loops while the outer loop has a FOPID controller with a voltage-controlling role. In that paper, the inner loop operates using feedback from the outer loop. The ant lion optimization was used for adjusting FOPID parameters. This optimization algorithm has significant advantages compared to other algorithms such as PSO. The results demonstrate that the proposed approach could strive for a high-rate variation of dynamic performance. In [97], an FO-based controller was investigated to increase the stability and decrease the effects of harmonics. In that paper, a FOPID-based buck converter on RLD is modeled. The ultimate figure of the continuous conduction mode in the steady-state situation shows that the parameters of the FO controller are reachable, stable, and robust.

2.3.2. AC-AC Inverters

The authors of [98] added a FOPID controller to an asymmetrical cascaded H-bridge multi-level inverter and compared it with conventional controllers. The results show an output voltage of the inverter with a considerable reduction in harmonic distortion. In [99], a multi-functional inverter was proposed for regulating the voltage and frequency of a photovoltaic energy storage system. In that paper, a shunt active power filter is utilized to mitigate harmonic content. Moreover, a virtual synchronous machine is considered to present the efficiency of the voltage–frequency regulation. This inverter has proper performance in the energy system of the photovoltaic panel and energy storage system on the AC power network. The authors of [100] presented a new approach for dealing with harmonic distortion that is delivered from the main grid. In this research, a FOPID controller is simulated in the LCL voltage-based inverter circuit, and it could decrease the disruptive effects of harmonics.

3. Fractional-Order Technique and Renewable Energy Sources

In this section, several papers that follow different objects are reviewed. In these papers, FOPID and conventional controllers are utilized to improve steady-state performance. This section focuses on FO controllers used for renewable energy systems.

Table 2 is mentioned to point the readers to an overview of papers that studied fractional-order techniques in renewable energy systems. This table summarizes the application of fractional-order methods in renewable energy systems. The types of the used controller, function, algorithm, and type of distributed generation unit of each paper are available in this table. Table 2 shows that most authors have used the FOPID controller for renewable energy systems. The TID controller is another practical method. The metaheuristic algorithms are utilized in most papers for optimizing the controller parameters. On the other hand, the MPPT is the most iterated function due to the role of the power amount without distortion. The major objective of the papers is to minimize the nonlinearity and uncertainty of the system, which causes unstable situations in the steady-state condition.

In the papers associated with fractional-order methods and renewable energy sources, FOPID controller, photovoltaic panel, and wind turbine are the most utilized keywords. In other words, the FO method has been utilized to control the most common types of renewable energy resources, mainly WTs and PVs. Different types of controllers, methods, and optimizations are other types of information that can be found in Table 2. For example, all the papers develop their controllers based on the FOPID controller. Based on the method, the MPPT is the most popular method in photovoltaic panels, while the SMES is the most popular approach in wind turbines. Moreover, one of the vital parts of implementing the suggested control method is to select an appropriate optimization approach, and the reviewed papers demonstrate that metaheuristic algorithms have extensive stocks in the papers.

In the first reviewed paper on fractional-order application in the RESs, the authors proposed a specific algorithm to reach maximum power [101]. In that study, due to uncertain irradiation, the maximum power point tracking (MPPT) approach was implemented in the photovoltaic panels (PVs) by combining step size variables. According to Figure 3, for eliminating the disturbance factor of MPPT, the novel FOPID controller was attached by an incremental conductance algorithm. The results illustrate enhanced steady-state performance [101]. In [63], focusing on MPPT in PV panels, a new novel approach is presented by combining incremental conductance and FOPID. According to the obtained results, the loss of output voltage is reduced considerably. In [64], a study to reduce perturbation in the output voltage of PV panels was conducted. In this research, MPPT through the perturb and observe (PO) technique under uncertain atmospheric situations was investigated. Reshaped output power by adding FOPID, which is tuned by the grey wolf optimization (GWO) technique, causes an improved transient factor in the result of the edited system. In [65], a novel high-energy tracking system from PV panels based on the perturb and observe technique was presented. In this paper, several events, such as inverter nonlinearity, the uncertainty of irradiation, and temperature, were adopted in the model. For compensating the PO technique, a FOPID controller is employed, and the optimal tuning of its parameters is obtained using the yin yang pair algorithm [102].

In the other paper, a geothermal power plant, dish Stirling, and high voltage DC system are connected to an uncertain environment. The specific scheme is presented for reducing the effects of harmonic distortion and having a robust system. The mixed FOPI-FOPID controller is replaced in the system and, using the sine cosine algorithm (SCA), the optimal adjustment of the parameters of the controller is conducted. The obtained results show improvements in the transient index such as overshoot and rise time [103]. Another application of the FOPID controller in the hybrid RES with combined energy storage was presented. The goal of the paper was to seek the objective frequency at which the system deviated from the equilibrium point, and after obtaining the result, the approach was validated with the mean square error method [104]. In [105], a fractional-order controller was presented to reduce tension and increase the robustness of the system. In this research, a novel controlling regime based on a fuzzy FOPID controller was proposed for hybrid renewable resources used for power generation and plentiful switching. The combination of the particle swarm optimization and chaotic map is utilized to extract the transient parameters of the controller [105].

Table 2. Summarization of applications of the fractional-order method in renewable energy systems.

Ref.	Type of Controller	Remark	Optimization/Analytical Method	Application
[94]	FOPID	Improving buck–boost efficiency with simulink	Analytical frequency domain design method	PV
[95]	FOPID	Tuning attached controller to DC-DC converter with simulink	Analytical frequency domain design method	PV
[97]	FOPID	Buck converter based on RLD	RLD	PV
[98]	FOPID	Asymmetrical cascaded H-bridge multi-level inverter with simulink	Analytical frequency domain design method	PV
[101]	FO controller	MPPT	Inc-Cond algorithm	PV
[102]	FOPID	MPPT	Yin yang pair algorithm	PV and WT
[103]	FOPI- FOPID	Deregulated AGC	Sine cosine algorithm	Geothermal plant
[104]	FOPID	Minimizing mean square error with simulink	Analytical frequency domain design method	RESs
[105]	Fuzzy FOPID	Chaos control	PSO	RESs
[106]	FO controller	Model control and space vector PWM	Analytical frequency domain design method	PV and WT
[107]	Fuzzy FOPID	Tuning WT inverter	TLBO	WT
[108]	FOPID	Pitch angle RBF neural network	Chaotic optimization	WT
[109]	FOPID	Generalized isodamping technique	Gain-scheduling algorithm	Solar system
[110]	FOPID	Yuning attached controller to DC-DC converter	PSO	HES
[111]	FOPI	Enhancing dynamic behavior	Metaheuristic algorithms	PV
[112]	FOPID-TID	load frequency control	Artificial ecosystem-based optimization	RESs
[113]	Fractional based TID	LFC and VIC	HGAPSO	RESs
[114]	FOPD-LFC	ITAE minimizing	SO algorithms	RESs
[115]	FOI-TD	Fitness-dependent optimizer	Hybrid sine cosine algorithm	RESs
[116]	FOPID	LFC and SEES controlling	Manta ray foraging optimization	RESs
[117]	FOPID	MPPT	Inc-Cond algorithm	PV
[118]	FOPID	MPPT	GWO	PV

In [106], research was presented for contrasting unstable situations and ensuring power quality in smart residential hybrid RESs. In this paper, a new FO controller implemented on an inverter's output break is controlled with the pulse-width modulation (PWM) method. A comparison of the proposed controller with the conventional controller shows a reduction in the tension of the output voltage [106]. Another paper about wind turbine (WT) output controllers was proposed in [107]. Fuzzy FOPID is replaced in the DC-AC converter section due to the uncertainty of wind speed and grid-connected wind power plants for reducing the effects of the dynamic situation. Rise time and fall time are improved by the tuning of controllers using the teaching–learning-based optimization (TLBO) algorithm overshoot [107]. In another paper, the pitch angle method was proposed for controlling the rotor speed and power production of the WT [108]. In this research, the FOPID controller is replaced with a radial basis function (RBF) neural network for improving system performance, and the chaotic optimization parameter of FOPID is tuned optimally. The result shows that by attaching the FOPID controller, the system's flexibility and robustness are increased in comparison to conventional controllers [108]. The authors of [109] designed a novel approach to deal with nonlinearity and increase the performance of the energy system. A FOPID controller is designed to cope with the destructive effects of the parameters. A numerical evaluation of the FOPID controller with a combined gain-scheduling algorithm and concentrated solar plant nonlinear model indicates its high performance in comparison with other controllers [109]. In another paper, the FOPID controller was presented due to the nonlinearity of the V-I characteristic of the PV panel [110]. In that paper, the boost DC-DC converter is utilized in the PV panel for regulating the output voltage. Moreover, particle swarm optimization is proposed for modifying the parameters of the FOPID controller [110].

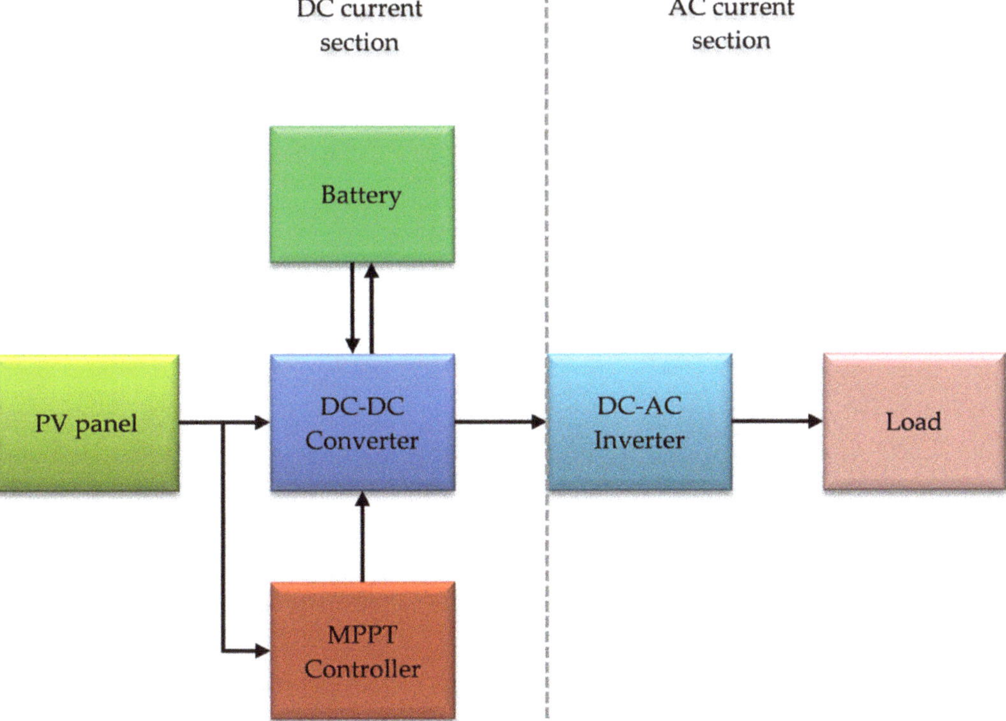

Figure 3. Diagram of the photovoltaic panel power generation.

Two controllers were presented for balancing the dynamic behavior of grid-connected PV panels in [111]. In this paper, metaheuristic algorithms such as cuckoo search (CS), GWO, the whale optimization algorithm (WOA), the mine blast algorithm (MBA), ABC, and the moth swarm algorithm (MSA) are used to optimize FOPI and PID controller parameters to reduce transient factors such as settling time and overshoot [111]. A study for mitigating the effects of renewable units and EVs present in the power system was proposed in [112]. In this paper, a controller consisting of both FOPID and TID controllers is presented to reduce frequency fluctuation and tie-line power deviation. The artificial ecosystem-based optimization (AEO) method for tuning parameters is used. Therefore, it can be said that the existence of different controllers with distinct performances increases the robustness and flexibility of the system [112]. A TID fractional-based controller was proposed for mitigating sensitive load and generation fluctuation. The power generation of RESs is penetrated to the power system due to the lowering of system inertia and unbalancing of the load and frequency. Therefore, load frequency control (LFC) and virtual inertia control (VIC) are proposed for compensating for RESs' penetration problems. Two TID controllers for each segment are placed using hybrid genetic PSO parameters of controllers, which use a case study combining RESs and conventional power production. In comparison to PSO and GA, the proposed algorithm is more effective and robust [113].

In [114], a new FOPD-LFC was proposed for covering the deficit of connecting RESs to the power grid due to power exchange and its violation. This paper aims to minimize the integral time absolute error (ITAE) with skill optimization. The proposed method was evaluated in two case studies, and the results were compared with the results of the jumping spider optimization algorithm and bonobo optimization. The proposed FO-based method has more reliability and stability than other algorithms [114]. A novel approach for controlling the nonlinearity of a system with RESs was proposed in [115]. The considered system has nonlinear loads and variable generation units. Therefore, the FOI-TD controller was designed for reducing the nonlinearity, and a hybrid sine cosine algorithm with a fitness-dependent optimizer was utilized to tune the controller's parameters. The result shows that the proposed algorithm outperforms other algorithms such as the fitness-dependent optimizer and PSO. Additionally, the designed controller reduces the overshoot and rise time [115]. The essential need for controlling the multi-area power system with renewable resources was investigated in [116]. In this paper, a new FOPID for controlling LFC and the superconducting magnetic energy storage system (SMESS) is designed. The optimum points of the FOPID parameters are determined using the manta ray foraging optimization (MRFO) algorithm. Robustness and flexibility against nonlinearity is a prominent characteristic of the proposed controller [116].

4. Fractional-Order Technique and Energy Storage Systems

In the last decade, energy storage systems have been widely utilized in MGs and EVs. Online and accurate modeling and estimation are essential to efficiently operate the storage systems and their upper energy system. The main purpose of studies on fractional-order control systems in ESSs is to improve the estimation rate of the SOC, state of energy (SOE), or other essential and uncertain parameters of the battery. In this section, the applications of FO techniques in the energy systems of different energy storage devices are presented and discussed based on various types of research.

Fractional-order method, lithium-ion battery, state of charge, Kalman filter, electro-chemical impedance spectroscopy, and parameter identification are the most repeated keywords in the papers on FO methods and ESSs. In Table 3, the details of articles on FO techniques and ESSs are presented. In this table, different parameters of each paper such as ESS models, FO-based structures and technologies, and main objectives are given.

Table 3. The details of articles on the fractional-order controller and energy storage systems.

Ref.	ESS Model					FO-Based Structures and Techniques															Main Objectives							
	LIB	Supercapacitor	LMB	SMESS	Battery	EIS	GA	KF	SMT	PSO	HGAPSO	LSA	PTM	HGPO	MCA	ANA	KHO	ACA	ITLO	MGSO	SOC	Temperature	Voltage	SOP	Electrode aging	Control costs	Solid phase diffusion	SOH
[119]	*					*	*	*	*												*							
[120]	*	*				*				*													*	*					
[121]	*										*										*				*			
[122]	*					*				*											*							
[123]			*			*															*		*					
[124]	*												*													*		
[125]	*					*															*		*					
[126]	*													*									*	*				
[127]		*																								*		
[128]	*						*	*													*							
[34]		*				*																	*	*				
[129]				*										*													*	
[130]	*					*	*	*													*							
[131]	*					*																				*	*	
[132]	*					*				*					*	*					*							
[133]	*					*															*		*	*				
[134]	*					*		*		*											*							
[135]	*							*													*	*						
[136]	*					*					*												*					
[137]	*						*	*													*							
[138]	*						*	*													*							
[22]	*						*	*													*							*
[139]	*																*				*	*						*
[140]	*							*		*								*			*		*					
[141]					*	*		*													*							
[142]	*					*				*													*					
[143]		*												*					*				*					
[144]	*					*										*												*
[145]	*					*					*												*					
[146]	*					*		*				*									*							
[147]	*					*															*					*		
[148]	*	*				*				*											*		*	*				
[149]					*																					*		
[150]					*																*		*					
[151]				*																	*		*					

While the LIB has been considered as the energy storage device in most of the studies, such as [119,125,137], in papers [141,149,150], the entire structure of the energy storage system has been studied. The particular type of energy device has not been considered. In [120,148], both lithium-ion and supercapacitor ESSs have been considered. In these articles, the performance of different storage devices and the application of the FO model have been compared and evaluated. The supercapacitor ESS has been investigated in Refs. [34,127,143]. The authors of [123] proposed fractional-order techniques for the LMB. Refs. [129,151] are also about superconducting magnetic energy storage. According to the literature, LIBs are more efficient for practical operation, and therefore, the main focus of most research is on this type of ESS.

In [119], the fractional-order impedance model was investigated to overcome the drawbacks of electrochemical and electrical circuit models. The fractional parameters and electrochemical impedance spectroscopy (EIS) data were utilized in the defined impedance model. The GA was used to identify the order of the fractional elements after achieving the state space equations of the model by the GLD. The performance of the FO model was improved by utilizing the Kalman filter (KF) and short memory techniques. The numerical results presented that the proposed fractional-order model can improve the estimation rate of the SOC of the battery so that its estimation error is around 3%. The application of the FO model in the energy system of LIBs and ultra-capacitors was experimentally evaluated in [120]. The EIS and temperature-compensating fractional models of batteries and capacitors were presented, and the PSO algorithm was considered to identify the online parameters. Based on the numerical results, the presented FO model can obtain a more accurate value of stochastic parameters with less than 4% error. The authors of [121] suggested FO estimation and control algorithms for the electrical storage system. The combination of the particle swarm optimization method and the GA was used to identify the parameters of the fractional model. Its structure was defined by evaluating a series of pulse tests at various levels of the battery's charge. Moreover, online charge estimation was used to increase the convergence speed and obtain more accurate results. In the proposed FO model, the behavior of the battery over different current and voltage rates was also researched. Numerical results, achieved under various load profiles, presented that the proposed FO-based estimation method has more accurate results than KF-based methods by up to 1.2%.

In [122], a factional-order model was presented to estimate the SOC of batteries, especially lithium-ion types. In the proposed model, the charging and discharging characteristics were described using a circuit model and then the model parameters were identified by the PSO algorithm. The proposed order-dependent model was evaluated utilizing real-time experimental data. The results showed the feasibility and validity of the FO model to observe the more accurate rate of battery charge. The authors of [123] investigated the fractional order model and EIS for liquid metal batteries (LMBs) as one of the practical ESSs in the last few years. The general electrochemical reaction process was considered to define the fractional-order circuit model. Moreover, the impedance spectra were analyzed to extract the parameters of the battery. The results, achieved from simulations and experiments, presented good performance and stability of the FO model for the battery management system. In [124], the relationship between the FO and electrode aging was investigated. The utilized battery for studying was the lithium-ion type. The system of the FO model was identified using the transient discharge dataset of the fully charged situation of the battery. The numerical results, which were achieved by applying the model to the actual data, presented that the FO can properly evaluate the degradation level of electrodes so that there is a steady relationship between the proposed model and the charging/discharging cycle of the battery. On the other hand, the battery life is terminated when the FO model tends to be stable.

The short- and long-time evaluation of the dynamic of ESSs was presented in [125] considering frequency parameters. In this study, the FO model was defined using modified electrochemical impedance spectroscopy. The modified version of the impedance model appropriately selected the internal dynamics in both short and long periods. Moreover, it could capture the low-frequency dynamics of the battery. Numerical results showed the high performance and proper adaptability of the proposed FO model to evaluate the dynamics of the battery in different operating conditions so that the maximum error was below 0.86%. The authors of [126] studied the application of the FO model to show the effect of thermal and electrolyte variations on LIBs. The polarization in electrolytes was utilized to modify the proposed FO model. In the proposed model, the heat absorption/generation of the battery cell was also described using the particle thermal model. The numerical evaluations presented high stability of the modified fractional-order model in various ranges of current, voltage, and temperature. In [127], the FO model was presented to explain the frequency-dependent behavior of electrical storage devices in energy systems. In this research, both the transient and steady modes of the storage system were considered to evaluate the electrical system of the fractional-order capacitance and inductance. The fractional-order coils and supercapacitors were utilized to verify the proposed FO model. The results expressed the proper performance of the proposed model to present the level of stored energy in commercial electrical storage devices. An FO model, which was modified using the Kalman filter, was presented in [128] to estimate the SOC of batteries. The LIBs were studied in that paper. In that paper, the FO model describes the physical behavior of the battery. The GA was utilized to identify the parameters of the model. The KF was also used to modify the proposed model in order to increase the stability of the estimation and better track the noise variance. The experimental data presented that the proposed FO-based model performs better and more accurately than the traditional KF-based models.

The authors of [34] developed an FO control method for the supercapacitor type of ESSs. In the defined model, the inherent physical characteristics of the storage system and the transient responses were considered and analyzed to obtain global control stability. Moreover, the sliding-mode control was also applied to the FO model to improve the robustness of the closed-loop system. The proposed control method was evaluated under different energy situations and the presence of various energy sources in the distribution system. The results presented more stability and feasibility of the FO-based model than the other control methods. Ref. [129] suggested a nonlinear control method utilizing FO control for analyzing the ESS. The ESS was a combination of battery and superconducting magnetic storage. The proposed FO control model has the ability to compensate for the nonlinearities and model the uncertainties through online estimation. Moreover, it improves the control performance of the ESS while only the voltage and current of the storage are measured instead of accurate modeling of the system. The numerical evaluation of the proposed method showed that the FO control method improves the storage system's performance and reduces the control costs of the system more than other control methods such as sliding-mode control and feedback linearization control. An FO model was proposed in [130] to estimate the electrochemistry dynamics of lithium-ion batteries. The GA was used to identify the parameters of the proposed model. Moreover, the KF was utilized to modify the FO control method in estimating the SOC. The proposed control method was simulated and compared with the Thevenin model. The numerical results showed that the FO control method is more accurate and robust than the other one. The authors of [131] presented a physics-based fractional-order model for simply analyzing the cycle of energy storage devices. This study focused on lithium-ion batteries. The dynamics of medium-high frequency were considered to define the full-cycle model. For this reason, it can be applied to the full-cycle operation of the battery. The proposed physical fractional-order model was evaluated using different loads. The obtained results presented that the defined model has suitable performance for online applications, so it has high stability in extracting the solid phase diffusion. In [132], a new technique based on the FO method was introduced to estimate the SOC of the battery. The proposed model was based on the open circuit voltage. The PSO method was utilized to select the parameters of the FO-equivalent model. To increase the accuracy and convergence rate of the estimation method, a particle filter, which was modified by an adaptive noise updating algorithm, was added to the proposed FO model. The numerical results, achieved by applying the proposed method to static and dynamic conditions, showed the appropriate convergence rate and high stability of the proposed charge estimation method. The state of power of LIBs was investigated in [133]. It is worth mentioning that in the evaluation of the SOP, it is considered that the battery is working with the maximum possible power rate. In this paper, the factional-order model was suggested to estimate the power level of the battery. In this model, the voltage, current, and SOC of the battery were modeled by FO calculus. The numerical results presented that the proposed estimation model with approximately 1.34% error has high accuracy for calculating the SOP of the battery in different operating conditions. The authors of [134] proposed another FO-based model for estimating the SOC of lithium-ion storage devices. In the presented method, the FO model was modified using a Kalman filter and EIS in order to reduce the error rate of the electrical equivalent circuit and increase the stability of the model. The parameter of the model was identified by the quantum particle swarm optimization algorithm. Moreover, the Grünwald–Letnikov fractional derivative and time-varying measurement error covariance were added to the model to promote the convergence speed of the estimation method. The simulation results showed a better performance of the FO model compared to other estimation methods to estimate the state of the charge of the battery.

In [135], to find the correct charge level of LIBs, a new fractional-order model was presented. The KF was utilized in this technique to increase the adaptive ability of the estimation method in a complex operating state. The Sigmoid function was added to the model to find the unknown parameters. Moreover, the augmented vector technique was utilized to better describe the nonlinear function of the storage device. Simulation and experimental data presented the high estimation stability of the FO model to estimate the SOC of the battery. Moreover, it has adequate ability in complex environments. A multi-parameter FO model, defined using 25 parameters, was presented for analyzing the battery in [136]. The considered battery type was lithium-ion. The combination of particle swarm optimization and the genetic algorithm was utilized to identify the parameters. Moreover, the proposed multi-domain model was defined using two domains including frequency and time. The frequency domain is based on EIS, and the time domain is based on the terminal voltage of the battery. The proposed FO model is robust and reliable for identifying and analyzing the storage device; this claim was pondered and proved by the numerical evaluations in the MATLAB environment. In [137], a new method for estimating the state of charge of LIBs was proposed. The suggested estimation method was based on the FO and KF. The parameters of the model were identified using the adaptive genetic algorithm. During the test condition, the root mean square error of the estimation method was less than 1%. Moreover, the FO model had more stability and robustness than other KF-based techniques in estimating the SOC of the battery. The authors of [138] proposed a hybrid method for estimating the state of charge and analyzing the ESS. This model combines the FO-equivalent model and the FO adaptive dual Kalman filter. The first part of the model is utilized to achieve the external electrical characteristics of the battery, while the second part of the proposed model estimates the battery's charge level. The evaluation of the proposed method in different experimental conditions presented more convergence, lower error, and higher stability of the proposed FO technique than other estimation techniques. In [22], an FO model was established to accurately estimate the state of charge and state of health of ESSs. In this second-order model, the adaptive genetic algorithm is utilized to identify the model parameters. Then, the multi-innovations unscented Kalman filter is applied to estimate the charge and health level of the battery. The performance of the estimation technique was pondered by evaluating the experimental results achieved from different cycles and operating conditions. By utilizing the proposed method, the estimation error of both SOC and SOH was lower than other methods, so its root mean square error was less than 1.2% and 0.007% in predicting the charge and health level of the battery, respectively. Thus, the proposed FO model had adequate accuracy. Another fractional-order model for estimating the SOC of the battery was introduced in [139]. The LIBs were studied in this paper. In the proposed FO model, various temperatures and operating conditions were modeled. The model parameters were optimized by using a Krill Herd optimizer in order to obtain an appropriate model. The validation of the proposed technique was evaluated using different operating conditions, temperatures, and SOC ranges. The numerical results, which were achieved from various tests such as hybrid pulse power characteristic and dynamic stress tests, presented high accuracy and reliability of the method in different situations. This method also had lower errors than other estimation methods. The authors of [140] proposed a multi-scale algorithm to estimate the accurate SOC of the batteries by focusing on lithium-ion technology. The estimation algorithm is based on the FO model. It was modified using the KF and variable forgetting factor recursive least squares. Indeed, the KF was used to enhance the estimation accuracy, and the variable forgetting factor algorithm was utilized to predict the internal resistance and capacity of the battery. The ant colony algorithm was utilized to extract the model parameters. The proposed model experimented with a fast variation of the SOC and a slow variation of the internal resistance of the battery. The results proved the strong power of the model in analyzing the energy storage device. The estimation error of the method was approximately 0.52%.

A new dual fractional-order model, which was extended using the KF and resistance–capacitor approximation method, was presented in [141] for simultaneous estimation of the SOC and fractional parameters of the battery. In this technique, the Grünwald–Letnikov definition was utilized to represent the discrete state space. Moreover, both frequency and time domains were investigated in this paper. The method was validated considering different operation levels. According to the simulation results, the proposed FO model with less than 0.28% root mean square error has high accuracy in SOC estimation, and therefore, it can be utilized in real operating conditions. In [142], a fractional-order-equivalent circuit model was suggested to analyze the battery energy system. To synchronously identify the parameters of the model and its order values, the PSO algorithm was utilized. In this study, the capability of the model was evaluated in various FO values, and therefore, the best values were selected. According to the numerical results, the proposed model has suitable performance for identifying the battery data. A robust fractional-order control method was proposed in [143] for controlling the parameters of the supercapacitor ESS. In the first step of the proposed technique, different parameters of the storage system such as uncertainties, disturbances, nonlinearities, and dynamics were predicted by the high-gain perturbation observer. Secondly, the FO controller estimated the online compensation rate. During the observation and control cycle, the interactive teaching and learning optimizer was utilized to achieve the gains of the observer and controller. The numerical results, which were achieved from different cases and control strategies, presented the high effectiveness of the proposed method for practical applications. In [144], an FO model was proposed to evaluate multiple groups of lithium-ion batteries with various states of health. Electrochemical impedance spectroscopy was utilized to extract the structure and parameters of the equivalent circuit model. Then, the P-type iterative learning algorithm was applied to optimize the selected parameters. To increase the reliability of the structure, the model was modified by pretest noise, correlative information criterion, and multiple correlations of parameters. The evaluation of the model considering different batteries with various states of health presented its high quality for estimating the health level and analyzing the situation of the energy storage devices. A fractional-order model was presented to evaluate the LIBs in [145]. The presented model was modified using the Randles model, equivalent circuit model, and free non-integer differentiation orders. Moreover, multi-objective particle swarm optimization was utilized to identify the parameters of the FO model. The efficiency and stability of the suggested method were proved by evaluating the storage system against the traditional resistance capacitor circuit model. The authors of [146] presented an FO model to estimate the charge situation of the battery in a storage management system. The LIB was the considered type of energy device for estimating. In the proposed method, the time and frequency domains were studied using the recursive least squares algorithm and recorded impedance spectroscopy, respectively. Moreover, the modified Kalman filter was used in the proposed method in order to estimate the SOC of the battery. According to the numerical evaluations, the proposed FO technique is more efficient and accurate than other methods such as the classical equivalent electric circuit.

In [147], an FO model was presented to implement the electrochemical impedance spectroscopy in the ESS. The main focus of this study was on LIBs. In the proposed method, the capacitive resistance circuit was utilized to extract the model parameters in both offline and online modes. The numerical evaluations presented high efficiency and stability of the proposed method for closely indicating the SOC, discharge rate, and aging degree of the storage device. A new FO-based technique was investigated in [148] to evaluate the remaining discharge time of the ESS. The LIBs and supercapacitors were studied in this paper. The combination of chaos theory and the PSO algorithm was utilized to extract the parameters of the model. Moreover, the Markov load trajectory prediction was applied to the method for increasing its accuracy. The reliability and robustness of the FO technique were proved by evaluating the method in a real operating environment. In [149], an FO model was presented to simultaneously evaluate the ESS and the demand response program. The method was investigated in a power network in the presence of thermal power plants, biogas units, wind turbines, and photovoltaic panels. Control of the frequency parameters of the system was the main goal of introducing this FO technique. The quasi-oppositional Harris hawk algorithm was utilized to optimize the considered coefficients. The simulation of the method in real-time conditions showed its appropriate feasibility to control the indices of the system. The authors of [150] presented another FO model for controlling the ESS in a distribution system in the presence of various energy sources such as wind turbines and PVs. In this study, the main goal of the controlling method was to stabilize the bus voltage of the system under different operation conditions. The numerical results presented that the proposed technique has more stability and reliability than other controllers such as sliding-mode control and PI control. The author of [151] proposed a control method based on FO controllers to evaluate and control the ESS in the power network. The considered indices of the system were optimized using metaheuristic algorithms. To achieve reliable and realistic results, the proposed model was modified using the governor dead band, generation rate constraint, and communication time delay. Moreover, the sensitivity analysis was utilized to verify the performance of the method. The numerical results presented that the suggested technique has more accuracy than common controlling methods so that the mean controlling error of the fractional-order method is considerably lower than PID and PI controllers.

The utilized electrical circuit for presenting the ESS is different in various papers. In Figure 4, the considered electrical circuits and their description are presented. In the presented circuits, different parameters such as dynamic characteristics, voltages, currents, temperatures, and storage capacities are considered and modeled using electronic devices such as resistances and capacitors and controlling blocks such as integrals and derivatives.

In the electric circuits of Figure 4, R_0 is the ohmic resistance. R_1 and R_2 show resistors that are modeled in parallel with constant phase elements. The CPE is considered to describe the charge transfer between the electrolyte interface and the double layer of the storage system. The impedance of CPE (Z_{CPE}) can be mathematically calculated by using Equation (12).

$$Z_{CPE}(S) = \frac{1}{C_{CPE}S^\alpha} \tag{12}$$

In Equation (12), C_{CPE} is an index similar to a capacitor, and S is the Laplace operator. The parameter α is the fractional order for describing the dispersion effect. It is a positive number between 0 and 1. It should be considered that the CPE behaves like the ideal capacitor when α is 1, and it behaves like the resistor when α is 0. The Warburg element (W), which represents the diffusion process in solid phases of the low-frequency band of the storage device, is like a constant phase element.

One of the steps for applying the FO model for controlling, estimating, or evaluating the ESS is the identification of the required parameters of the proposed model. Optimization algorithms and electrochemical tests are the most used methods for identifying the model parameters in different research studies. In Table 4, the utilized methods in various papers for selecting the required parameters of the FO models of ESSs are presented. As can be seen in this table, electrochemical tests are the most utilized method for identifying the parameters. In contrast, the PSO method is the most applied intelligent algorithm to extract the model parameters.

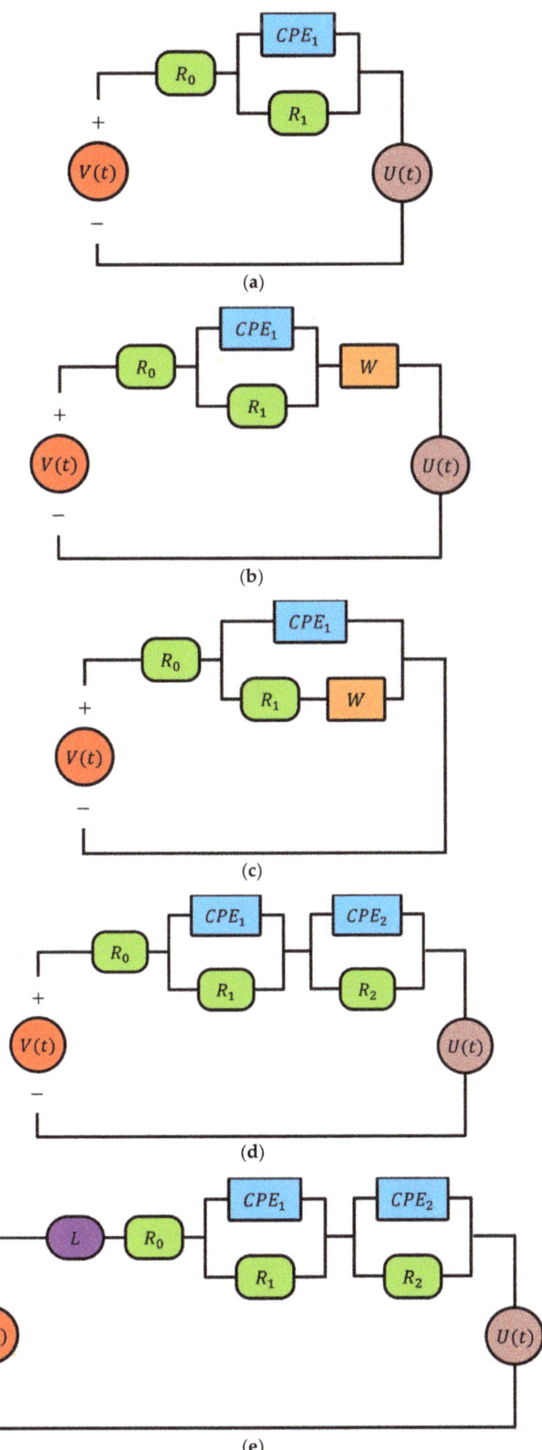

Figure 4. *Cont.*

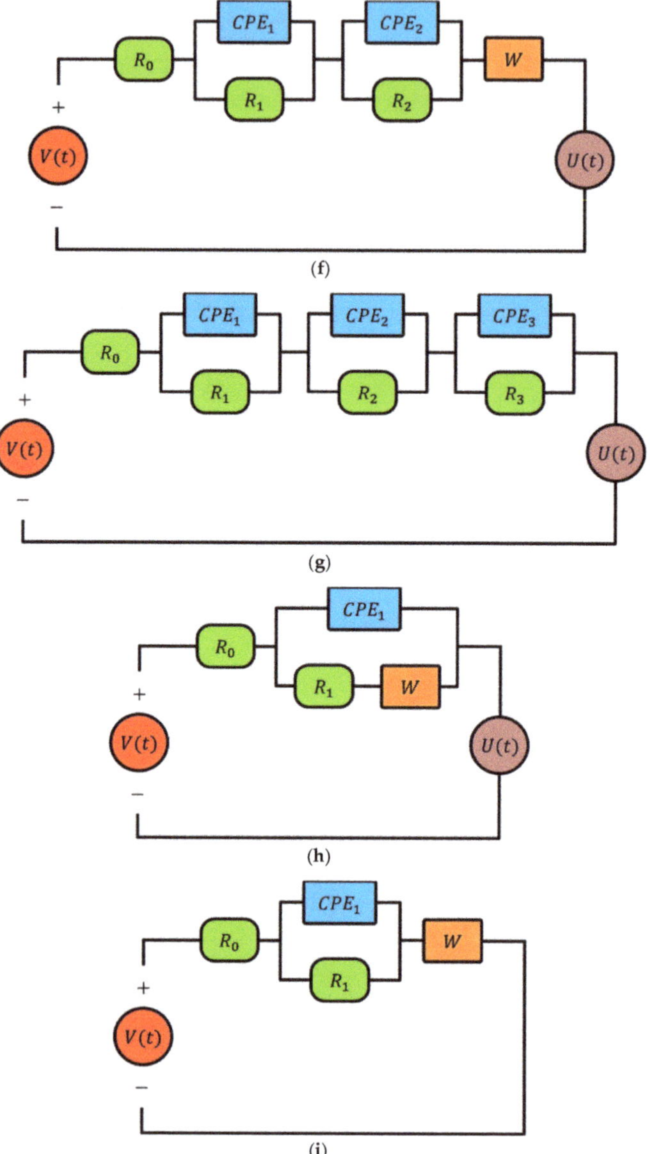

Figure 4. The equivalent electrical circuits for energy storage systems. (**a**) Lithium-ion battery [119,122,124,132,134,135,138]; (**b**) Lithium-ion battery [119,120,144]; (**c**) Supercapacitor storage [120]; (**d**) Lithium-ion battery [22,121,128,130,137,140,142,146,148]; (**e**) Liquid metal battery [123]; (**f**) Lithium-ion battery [133,136,139]; (**g**) Lithium-ion battery [141]; (**h**) Lithium-ion battery [145]; (**i**) Supercapacitor storage [148].

One of the techniques for the practical implementation of the FO model in storage systems is EIS. It is a powerful method for separating the electrochemical reactions at electrode surfaces [152]. The measured electrochemical impedance spectra are usually divided into three parts based on the frequency. High frequency, middle frequency, and low frequency are the divided parts of the impedance spectra [153]. In Figure 5, the sample electrochemical impedance spectra of the two most utilized types of ESSs, lithium-ion batteries and liquid metal batteries, are demonstrated [119–123]. In the FO method, the impedance spectra are utilized to model the dynamics between the electrodes of the storage system in different operating situations.

Table 4. The identifier of model parameters of the fractional-order technique in energy storage systems.

Identification Method	Refs.
Hybrid of genetic algorithm and particle swarm optimization	[119,136]
Global optimization	[22,120]
Hybrid pulse tests	[121]
Particle swarm optimization	[122,133,134,142,145,148]
Electrochemical impedance spectroscopy	[123,125,141,144,146,147]
Least squares method	[124]
Pseudo-two-dimensional electrochemical method	[126]
Specific current condition test	[128]
Genetic algorithm	[130,137]
Decoupling the dynamics in frequency and spatial domain	[131]
Dynamic stress test	[132,139]
Augmented vector method	[135]
Forgetting factor recursive least squares method	[138]
Ant colony algorithm	[140]

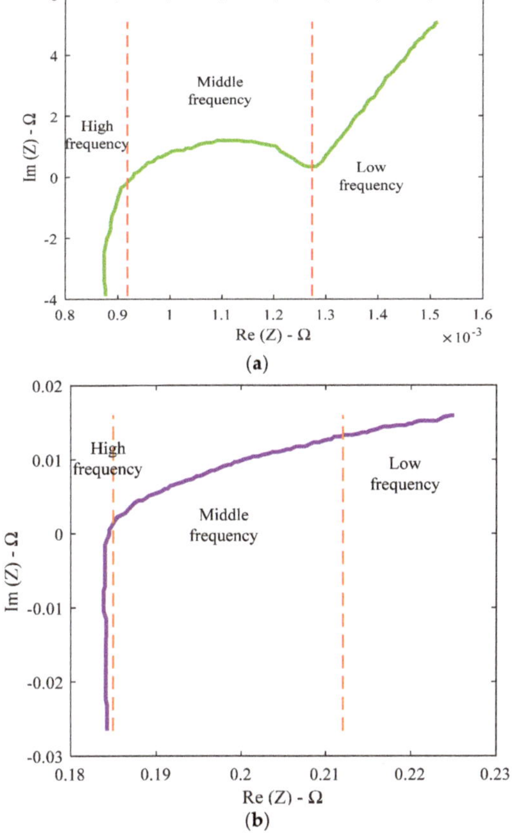

Figure 5. The sample electrochemical impedance spectra of energy storage systems. (**a**) Lithium-ion batteries; (**b**) Liquid metal batteries.

The estimation of the battery's parameters such as the SOC during real operation of the power network and also EV operation is a major and complex challenge due to the nonlinear properties of the storage systems. The FO technique is useful for analyzing and achieving the proper parameters. The FO models have unique advantages and more accuracy compared with the conventional electrochemical models and equivalent circuit models [120]. The fractional-order technique has been utilized to estimate the state of charge of ESSs used in EVs and MGs. In [122], the estimation power of the FO model was compared with the real-time data and the extended Kalman filter method. Figure 6 shows the SOC of the battery considering different estimation methods and the actual data. According to the results, the maximum estimation error of the FO method is 1.50%, while the maximum estimation error of the extended Kalman filter is 5.11%. Therefore, the fractional-order model reduces the estimation error by up to 71% [122].

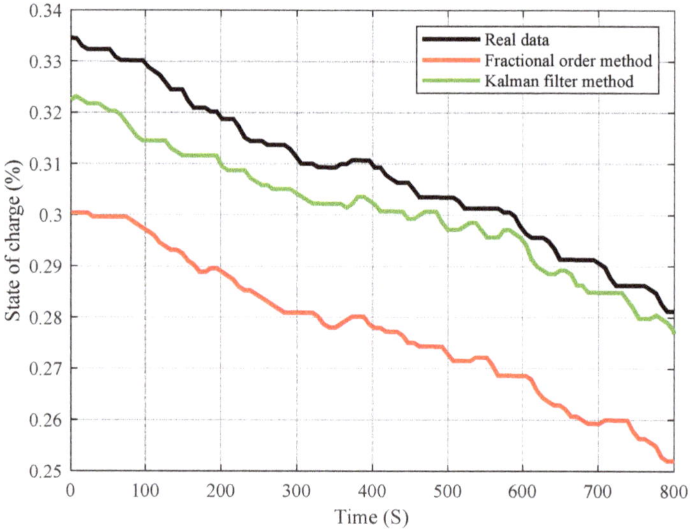

Figure 6. The state of charge of the energy storage system considering different methods.

In another study, experimental evaluations proved that the FO method has high stability and proper adaptability so that it appropriately identifies the electrochemical dynamics of the ESS with a 0.86% maximum relative absolute error [125].

The estimation error of the voltage of the ESS during the operation considering two methods is shown in Figure 7. As can be seen in this figure, the voltage estimation error of the FO model is lower than that of the integer-order model. The error rate of the fractional-order method is about 0.13%, while the estimation voltage of the integer-order method has an error of approximately 0.16%. Therefore, the fractional-order technique has more accuracy in estimating the voltage of the energy storage system [128].

The FO method also has more effectiveness and stability in controlling the nonlinear and uncertain parameters of the ESS. According to the numerical evaluations, the error rate of the fractional-order-based technique is 8.79%. At the same time, the feedback linearization control, the proportional integral derivative control, and the sliding-mode control have 31.86%, 14.54%, and 11.31% error rates, respectively. Thus, the FO method has more stability and efficiency than other methods [129].

In another study, the performance of the fractional-order method in estimating the SOC of the ESS was compared with the KF [137]. Experimental results presented that the average error of the Kalman filter is 0.73%, while the average error of the fractional-order model is 0.55%. Therefore, the estimated SOC by the fractional-order method has approximately 25% more accuracy than the result obtained by using the KF [137].

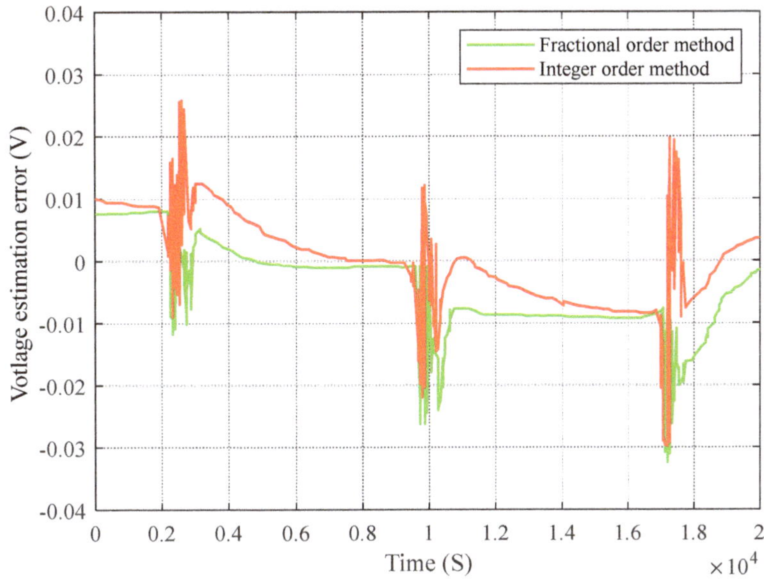

Figure 7. The voltage estimation error of energy storage system considering different methods.

The fractional-order methods have more stability than other methods in controlling systems. Table 5 presents the details of the energy system after applying a fault considering FOPID, TID, and PID controllers. In this table, the settling time (ST), the peak time (PT), the magnitude of the peak (MP), the integral of squared error (ISE), and the integral of time-multiplied absolute error (ITMAE) of the energy system after applying a fault to the system are given. As can be shown in this table, the FOPID method is about 31% faster than PID and TID controllers in controlling any disturbance in the system. Moreover, it has 33–50% and 45–57% lower ISE and ITMAE than classical controllers, respectively. The numerical results show that the fractional-order controller has more stability and lower error than other ones [151].

Table 5. The details of different controllers.

Control Strategy	Control Parameters				
	ST (s)	PT (s)	MP (Pu)	ISE	ITMAE
PID	25.85	1.76	0.07	0.04	7.67
TID	25.23	1.74	0.06	0.03	6.04
FOPID	17.81	1.74	0.04	0.02	3.31

5. Future Perspectives

Fractional-order systems have attracted the attention of many researchers in recent years due to their wide application in various branches of engineering, such as renewable energy systems, energy storage systems, secure communications, nonlinear control, information processing, biological systems, etc. The most important goals and challenges in applications of fractional-order systems that require more extensive investigation are as follows:

1. Utilizing the combination of fractional-order techniques and intelligent estimation methods in order to model uncertain and stochastic dynamics of RESs and ESSs in different operational conditions.

2. Considering fractional-order methods and training algorithms to design self-regulated systems for RESs and ESSs in order to respond to various practical faults in the distribution systems appropriately.

3. Studying fractional-order controllers of energy systems of RESs and ESSs considering the delay of measurement devices, which causes a delay in the output controlling signal, and investigating the effect of this delay on the practical performance of RESs and ESSs.

4. Studying the effects of estimations errors, the uncertainty of the system, and external perturbations on the modeling, controlling, and stability of fractional-order methods in the energy systems of RESs and ESSs.

6. Conclusions

A fractional-order system is a dynamical system that is modeled by fractional differential equations and a non-integer derivative. In other words, the fractional-order technique utilizes an impedance model based on the FO theory to identify, estimate, and control the energy system. In this paper, a comprehensive review of the energy system of renewable energy units and energy storage devices was presented. The mathematical fundamentals of the FO method were mentioned, and the various studies were categorized based on different parameters. The FO formulations were presented, and its most utilized definitions were formulated. Additionally, its applications in inverters and converters were investigated. Different studies and numerical evaluations present appropriate efficiency and stability of the FO techniques for estimating, controlling, and improving the performance of energy systems in various operational conditions. According to the different studies, the FO method has appropriate accuracy for estimating uncertain parameters. Its estimation error is considerably lower than that of other classical methods in practical systems. The fractional-order technique has more stability and lower steady-state error than other methods such as integer-order and electrochemical models, so the estimation and control errors of the FO technique are considerably lower than those of other ones. Moreover, it is also faster than other techniques. Therefore, it has appropriate performance in online, real-time, and complex operating conditions. Although fractional-order methods have attracted the attention of many researchers in recent years, modeling the uncertainties and stochastic dynamics of RESs and ESSs, designing self-regulated systems, considering the delay of measurement devices, and evaluating the effects of delays and estimation errors on the output controlling signals of FO-based controllers are the most important challenges in applications of FO systems in RESs and ESSs, which can be considered for more investigations in future projects about fractional-order methods.

Author Contributions: Conceptualization, M.A. and H.A.; methodology, M.A.; software, M.A.; validation, M.A. and A.O.; formal analysis, A.O.; investigation, B.M.-I.; resources, M.A. and H.A.; data curation, M.A.; writing—original draft preparation, M.A. and H.A.; writing—review and editing, A.O., B.M.-I. and R.T.; visualization, R.T.; supervision, B.M.-I.; project administration, M.A.; funding acquisition, B.M.-I. All authors have read and agreed to the published version of the manuscript.

Funding: This research is supported by the research grant of the University of Tabriz (number S-10).

Data Availability Statement: Not applicable.

Conflicts of Interest: The authors declare no conflict of interest.

Nomenclatures and Abbreviations (In Alphabetical Order)

ABC	Artificial bee colony
ACA	Ant colony algorithm
AEO	Artificial ecosystem-based optimization
ANA	Adaptive noise algorithm
ANN	Artificial neural network
BEL	Brain emotional learning
CD	Caputo definition

CPE	Constant phase element
CS	Cuckoo search
EIS	Electrochemical impedance spectroscopy
ESS	Energy storage system
EV	Electric vehicle
FO	Fractional order
FOPID	Fractional-order proportional integral derivative
GA	Genetic algorithm
GLD	Grünwald–Letnikov definition
GWO	Grey wolf optimization
HGAPSO	Hybrid of GA and PSO
HGPO	High-gain perturbation observer
ISE	Integral of squared error
ITAE	Integral time absolute error
ITLO	Interactive teaching–learning optimization
ITMAE	Integral of time multiplied absolute error
KF	Kalman filter
KHO	Krill herd optimization
LFC	Load frequency control
LIB	Lithium-ion battery
LMB	Liquid metal battery
LSA	Least square algorithm
MBA	Mine blast algorithm
MCA	Monte Carlo algorithm
MG	Microgrid
MGSO	Modified group search optimization
MP	Magnitude of the peak
MPC	Model predictive control
MPPT	Maximum power point tracking
MRFO	Manta ray foraging optimization
MSA	Moth swarm algorithm
OA	Optimization algorithm
OSD	Oldham and Spanier definition
PD	Proportional derivative
PI	Proportional integral
PID	Proportional integral derivative
PO	Perturb and observe technique
PSO	Particle swarm optimization
PT	peak time
PTM	Particle thermal method
PV	Photovoltaic panel
PWM	Pulse-width modulation
RBF	Radial basis function
RES	Renewable energy source
RLD	Riemann–Liouville definition
SCA	Sine cosine algorithm
SMESS	Superconducting magnetic energy storage system
SMT	Short memory technique
ST	Settling time
SG	Smart grid
SOC	State of charge
SOH	State of health
SOP	State of power
TID	Tilted integral derivative
TLBO	Teaching–learning-based optimization
VIC	Virtual inertia control
WOA	Whale optimization algorithm
WT	Wind turbine

References

1. Mohamed, M.A. A relaxed consensus plus innovation based effective negotiation approach for energy cooperation between smart grid and microgrid. *Energy* **2022**, *252*, 123996. [CrossRef]
2. Alilou, M.; Tousi, B.; Shayeghi, H. Multi-objective energy management of smart homes considering uncertainty in wind power forecasting. *Electr. Eng.* **2021**, *103*, 1367–1383. [CrossRef]
3. Nazari-Heris, M.; Abapour, M.; Mohammadi-Ivatloo, B. An Updated Review and Outlook on Electric Vehicle Aggregators in Electric Energy Networks. *Sustainability* **2022**, *14*, 5747. [CrossRef]
4. Rezaei, H.; Abdollahi, S.E.; Abdollahi, S.; Filizadeh, S. Energy management strategies of battery-ultracapacitor hybrid storage systems for electric vehicles: Review, challenges, and future trends. *J. Energy Storage* **2022**, *53*, 105045. [CrossRef]
5. Aliasghari, P.; Mohammadi-Ivatloo, B.; Alipour, M.; Abapour, M.; Zare, K. Optimal scheduling of plug-in electric vehicles and renewable micro-grid in energy and reserve markets considering demand response program. *J. Clean. Prod.* **2018**, *186*, 293–303. [CrossRef]
6. Erdinç, F.G.; Çiçek, A.; Erdinç, O.; Yumurtacı, R.; Oskouei, M.Z.; Mohammadi-Ivatloo, B. Decision-making framework for power system with RES including responsive demand, ESSs, EV aggregator and dynamic line rating as multiple flexibility resources. *Electr. Power Syst. Res.* **2022**, *204*, 107702. [CrossRef]
7. Alilou, M.; Gharehpetian, G.B.; Ahmadiahangar, R.; Rosin, A.; Anvari-Moghaddam, A. Day-Ahead Scheduling of Electric Vehicles and Electrical Storage Systems in Smart Homes Using a Novel Decision Vector and AHP Method. *Sustainability* **2022**, *14*, 11773. [CrossRef]
8. Zhang, G.; Ge, Y.; Ye, Z.; Al-Bahrani, M. Multi-objective planning of energy hub on economic aspects and resources with heat and power sources, energizable, electric vehicle and hydrogen storage system due to uncertainties and demand response. *J. Energy Storage* **2023**, *57*, 106160. [CrossRef]
9. Sun, H.; Zhang, Y.; Baleanu, D.; Chen, W.; Chen, Y. A new collection of real world applications of fractional calculus in science and engineering. *Commun. Nonlinear Sci. Numer. Simul.* **2018**, *64*, 213–231. [CrossRef]
10. Shah, P.; Agashe, S. Review of fractional PID controller. *Mechatronics* **2016**, *38*, 29–41. [CrossRef]
11. Hidalgo-Reyes, J.I.; Gómez-Aguilar, J.F.; Escobar-Jiménez, R.F.; Alvarado-Martínez, V.M.; López-López, M.G. Classical and fractional-order modeling of equivalent electrical circuits for supercapacitors and batteries, energy management strategies for hybrid systems and methods for the state of charge estimation: A state of the art review. *Microelectron. J.* **2019**, *85*, 109–128. [CrossRef]
12. Huang, S.; Zhou, B.; Li, C.; Wu, Q.; Xia, S.; Wang, H.; Yang, H. Fractional-order modeling and sliding mode control of energy-saving and emission-reduction dynamic evolution system. *Int. J. Electr. Power Energy Syst.* **2018**, *100*, 400–410. [CrossRef]
13. Wang, X.; Wei, X.; Meng, Y. Experiment on Grid-Connection Process of Wind Turbines in Fractional Frequency Wind Power System. *IEEE Trans. Energy Convers.* **2015**, *30*, 22–31. [CrossRef]
14. Sondhi, S.; Hote, Y. V Fractional order PID controller for perturbed load frequency control using Kharitonov's theorem. *Int. J. Electr. Power Energy Syst.* **2016**, *78*, 884–896. [CrossRef]
15. Aghababa, M.P.; Haghighi, A.R.; Roohi, M. Stabilisation of unknown fractional-order chaotic systems: An adaptive switching control strategy with application to power systems. *IET Gener. Transm. Distrib.* **2015**, *9*, 1883–1893. [CrossRef]
16. Aghababa, M.P. Fractional modeling and control of a complex nonlinear energy supply-demand system. *Complexity* **2015**, *20*, 74–86. [CrossRef]
17. Borah, M.; Roy, B.K. Dynamics of the fractional-order chaotic PMSG, its stabilisation using predictive control and circuit validation. *IET Electr. Power Appl.* **2017**, *11*, 707–716. [CrossRef]
18. Mohanty, A.; Viswavandya, M.; Mohanty, S. An optimised FOPID controller for dynamic voltage stability and reactive power management in a stand-alone micro grid. *Int. J. Electr. Power Energy Syst.* **2016**, *78*, 524–536. [CrossRef]
19. Liu, L.; Qian, J.; Hua, L.; Zhang, B. System estimation of the SOFCs using fractional-order social network search algorithm. *Energy* **2022**, *255*, 124516. [CrossRef]
20. Ye, L.; Peng, D.; Xue, D.; Chen, S.; Shi, A. Co-estimation of lithium-ion battery state-of-charge and state-of-health based on fractional-order model. *J. Energy Storage* **2023**, *65*, 107225. [CrossRef]
21. Chen, L.; Wang, S.; Jiang, H.; Fernandez, C. A novel combined estimation method for state of energy and predicted maximum available energy based on fractional-order modeling. *J. Energy Storage* **2023**, *62*, 106930. [CrossRef]
22. Ma, L.; Xu, Y.; Zhang, H.; Yang, F.; Wang, X.; Li, C. Co-estimation of state of charge and state of health for lithium-ion batteries based on fractional-order model with multi-innovations unscented Kalman filter method. *J. Energy Storage* **2022**, *52*, 104904. [CrossRef]
23. Liu, S.; Sun, H.; Yu, H.; Miao, J.; Zheng, C.; Zhang, X. A framework for battery temperature estimation based on fractional electro-thermal coupling model. *J. Energy Storage* **2023**, *63*, 107042. [CrossRef]
24. Raheem, A.; Afreen, A.; Khatoon, A. Multi-term time-fractional stochastic system with multiple delays in control. *Chaos Solitons Fractals* **2023**, *167*, 112979. [CrossRef]
25. Mousavi, Y.; Bevan, G.; Kucukdemiral, I.B.; Fekih, A. Sliding mode control of wind energy conversion systems: Trends and applications. *Renew. Sustain. Energy Rev.* **2022**, *167*, 112734. [CrossRef]

26. Darvish Falehi, A.; Torkaman, H. Promoted supercapacitor control scheme based on robust fractional-order super-twisting sliding mode control for dynamic voltage restorer to enhance FRT and PQ capabilities of DFIG-based wind turbine. *J. Energy Storage* **2021**, *42*, 102983. [CrossRef]
27. Girgis, M.E.; Fahmy, R.A.; Badr, R.I. Optimal fractional-order PID control for plasma shape, position, and current in Tokamaks. *Fusion Eng. Des.* **2020**, *150*, 111361. [CrossRef]
28. Gao, Q.; Cai, J.; Liu, Y.; Chen, Y.; Shi, L.; Xu, W. Power mapping-based stability analysis and order adjustment control for fractional-order multiple delayed systems. *ISA Trans.* 2023, *Online ahead of print*. [CrossRef]
29. Lusekelo, E.; Helikumi, M.; Kuznetsov, D.; Mushayabasa, S. Dynamic modelling and optimal control analysis of a fractional order chikungunya disease model with temperature effects. *Results Control Optim.* **2023**, *10*, 100206. [CrossRef]
30. Barry, L.E.; O'Neill, C.; Butler, C.; Chaudhuri, R.; Heaney, L.G. Cost-Effectiveness of Fractional Exhaled Nitric Oxide Suppression Testing as an Adherence Screening Tool Among Patients With Difficult-to-Control Asthma. *J. Allergy Clin. Immunol. Pract.* 2023, *Online ahead of print*. [CrossRef]
31. Huang, S.; Wang, J.; Huang, C.; Zhou, L.; Xiong, L.; Liu, J.; Li, P. A fixed-time fractional-order sliding mode control strategy for power quality enhancement of PMSG wind turbine. *Int. J. Electr. Power Energy Syst.* **2022**, *134*, 107354. [CrossRef]
32. Hao, J.; Wang, J.; Yu, D.; Zhu, J.; Moattari, M. Fractional-order pathfinder algorithm for optimum design of an HRES based on photovoltaic and proton exchange membrane fuel cell: A case study. *Int. J. Hydrogen Energy*, 2023; in press. [CrossRef]
33. Zhang, Q.; Shang, Y.; Li, Y.; Cui, N.; Duan, B.; Zhang, C. A novel fractional variable-order equivalent circuit model and parameter identification of electric vehicle Li-ion batteries. *ISA Trans.* **2020**, *97*, 448–457. [CrossRef]
34. Yang, B.; Wang, J.; Sang, Y.; Yu, L.; Shu, H.; Li, S.; He, T.; Yang, L.; Zhang, X.; Yu, T. Applications of supercapacitor energy storage systems in microgrid with distributed generators via passive fractional-order sliding-mode control. *Energy* **2019**, *187*, 115905. [CrossRef]
35. Statistical Review of World Energy. 2022. Available online: www.bp.com (accessed on 1 June 2022).
36. François, B.; Puspitarini, H.D.; Volpi, E.; Borga, M. Statistical analysis of electricity supply deficits from renewable energy sources across an Alpine transect. *Renew. Energy* **2022**, *201*, 1200–1212. [CrossRef]
37. Alilou, M.; Toosi, B.; Shayeghi, H. Multi-objective unit and load commitment in smart homes considering uncertainties. *Int. Trans. Electr. Energy Syst.* **2020**, *30*, e12614. [CrossRef]
38. Singh, S.; Tayal, V.K.; Singh, H.P.; Yadav, V.K. Fractional Control Design of Renewable Energy Systems. In Proceedings of the ICRITO 2020 8th International Conference on Reliability, Infocom Technologies and Optimization (Trends and Future Directions), Noida, India, 4–5 June 2020; pp. 1246–1251. [CrossRef]
39. Azami, H.; Alizadeh, B.A.M.; Abapour, M. Optimal Smart Home Scheduling with Considering Hybrid Resource Management. In Proceedings of the 2021 11th Smart Grid Conference (SGC), Tabriz, Iran, 7–9 December 2021. [CrossRef]
40. Thangam, T.; Muthuvel, M.K. Passive Fractional-order Proportional-Integral-Derivative control design of a Grid-connected Photovoltaic inverter for Maximum Power Point Tracking. *Comput. Electr. Eng.* **2022**, *97*, 107657. [CrossRef]
41. Oshnoei, S.; Oshnoei, A.; Mosallanejad, A.; Haghjoo, F. Contribution of GCSC to regulate the frequency in multi-area power systems considering time delays: A new control outline based on fractional order controllers. *Int. J. Electr. Power Energy Syst.* **2020**, *123*, 106197. [CrossRef]
42. Oshnoei, S.; Aghamohammadi, M.; Oshnoei, S.; Oshnoei, A.; Mohammadi-Ivatloo, B. Provision of Frequency Stability of an Islanded Microgrid Using a Novel Virtual Inertia Control and a Fractional Order Cascade Controller. *Energies* **2021**, *14*, 4152. [CrossRef]
43. Suo, J.; Shi, M.; Li, Y.; Yang, Y. Proportional-integral control for synchronization of complex dynamical networks under dynamic event-triggered mechanism. *J. Frankl. Inst.* **2023**, *360*, 1436–1453. [CrossRef]
44. Çelik, E.; Öztürk, N. Novel fuzzy 1PD-TI controller for AGC of interconnected electric power systems with renewable power generation and energy storage devices. *Eng. Sci. Technol. Int. J.* **2022**, *35*, 101166. [CrossRef]
45. Aryan, P.; Raja, G.L. Restructured LFC Scheme with Renewables and EV Penetration using Novel QOEA Optimized Parallel Fuzzy I-PID Controller. *IFAC-Pap.* **2022**, *55*, 460–466. [CrossRef]
46. Oberlin, P.; Rathinam, S.; Darbha, S. A transformation for a Heterogeneous, Multiple Depot, Multiple Traveling Salesman Problem. In Proceedings of the 2009 American Control Conference, Saint Louis, Mo, USA, 10–12 June 2009; IEEE: New York, NY, USA, 2009; pp. 1292–1297.
47. Oshnoei, A.; Sadeghian, O.; Anvari-Moghaddam, A. Intelligent Power Control of Inverter Air Conditioners in Power Systems: A Brain Emotional Learning-Based Approach. *IEEE Trans. Power Syst.* **2022**, 1–15. [CrossRef]
48. Yang, S.; Wan, M.P.; Chen, W.; Ng, B.F.; Dubey, S. Model predictive control with adaptive machine-learning-based model for building energy efficiency and comfort optimization. *Appl. Energy* **2020**, *271*, 115147. [CrossRef]
49. Oshnoei, A.; Kheradmandi, M.; Muyeen, S.M. Robust Control Scheme for Distributed Battery Energy Storage Systems in Load Frequency Control. *IEEE Trans. Power Syst.* **2020**, *35*, 4781–4791. [CrossRef]
50. Oshnoei, A.; Khezri, R.; Muyeen, S.M.; Oshnoei, S.; Blaabjerg, F. Automatic Generation Control Incorporating Electric Vehicles. *Electr. Power Compon. Syst.* **2019**, *47*, 720–732. [CrossRef]
51. Oshnoei, S.; Oshnoei, A.; Mosallanejad, A.; Haghjoo, F. Novel load frequency control scheme for an interconnected two-area power system including wind turbine generation and redox flow battery. *Int. J. Electr. Power Energy Syst.* **2021**, *130*, 107033. [CrossRef]

52. Al-Dhaifallah, M.; Nassef, A.M.; Rezk, H.; Nisar, K.S. Optimal parameter design of fractional order control based INC-MPPT for PV system. *Sol. Energy* **2018**, *159*, 650–664. [CrossRef]

53. Ahmed, E.M.; Selim, A.; Mohamed, E.A.; Aly, M.; Alnuman, H.; Ramadan, H.A. Modified manta ray foraging optimization algorithm based improved load frequency controller for interconnected microgrids. *IET Renew. Power Gener.* **2022**, *16*, 3587–3613. [CrossRef]

54. Ahmed, E.M.; Elmelegi, A.; Shawky, A.; Aly, M.; Alhosaini, W.; Mohamed, E.A. Frequency Regulation of Electric Vehicle-Penetrated Power System Using MPA-Tuned New Combined Fractional Order Controllers. *IEEE Access* **2021**, *9*, 107548–107565. [CrossRef]

55. Song, K.; Hu, D.; Tong, Y.; Yue, X. Remaining life prediction of lithium-ion batteries based on health management: A review. *J. Energy Storage* **2023**, *57*, 106193. [CrossRef]

56. Bhat, M.Y.; Hashmi, S.A.; Khan, M.; Choi, D.; Qurashi, A. Frontiers and recent developments on supercapacitor's materials, design, and applications: Transport and power system applications. *J. Energy Storage* **2023**, *58*, 106104. [CrossRef]

57. Elkasem, A.H.A.; Khamies, M.; Hassan, M.H.; Agwa, A.M.; Kamel, S. Optimal Design of TD-TI Controller for LFC Considering Renewables Penetration by an Improved Chaos Game Optimizer. *Fractal Fract.* **2022**, *6*, 220. [CrossRef]

58. Chen, Y.; Li, R.; Sun, Z.; Zhao, L.; Guo, X. SOC estimation of retired lithium-ion batteries for electric vehicle with improved particle filter by H-infinity filter. *Energy Rep.* **2023**, *9*, 1937–1947. [CrossRef]

59. Shrivastava, P.; Soon, T.K.; Bin Idris, M.Y.I.; Mekhilef, S.; Adnan, S.B.R.S. Comprehensive co-estimation of lithium-ion battery state of charge, state of energy, state of power, maximum available capacity, and maximum available energy. *J. Energy Storage* **2022**, *56*, 106049. [CrossRef]

60. Sheng, C.; Fu, J.; Li, D.; Jiang, C.; Guo, Z.; Li, B.; Lei, J.; Zeng, L.; Deng, Z.; Fu, X.; et al. Energy management strategy based on health state for a PEMFC/Lithium-ion batteries hybrid power system. *Energy Convers. Manag.* **2022**, *271*, 116330. [CrossRef]

61. Chen, L.; Wu, X.; Tenreiro Machado, J.A.; Lopes, A.M.; Li, P.; Dong, X. State-of-Charge Estimation of Lithium-Ion Batteries Based on Fractional-Order Square-Root Unscented Kalman Filter. *Fractal Fract.* **2022**, *6*, 52. [CrossRef]

62. Chen, J.; Feng, X.; Jiang, L.; Zhu, Q. State of charge estimation of lithium-ion battery using denoising autoencoder and gated recurrent unit recurrent neural network. *Energy* **2021**, *227*, 120451. [CrossRef]

63. Wang, X.; Sun, Q.; Kou, X.; Ma, W.; Zhang, H.; Liu, R. Noise immune state of charge estimation of li-ion battery via the extreme learning machine with mixture generalized maximum correntropy criterion. *Energy* **2022**, *239*, 122406. [CrossRef]

64. Li, X.; Wang, Z.; Zhang, L. Co-estimation of capacity and state-of-charge for lithium-ion batteries in electric vehicles. *Energy* **2019**, *174*, 33–44. [CrossRef]

65. Ma, Y.; Duan, P.; Sun, Y.; Chen, H. Equalization of Lithium-Ion Battery Pack Based on Fuzzy Logic Control in Electric Vehicle. *IEEE Trans. Ind. Electron.* **2018**, *65*, 6762–6771. [CrossRef]

66. Xing, Y.; He, W.; Pecht, M.; Tsui, K.L. State of charge estimation of lithium-ion batteries using the open-circuit voltage at various ambient temperatures. *Appl. Energy* **2014**, *113*, 106–115. [CrossRef]

67. Xu, J.; Mi, C.C.; Cao, B.; Deng, J.; Chen, Z.; Li, S. The State of Charge Estimation of Lithium-Ion Batteries Based on a Proportional-Integral Observer. *IEEE Trans. Veh. Technol.* **2014**, *63*, 1614–1621. [CrossRef]

68. Hubert, A.; Forgez, C.; Yvars, P.-A. Designing the architecture of electrochemical energy storage systems. A model-based system synthesis approach. *J. Energy Storage* **2022**, *54*, 105351. [CrossRef]

69. Mohammed, A.; Ghaithan, A.M.; Al-Hanbali, A.; Attia, A.M. A multi-objective optimization model based on mixed integer linear programming for sizing a hybrid PV-hydrogen storage system. *Int. J. Hydrogen Energy* **2022**, *48*, 9748–9761. [CrossRef]

70. Shi, Q.; Guo, Z.; Wang, S.; Yan, S.; Zhou, X.; Li, H.; Wang, K.; Jiang, K. Physics-based fractional-order model and parameters identification of liquid metal battery. *Electrochim. Acta* **2022**, *428*, 140916. [CrossRef]

71. Bingi, K.; Rajanarayan Prusty, B.; Pal Singh, A. A Review on Fractional-Order Modelling and Control of Robotic Manipulators. *Fractal Fract.* **2023**, *7*, 77. [CrossRef]

72. Daraz, A.; Malik, S.A.; Basit, A.; Aslam, S.; Zhang, G. Modified FOPID Controller for Frequency Regulation of a Hybrid Interconnected System of Conventional and Renewable Energy Sources. *Fractal Fract.* **2023**, *7*, 89. [CrossRef]

73. Ali, F.; Iftikhar, M.; Khan, I.; Sheikh, N.A.; Aamina; Nisar, K.S. Time fractional analysis of electro-osmotic flow of Walters's-B fluid with time-dependent temperature and concentration. *Alex. Eng. J.* **2020**, *59*, 25–38. [CrossRef]

74. Kaleem, M.M.; Usman, M.; Asjad, M.I.; Eldin, S.M. Magnetic Field, Variable Thermal Conductivity, Thermal Radiation, and Viscous Dissipation Effect on Heat and Momentum of Fractional Oldroyd-B Bio Nano-Fluid within a Channel. *Fractal Fract.* **2022**, *6*, 712. [CrossRef]

75. Eyebe, G.J.; Betchewe, G.; Mohamadou, A.; Kofane, T.C. Nonlinear Vibration of a Nonlocal Nanobeam Resting on Fractional-Order Viscoelastic Pasternak Foundations. *Fractal Fract.* **2018**, *2*, 21. [CrossRef]

76. Alazman, I.; Alkahtani, B.S.T. Investigation of Novel Piecewise Fractional Mathematical Model for COVID-19. *Fractal Fract.* **2022**, *6*, 661. [CrossRef]

77. Sajid, M.; Chaudhary, H.; Allahem, A.; Kaushik, S. Chaos Controllability in Fractional-Order Systems via Active Dual Combination–Combination Hybrid Synchronization Strategy. *Fractal Fract.* **2022**, *6*, 717. [CrossRef]

78. Bertsias, P.; Psychalinos, C.; Minaei, S.; Yesil, A.; Elwakil, A.S. Fractional-order inverse filters revisited: Equivalence with fractional-order controllers. *Microelectron. J.* **2023**, *131*, 105646. [CrossRef]

79. Meng, X.; Jiang, B.; Karimi, H.R.; Gao, C. An event-triggered mechanism to observer-based sliding mode control of fractional-order uncertain switched systems. *ISA Trans.* **2022**, *135*, 115–129. [CrossRef] [PubMed]

80. Xu, Y.; Zhang, H.; Zhang, J.; Yang, F.; Tong, L.; Yan, D.; Yang, H.; Wang, Y. State of charge estimation under different temperatures using unscented Kalman filter algorithm based on fractional-order model with multi-innovation. *J. Energy Storage* **2022**, *56*, 106101. [CrossRef]

81. Jin, X.-C.; Lu, J.-G.; Zhang, Q.-H. Delay-dependent and order-dependent conditions for stability and stabilization of fractional-order memristive neural networks with time-varying delays. *Neurocomputing* **2023**, *522*, 53–63. [CrossRef]

82. Ortigueira, M.D.; Rodríguez-Germá, L.; Trujillo, J.J. Complex Grünwald–Letnikov, Liouville, Riemann–Liouville, and Caputo derivatives for analytic functions. *Commun. Nonlinear Sci. Numer. Simul.* **2011**, *16*, 4174–4182. [CrossRef]

83. Scherer, R.; Kalla, S.L.; Tang, Y.; Huang, J. The Grünwald–Letnikov method for fractional differential equations. *Comput. Math. Appl.* **2011**, *62*, 902–917. [CrossRef]

84. Haq, A.; Sukavanam, N. Existence and partial approximate controllability of nonlinear Riemann–Liouville fractional systems of higher order. *Chaos Solitons Fractals* **2022**, *165*, 112783. [CrossRef]

85. Turkyilmazoglu, M.; Altanji, M. Fractional models of falling object with linear and quadratic frictional forces considering Caputo derivative. *Chaos Solitons Fractals* **2023**, *166*, 112980. [CrossRef]

86. Shukla, K.; Sapra, P. Fractional Calculus and Its Applications for Scientific Professionals: A Literature Review. *Int. J. Mod. Math. Sci.* **2019**, *17*, 111–137.

87. Frikh, M.L.; Soltani, F.; Bensiali, N.; Boutasseta, N.; Fergani, N. Fractional order PID controller design for wind turbine systems using analytical and computational tuning approaches. *Comput. Electr. Eng.* **2021**, *95*, 107410. [CrossRef]

88. Dudhe, S.; Kumar Dheer, D.; Lloyds Raja, G. Modeling and control of suction pressure in portable meconium aspirator system using fractional order IMC-PID controller and RDR techniques. *Mater. Today Proc.* **2023**, *80*, 320–326. [CrossRef]

89. Xuet, D.; Chen, Y.; Systems, I. A comparative introduction of four fractional order controllers. In Proceedings of the 4th World Congress on Intelligent Control and Automation, online, 10–14 June 2002; pp. 3228–3235.

90. Hasan, R.; Masud, M.S.; Haque, N.; Abdussami, M.R. Frequency control of nuclear-renewable hybrid energy systems using optimal PID and FOPID controllers. *Heliyon* **2022**, *8*, e11770. [CrossRef] [PubMed]

91. Yumuk, E.; Güzelkaya, M.; Eksin, İ. Analytical fractional PID controller design based on Bode's ideal transfer function plus time delay. *ISA Trans.* **2019**, *91*, 196–206. [CrossRef]

92. Kumar, A.; Kumar, V. Hybridized ABC-GA optimized fractional order fuzzy pre-compensated FOPID control design for 2-DOF robot manipulator. *AEU-Int. J. Electron. Commun.* **2017**, *79*, 219–233. [CrossRef]

93. Dastjerdi, A.A.; Vinagre, B.M.; Chen, Y.Q.; HosseinNia, S.H. Linear fractional order controllers; A survey in the frequency domain. *Annu. Rev. Control* **2019**, *47*, 51–70. [CrossRef]

94. Vanitha, D.; Rathinakumar, M. Fractional order PID controlled PV buck boost converter with coupled inductor. *Int. J. Power Electron. Drive Syst.* **2017**, *8*, 1401–1407.

95. Soriano-Sánchez, A.G.; Rodríguez-Licea, M.A.; Pérez-Pinal, F.J.; Vázquez-López, J.A. Fractional-order approximation and synthesis of a PID controller for a buck converter. *Energies* **2020**, *13*, 629. [CrossRef]

96. Mollaee, H.; Ghamari, S.M.; Saadat, S.A.; Wheeler, P. A novel adaptive cascade controller design on a buck–boost DC–DC converter with a fractional-order PID voltage controller and a self-tuning regulator adaptive current controller. *IET Power Electron.* **2021**, *14*, 1920–1935. [CrossRef]

97. Wei, Z.; Zhang, B.; Jiang, Y. Analysis and Modeling of Fractional-Order Buck Converter Based on Riemann-Liouville Derivative. *IEEE Access* **2019**, *7*, 162768–162777. [CrossRef]

98. Kumar, V.A.; Mouttou, A. Improved performance with fractional order control for asymmetrical cascaded h-bridge multilevel inverter. *Bull. Electr. Eng. Inform.* **2020**, *9*, 1335–1344. [CrossRef]

99. Silva Júnior, D.C.; Oliveira, J.G.; de Almeida, P.M.; Boström, C. Control of a multi-functional inverter in an AC microgrid–Real-time simulation with control hardware in the loop. *Electr. Power Syst. Res.* **2019**, *172*, 201–212. [CrossRef]

100. Lai, J.; Yin, X.; Yin, X.; Jiang, L. Fractional order harmonic disturbance observer control for three-phase LCL-type inverter. *Control Eng. Pract.* **2021**, *107*, 104697. [CrossRef]

101. Arulmurugan, R.; Suthanthiravanitha, N. Improved Fractional Order VSS Inc-Cond MPPT Algorithm for Photovoltaic Scheme. *Int. J. Photoenergy* **2014**, *2014*, 1–10. [CrossRef]

102. Yang, B.; Yu, T.; Shu, H.; Zhu, D.; Zeng, F.; Sang, Y.; Jiang, L. Perturbation observer based fractional-order PID control of photovoltaics inverters for solar energy harvesting via Yin-Yang-Pair optimization. *Energy Convers. Manag.* **2018**, *171*, 170–187. [CrossRef]

103. Tasnin, W.; Saikia, L.C.; Raju, M. Deregulated AGC of multi-area system incorporating dish-Stirling solar thermal and geothermal power plants using fractional order cascade controller. *Int. J. Electr. Power Energy Syst.* **2018**, *101*, 60–74. [CrossRef]

104. Nosrati, K.; Mansouri, H.R.; Saboori, H. Fractional-order PID controller design of frequency deviation in a hybrid renewable energy generation and storage system. *CIRED-Open Access Proc. J.* **2017**, *2017*, 1148–1152. [CrossRef]

105. Pan, I.; Das, S. Fractional order fuzzy control of hybrid power system with renewable generation using chaotic PSO. *ISA Trans.* **2016**, *62*, 19–29. [CrossRef]

106. Gül, O.; Tan, N. Application of fractional-order voltage controller in building-integrated photovoltaic and wind turbine system. *Meas. Control* **2019**, *52*, 1145–1158. [CrossRef]

107. Pathak, D.; Gaur, P. A fractional order fuzzy-proportional-integral-derivative based pitch angle controller for a direct-drive wind energy system. *Comput. Electr. Eng.* **2019**, *78*, 420–436. [CrossRef]
108. Asgharnia, A.; Jamali, A.; Shahnazi, R.; Maheri, A. Load mitigation of a class of 5-MW wind turbine with RBF neural network based fractional-order PID controller. *ISA Trans.* **2020**, *96*, 272–286. [CrossRef] [PubMed]
109. Beschi, M.; Padula, F.; Visioli, A. Fractional robust PID control of a solar furnace. *Control Eng. Pract.* **2016**, *56*, 190–199. [CrossRef]
110. Sahin, E. A PSO Optimized Fractional-Order PID Controller for a PV System with DC-DC Boost Converter. In Proceedings of the 2014 16th International Power Electronics and Motion Control Conference and Exposition, Antalya, Turkey, 21–24 September 2014; pp. 477–481.
111. Ramadan, H.S. Optimal fractional order PI control applicability for enhanced dynamic behavior of on-grid solar PV systems. *Int. J. Hydrogen Energy* **2017**, *42*, 4017–4031. [CrossRef]
112. Ahmed, E.M.; Member, S.; Mohamed, E.A.; Elmelegi, A. Optimum Modified Fractional Order Controller for Future Electric Vehicles and Renewable Energy-Based Interconnected Power Systems. *IEEE Access* **2021**, *9*, 29993–30010. [CrossRef]
113. Elmelegi, A.; Mohamed, E.A.; Aly, M.; Ahmed, E.M.; Mohamed, A.A.A.; Elbaksawi, O. Optimized Tilt Fractional Order Cooperative Controllers for Preserving Frequency Stability in Renewable Energy-Based Power Systems. *IEEE Access* **2021**, *9*, 8261–8277. [CrossRef]
114. Fathy, A.; Rezk, H.; Ferahtia, S.; Ghoniem, R.M.; Alkanhel, R.; Ghoniem, M.M. A New Fractional-Order Load Frequency Control for Multi-Renewable Energy Interconnected Plants Using Skill Optimization Algorithm. *Sustainability* **2022**, *14*, 14999. [CrossRef]
115. Daraz, A.; Malik, S.A.; Azar, A.T.; Aslam, S.; Alkhalifah, T.; Alturise, F. Optimized Fractional Order Integral-Tilt Derivative Controller for Frequency Regulation of Interconnected Diverse Renewable Energy Resources. *IEEE Access* **2022**, *10*, 43514–43527. [CrossRef]
116. Mohamed, E.A.; Ahmed, E.M.; Elmelegi, A.; Aly, M.; Elbaksawi, O.; Mohamed, A.A.A. An Optimized Hybrid Fractional Order Controller for Frequency Regulation in Multi-Area Power Systems. *IEEE Access* **2020**, *8*, 213899–213915. [CrossRef]
117. Yu, K.N.; Liao, C.K.; Yau, H.T. A New Fractional-Order Based Intelligent Maximum Power Point Tracking Control Algorithm for Photovoltaic Power Systems. *Int. J. Photoenergy* **2015**, *2015*, 1–8. [CrossRef]
118. Yang, B.; Yu, T.; Shu, H.; Zhu, D.; An, N.; Sang, Y.; Jiang, L. Energy reshaping based passive fractional-order PID control design and implementation of a grid-connected PV inverter for MPPT using grouped grey wolf optimizer. *Sol. Energy* **2018**, *170*, 31–46. [CrossRef]
119. Mu, H.; Xiong, R.; Zheng, H.; Chang, Y.; Chen, Z. A novel fractional order model based state-of-charge estimation method for lithium-ion battery. *Appl. Energy* **2017**, *207*, 384–393. [CrossRef]
120. Wang, Y.; Li, M.; Chen, Z. Experimental study of fractional-order models for lithium-ion battery and ultra-capacitor: Modeling, system identification, and validation. *Appl. Energy* **2020**, *278*, 115736. [CrossRef]
121. Guo, R.; Shen, W. Online state of charge and state of power co-estimation of lithium-ion batteries based on fractional-order calculus and model predictive control theory. *Appl. Energy* **2022**, *327*, 120009. [CrossRef]
122. Chen, L.; Guo, W.; Lopes, A.M.; Wu, R.; Li, P.; Yin, L. State-of-charge estimation for lithium-ion batteries based on incommensurate fractional-order observer. *Commun. Nonlinear Sci. Numer. Simul.* **2023**, *118*, 107059. [CrossRef]
123. Xu, C.; Cheng, S.; Wang, K.; Jiang, K. A Fractional-order Model for Liquid Metal Batteries. *Energy Procedia* **2019**, *158*, 4690–4695. [CrossRef]
124. Lu, X.; Li, H.; Chen, N. An indicator for the electrode aging of lithium-ion batteries using a fractional variable order model. *Electrochim. Acta* **2019**, *299*, 378–387. [CrossRef]
125. Ruan, H.; Sun, B.; Jiang, J.; Zhang, W.; He, X.; Su, X.; Bian, J.; Gao, W. A modified-electrochemical impedance spectroscopy-based multi-time-scale fractional-order model for lithium-ion batteries. *Electrochim. Acta* **2021**, *394*, 139066. [CrossRef]
126. Zhu, G.; Kong, C.; Wang, J.V.; Kang, J.; Yang, G.; Wang, Q. A fractional-order model of lithium-ion battery considering polarization in electrolyte and thermal effect. *Electrochim. Acta* **2023**, *438*, 141461. [CrossRef]
127. Fouda, M.E.; Elwakil, A.S.; Radwan, A.G.; Allagui, A. Power and energy analysis of fractional-order electrical energy storage devices. *Energy* **2016**, *111*, 785–792. [CrossRef]
128. Zhu, Q.; Xu, M.; Liu, W.; Zheng, M. A state of charge estimation method for lithium-ion batteries based on fractional order adaptive extended kalman filter. *Energy* **2019**, *187*, 115880. [CrossRef]
129. Yang, B.; Zhu, T.; Zhang, X.; Wang, J.; Shu, H.; Li, S.; He, T.; Yang, L.; Yu, T. Design and implementation of Battery/SMES hybrid energy storage systems used in electric vehicles: A nonlinear robust fractional-order control approach. *Energy* **2020**, *191*, 116510. [CrossRef]
130. He, L.; Wang, Y.; Wei, Y.; Wang, M.; Hu, X.; Shi, Q. An adaptive central difference Kalman filter approach for state of charge estimation by fractional order model of lithium-ion battery. *Energy* **2022**, *244*, 122627. [CrossRef]
131. Guo, D.; Yang, G.; Feng, X.; Han, X.; Lu, L.; Ouyang, M. Physics-based fractional-order model with simplified solid phase diffusion of lithium-ion battery. *J. Energy Storage* **2020**, *30*, 101404. [CrossRef]
132. Li, S.; Li, Y.; Zhao, D.; Zhang, C. Adaptive state of charge estimation for lithium-ion batteries based on implementable fractional-order technology. *J. Energy Storage* **2020**, *32*, 101838. [CrossRef]
133. Liu, C.; Hu, M.; Jin, G.; Xu, Y.; Zhai, J. State of power estimation of lithium-ion battery based on fractional-order equivalent circuit model. *J. Energy Storage* **2021**, *41*, 102954. [CrossRef]

134. Solomon, O.O.; Zheng, W.; Chen, J.; Qiao, Z. State of charge estimation of Lithium-ion battery using an improved fractional-order extended Kalman filter. *J. Energy Storage* **2022**, *49*, 104007. [CrossRef]
135. Miao, Y.; Gao, Z. Estimation for state of charge of lithium-ion batteries by adaptive fractional-order unscented Kalman filters. *J. Energy Storage* **2022**, *51*, 104396. [CrossRef]
136. Zhang, L.; Wang, X.; Chen, M.; Yu, F.; Li, M. A fractional-order model of lithium-ion batteries and multi-domain parameter identification method. *J. Energy Storage* **2022**, *50*, 104595. [CrossRef]
137. Wu, J.; Fang, C.; Jin, Z.; Zhang, L.; Xing, J. A multi-scale fractional-order dual unscented Kalman filter based parameter and state of charge joint estimation method of lithium-ion battery. *J. Energy Storage* **2022**, *50*, 104666. [CrossRef]
138. Liu, Z.; Chen, S.; Jing, B.; Yang, C.; Ji, J.; Zhao, Z. Fractional variable-order calculus based state of charge estimation of Li-ion battery using dual fractional order Kalman filter. *J. Energy Storage* **2022**, *52*, 104685. [CrossRef]
139. Abdullaeva, B.; Opulencia, M.J.C.; Borisov, V.; Uktamov, K.F.; Abdelbasset, W.K.; Al-Nussair, A.K.J.; Abdulhasan, M.M.; Thangavelu, L.; Jabbar, A.H. Optimal variable estimation of a Li-ion battery model by fractional calculus and bio-inspired algorithms. *J. Energy Storage* **2022**, *54*, 105323. [CrossRef]
140. Guo, H.; Han, X.; Yang, R.; Shi, J. State of charge estimation for lithium-ion batteries based on fractional order multiscale algorithm. *J. Energy Storage* **2022**, *55*, 105630. [CrossRef]
141. Rodríguez-Iturriaga, P.; Alonso-del-Valle, J.; Rodríguez-Bolívar, S.; Anseán, D.; Viera, J.C.; López-Villanueva, J.A. A novel Dual Fractional-Order Extended Kalman Filter for the improved estimation of battery state of charge. *J. Energy Storage* **2022**, *56*, 105810. [CrossRef]
142. Mao, S.; Yu, Z.; Zhang, Z.; Lv, B.; Sun, Z.; Huai, R.; Chang, L.; Li, H. Parameter identification method for the variable order fractional-order equivalent model of lithium-ion battery. *J. Energy Storage* **2023**, *57*, 106273. [CrossRef]
143. Yang, B.; Wang, J.; Wang, J.; Shu, H.; Li, D.; Zeng, C.; Chen, Y.; Zhang, X.; Yu, T. Robust fractional-order PID control of supercapacitor energy storage systems for distribution network applications: A perturbation compensation based approach. *J. Clean. Prod.* **2021**, *279*, 123362. [CrossRef]
144. Yu, M.; Li, Y.; Podlubny, I.; Gong, F.; Sun, Y.; Zhang, Q.; Shang, Y.; Duan, B.; Zhang, C. Fractional-order modeling of lithium-ion batteries using additive noise assisted modeling and correlative information criterion. *J. Adv. Res.* **2020**, *25*, 49–56. [CrossRef] [PubMed]
145. Wang, B.; Li, S.E.; Peng, H.; Liu, Z. Fractional-order modeling and parameter identification for lithium-ion batteries. *J. Power Sources* **2015**, *293*, 151–161. [CrossRef]
146. Mawonou, K.S.R.; Eddahech, A.; Dumur, D.; Beauvois, D.; Godoy, E. Improved state of charge estimation for Li-ion batteries using fractional order extended Kalman filter. *J. Power Sources* **2019**, *435*, 226710. [CrossRef]
147. Sun, Y.; Li, Y.; Yu, M.; Zhou, Z.; Zhang, Q.; Duan, B.; Shang, Y.; Zhang, C. Variable fractional order-A comprehensive evaluation indicator of lithium-ion batteries. *J. Power Sources* **2020**, *448*, 227411. [CrossRef]
148. Wang, Y.; Gao, G.; Li, X.; Chen, Z. A fractional-order model-based state estimation approach for lithium-ion battery and ultra-capacitor hybrid power source system considering load trajectory. *J. Power Sources* **2020**, *449*, 227543. [CrossRef]
149. Saxena, A.; Shankar, R. Improved load frequency control considering dynamic demand regulated power system integrating renewable sources and hybrid energy storage system. *Sustain. Energy Technol. Assess.* **2022**, *52*, 102245. [CrossRef]
150. Prasad, E.N.V.D.V.; Sahani, M.; Dash, P.K. A new adaptive integral back stepping fractional order sliding mode control approach for PV and wind with battery system based DC microgrid. *Sustain. Energy Technol. Assess.* **2022**, *52*, 102261. [CrossRef]
151. Morsali, J. Fractional order control strategy for superconducting magnetic energy storage to take part effectually in automatic generation control issue of a realistic restructured power system. *J. Energy Storage* **2022**, *55*, 105764. [CrossRef]
152. Saha, S.K.; Takano, T.; Fushimi, K.; Sakairi, M.; Saito, R. Passivity of iron surface in curing cement paste environment investigated by electrochemical impedance spectroscopy and surface characterization techniques. *Surf. Interfaces* **2023**, *36*, 102549. [CrossRef]
153. Wang, L.; Zhao, X.; Deng, Z.; Yang, L. Application of electrochemical impedance spectroscopy in battery management system: State of charge estimation for aging batteries. *J. Energy Storage* **2023**, *57*, 106275. [CrossRef]

fractal and fractional

Article

Fractional-Order Model-Free Predictive Control for Voltage Source Inverters

Hani Albalawi [1,2,*], Abualkasim Bakeer [3,*], Sherif A. Zaid [1], El-Hadi Aggoune [1], Muhammad Ayaz [4], Ahmed Bensenouci [5] and Amir Eisa [1]

[1]   Electrical Engineering Department, Faculty of Engineering, University of Tabuk, Tabuk 47913, Saudi Arabia; shfaraj@ut.edu.sa (S.A.Z.)
[2]   Renewable Energy and Energy Efficiency Centre (REEEC), University of Tabuk, Tabuk 47913, Saudi Arabia
[3]   Electrical Engineering Department, Faculty of Engineering, Aswan University, Aswan 81542, Egypt
[4]   Sensor Networks and Cellular Systems (SNCS) Research Center, University of Tabuk, Tabuk 71491, Saudi Arabia
[5]   College of Engineering, Effat University, Jeddah 21478, Saudi Arabia
*   Correspondence: halbala@ut.edu.sa (H.A.); abualkasim.bakeer@aswu.edu.eg (A.B.)

Abstract: Currently, a two-level voltage source inverter (2L-VSI) is regarded as the cornerstone of modern industrial applications. However, the control of VSIs is a challenging task due to their nonlinear and time-varying nature. This paper proposes employing the fractional-order controller (FOC) to improve the performance of model-free predictive control (MFPC) of the 2L-VSI voltage control in uninterruptible power supply (UPS) applications. In the conventional MFPC based on the ultra-local model (ULM), the unknown variable that includes all the system disturbances is estimated using algebraic identification, which is insufficient to improve the prediction accuracy in the predictive control. The proposed FO-MFPC uses fractional-order proportional-integral control (FOPI) to estimate the unknown function associated with the MFPC. To get the best performance from the FOPI, its parameters are optimally designed using the grey wolf optimization (GWO) approach. The number of iterations of the GWO is 100, while the grey wolf's number is 20. The proposed GWO algorithm achieves a small fitness function value of approximately 0.156. In addition, the GWO algorithm nearly finds the optimal parameters after 80 iterations for the defined objective function. The performance of the proposed FO-MFPC controller is compared to that of conventional MFPC for the three loading cases and conditions. Using MATLAB simulations, the simulation results indicated the superiority of the proposed FO-MFPC controller over the conventional MFPC in steady state and transient responses. Moreover, the total harmonic distortion (THD) of the output voltage at different sampling times proves the excellent quality of the output voltage with the proposed FO-MFPC controller over the conventional MFPC controller. The results confirm the robustness of the two control systems against parameter mismatches. Additionally, using the TMS320F28379D kit, the experimental verification of the proposed FO-MFPC control strategy is implemented for 2L-VSI on the basis of the Hardware-in-the-Loop (HIL) simulator, demonstrating the applicability and effective performance of our proposed control strategy under realistic circumstances.

Keywords: uninterruptible power supply (UPS); model predictive control (MPC); ultra-local model (ULM); model-free predictive control (MFPC); fractional-order control (FOC)

Citation: Albalawi, H.; Bakeer, A.; Zaid, S.A.; Aggoune, E.-H.; Ayaz, M.; Bensenouci, A.; Eisa, A. Fractional-Order Model-Free Predictive Control for Voltage Source Inverters. *Fractal Fract.* **2023**, *7*, 433. https://doi.org/10.3390/fractalfract7060433

Academic Editor: Da-Yan Liu

Received: 7 April 2023
Revised: 20 May 2023
Accepted: 25 May 2023
Published: 27 May 2023

1. Introduction

Voltage source inverters (VSIs) are widely used in power electronic systems for applications, such as renewable energy systems, electric vehicles, and industrial drives [1–3]. However, controlling VSIs is challenging due to their nonlinear and time-varying nature [4–7]. Additionally, the reliability of the VSI is at risk as a result of the likelihood of a short-circuit occurring between the two switches located on the same leg. This potential problem could compromise the overall functioning and performance of the VS.

Conventional VSIs, also called two-level inverters, are limited to only two output levels and require particular features to achieve high-quality output [8]. Although it has the merit of simplicity, two-level VSI has the drawbacks of high switching frequency, high switching stresses, power losses, and electromagnetic interference. There is now multilevel architecture, which overcomes the disadvantages of conventional inverters. The famous multilevel VSI topologies are the cascaded [9], flying capacitor [10], and neutral point clamped multilevel inverters [11]. The number of output voltage levels is the primary distinction between multilevel inverters and conventional VSI topologies. Many control techniques have been implemented in the literature, such as internal model controllers, hysteresis controllers, proportional-resonant controllers, proportional-integral controllers, and deadbeat controllers [12]. The finite control set-model predictive control (FCS-MPC) has several advantages over other control methods: its simplicity, ability to handle nonlinearity, and fast response during transients [13–15]. However, the FCS-MPC's performance depends on the system model's accuracy [16]. In recent years, model-free predictive control (MFPC) has emerged as a promising approach to VSI control. It has been widely used in many applications, such as energy management and intelligent transportation [17].

MFPC is a control strategy that uses historical data to predict the future behavior of a system, then uses this information to determine the control action. Unlike traditional model-based control methods, MFPC does not require a detailed system dynamics model. This makes it suitable for complex or uncertain dynamic systems, such as VSIs [18–21]. One of the main advantages of MFPC for VSI control is its ability to handle nonlinear and time-varying system dynamics. MFPC can handle these dynamics using a prediction model updated with real-time data. This allows the control algorithm to adapt to changes in the system dynamics, resulting in improved control performance. Another advantage of MFPC for VSI control is its ability to handle constraints. MFPC can consider constraints, such as voltage, current, and power limits, and use this information to determine the optimal control action. This improves the robustness of the control algorithm and reduces the risk of system failures.

Several studies have been conducted on the application of MFPC to VSI control. For example, a study has proposed an MPC-based control strategy for a VSI in a wind energy system [22]. The authors used a prediction model based on historical data to predict the wind speed and power output of the wind turbine. The control algorithm then used this information to determine the optimal VSI control action. The authors found that the MPC-based control strategy improved the performance of the VSI compared to a traditional model-based control strategy. Another study [23] proposed a neural network-based MFPC controller for the rigorous performance of the power converters. The authors utilized a new framework named the state-space neural network to implement the MFPC controller for the 3-Φ VSI converters. Though the proposed system was robust, the architecture of the neural network structure is unavoidably affected by the nonlinearities in the system. An innovative MFPC controller has been introduced [24] for three-level grid-connected inverters. The proposal was amazing; however, the system was complex. In [25], a modified MFPC technique has been introduced for pulse width modulation (PWM) converters. To achieve excellent performance, the technique has utilized two successive current samples. A new MFPC strategy has been implemented for the DC choppers; however, it does not apply to 3-Φ converters [26]. The observer has been built to enhance the performance of the MPC against parameter uncertainty.

Fractional-order control (FOC) is a relatively new control technique that has been applied to various systems, including voltage source inverters (VSIs) [27–29]. FOC is an extension of traditional integer order control and offers several advantages over conventional control techniques, such as improved performance, better robustness, and increased flexibility. Additionally, FOC can improve the VSI output's power quality, reducing harmonic and total harmonic distortion (THD). It can control current and voltage in a VSI, whereas traditional integer order control is typically used to control only one of these variables. Despite these advantages, there are also some disadvantages to FOC. The main disadvantages are the complexity, difficulty of implementation, and computationally intensive nature.

Several studies have been conducted on applying FOC to VSIs, and the results have been promising. A controller that utilizes FOC and repetition control principles has been proposed to eliminate harmonics and steady-state errors in power converters [30]. In [31], a robust FOC for VSIs was utilized in microgrid applications. Although the performance of the control system has improved, the presence of load variations has affected its robustness.

Despite the significant reduction in VSI-dependent parameters, finding the appropriate function in the MFPC's input-output relationship still poses a challenge. This paper introduces using FOC and MFPC controllers with 2L-VSI for UPS applications. Combining these controllers allows for more accurate and efficient operation of the UPS system. The basic goal of the control system is to keep the output voltage on the load terminals sinusoidal with low harmonic distortions. The fractional-order proportional integral (FOPI) is a numerical method used to calculate the unknown function in the MFPC, representing the total disturbances of the system. Consequently, the MFPC can predict the output voltage at different voltage vectors and choose the one that results in the best performance. Moreover, the FOPI gains are optimally selected using the GWO approach. The main contributions of this study can be summarized as follows:

- The FOPI controller and the MFPC controllers have been integrated to improve the performance of the 2L-VSI. This has been carried out by accurately estimating the unknown function of the MFPC for the voltage control of the 2L-VSI.
- The metaheuristic optimization approach (GWO) has been implemented to find the optimal gains of the proposed FO-MFPC controller.
- The performance of the proposed system utilizing the FO-MFPC controller and the conventional MFPC has been compared. The controller's performance has been tested under linear and nonlinear load disturbances.
- The robustness of the proposed control system under parameter uncertainty has been discussed.
- The effect of changing the sampling period on the system performance has been studied and compared for the proposed FO-MFPC controller and the conventional MFPC.

The manuscript is arranged as follows. First, the conventional model-free predictive control based on the ultra-local model is explained in Section 2. Then, in Section 3, the proposed fractional-order model-free predictive control is described. Next, Section 4 discusses the simulation results. Finally, Section 5 presents the research conclusions.

2. Conventional Model-Free Predictive Control of UPS Based on an Ultra-Local Model

Figure 1 shows a 3-Φ VSI power circuit with the conventional MFPC controller. The converter is connected to a load via an LC filter to eliminate the current's low-order harmonics and provide a sinusoidal 3-Φ voltage at the load terminals. All of the circuit 3-Φ variables, such as (v_a, v_b, and v_c), are represented by the space vector ($V_{x,\alpha\beta}$) notation:

$$V_{x,\alpha\beta} = 2/3(v_a + e^{j(2\pi/3)}v_b + e^{j(4\pi/3)}v_c) \tag{1}$$

The three-phase 2L-VSI has six switched devices ($S_1 : S_6$) with eight possible switching states (i.e., 2^3), as listed in Table 1, in which Vdc is the value of the input dc source. The space vectors of the inverter output voltage ($V_{x,\alpha\beta}$) during the eight switching states ($x \in [0, 7]$) are shown in Figure 2. The space vector diagram is evidently comprised of six distinct sectors. In this space vector modeling, there are a total of eight vectors, out of which two are zero vectors, and the remaining six are referred to as active vectors. During the active vectors, the DC source and load are exclusively connected through a direct path. More details about the conventional MPC for the three-phase 2L-VSI in UPS applications can be found in [32].

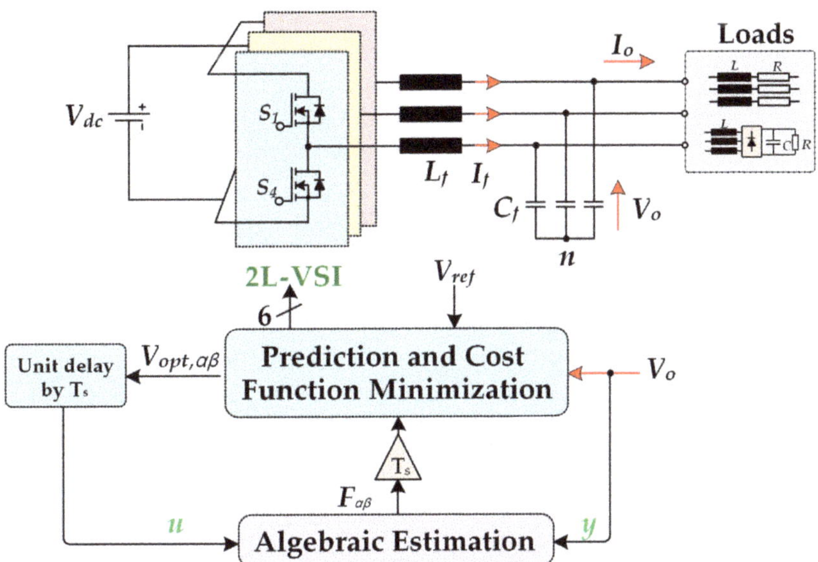

Figure 1. The UPS-based 2L-VSI power circuit with the conventional MFPC controller.

Table 1. Switching states of the 2L-VSI for UPS applications.

x	V_x	Output Voltage $V_{x,\alpha\beta}$	S_1	S_2	S_3	S_4	S_5	S_6
0	V_0	0	0	0	0	1	1	1
1	V_1	$\frac{2}{3}V_{dc}$	1	0	0	0	1	1
2	V_2	$\frac{1}{3}V_{dc} + j\frac{\sqrt{3}}{3}V_{dc}$	1	1	0	0	0	1
3	V_3	$-\frac{1}{3}V_{dc} + j\frac{\sqrt{3}}{3}V_{dc}$	0	1	0	1	0	1
4	V_4	$-\frac{2}{3}V_{dc}$	0	1	1	1	0	0
5	V_5	$-\frac{1}{3}V_{dc} - j\frac{\sqrt{3}}{3}V_{dc}$	0	0	1	1	1	0
6	V_6	$\frac{1}{3}V_{dc} - j\frac{\sqrt{3}}{3}V_{dc}$	1	0	1	0	1	0
7	V_7	0	1	1	1	0	0	0

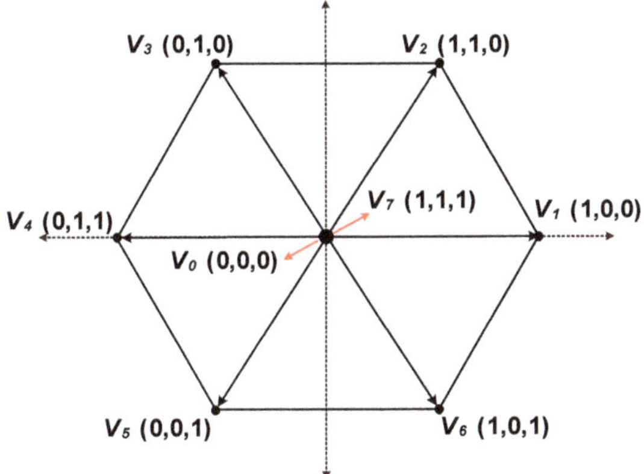

Figure 2. Space vectors of the output voltage at the 2L-VSI terminals.

Figure 3 depicts the fundamental building blocks of the ULM. The symbol F is the unknown function or variable in the ULM that includes the system's overall uncertainty and disruption [33]. The system output and preceding control input are measured in order to define this unknown function F. In addition, the ULM principle can be expressed as follows:

$$y^{(n)} = F + \alpha u \tag{2}$$

where $y^{(n)}$ denotes the nth derivative of y (i.e., in most cases, the practitioner chooses either 1 or 2, with 1 being the most frequently chosen option in all actual circumstances) [20], u indicates the input of the controlled plant, y denotes the plant output, and $\alpha \in \mathrm{R}$ stands for a non-physical parameter.

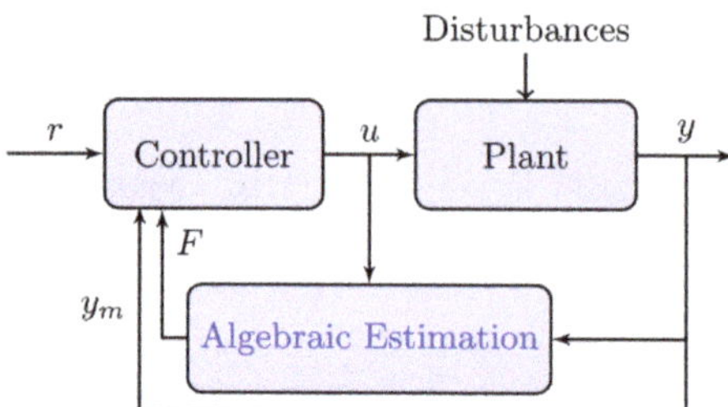

Figure 3. The basic implementation of the ULM.

When using algebraic identification approaches, the value of F can be substituted with a more exact number in place of the estimate by using the letter $\hat{F}$. Finally, the value of $\hat{F}$ may be determined using the Heun technique as follows [34]:

$$\hat{F} = -\frac{3}{N_f{}^3 T_s} \sum_{i=1}^{N_f}(F_1 + F_2) \tag{3}$$

where

$$F_1 = (N_f - 2(i-1))y(k-1) + (N_f - 2i)y(k)$$

$$F_2 = \left(\alpha(i-1)T_s\left(N_f - (i-1)\right)\right)u(k-1) + \alpha i T_s(N_f - i)u(k)$$

where N_f is the length of the window and k is the current instant of the variable.

More specifically, in the case of the UPS, the control target is the output voltage, so the ULM for the VSI with the UPS is given by:

$$\frac{dV_{o,\alpha\beta}(k)}{dt} = \hat{F}_{\alpha\beta} + \alpha u_{\alpha\beta} \tag{4}$$

where $V_{o,\alpha\beta}(k)$ is the output voltage in the $(\alpha\beta)$ coordination frame at kth instant; $u_{\alpha\beta}$ is the optimal voltage vector from Table 1, which is applied at the instant k in the $(\alpha\beta)$ coordination frame; and $\hat{F}_{\alpha\beta}$ is the $(\alpha\beta)$ component of the approximated unknown function $\hat{F}$.

The MFPC model can predict the output voltage at different voltage vectors V_x when applied in the next sampling interval. Euler theory can be used to solve the differential term in Equation (4) and obtain the discrete equation that can be used to predict the output voltage at any given voltage vector as below:

$$V_{o,\alpha\beta}(k+1) = V_{o,\alpha\beta}(k) + T_s\left(\hat{F}_{\alpha\beta} + \alpha V_{x,\alpha\beta}(k+1)\right) \tag{5}$$

where $V_{o,\alpha\beta}(k+1)$ is the predicted voltage across the capacitor C_f in the $(\alpha\beta)$ coordination frame, $V_{o,\alpha\beta}(k)$ is the measured output voltage, $V_{x,\alpha\beta}(k+1)$ is the voltage vector from Table 1 and Equation (1), and T_s is the sampling period.

The multi-objective optimization of the MFPC aims to minimize the total cost functional at any voltage vector x from Table 1, which includes two terms with equal priority as in Equation (6). Consequently, the employed cost function does not need to use weighting factors as we only have one objective: the inverter output voltage. This introduces a flexible algorithm with enhanced power quality.

$$g(x) = (V_{ref,\alpha}(k+1) - V_{o,\alpha}(k+1))^2 + (V_{ref,\beta}(k+1) - V_{o,\beta}(k+1))^2 \tag{6}$$

where $V_{ref,\alpha}$ $(k + 1)$ and $V_{ref,\beta}(k + 1)$ are the reference voltages in the $(\alpha\beta)$ coordination frame and $V_{o,\alpha}$ $(k + 1)$ and $V_{o,\beta}$ $(k + 1)$ are the predicted output voltages in the $(\alpha\beta)$ coordination frame.

3. Proposed Fractional-Order Model-Free Predictive Control

3.1. Fractional-Order Calculus

When using fractional operators in the controller, every real number may be represented as a generic differential or integral notation [34]. The fundamental mathematical relationship of the FO differential or integral operators can be written as follows:

$$D_{lb,ub}^{q} f(t) = \begin{cases} \frac{d^q}{dt^q} f(t) & q > 0 \\ f(t) & q = 0 \\ \int_{lb}^{ub} f(t) d\tau^{-q} & q < 0 \end{cases} \tag{7}$$

where q is the order of the FO calculus, *lb* and *ub* denote the lower and upper bands, respectively. It is clear that when the order is positive (i.e., $q > 0$), it is considered FO differential. On the other hand, when the order is negative (i.e., $q < 0$), it is considered FO integral. There are two different ways to figure out the principle of the FO. One is to use the Riemann–Liouville (R-L), which helps to derive the order derivative of a function $f(t)$ [35]:

$$D_{lb,ub}^{q} f(t) = \frac{1}{\Gamma(n-q)} \left(\frac{d}{dt} \right)^n \int_{lb}^{ub} \frac{f(\tau)}{(t-\tau)^{q-n+1}} d\tau \tag{8}$$

where $\Gamma(w) = \int_0^\infty t^{w-1} e^{-t} dt$ is the Gamma function, $n \in N$, and $n - 1 < q < n$.

The Laplace technique may be used to translate the fractional derivative of R-L found in Equation (8) to obtain the solution in Equation (9) [34]. We may also express the time domain representation of the q order of the function $f(t)$ by using the definition of Caputo, which is a second definition connected to the idea of FO, as indicated in Equation (10).

$$\mathcal{L}\left\{ D_0^q f(t) \right\} = s^q F(s) - \sum_{z=0}^{n-1} s^z \left(D_0^{q-z-1} f(t) \right) \Big|_{t=0} \tag{9}$$

$$D_{lb,ub}^{q} f(t) = \begin{cases} \frac{1}{\Gamma(n-q)} \left(\int_{lb}^{ub} \frac{f^n(\tau)}{(t-\tau)^{1-n+q}} d\tau \right) & n - 1 < q < n \\ \left(\frac{d}{dt} \right)^n f(t) & q = n \end{cases} \tag{10}$$

Applying the Laplace transformation to Equation (10), the integral order has an initial condition, which indicates its physical meaning, as described in Equation (11):

$$\mathcal{L}\left\{ D_0^q f(t) \right\} = s^q F(s) - \sum_{z=0}^{n-1} s^{q-z-1} f^{(z)}(0) \tag{11}$$

where s is the Laplace operator.

A FOPI controller has three parameters: the proportional gain K_p, integral gain K_i, and integral fractional order λ, as presented in Figure 4. In addition, the complete transfer function of the FOPI in Laplace form, $G_c(s)$, is given in Equation (12). It has been found that controllers built using these specific parameters can have improved transient time, stability, and overall accuracy compared to traditional PI controllers. Additionally, the controller provides more flexibility and resilience when dealing with system disturbances. This allows it to handle a wide range of disturbances.

$$G_c(s) = K_p + K_i \left(\frac{1}{s}\right)^{\lambda} \tag{12}$$

where λ is frequently in the range of [0, 1].

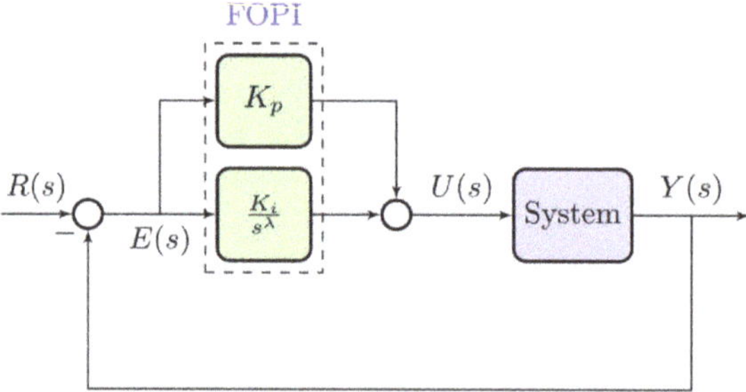

Figure 4. The FOPI controller's fundamental structure.

3.2. Proposed FO-MFPC for 2L-VSI in UPS Applications

The main idea behind the proposed FO-MFPC is to enhance the calculations of the unknown function $\hat{F}$ compared to the algebraic identification with the conventional MFPC. The algebraic estimation for the function $\hat{F}$ in the conventional MFPC will be added to the output of the FOPI controller, resulting in a modified $\hat{F}$ (i.e., $\hat{F}_{m,\alpha\beta}$) as in Equation (13). This could help improve the rejection of disturbances caused by load changes and parameter mismatches.

$$\hat{F}_{m,\alpha\beta} = T_s \hat{F}_{\alpha\beta} + \left(V_{ref,\alpha\beta}(k) - V_{o,\alpha\beta}(k)\right) \times \left(K_p + K_i \left(\frac{1}{s}\right)^{\lambda}\right) \tag{13}$$

Then, the value of future output voltage across the capacitor of the filter by which the trajectory of the load voltage could be predicted is given as:

$$V_{o,\alpha\beta}(k+1) = V_{o,\alpha\beta}(k) + T_s\left(\hat{F}_{m,\alpha\beta} + \alpha V_{x,\alpha\beta}(k+1)\right) \tag{14}$$

The complete structure of the proposed FO-MFPC is shown in Figure 5. First, the algebraic estimation of F is obtained using Equation (3) and updated every sampling interval T_s. Using this value, the predicted value of 2L-VSI at different possible voltage vectors can be calculated with Equation (5). Then, the cost function is evaluated to select the switching vector that provides the minimum value. Implementing the proposed FO-MFPC can be time-consuming, but more feasible as digital signal processors (DSPs) become more powerful. Additionally, a multiple-step prediction can decrease the influence of computational delay on control performance [36].

The complete flowchart of the proposed FO-MFPC for 2L-VSI is depicted in Figure 6. The entire procedure of the proposed FO-MFPC for 2L-VSI can be described step-by-step as follows:

(1) At sampling instant k, the controlled variables ($V_{o,\alpha\beta}(k)$) should be measured.
(2) Those controlled variables are then predicted at instant $k + 1$ based on the discrete model of the converter given in Equation (14).
(3) After defining a proper cost function $g(x)$, as in Equation (6), it should be calculated for the current switching states (x) based on the desired value of the controlled variable.

(4)   As the main objective of the optimization problem is to find the optimum switching state that minimizes the cost fuction, the cost function of the current switching state $g(x)$ is compared with the smallest previous value.

(5)   Steps (2) to (4) are repeated for all possible switching states given in Table 1.

(6)   Finally, the optimum switching state is applied at the next sampling instant.

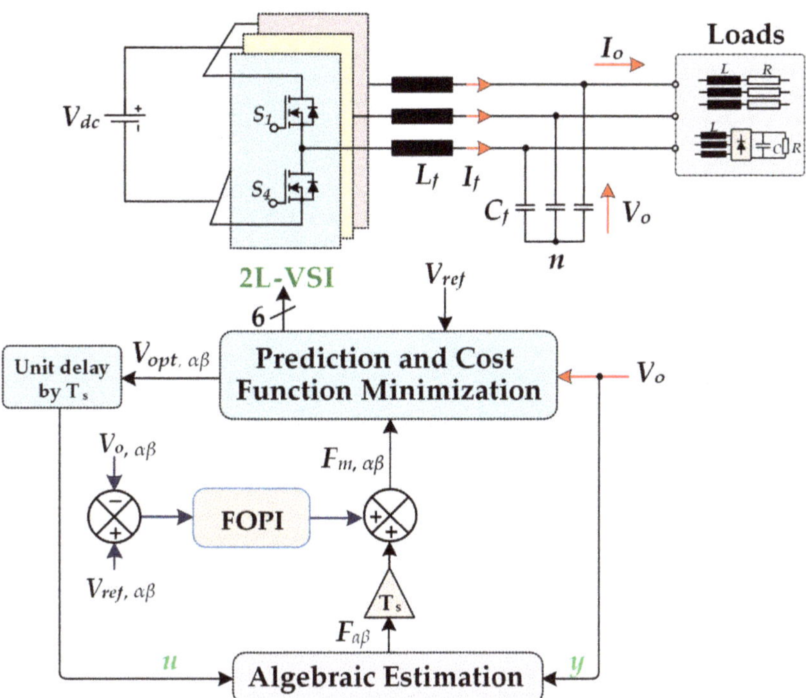

Figure 5. Structure of the proposed FO-MFPC of the 2L-VSI for UPS applications.

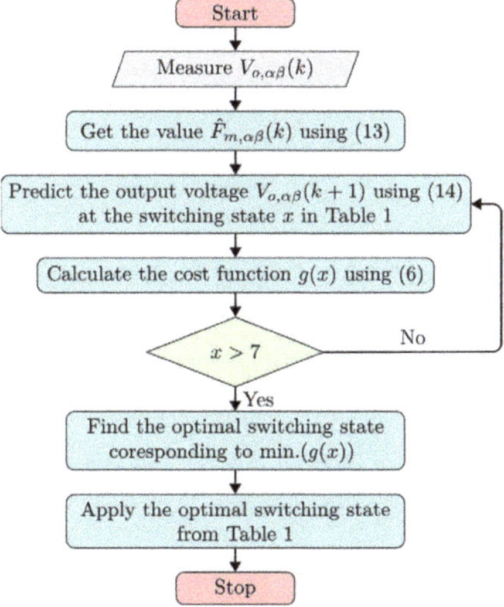

Figure 6. Flowchart of the proposed FO-MFPC for one sampling interval.

The FOPI parameters can be established by trial and error, which can be challenging and dependent on the practitioner's experience. Finding the right values for the proposed FO-MFPC parameters can be challenging. Still, it is crucial to carry it out in a manner that improves system performance and guarantees system stability against interruptions. A metaheuristic optimization technique based on GWO is utilized to determine the optimal value for the parameters of the FOPI controller.

Figure 7 depicts the FOPI parameters' tuning procedure. The GWO algorithm runs on a personal computer employing an Intel© Core™ i5-8265U processor operating at 1.60 GHz and 16 GB of RAM. The GWO will keep going around 100 times, and the grey wolf's number will be 20. The minimum range of the parameters is $[-1,-1,0.1]$, while the maximum range is $[1,1,1]$. The employed fitness function for the GWO is the integral square error (*ISE*) as in Equation (15). The convergence curve of the employed GWO is shown in Figure 8, and the optimal parameters of the FOPI are summarized in Table 2. The proposed GWO algorithm achieves a small fitness function value of approximately 0.156. In addition, the GWO algorithm nearly finds the optimal parameters after 80 iterations for the *ISE* objective function. The parameter α is selected before running the GWO algorithm in order to ensure optimal FOPI gains at the current system parameter setting of the ULM.

$$ISE = \int_0^{tsim} \left(\left(V_{ref,\alpha}(k) - V_{o,\alpha}(k) \right)^2 + \left(V_{ref,\beta}(k) - V_{o,\beta}(k) \right)^2 \right) dt \tag{15}$$

where t_{sim} is the simulation time.

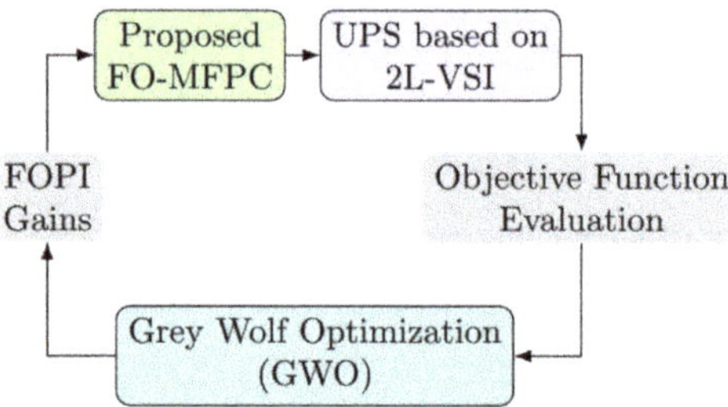

Figure 7. The tuning procedure of the proposed FO-MFPC for UPS.

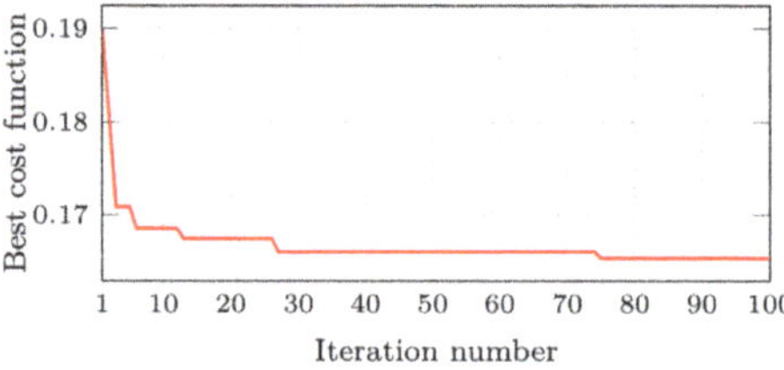

Figure 8. Convergence curve of the employed GWO to tune the FOPI gains of the proposed FO-MFPC.

Table 2. The optimal parameters of the FO-MFPC using GWO.

Parameter	Value
K_p	0.360
K_i	0.034
λ	0.605

4. Simulation Results

The proposed VSI with the investigated FO-MFPC controller, as shown in Figure 5, is simulated using MATLAB. The proposed system's technical parameters are presented in Table 3, while Figure 9 shows the Simulink modeling for the simulation results. The proposed system has been tested under three circumstances to investigate the benefits of the control system. The first case tests the steady-state response of the proposed system under a linear resistive load. The system's transient response under a step resistive load change is verified in the second case. In the third case, the steady-state response of the proposed system under nonlinear load has been tested. The performance of the proposed FO-MFPC controller is compared to that of the conventional MFPC for the three loading cases. Discussions and comparisons of the results are presented in the following paragraphs.

Table 3. Parameters of the studied 2L-VSI for UPS applications.

Parameter	Symbol	Value
Input voltage	V_{dc}	500 V
Filter inductance	L_f	1.5 mH
Filter capacitance	C_f	150 μF
Nominal RMS output voltage (L-L)	$V_{o,ref}$	200 V
Sampling time	T_s	20 μs

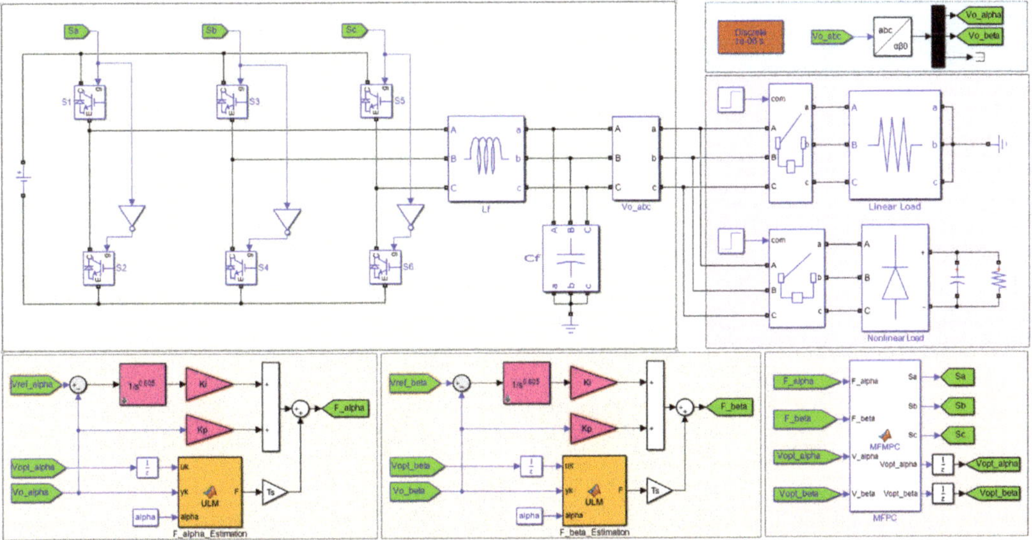

Figure 9. Simulink model of the proposed FO-MFPC for 2L-VSI.

4.1. Case 1: Steady-State Response @ Linear Resistive Load

In this case, the system's steady-state performance is demonstrated by the inverter load, which is a linear resistive load. Figure 10 displays the system's steady-state performance utilizing the proposed FO-MFPC controller and a traditional MFPC controller. For both controllers, the 3-Φ load currents are shown in Figure 10a,b. It is seen that the currents for the two controllers are sinusoidal and balanced. Additionally, as illustrated in Figure 10c,d, the output 3-Φ voltage for both controllers is sinusoidal and balanced. However, the typical MFPC controller's output voltage has a little bit more ripple. The $\alpha\beta$ components of the output voltage compared to their reference values are presented in Figure 10e,f. Additionally, the performance of the proposed FO-MFPC controller is better than that of the conventional one in tracking the reference signals. The controller's measured unknown function, which contains all the system disturbances [37,38], is presented in Figure 10g,h. It is noted that the function has serious disturbances and noise in the case of the conventional MFPC controller. Figure 10i,j present the voltage harmonic spectrum for the two controllers. The harmonic spectrum and the THD of the proposed controller are the best. Therefore, the overall response of the VSI with the proposed FO-MFPC controller is better than that with the conventional MFPC controller.

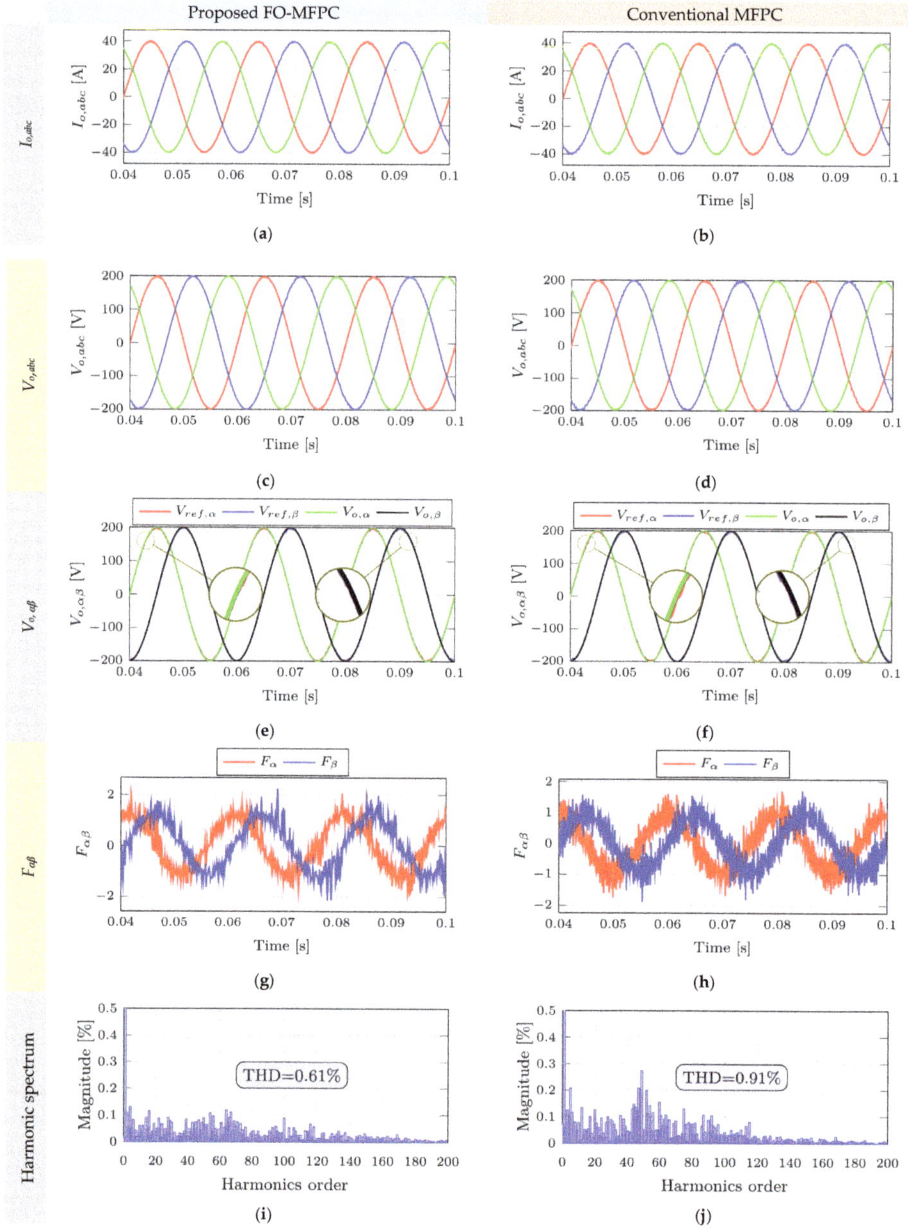

Figure 10. Steady-state response @ fixed resistive load for the UPS with the proposed FO-MFPC and conventional MFPC.

4.2. Case 2: Transient Response @ Step Resistive Load Change

In this case, the inverter load is a linear resistive load with a step change to present the transient state performance of the system. The transient performance of the system using the proposed FO-MFPC controller and the conventional MFPC controller is shown in Figure 11. The load step is applied at 0.07 s. Figure 11a,b show the 3-Φ load currents for both controllers. It is noticed that the currents are sinusoidal and balanced for the two controllers. The currents encounter some transients with each controller. However, the transients have a lower amplitude, ~50%, and shorter time, ~30%, in the case of the proposed controller. Additionally, the output 3-Φ voltages for both controllers have a sinusoidal and balanced nature, as shown in Figure 11c,d. As a result of the presence of the filter inductance, the current transients produce a transient distortion in the output voltage waves. Nevertheless, the output voltage in the case of the conventional MFPC controller has slightly higher transient distortions. The transient responses of the $\alpha\beta$ components of the output voltage compared to their reference values are presented in Figure 11e,f. Additionally, the performance of the proposed FO-MFPC controller is better than that of the conventional one in tracking the reference signals and the transient response. The error between the output voltage and its reference value in the $\alpha\beta$ frame is shown in Figure 11g,h. It is clear that the proposed FO-MFPC achieves the minimum error compared to the conventional MFPC. The controller's unknown functions are presented in Figure 11i,j. It is noted that the function has high noise and spikes in the case of the conventional MFPC controller. The overall transient response of the VSI with the proposed FO-MFPC controller is better than that with the conventional MFPC controller.

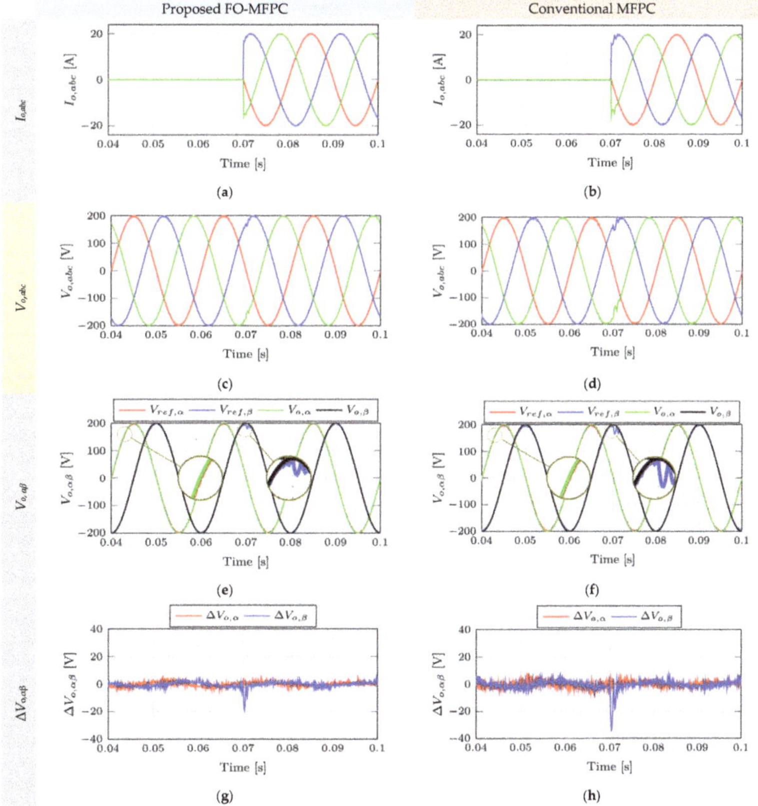

Figure 11. *Cont.*

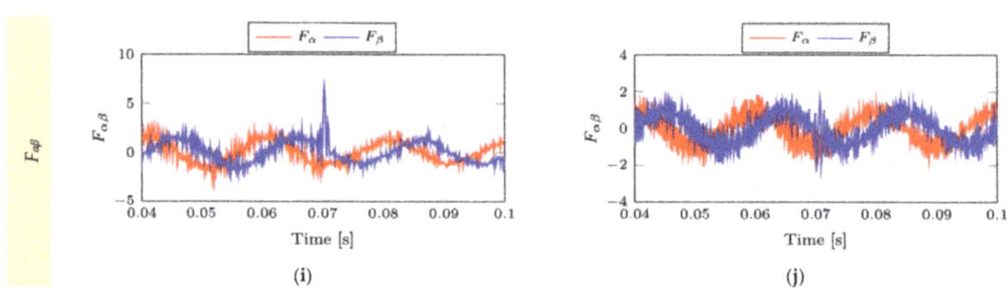

Figure 11. Transient response @ step changes from no-load to 20 A loading with the proposed FO-MFPC and conventional MFPC.

4.3. Case 3: Steady-State Response @ Nonlinear Load

In this case, the inverter load is nonlinear and consists of a three-phase rectifier and a filtered capacitor at the output terminal with a 200 µF capacitance. The load resistance, in this case, is 100 Ω. The system steady state performance using the proposed FO-MFPC controller and conventional MFPC controller is shown in Figure 12. Figure 12a,b show the 3-Φ load currents for both controllers. It is noticed that the currents are highly distorted, far from sinusoidal waves, and unbalanced for the two controllers. However, the output 3-Φ voltages for both controllers have a sinusoidal and balanced nature, as shown in Figure 12c,d. Nevertheless, the output voltage in the case of the conventional MFPC controller has slightly higher ripples. The unknown function in the controller that contains all the disturbances in the system is presented in Figure 12e,f. It is noted that the function has higher noise in the case of the conventional MFPC controller. Therefore, the overall response of the VSI, which supplies the nonlinear load, with the proposed FO-MFPC controller is better than that with the conventional MFPC controller.

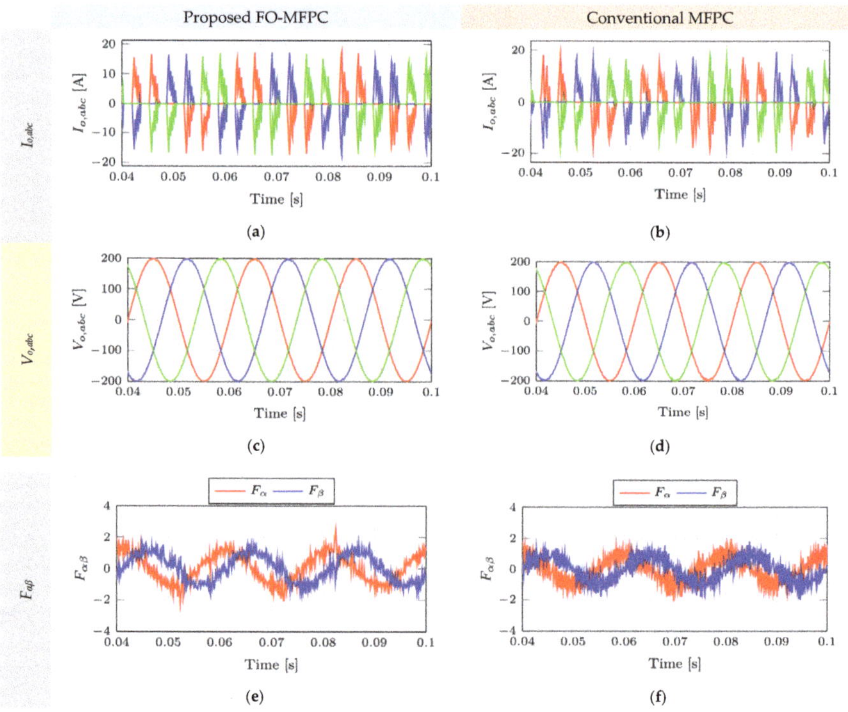

Figure 12. Steady-state response @ nonlinear load with the proposed FO-MFPC and conventional MFPC.

4.4. Case 4: Parameter Mismatch

To check the robustness of the control system under parameter uncertainty, the effect of a 50% change in the filter capacitor value (C_f) on VSI performance with the two controllers is presented in Figure 13. It is noted that the two controllers track the reference signals well. This shows the robustness of the two control systems against parameter mismatches. On the other hand, the THD of the output voltage shows a small relative increase between the two controllers.

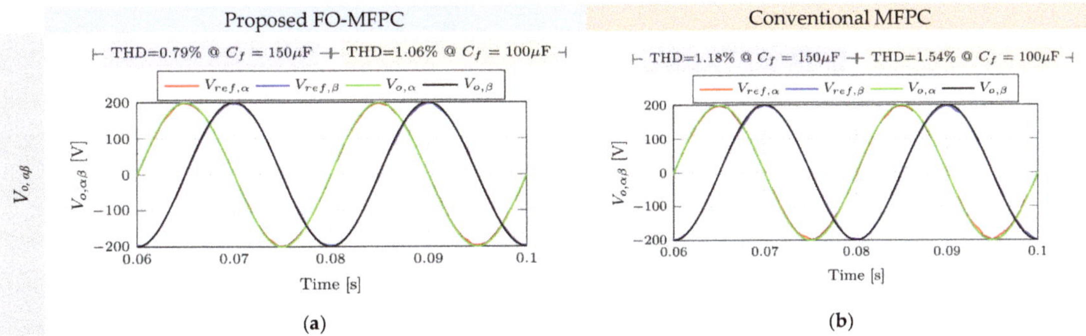

Figure 13. Mismatch of C_f with the proposed FO-MFPC and conventional MFPC.

4.5. THD Evaluation at Different Sampling Intervals

The effect of varying the sampling period on the performance of the VSI controlled using the proposed FO-MFPC and the conventional MFPC has been studied. Figure 14 compares the output voltage THD for the two controllers at different sampling periods. The THDs using the two controllers are lower than the standard recommended values [39]. As expected, the THD using the two controllers increased with the sampling period. However, it is clear that the proposed FO-MFPC controller usually has the lowest THD for any sampling time. The minimum decrease in the THD when using the proposed controller is 10% and the maximum is 48%.

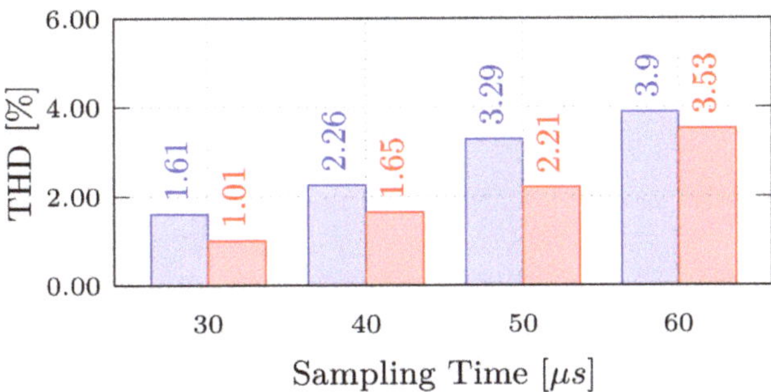

Figure 14. Comparisons of the THD at different sampling times of the proposed FO-MFPC and the conventional MFPC.

4.6. HIL Validation Results

The C2000TM-microcontroller-LaunchPadXL TMS320F28379D kit has been constructed as a Hardware-in-the-Loop (HIL) emulator to test the proposed system and confirm the researched simulation findings. The HIL emulator works by hosting a particular system component—typically the power component—in the computer as a MATLAB model. The

MATLAB application simulates and hosts the planned system power units, such as the power converters and filters. On the other side, the micro-controller kit implements the control algorithms, namely, the proposed FO-MFPC. The virtual serial COM ports [6] facilitate the communication between the PC and the kit. It enables MATLAB to provide measured signals from the power circuit to the kit, including the DC bus voltage, load voltages, and load currents. To produce the 2L-VSI switching signals, the kit performs the control algorithms. Figure 15a shows a schematic diagram of the HIL implementation of the proposed 2L-VSI.

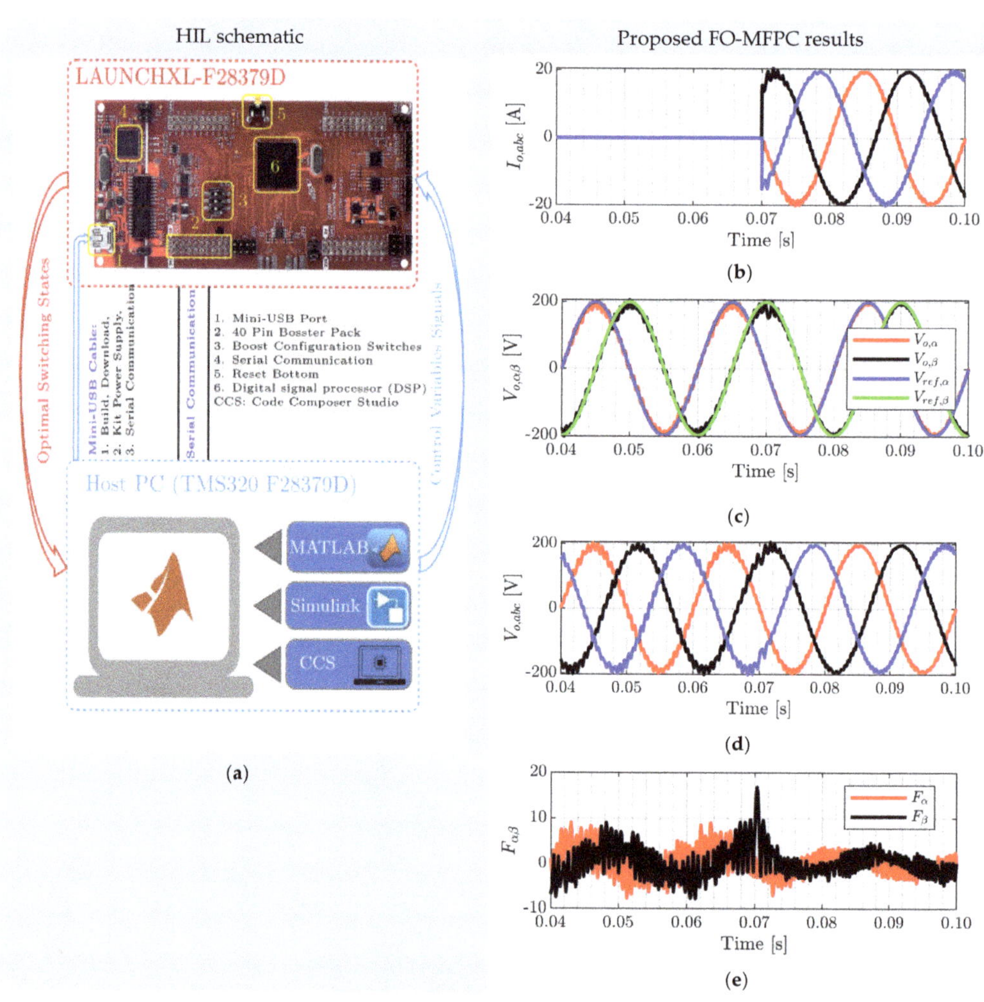

Figure 15. The HIL validation of the proposed 2L-VSI: (**a**) the schematic diagram (**b–e**) the transient response @ step changes from no-load to 20 A loading using the proposed FO-MFPC.

Figure 15b–e presents the results of the HIL validation of the proposed 2L-VSI with FO-MFPC for case (2): the inverter load is a linear resistive load with a step change to present the transient state performance of the system. The load currents and voltages are very close to the simulations results except for some contaminated noise on the waveforms. It is noted that the signals have higher noise and spikes in the case of the HIL implementation than the simulation results of Figure 11.

5. Conclusions

This paper proposes employing the fractional-order controller to improve the performance of the MFPC of the 2L-VSI voltage control in UPS applications. The proposed FO-MFPC uses fractional-order proportional-integral control (FOPI) to estimate the unknown function associated with the MFPC. To get the best performance from FOPI, its parameters are optimally designed using the GWO approach. For three loading cases and conditions, the performance of the proposed FO-MFPC controller is compared to that of the conventional MFPC. Using MATLAB simulations, the simulation results indicated the superiority of the proposed FO-MFPC controller over the conventional MFPC in steady state and transient responses. The results indicated that the THD of the output voltage for the two controllers is much lower than the recommended standard. However, the THD with the proposed FO-MFPC controller is lower than that with the conventional MFPC controller. Additionally, it has been noticed that the proposed FO-MFPC controller usually has the lowest THD. The suggested controller can reduce the THD by as little as 10% and as much as 48%. To check the robustness of the control system under parameter uncertainty, the effect of a 50% change in the filter capacitor value on the performance of the VSI has been determined. The results prove the robustness of the two control systems against parameter mismatches. Moreover, the effect of varying the sampling period on the performance of the VSI controlled using the proposed FO-MFPC and the conventional MFPC has been studied. As expected, the THD using the two controllers increased with the sampling period increase and the proposed FO-MFPC controller has the lowest THD for any sampling time. The future work of the paper could focus on employing the fuzzy logic controller to enhance the calculations of the disturbance function associated with the ULM. Additionally, using the TMS320F28379D kit, the experimental verification of the proposed FO-MFPC control strategy is implemented for 2L-VSI on the basis of the HIL simulator, demonstrating the applicability and effective performance of our proposed control strategy under realistic circumstances.

Author Contributions: A.B. (Abualkasim Bakeer) designed the system, and S.A.Z. derived the model and analyzed the results. H.A. helped in writing the paper. E.-H.A., A.E., A.B. (Ahmed Bensenouci), and M.A. supported the funding process. All authors have read and agreed to the published version of the manuscript.

Funding: This research was funded by Deputyship for Research and Innovation, Ministry of Education in Saudi Arabia, Grant Number 0029-1442-S.

Data Availability Statement: Data are available from the authors upon reasonable request.

Acknowledgments: The authors extend their appreciation to the Deputyship for Research and Innovation, Ministry of Education in Saudi Arabia for funding this research work through the project number (0029-1442-S). The authors also acknowledge the support of the Deanship of Scientific Research at the University of Tabuk.

Conflicts of Interest: The authors declare no conflict of interest.

Nomenclature

2L-VSI	Two-level voltage source inverter
FOC	Fractional-order controller
MFPC	Model-free predictive control
UPS	Uninterruptable power supply
ULM	Ultra-local model
FOPI	Fractional-order proportional-integral
GWO	Grey wolf optimization
THD	Total harmonics distortion
FCS-MPC	Finite control set-model predictive control
PWM	Pulse width modulation
LC	Inductor-capacitor

V_x	Space voltage vector
F	Unknown function associated with MFPC
u	Plant input
y	Plant output
α	Non-physical parameter
T_s	Sampling time
N_f	Length of the window
$\hat{F}$	Approximated value of the unknown function F
C_f	Filter capacitor
x	Voltage vector number in Table 1
FO	Fractional-order
q	Order of the FO calculus
lb	Lower band of the FO integrator
ub	Upper band of the FO integrator
R-L	Riemann–Liouville
K_p	Proportional gain of the FOPI
K_i	Integral gain of the FOPI
λ	Integral fractional order
PI	Proportional integral
$G_c(s)$	FOPI transfer function
s	Laplace operator
$\hat{F}_{m,\alpha\beta}$	Modified value of the unknown function $\hat{F}_{\alpha\beta}$
ISE	Integral square error

References

1. Yan, S.; Yang, Y.; Hui, S.Y.; Blaabjerg, F. A Review on Direct Power Control of Pulsewidth Modulation Converters. *IEEE Trans. Power Electron.* **2021**, *36*, 11984–12007. [CrossRef]
2. Padmanaban, S.; Samavat, T.; Nasab, M.A.; Nasab, M.A.; Zand, M.; Nikokar, F. Electric Vehicles and IoT in Smart Cities. *Artif. Intell.-Based Smart Power Syst.* **2023**, 273–290.
3. Khalili, M.; Dashtaki, M.A.; Nasab, M.A.; Hanif, H.R.; Padmanaban, S.; Khan, B. Optimal instantaneous prediction of voltage instability due to transient faults in power networks taking into account the dynamic effect of generators. *Cogent. Eng.* **2022**, *9*, 2072568. [CrossRef]
4. Zaid, S.A.; Albalawi, H.; AbdelMeguid, H.; Alhmiedat, T.A.; Bakeer, A. Performance Improvement of H8 Transformerless Grid-Tied Inverter Using Model Predictive Control Considering a Weak Grid. *Processes* **2022**, *10*, 1243. [CrossRef]
5. Ahmed, A.; Biswas, S.P.; Anower, S.; Islam, R.; Mondal, S.; Muyeen, S.M. A Hybrid PWM Technique to Improve the Performance of Voltage Source Inverters. *IEEE Access* **2023**, *11*, 4717–4729. [CrossRef]
6. Zaid, S.A.; Mohamed, I.S.; Bakeer, A.; Liu, L.; Albalawi, H.; Tawfiq, M.E.; Kassem, A.M. From *MPC*-Based to End-to-End (*E2E*) Learning-Based Control Policy for Grid-Tied *3L-NPC* Transformerless Inverter. *IEEE Access* **2022**, *10*, 57309–57326. [CrossRef]
7. Albalawi, H.; Zaid, S.A. Performance Improvement of a Grid-Tied Neutral-Point-Clamped 3-φ Transformerless Inverter Using Model Predictive Control. *Processes* **2019**, *7*, 856. [CrossRef]
8. De Bén, A.A.E.; Alvarez-Diazcomas, A.; Rodriguez-Resendiz, J. Transformerless Multilevel Voltage-Source Inverter Topology Comparative Study for PV Systems. *Energies* **2020**, *13*, 3261. [CrossRef]
9. Yuan, W.; Wang, T.; Diallo, D.; Delpha, C. A Fault Diagnosis Strategy Based on Multilevel Classification for a Cascaded Photovoltaic Grid-Connected Inverter. *Electronics* **2020**, *9*, 429. [CrossRef]
10. Rana, R.A.; Patel, S.A.; Muthusamy, A.; Lee, C.W.; Kim, H.-J. Review of Multilevel Voltage Source Inverter Topologies and Analysis of Harmonics Distortions in FC-MLI. *Electronics* **2019**, *8*, 1329. [CrossRef]
11. Anwar, M.A.; Abbas, G.; Khan, I.; Awan, A.B.; Farooq, U.; Khan, S.S. An Impedance Network-Based Three Level Quasi Neutral Point Clamped Inverter with High Voltage Gain. *Energies* **2020**, *13*, 1261. [CrossRef]
12. Heredero-Peris, D.; Chillón-Antón, C.; Sánchez-Sánchez, E.; Montesinos-Miracle, D. Fractional proportional-resonant current controllers for voltage source converters. *Electr. Power Syst. Res.* **2018**, *168*, 20–45. [CrossRef]
13. Carlet, P.G.; Tinazzi, F.; Bolognani, S.; Zigliotto, M. An Effective Model-Free Predictive Current Control for Synchronous Reluctance Motor Drives. *IEEE Trans. Ind. Appl.* **2019**, *55*, 3781–3790. [CrossRef]
14. Mohamed, I.S.; Rovetta, S.; Do, T.D.; Dragicevi´c, T.; Diab, A.A.Z. A neural-network-based model predictive control of three-phase inverter with an output LC filter. *IEEE Access* **2019**, *7*, 124737–124749. [CrossRef]
15. Mohamed-Seghir, M.; Krama, A.; Refaat, S.S.; Trabelsi, M.; Abu-Rub, H. Artificial Intelligence-Based Weighting Factor Autotuning for Model Predictive Control of Grid-Tied Packed U-Cell Inverter. *Energies* **2020**, *13*, 3107. [CrossRef]
16. Lucia, S.; Navarro, D.; Karg, B.; Sarnago, H.; Lucia, O. Deep Learning-Based Model Predictive Control for Resonant Power Converters. *IEEE Trans. Ind. Inform.* **2020**, *17*, 409–420. [CrossRef]
17. Fliess, M.; Join, C. Model-free control. *Int. J. Control* **2013**, *86*, 2228–2252. [CrossRef]

18. Stenman, A. Model-free predictive control. *Proc. 38th IEEE Conf. Decis. Control.* **1999**, *4*, 3712–3717.
19. Rodriguez, J.; Heydari, R.; Rafiee, Z.; Young, H.A.; Flores-Bahamonde, F.; Shahparasti, M. Model-Free Predictive Current Control of a Voltage Source Inverter. *IEEE Access* **2020**, *8*, 211104–211114. [CrossRef]
20. Zhang, Y.; Liu, X.; Liu, J.; Rodriguez, J.; Garcia, C. Model-Free Predictive Current Control of Power Converters Based on Ultra-Local Model. *IEEE Int. Conf. Industrial Technol. (ICIT)* **2020**, 1089–1093. [CrossRef]
21. Khalilzadeh, M.; Vaez-Zadeh, S.; Rodriguez, J.; Heydari, R. Model-Free Predictive Control of Motor Drives and Power Converters: A Review. *IEEE Access* **2021**, *9*, 105733–105747. [CrossRef]
22. Wang, S.; Li, J.; Hou, Z.; Meng, Q.; Li, M. Composite Model-free Adaptive Predictive Control for Wind Power Generation Based on Full Wind Speed. *CSEE J. Power Energy Syst.* **2022**, *8*, 1659–1669. [CrossRef]
23. Sabzevari, S.; Heydari, R.; Mohiti, M.; Savaghebi, M.; Rodriguez, J. Model-Free Neural Network-Based Predictive Control for Robust Operation of Power Converters. *Energies* **2021**, *14*, 2325. [CrossRef]
24. Yin, Z.; Hu, C.; Luo, K.; Rui, T.; Feng, Z.; Lu, G.; Zhang, P. A Novel Model-Free Predictive Control for T-Type Three-Level Grid-Tied Inverters. *Energies* **2022**, *15*, 6557. [CrossRef]
25. Zhang, Y.; Liu, J. An improved model-free predictive current control of pwm rectifiers. In Proceedings of the 20th International Conference on Electrical Machines and Systems (ICEMS), Sydney, Australia, 11–14 August 2017; pp. 1–5.
26. Fliess, M.; Join, C. Model-free control and intelligent pid controllers: Towards a possible trivialization of nonlinear control? *IFAC Proc. Vol.* **2009**, *42*, 1531–1550. [CrossRef]
27. Birs, I.; Muresan, C.; Nascu, I.; Ionescu, C. A Survey of Recent Advances in Fractional Order Control for Time Delay Systems. *IEEE Access* **2019**, *7*, 30951–30965. [CrossRef]
28. Traver, J.E.; Nuevo-Gallardo, C.; Tejado, I.; Fernández-Portales, J.; Ortega-Morán, J.F.; Pagador, J.B.; Vinagre, B.M. Cardiovascular Circulatory System and Left Carotid Model: A Fractional Approach to Disease Modeling. *Fractal Fract.* **2022**, *6*, 64. [CrossRef]
29. Shah, P.; Agashe, S. Review of fractional PID controller. *Mechatronics* **2016**, *38*, 29–41. [CrossRef]
30. Nazir, R. Taylor series expansion based repetitive controllers for power converters, subject to fractional delays. *Control Eng. Pract.* **2017**, *64*, 140–147. [CrossRef]
31. Pullaguram, D.; Mishra, S.; Senroy, N.; Mukherjee, M. Design and Tuning of Robust Fractional Order Controller for Autonomous Microgrid VSC System. *IEEE Trans. Ind. Appl.* **2017**, *54*, 91–101. [CrossRef]
32. Bakeer, A.; Alhasheem, M.; Peyghami, S. Efficient Fixed-Switching Modulated Finite Control Set-Model Predictive Control Based on Artificial Neural Networks. *Appl. Sci.* **2022**, *12*, 3134. [CrossRef]
33. Bakeer, A.; Magdy, G.; Chub, A.; Jurado, F.; Rihan, M. Optimal Ultra-Local Model Control Integrated with Load Frequency Control of Renewable Energy Sources Based Microgrids. *Energies* **2022**, *15*, 9177. [CrossRef]
34. Zaid, S.A.; Bakeer, A.; Magdy, G.; Albalawi, H.; Kassem, A.M.; El-Shimy, M.E.; AbdelMeguid, H.; Manqarah, B. A New Intelligent Fractional-Order Load Frequency Control for Interconnected Modern Power Systems with Virtual Inertia Control. *Fractal Fract.* **2023**, *7*, 62. [CrossRef]
35. Morsali, J.; Zare, K.; Hagh, M.T. Applying fractional order PID to design TCSC-based damping controller in coordination with automatic generation control of interconnected multi-source power system. *Eng. Sci. Technol. Int. J.* **2017**, *20*, 1–17. [CrossRef]
36. Fawzy, A.; Bakeer, A.; Magdy, G.; Atawi, I.E.; Roshdy, M. Adaptive Virtual Inertia-Damping System Based on Model Predictive Control for Low-Inertia Microgrids. *IEEE Access* **2021**, *9*, 109718–109731. [CrossRef]
37. Peng, Y.; Tang, S.; Huang, J.; Tang, C.; Wang, L.; Liu, Y. Fractal Analysis on Pore Structure and Modeling of Hydration of Magnesium Phosphate Cement Paste. *Fractal Fract.* **2022**, *6*, 337. [CrossRef]
38. Tartaglione, V.; Farges, C.; Sabatier, J. Fractional Behaviours Modelling with Volterra Equations: Application to a Lithium-Ion Cell and Comparison with a Fractional Model. *Fractal Fract.* **2022**, *6*, 137. [CrossRef]
39. ILangella, R.; Testa, A. *AliiIEEE Recommended Practice and Requirements for Harmonic Control in Electric Power Systems*; IEEE Std 519-2014 (Revision of IEEE Std 519-1992); IEEE: Piscataway, NJ, USA, 2014; pp. 1–29. [CrossRef]

 *fractal and fractional*

Article

Theoretical Analysis of a Fractional-Order LLCL Filter for Grid-Tied Inverters

Xiaogang Wang [1,*], Ruidong Zhuang [1] and Junhui Cai [2]

[1] School of Mechanical and Electrical Engineering, Guangzhou University, Guangzhou 510006, China
[2] School of Electronics and Communication Engineering, Guangzhou University, Guangzhou 510006, China
*   Correspondence: wxg@gzhu.edu.cn

Abstract: The LLCL-filter-based grid-tied inverter performs better than the LCL-type grid-tied inverter due to its outstanding switching-frequency current harmonic elimination capability, but the positive resonance peak must be suppressed by passive or active damping methods. This paper proposes a class of fractional-order LLCL (FOLLCL) filters, which provides rich features by adjusting the orders of three inductors and one capacitor of the filter. Detailed analyses are performed to reveal the frequency characteristics of the FOLLCL filter; the orders must be selected reasonably to damp the positive resonance peak while reserving the negative resonance peak to attenuate the switching-frequency harmonics. Furthermore, the control system of the grid-tied inverter based on the FOLLCL filter is studied. When the positive resonance is suppressed by the intrinsic damping effect of the FOLLCL filter, the passive or active damper can be avoided; the grid current single close-loop is adequate to control the grid-tied inverter. For low-frequency applications, proportional-resonant (PR) controller is more suitable for the FOLLCL-type grid-tied inverter compared with the proportional-integral (PI) and fractional-order PI controllers due to its overall performance. Simulation results are consistent with theoretical expectations.

Keywords: LLCL filter; active damping; fractional-order; grid-tied inverter; proportional-resonant (PR) control

Citation: Wang, X.; Zhuang, R.; Cai, J. Theoretical Analysis of a Fractional-Order LLCL Filter for Grid-Tied Inverters. *Fractal Fract.* **2023**, *7*, 135. https://doi.org/10.3390/fractalfract7020135

Academic Editors: Behnam Mohammadi-Ivatloo and Arman Oshnoei

Received: 9 January 2023
Revised: 25 January 2023
Accepted: 29 January 2023
Published: 31 January 2023

1. Introduction

The grid-tied inverter is widely used in renewable energy generation; the voltage-source inverter (VSI) interfaces with the grid through a low-pass filter to limit the excessive current harmonics. A third-order LCL filter is the most popular solution over a first-order L filter due to its smaller size, lower cost, and better harmonic attenuation capability [1–5]. However, large inductance should be selected for an LCL filter in low-frequency applications to suppress the more abundant current harmonics. To solve this problem, a high-order LLCL filter has been proposed in [6] and further developed in [7–15]. Based on the LCL filter, a small inductor is inserted in series with the capacitor to form a series resonant branch. The series resonant frequency is thus designed to further attenuate the switching harmonics. The total harmonic distortion (THD) of the grid current will be much lower with LLCL-type inverters compared with LCL-type ones in low-frequency applications.

However, the LLCL filter retains the positive resonant feature of the LCL filter, which causes system instability. Passive or active dampers are used to mitigate the impact of the positive resonance, leading to power loss or control complexity. It is even worse that the capacitor current feedback, the most commonly used active damping method, may introduce a negative resistance and cause instability due to the control delay [8].

In recent years, the fractional-order modeling of power converters has been paid much attention because the inductors and capacitors, the key components of power converters, have fractional-order characteristics, or can be specially designed as fractional-order components. The research of fractional-order power converters began from the modeling of

DC–DC converters. In [16], fractional calculus and the circuit-averaging technique are used to model the buck converter. This technique is also used to build the model of the fractional-order magnetic coupled boost converter [17]. The fractional-order model of the buck converter based on the Caputo–Fabrizio derivative is presented in [18]. The Riemann–Liouville derivative is also used to obtain more accurate models of the fractional-order buck converter [19] and fractional-order buck–boost converter [20]. Instead of considering the complex definitions of fractional calculus, the harmonic balance principle and equivalent small parameter method are used to describe the fractional-order DC–DC converters [21]. Different from the above studies, time domain expressions for fractional-order DC–DC converters are derived in [22]. The modeling methods for fractional-order DC–AC converters are also reported in the literature. In [23], the Caputo derivative method is used to build the model of the voltage source converter, and small-signal analysis and averaging state-space model-based analysis are developed. The fractional-order model of the three-phase voltage source PWM rectifier is constructed in [24]; the Caputo fractional calculus operator is used to describe the fractional-order characteristics of the inductor and capacitor. In addition, the influence of the orders of the inductors and the capacitor on the operating characteristics of the PWM rectifier is studied. In [25], an LCαL filter-based grid-connected inverter is modeled and a filter design example is given. However, the above literature only focuses on the modeling methods; the control strategies are not considered.

On the other hand, fractional-control theories are developed to control the power converters. The fractional-order PID control method is employed to regulate DC–DC converters [26,27]; the results show that the method achieves less overshoot and a faster recovery time compared to the integer-order PID regulator. In [28], the factional-order adaptive sliding mode control approach is proposed for fractional-order buck–boost converters, which shows stronger robustness under various disturbances. For LCL-type grid-tied inverters, an active damping method based on fractional-order proportional-derivative (PD) grid current feedback is presented in [29], which shows better performance compared to the integer-order PD damping method. In [30], a capacitive current fractional proportional-integral feedback strategy is proposed to increase the limit of the damping region of the LCL grid-tied inverter under the weak grid condition. A fractional-order LCL (FOLCL) filter-based grid-tied inverter is studied in [31]; the capacitor current feedback loop can be omitted by only changing the orders of the passive components. Especially, PI and fractional-order PI controllers especially are designed for this grid-tied inverter.

Considering the advantages of fractional-order converters, this paper proposes a fractional-order LLCL-type grid-tied inverter, which can avoid the use of an active damper. The contributions of this paper include the following points:

i.   The characteristics of the FOLLCL filter is analyzed, including the condition of resonance, magnitude–frequency characteristic, phase–frequency characteristic, and the impacts of inductor and capacitor orders on the characteristics.

ii.   The control system of the FOLLCL-type grid-tied inverter is given. Active damping can be avoided, thus improving the ease of control and saving the cost of the control system.

iii.   The performances of the FOLLCL-type grid-tied inverter based on PI, PI$^\lambda$, and PR control are analyzed through four cases. Among these three control methods, the most suitable one for the FOLLCL-type grid-tied inverter without an active damper is determined.

The remainder of the paper is organized as follows: Section 2 introduces the integer-order LLCL (IOLLCL) filter and makes a comparison between the IOLLCL filter and the IOLCL filter. Section 3 analyzes the characteristics of the FOLLCL filter, including resonant frequency, magnitude–frequency characteristic, and phase–frequency characteristic. In Section 4, the structure of the control system of the FOLLCL-type grid-tied inverter is described. Based on the expression of loop gain, the system performance is analyzed. Four cases are presented to discuss the performance of the FOLLCL-type grid-tied inverter.

The simulation results are given in Section 5 to validate the theoretical analysis. Finally, Section 6 concludes this paper.

2. Integer-Order LLCL Filter

An VSI can connect the power grid through an LLCL filter to form a grid-tied inverter. The equivalent circuit of a single-phase integer-order LLCL-filter-based grid-tied inverter is shown in Figure 1, where L_1 and L_2 are the inverter-side and grid-side inductors, a small inductor L_f and a capacitor C_f composing a series resonant circuit, u_i and i_1 are the inverter output voltage and current, u_g and i_g are the grid voltage and current, and i_c is the capacitor current.

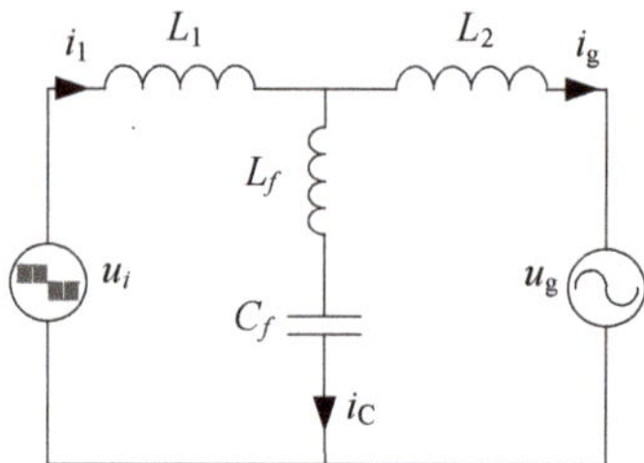

Figure 1. Equivalent circuit of a single-phase integer-order LLCL-filter-based grid-tied inverter.

The transfer function $i_g(s)/u_i(s)$ of the IOLLCL filter can be derived as

$$G_{IO} = \left.\frac{i_g(s)}{u_i(s)}\right|_{u_g(s)=0} = \frac{L_f C_f s^2 + 1}{\left(L_1 L_2 C_f + (L_1 + L_2)C_f L_f\right)s^3 + (L_1 + L_2)s} \tag{1}$$

Figure 2 illustrates the bode diagrams of $i_g(s)/u_i(s)$ for both the IOLLCL filter and IOLCL filter while all the other parameters are the same except for L_f. The specific parameters of the filters are given in Table 1. Unlike the IOLCL filter, the IOLLCL filter has two resonance peaks: a negative one and a positive one; the resonance frequencies are f_{rp1} (ω_{rp1}) and f_{rp2} (ω_{rp2}), respectively. When the VSI operates under the condition of the dual-carrier sine-wave PWM, the uppermost harmonics of i_g are around the switching frequency $2f_s$. Therefore, f_{rp1} is designed to be equal to $2f_s$ to attenuate such harmonics. The positive resonance peak at f_{rp2}, as in the resonance peak at f_{rp} for the IOLCL filter, would lead to system instability in grid-tied inverter applications and should be damped by passive or active methods.

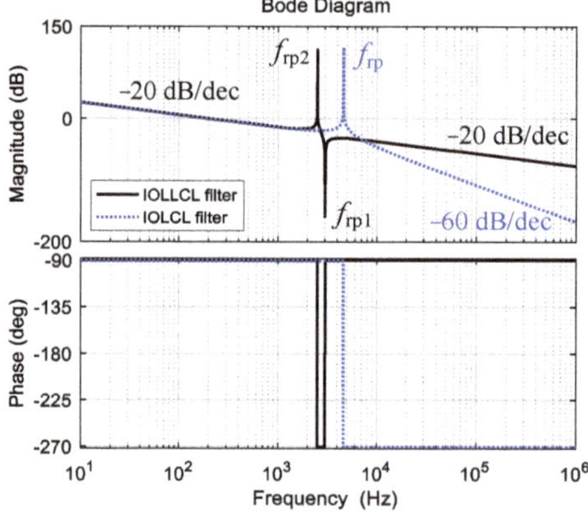

Figure 2. Frequency-response characteristic of the IOLLCL filter.

Table 1. Filter Parameters.

Parameter	Symbol	Value
inverter-side inductor	L_1	600 µH
grid-side inductor	L_2	150 µH
series resonant circuit inductor	L_f	70.362 µH
series resonant circuit capacitor	C_f	10 µF

The negative resonant frequency of the IOLLCL filter is

$$\omega_{\mathrm{rp1}} = \sqrt{\tfrac{1}{L_f C_f}} \tag{2}$$

The positive resonant frequency of the IOLLCL filter is

$$\omega_{\mathrm{rp2}} = \sqrt{\tfrac{L_1+L_2}{L_1 L_2 C_f + L_f C_f (L_1+L_2)}} \tag{3}$$

It can also be seen from Figure 2 that the IOLLCL and IOLCL filters have similar low-frequency magnitude characteristics, while the IOLCL filter exhibits a better attenuation ability at a high-frequency band than the IOLLCL filter. However, overall, compared with the IOLCL filter, the grid current can obtain lower total harmonic distortion with the IOLLCL filter.

3. Fractional-Order LLCL Filter

The IOLLCL filter in the grid-tied inverter can be replaced by an FOLLCL filter to achieve better performance. The FOLLCL filter consists of four components: three inductors and a capacitor, as shown in Figure 3. In this paper, an LLCL filter can be called a fractional-order LLCL filter, with all or part of its components being fractional-order ones.

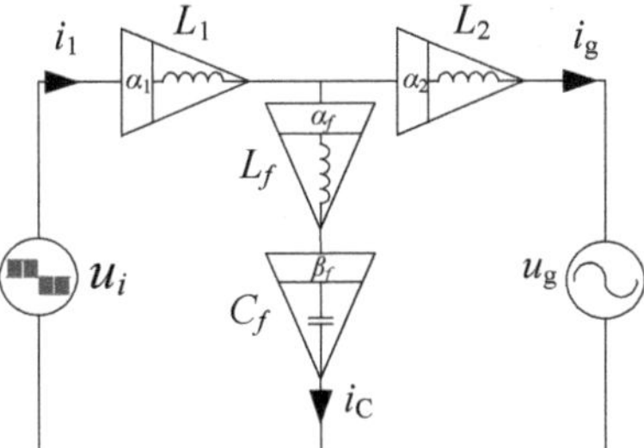

Figure 3. Equivalent circuit of a single-phase fractional-order LLCL-filter-based grid-tied inverter.

The transfer function from input (inverter output voltage u_i) to output (grid current i_g) is expressed as

$$G_{\mathrm{FO}} = \frac{i_g(s)}{u_i(s)} = \frac{L_f C_f s^{\alpha_f+\beta_f} + 1}{L_1 L_2 C_f s^{\alpha_1+\alpha_2+\beta_f} + C_f L_f L_1 s^{\alpha_1+\alpha_f+\beta_f} + C_f L_f L_2 s^{\alpha_2+\alpha_f+\beta_f} + L_1 s^{\alpha_1} + L_2 s^{\alpha_2}} \tag{4}$$

where α_1, α_2, α_f, and β_f are the orders of L_1, L_2, L_f, and C_f, respectively. The magnitude–frequency and phase–frequency characteristic expressions obtained from (4) are quite complex. To simplify the analysis, set $\alpha_1 = \alpha_2 = \alpha$; (4) is rewritten as

$$G_{\mathrm{FO}} = \frac{i_g(s)}{u_i(s)} = \frac{L_f C_f s^{\alpha_f+\beta_f} + 1}{L_1 L_2 C_f s^{2\alpha+\beta_f} + C_f L_f (L_1+L_2) s^{\alpha+\alpha_f+\beta_f} + (L_1+L_2)s^{\alpha}} = \frac{L_f C_f s^{\alpha_f+\beta_f} + 1}{L_1 L_2 C_f s^{\alpha}\left[s^{\alpha+\beta_f} + B s^{\alpha_f+\beta_f} + A\right]} \tag{5}$$

where $A = (L_1 + L_2)/(L_1 L_2 C_f)$ and $B = L_f (L_1 + L_2)/(L_1 L_2)$. Substitute $(j\omega)^\alpha = \omega^\alpha \cos(\alpha\pi/2) + j\omega^\alpha \sin(\alpha\pi/2)$ into (5); the mathematical model in frequency domain can be obtained as

$$G_{FO}(j\omega) = \frac{L_f C_f \omega^{\alpha_f + \beta_f}\left[\cos\frac{(\alpha_f+\beta_f)\pi}{2} + j\sin\frac{(\alpha_f+\beta_f)\pi}{2}\right] + 1}{L_1 L_2 C_f \omega^\alpha \left(\cos\frac{\alpha\pi}{2} + j\sin\frac{\alpha\pi}{2}\right)\left[\omega^{\alpha+\beta_f}\cos\frac{(\alpha+\beta_f)\pi}{2} + A + j\omega^{\alpha+\beta_f}\sin\frac{(\alpha+\beta_f)\pi}{2} + B\omega^{\alpha_f+\beta_f}\cos\frac{(\alpha_f+\beta_f)\pi}{2} + jB\omega^{\alpha_f+\beta}\sin\frac{(\alpha_f+\beta_f)\pi}{2}\right]} \tag{6}$$

The magnitude–frequency characteristic of G_{FO} is expressed as

$$|G_{FO}(j\omega)| = \frac{1}{L_1 L_2 C_f \omega^\alpha}\frac{\sqrt{\left[L_f C_f \omega^{\alpha_f+\beta_f}\cos\frac{(\alpha_f+\beta_f)\pi}{2} + 1\right]^2 + \left(\sin\frac{(\alpha_f+\beta_f)\pi}{2}\right)^2}}{\sqrt{\left(\omega^{\alpha+\beta_f}\cos\frac{(\alpha+\beta_f)\pi}{2} + A + B\omega^{\alpha_f+\beta_f}\cos\frac{(\alpha_f+\beta_f)\pi}{2}\right)^2 + \left(\omega^{\alpha+\beta_f}\sin\frac{(\alpha+\beta_f)\pi}{2} + B\omega^{\alpha_f+\beta}\sin\frac{(\alpha_f+\beta_f)\pi}{2}\right)^2}} \tag{7}$$

3.1. Resonant Frequencies

Define angular frequency ω_{r1} as follows:

$$\omega_{r1} = \left[-\frac{1}{L_f C_f \cos[(\alpha_f+\beta_f)\pi/2]}\right]^{\frac{1}{\alpha_f+\beta_f}} \tag{8}$$

Then, the numerator of (7) can be expressed as

$$\text{num}(|G_{FO}|) = \sqrt{\left[-\left(\frac{\omega}{\omega_{r1}}\right)^{\alpha_f+\beta_f} + 1\right]^2 + \left(\sin\frac{(\alpha_f+\beta_f)\pi}{2}\right)^2} \tag{9}$$

When $\omega = \omega_{r1}$, (9) can be reduced to

$$\text{num}(|G_{FO}|) = \left|\sin\frac{(\alpha_f+\beta_f)\pi}{2}\right| \tag{10}$$

If $\alpha_f + \beta_f = 2n$ (n is an integer), $\sin[(\alpha_f + \beta_f) n\pi/2] = 0$, and $|G_{FO}(j\omega_{r1})|$ as shown in (7) is zero. It means that the magnitude–frequency characteristic of the FOLLCL filter has a negative resonance (series resonance) peak at $\omega = \omega_{r1}$. According to the present literature, the orders of the actually realizable fractional-order inductors and capacitors are greater than 0 and less than 2, so n is set to 1 in this research. Therefore, to attenuate the switching-frequency current ripple in grid-tied inverter applications, the sum of α_f and β_f must equal 2. Substitute $\alpha_f + \beta_f = 2$ back into (8); the negative resonant frequency can be obtained as

$$\omega_{rp1} = \sqrt{\frac{1}{L_f C_f}} \tag{11}$$

It can be seen from (8) that the series resonance peak of the FOLLCL filter has the same form as the IOLLCL filter when $\alpha_f + \beta_f = 2$. Series resonance is the most critical feature for the FOLLCL filter, so the following analysis is based on the relationship of $\alpha_f + \beta_f = 2$. Substitute $\alpha_f + \beta_f = 2$ to (7); the denominator of (7) is expressed as

$$\text{den}(|G_{FO}|) = L_1 L_2 C_f \omega^\alpha \sqrt{\left(\omega^{\alpha+\beta_f}\cos\frac{(\alpha+\beta_f)\pi}{2} + A - B\omega^2\right)^2 + \left(\omega^{\alpha+\beta_f}\sin\frac{(\alpha+\beta_f)\pi}{2}\right)^2} \tag{12}$$

Define angular frequency ω_{r2} as follows:

$$\omega_{r2} = \left[-\frac{A - B\omega^2}{\cos[(\alpha+\beta_f)\pi/2]}\right]^{\frac{1}{\alpha+\beta_f}} \tag{13}$$

Therefore, (12) can be expressed as

$$\text{den}(|G_{FO}|) = L_1 L_2 C_f \omega^\alpha \sqrt{\left[-\left(\frac{\omega}{\omega_{r2}}\right)^{\alpha+\beta_f}(A - B\omega^2) + A - B\omega^2\right]^2 + \left(\omega^{\alpha+\beta_f}\sin\frac{(\alpha+\beta_f)\pi}{2}\right)^2} \tag{14}$$

When $\omega = \omega_{r2}$, (14) can be reduced to

$$\text{den}(|G_{FO}|) = L_1 L_2 C_f \omega^\alpha \sqrt{\left(\omega_{r2}^{\alpha+\beta_f}\sin\frac{(\alpha+\beta_f)\pi}{2}\right)^2} \tag{15}$$

If $\alpha + \beta_f = 2$, $\sin[(\alpha + \beta_f)\pi/2] = 0$, and $|G_{FO}(j\omega_{r2})|$ as shown in (7) is positive infinity. It means that the magnitude–frequency characteristic of the FOLLCL filter has a positive resonance peak at $\omega = \omega_{r2}$. Substitute $\alpha + \beta_f = 2$ back into (13); the positive resonant frequency can be obtained as

$$\omega_{rp2} = \sqrt{\frac{L_1 + L_2}{L_1 L_2 C_f + L_f C_f(L_1 + L_2)}} \tag{16}$$

It can be seen from (16) that the FOLLCL filter has the same expression of positive resonance peak as the IOLLCL filter when $\alpha_f + \beta_f = 2$ and $\alpha + \beta_f = 2$.

Theorem 1. *When $\alpha_1 = \alpha_2 = \alpha$, the negative resonance (series resonance) peak of the FOLLCL filter exists only when $\alpha_f + \beta_f = 2$; the resonant frequency is $\omega_{rp1} = \sqrt{1/L_f C_f}$. The positive resonance peak of the FOLLCL filter exists only when $\alpha_f + \beta_f = 2$ as well as $\alpha + \beta_f = 2$; the resonant frequency is $\omega_{rp2} = \sqrt{(L_1 + L_2)/\left[L_1 L_2 C_f + L_f C_f(L_1 + L_2)\right]}$. The positive resonant frequency ω_{rp2} is always less than the negative resonant frequency ω_{rp1}.*

Theorem 1 essentially reveals the resonant conditions for FOLLCL filters and provides a criterion to estimate whether an FOLLCL filter has resonance peaks. Orders α, α_f, and β_f of a conventional IOLLCL filter are all equal to 1. Both conditions $\alpha_f + \beta_f = 2$ and $\alpha + \beta_f = 2$ are satisfied; therefore, the IOLLCL filter is just a special case of the FOLLCL filter. Moreover, the positive resonance peak can be avoided according to Theorem 1 by choosing reasonable orders for the inverter-side inductor L_1, grid-side inductor L_2, and capacitor C_f. Passive or active damping approaches used in an IOLLCL filter can be omitted for an FOLLCL filter.

3.2. Magnitude–Frequency Characteristic

As previously mentioned, $\alpha_f + \beta_f = 2$ must be satisfied for the FOLLCL filter, so (7) can be arranged as

$$|G_{FO}(j\omega)| = \frac{1}{L_1 L_2 C_f \omega^\alpha} \frac{\left|-L_f C_f \omega^2 + 1\right|}{\sqrt{\left(\omega^{\alpha+\beta_f} + (A - B\omega^2)\cos\frac{(\alpha+\beta_f)\pi}{2}\right)^2 + (A - B\omega^2)^2 \sin^2\frac{(\alpha+\beta_f)\pi}{2}}} \tag{17}$$

(1) When $\omega \ll \omega_{rp2}$, $L_f C_f \omega^2 \ll 1$, and $A - B\omega^2 \approx A$, (17) can be further expressed as

$$|G_{FO}(j\omega)| = \frac{1}{L_1 L_2 C_f \omega^\alpha} \frac{1}{\sqrt{\left(\omega^{\alpha+\beta_f} + A\cos\frac{(\alpha+\beta_f)\pi}{2}\right)^2 + A^2 \sin^2\frac{(\alpha+\beta_f)\pi}{2}}} \tag{18}$$

Define intermediate variables ω_{t1} and ω_{t2} as follows:

$$\omega_{t1} = \left| A \cos \frac{(\alpha+\beta_f)\pi}{2} \right|^{\frac{1}{\alpha+\beta_f}} \tag{19}$$

$$\omega_{t2} = \left| A \sin \frac{(\alpha+\beta_f)\pi}{2} \right|^{\frac{1}{\alpha+\beta_f}} \tag{20}$$

Substitute (19) and (20) into (18); the magnitude–frequency characteristic can be derived as

$$|G_{\text{FO}}(j\omega)| = \frac{1}{L_1 L_2 C_f \omega^{\alpha}} \frac{1}{\sqrt{\left(\omega^{\alpha+\beta_f} + \tau \omega_{t1}^{\alpha+\beta_f}\right)^2 + \omega_{t2}^{2(\alpha+\beta_f)}}} \tag{21}$$

where $\tau = 1$ ($\alpha + \beta_f \in (0, 1] \cup [3, 4)$) or $\tau = -1$ ($\alpha + \beta_f \in (1, 3)$).

When $\omega << \omega_{t1}$, $(\omega/\omega_{t1})^{\alpha+\beta_f} \approx 0$ and $\tau^2 = 1$, (21) can be simplified as

$$\begin{aligned}
|G_{\text{FO}}(j\omega)| &= \frac{1}{L_1 L_2 C_f \omega^{\alpha} \omega_{t1}^{\alpha+\beta_f}} \frac{1}{\sqrt{\left((\omega/\omega_{t1})^{\alpha+\beta_f} + \tau\right)^2 + (\omega_{t2}/\omega_{t1})^{2(\alpha+\beta_f)}}} \\
&\approx \frac{1}{L_1 L_2 C_f \omega^{\alpha}} \frac{1}{\sqrt{\omega_{t1}^{2(\alpha+\beta_f)} + \omega_{t2}^{2(\alpha+\beta_f)}}} = \frac{1}{L_1 L_2 C_f A \omega^{\alpha}}
\end{aligned} \tag{22}$$

The log magnitude–frequency characteristic and the slope of its asymptote are expressed as (23) and (24).

$$L(\omega) \approx -20\lg(L_1 L_2 C_f A) - 20\alpha\lg\omega \tag{23}$$

$$\frac{dL(\omega)}{d\lg\omega} \approx -20\alpha \text{ dB/dec}, \omega << \omega_{t1} \tag{24}$$

(2) When $\omega >> \omega_{\text{rp1}}$, $L_f C_f \omega^2 >> 1$, and $A - B\omega^2 \approx B\omega^2$, (17) can be expressed as

$$|G_{\text{FO}}(j\omega)| = \frac{1}{L_1 L_2 C_f \omega^{\alpha}} \frac{L_f C_f \omega^2}{\sqrt{\left(\omega^{\alpha+\beta_f} + B\omega^2 \cos \frac{(\alpha+\beta_f)\pi}{2}\right)^2 + B^2\omega^4 \sin^2 \frac{(\alpha+\beta_f)\pi}{2}}} \tag{25}$$

$$\omega_{t3} = \left| B \cos \frac{(\alpha+\beta_f)\pi}{2} \right|^{\frac{1}{\alpha+\beta_f}} \tag{26}$$

Define intermediate variables ω_{t3} and ω_{t4} as follows:

$$\omega_{t4} = \left| B \sin \frac{(\alpha+\beta_f)\pi}{2} \right|^{\frac{1}{\alpha+\beta_f}} \tag{27}$$

Substitute (26) and (27) into (25); the magnitude–frequency characteristic can be derived as

$$|G_{\text{FO}}(j\omega)| = \frac{1}{L_1 L_2 C_f \omega^{\alpha}} \frac{L_f C_f \omega^2}{\sqrt{\left(\omega^{\alpha+\beta_f} + \tau\omega^2 \omega_{t3}^{\alpha+\beta_f}\right)^2 + \omega^4 \omega_{t4}^{2(\alpha+\beta_f)}}} \tag{28}$$

When $\omega >> \omega_{\text{rp1}}$ and $\alpha + \beta_f \in [2, 4)$, $\omega^2(\omega_{t3}/\omega)^{\alpha+\beta_f} \approx 0$, and $\omega^2(\omega_{t4}/\omega)^{\alpha+\beta_f} \approx 0$, (17) can be expressed as

$$|G_{\text{FO}}(j\omega)| = \frac{1}{L_1 L_2 C_f \omega^{2\alpha+\beta_f}} \frac{L_f C_f \omega^2}{\sqrt{\left(1 + \tau\omega^2(\omega_{t3}/\omega)^{\alpha+\beta_f}\right)^2 + \omega^4(\omega_{t4}/\omega)^{2(\alpha+\beta_f)}}} \approx \frac{L_f}{L_1 L_2 \omega^{2\alpha+\beta_f-2}} \tag{29}$$

The log magnitude–frequency characteristics and the slope of its asymptote are expressed as (30) and (31).

$$L(\omega) \approx 20\lg L_f - 20\lg(L_1 L_2) - 20\left(2\alpha + \beta_f - 2\right)\lg\omega \tag{30}$$

$$\frac{dL(\omega)}{d\lg\omega} \approx -20\left(2\alpha + \beta_f - 2\right) \text{ dB/dec}, \, \omega >> \omega_{\text{rp1}} \tag{31}$$

Similarly, when $\omega >> \omega_{\text{rp1}}$ and $\alpha + \beta_f \in (0,2)$, the slope of the asymptote is -20α dB/dec.

Theorem 2. *For FOLLCL, when $\omega << \omega_{rp2}$, the asymptote slope of the low-frequency log magnitude–frequency characteristic is -20α dB/dec. When $\omega >> \omega_{rp1}$, $\alpha + \beta_f \in [2, 4)$, the asymptote slope of the high-frequency log magnitude–frequency characteristics is $-20(2\alpha + \beta_f - 2)$ dB/dec; if $\alpha + \beta_f \in (0, 2)$, the asymptote slope is -20α dB/dec.*

3.3. Phase–Frequency Characteristic

According to (6), when $\alpha_f + \beta_f = 2$, the phase model can be expressed as

$$\angle G_{\text{FO}}(j\omega) = -\arctan\left(\tan\frac{\pi\alpha}{2}\right) - \arctan\frac{\omega^{\alpha+\beta_f}\sin\dfrac{\left(\alpha+\beta_f\right)\pi}{2}}{\omega^{\alpha+\beta_f}\cos\dfrac{\left(\alpha+\beta_f\right)\pi}{2} + A - B\omega^2} \tag{32}$$

(1) When $\omega << \omega_{tl}, A - B\omega^2 \approx A$ and $\omega^{\alpha+\beta_f}\sin[\left(\alpha + \beta_f\right)\pi/2] << \omega^{\alpha+\beta_f}\cos[\left(\alpha + \beta_f\right)\pi/2] + A$, so the low-frequency phase is expressed as

$$\angle G_{\text{FO}}(j\omega) \approx -\arctan\left(\tan\tfrac{\pi\alpha}{2}\right) - \arctan\frac{\omega^{\alpha+\beta_f}\sin\dfrac{\left(\alpha+\beta_f\right)\pi}{2}}{\omega^{\alpha+\beta_f}\cos\dfrac{\left(\alpha+\beta_f\right)\pi}{2} + A} \approx -\frac{\pi\alpha}{2} \tag{33}$$

(2) When $\omega >> \omega_{\text{rp1}}$, $A - B\omega^2 \approx -B\omega^2$ and $B\omega^2 << |\omega^{\alpha+\beta_f}\cos[\left(\alpha + \beta_f\right)\pi/2]|$. Moreover, when $\alpha + \beta_f \in [2, 4)$, $\arctan\left\{\tan[\left(\alpha + \beta_f\right)\pi/2]\right\} = \left(\alpha + \beta_f\right)\pi/2 - 2\pi$, so the high-frequency phase is expressed as

$$\angle G_{\text{FO}}(j\omega) \approx \pi - \arctan\left(\tan\tfrac{\pi\alpha}{2}\right) - \arctan\frac{\omega^{\alpha+\beta_f}\sin\dfrac{\left(\alpha+\beta_f\right)\pi}{2}}{\omega^{\alpha+\beta_f}\cos\dfrac{\left(\alpha+\beta_f\right)\pi}{2} - B\omega^2}$$

$$\approx \pi - \arctan\left(\tan\frac{\pi\alpha}{2}\right) - \arctan\frac{\omega^{\alpha+\beta_f}\sin\dfrac{\left(\alpha+\beta_f\right)\pi}{2}}{\omega^{\alpha+\beta_f}\cos\dfrac{\left(\alpha+\beta_f\right)\pi}{2}} = -\pi\left(\alpha + \beta_f/2\right) + \pi \tag{34}$$

When $\alpha + \beta_f \in (0,2)$, $\omega^{\alpha+\beta_f}\sin[(\alpha+\beta_f)\pi/2] << \left|\omega^{\alpha+\beta_f}\cos[(\alpha+\beta_f)\pi/2] - B\omega^2\right|$, so the high-frequency phase can be expressed as

$$\angle G_{FO}(j\omega) \approx -\arctan\left(\tan\frac{\pi\alpha}{2}\right) - \arctan\frac{\omega^{\alpha+\beta_f}\sin\dfrac{(\alpha+\beta_f)\pi}{2}}{\omega^{\alpha+\beta_f}\cos\dfrac{(\alpha+\beta_f)\pi}{2}-B\omega^2} \approx -\frac{\pi\alpha}{2} \tag{35}$$

Theorem 3. *For FOLLCL, when $\omega << \omega_{rp2}$, the low-frequency phase is $-\pi\alpha/2$. When $\omega >> \omega_{rp1}$, if $\alpha + \beta_f \in [2, 4)$, the high-frequency phase is $-\pi(\alpha + \beta_f/2) + \pi$; if $\alpha + \beta_f \in (0, 2)$, the high-frequency phase is $-\pi\alpha/2$.*

It can be seen from Theorem 2 and Theorem 3 that the low-frequency characteristics only depend on the orders of L_1 and L_2, and are independent of the orders of L_f and C_f. The high-frequency characteristics are determined by the orders of L_1, L_2, and C_f when $\alpha + \beta_f \in [2, 4)$, and only depend on the orders of L_1 and L_2 when $\alpha + \beta_f \in (0, 2)$.

3.4. Simulation Analyses

The bode plots of the FOLLCL filter are shown in Figure 4. The specific parameters of the FOLLCL filter are given in Table 1. Two cases are considered, namely, $\alpha + \beta_f \leq 2$ and $\alpha + \beta_f \geq 2$. The values of the asymptote slopes and phases are marked in the plots; it can be seen that the results are consistent with the theoretical analyses. In particular, the positive resonance peak is suppressed when $\alpha + \beta_f \neq 2$. Furthermore, it is shown in Figure 4b that when $\alpha + \beta_f > 2$, the phase–frequency characteristic curves do not pass through $-180°$, which means that the phase crossover frequency does not exist. Therefore, $\alpha + \beta_f$ must be less than or equal to 2 to guarantee the stability.

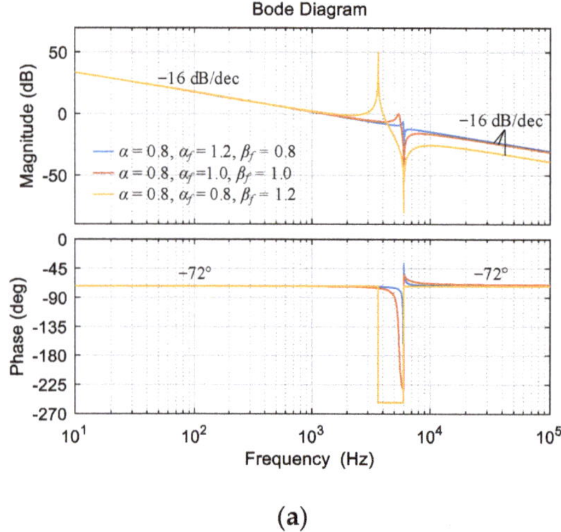

(a)

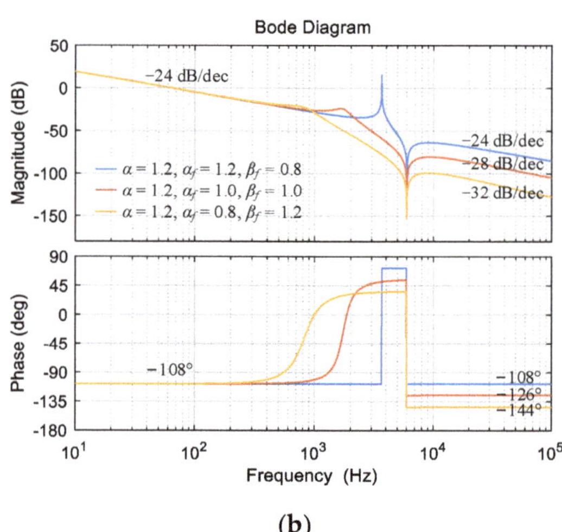

(b)

Figure 4. Bode diagrams of the FOLLCL filter. (**a**) $\alpha = 0.8$, $\alpha + \beta_f \leq 2$, (**b**) $\alpha = 1.2$, $\alpha + \beta_f \geq 2$.

4. Grid-Tied Inverter Based on Fractional-Order LLCL Filter

An FOLLCL filter and a VSI can be combined to form a grid-tied inverter. Figure 5 shows the single-phase FOLLCL-type grid-tied inverter and its control system. The primary objective of the grid-tied inverter is to control the grid-side current i_g to be synchronized with the grid voltage, which is denoted by u_g. I^* is the reference amplitude of the grid-side current, θ is the phase of grid voltage obtained by the phase-locked loop (PLL), and

i_g^* is the reference of the grid-side current. i_g is sensed with the sensor gain of H_{ig} and compared with i_g^*. The current error is sent to the current regulator G_i; $G_i = K_p + K_i/s^\lambda$, and K_p, K_i, and λ are the proportional coefficient, integral coefficient, and order of the integrator, respectively. For the FOLLCL filter with $\alpha + \beta_f$ equaling or very close to 2, an active damping method is used to attenuate the positive resonance. The output of G_i is sent to the PWM generator after subtracting the capacitor current i_C, which is sensed with the sensor gain of H_{iC}. For the FOLLCL filter with $\alpha + \beta_f$ deviating from 2, the output of G_i is sent to the PWM generator directly; the active damping can be avoided.

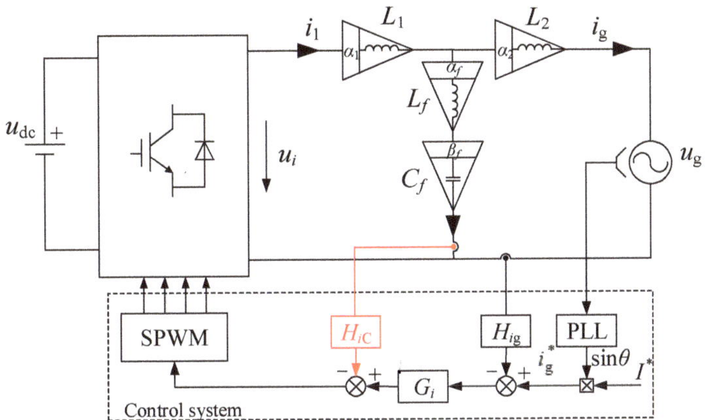

Figure 5. Single-phase FOLLCL-type grid-tied inverter and its control system.

4.1. Structure of the Control System

According to Figure 5, the control block diagram of the single-phase FOLLCL-type grid-tied inverter when $\alpha + \beta_f$ equals to or is very close to 2 is shown in Figure 6, where K_{PWM} is the transfer function from the modulation signal to the inverter output voltage, expressed as $K_{PWM} = u_{dc}/V_{tri}$, and V_{tri} is the amplitude of triangular carrier. $Z_{L1}(s)$, $Z_{L2}(s)$, $Z_{Lf}(s)$, and $Z_{Cf}(s)$ are the impedance of L_1, L_2, L_f, and C_f, respectively, which are expressed as

$$Z_{L1}(s) = s^\alpha L_1, \quad Z_{L2}(s) = s^\alpha L_2, \quad Z_{Lf}(s) = s^{\alpha_f} L_1, \quad Z_{Cf}(s) = 1/s^{\beta_f} C_f \tag{36}$$

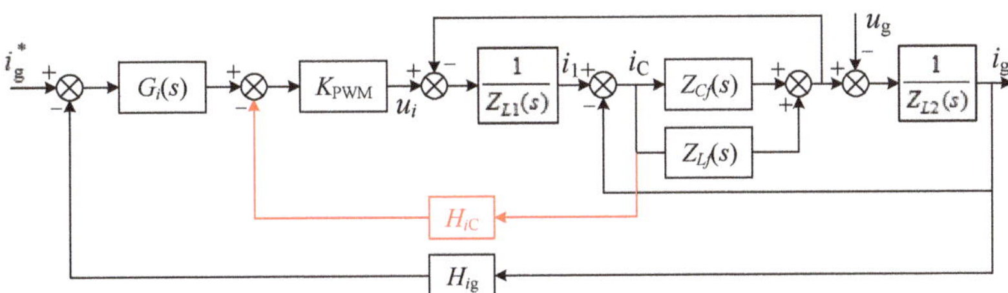

Figure 6. Control block diagram of single-phase FOLLCL-type grid-tied inverter when $\alpha + \beta_f$ equals or very close to 2.

The control block diagram of the single-phase FOLLCL-type grid-tied inverter when $\alpha + \beta_f$ deviates from 2 is shown in Figure 7. Compared with Figure 6, the capacitor current feedback loop is removed.

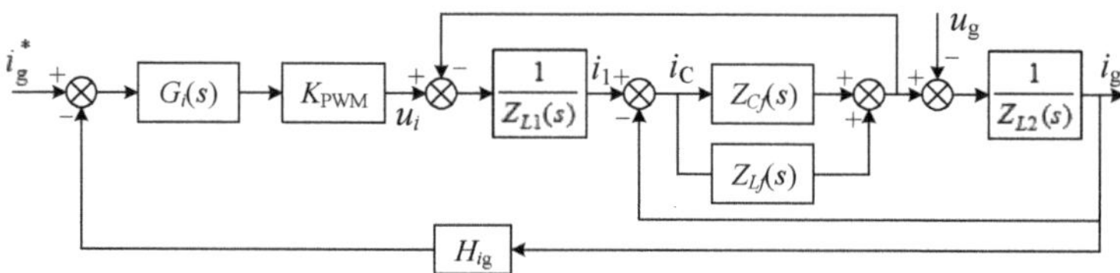

Figure 7. Control block diagram of single-phase FOLLCL-type grid-tied inverter when $\alpha + \beta_f$ deviates from 2.

The control block diagrams in Figures 6 and 7 can be equivalently transformed into the block diagram in Figure 8. The transfer functions G_{x1} and G_{x2} are expressed as (37) and (38), respectively:

$$G_{x1}(s) = \frac{K_{\text{PWM}} G_i(s) \left[Z_{Lf}(s) + Z_{Cf}(s) \right]}{Z_{L1}(s) + Z_{Lf}(s) + Z_{Cf}(s) + H_{iC} K_{\text{PWM}}} \tag{37}$$

$$G_{x2}(s) = \frac{Z_{L1}(s) + Z_{Lf}(s) + Z_{Cf}(s) + H_{iC} K_{\text{PWM}}}{Z_{L1}(s) Z_{L2}(s) + [Z_{L1}(s) + Z_{L2}(s)] \left[Z_{Lf}(s) + Z_{Cf}(s) \right] + H_{iC} K_{\text{PWM}} Z_{L2}(s)} \tag{38}$$

where $H_{iC} = 0$ when $\alpha + \beta_f \neq 2$.

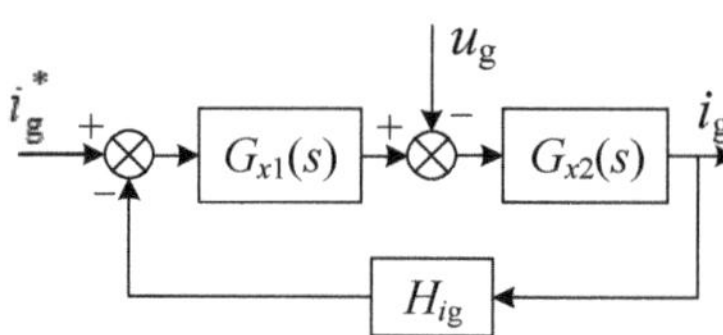

Figure 8. Equivalent block diagram of Figures 6 and 7.

According to the equivalent block diagram in Figure 8 and (36)~(38), the expression of the loop gain can be derived as

$$T_A(s) = G_{x1}(s) G_{x2}(s) H_{ig}(s) = \frac{H_{ig} K_{\text{PWM}} G_i(s) \left(L_f C_f s^{\alpha_f + \beta_f} + 1 \right)}{L_1 L_2 C_f s^{2\alpha + \beta_f} + (L_1 + L_2) L_f C_f s^{\alpha + \alpha_f + \beta_f} + L_2 C_f H_{iC} K_{\text{PWM}} s^{\alpha + \beta_f} + (L_1 + L_2) s^{\alpha}} \tag{39}$$

As discussed in Part 3, $\alpha_f + \beta_f = 2$ must be satisfied in grid-tied inverter applications, so (39) is rewritten as

$$T_A(s) = \frac{H_{ig} K_{\text{PWM}} G_i(s) \left(L_f C_f s^2 + 1 \right)}{L_1 L_2 C_f s^{2\alpha + \beta_f} + (L_1 + L_2) L_f C_f s^{\alpha + 2} + L_2 C_f H_{iC} K_{\text{PWM}} s^{\alpha + \beta_f} + (L_1 + L_2) s^{\alpha}} \tag{40}$$

The grid current i_g can be expressed as

$$i_g(s) = \frac{1}{H_{ig}} \frac{T_A(s)}{1 + T_A(s)} i_g^*(s) - \frac{G_{x2}(s)}{1 + T_A(s)} u_g(s) = i_{g1}(s) + i_{g2}(s) \tag{41}$$

It can be seen from (41) that $i_g(s)$ consists of two parts: the reference tracking component $i_{g1}(s)$ and the disturbance component $i_{g2}(s)$ caused by the grid voltage, which can be expressed as (42) and (43), respectively.

$$i_{g1}(s) = \frac{1}{H_{ig}} \frac{T_A(s)}{1 + T_A(s)} i_g^*(s) \tag{42}$$

$$i_{g2}(s) = -\frac{G_{x2}(s)}{1 + T_A(s)} u_g(s) \tag{43}$$

4.2. System Performance Analysis

The loop gain at the fundamental frequency is often much greater than one, so $1 + T_A(s) \approx T_A(s)$, and (42) can be rewritten as $i_{g1}(s) \approx i_g^*(s)/H_{ig}$. Therefore, $i_{g1}(s)$ and $i_g^*(s)$ are almost in phase. For $f \leq f_c$, the $L_f - C_f$ branch can be considered open. According to (38) and (40), the expression of $G_{x2}(s)$ and $T_A(s)$ at the fundamental frequency can be obtained as follows:

$$G_{x2}(j2\pi f_o) \approx \frac{1}{(j2\pi f_o)^{\alpha}(L_1 + L_2)} \tag{44}$$

$$T_A(j2\pi f_o) \approx \frac{H_{ig}K_{PWM}G_i(j2\pi f_o)}{(j2\pi f_o)^{\alpha}(L_1 + L_2)} \tag{45}$$

where f_o is the fundamental frequency. Substitute (44) and (45) into (43), and for PI regulator, $G_i(j2\pi f_o) \approx K_i/(j2\pi f_o)$, so we have

$$i_{g2} \approx -\frac{u_g}{H_{ig}K_{PWM}G_i(j2\pi f_o)} \approx -\frac{j2\pi f_o u_g}{H_{ig}K_{PWM}K_i} \tag{46}$$

From (46), it can be seen that i_{g2} lags behind u_g by $90°$; a small i_{g2} is expected to reduce the amplitude and phase tracking errors for i_g. From (45) and (46), the RMS value of i_{g2} can be expressed as

$$I_{g2} \approx \frac{U_g}{H_{ig}K_{PWM}|G_i(j2\pi f_o)|} \approx \frac{U_g}{(2\pi f_o)^{\alpha}(L_1 + L_2)|T_A(j2\pi f_o)|} \tag{47}$$

The magnitude of the loop gain at f_o is expressed as

$$T_{fo} = 20lg\left|T_A(j2\pi f_o)\right| \approx 20lg\frac{U_g}{(2\pi f_o)^{\alpha}(L_1 + L_2)I_{g2}} \tag{48}$$

where the unit of T_{fo} is dB. Thus, the steady-state error requirement for I_{g2} is converted to the requirement for T_{fo}. Obviously, for a given value of I_{g2}, a smaller-order α means bigger T_{fo}.

Compared to the PI regulator, the PR regulator can significantly increase T_{fo}, and thus decrease the steady-state error of the grid current. The expression of the PR regulator is

$$G_i(s) = K_p + \frac{2K_r\omega_i s}{s^2 + 2\omega_i s + \omega_0^2} \tag{49}$$

where K_p is the proportional coefficient, K_r is the resonant coefficient, ω_i is the bandwidth concerning the -3 dB cutoff frequency of the resonant compensator, and $\omega_o = 2\pi f_o$ is the fundamental angular frequency. The design criteria of the PR regulator have been reported in many works in the literature and will not be repeated here.

In order to demonstrate the control system design criteria of the FOLLCL-type grid-tied inverter, four cases are presented based on the system parameters listed in Tables 1 and 2.

Case I ($\alpha + \beta_f = 2$, PI control): For $\alpha + \beta_f = 2$ (taking $(\alpha, \alpha_f, \beta_f) = (1.2, 1.2, 0.8)$ and $(\alpha, \alpha_f, \beta_f) = (1.1, 1.1, 0.9)$ as examples), the bode diagrams of the loop gain before compensation ($G_i(s) = 1$) are drawn in Figure 9 according to (40), where f_c is the cut-off frequency of the loop gain. As shown in Figure 9, the capacitor current feedback can effectively suppress the positive resonance peak of the FOLLCL filter, and the resonance damping capability becomes stronger with the increase of H_{iC}. As with the application in conventional IOLCL-type grid-tied inverters, this well-known active damping method only changes the magnitude–frequency characteristics around the resonant frequency f_{rp2}. However, the phase–frequency characteristics vary observably; they decrease from $-(90\alpha)°$ when $f < f_{rp2}$.

Table 2. System Parameters.

Parameter	Symbol	Value
DC voltage	u_{dc}	360 V
grid voltage (RMS)	U_g	220 V
fundamental frequency	f_o	50 Hz
switching frequency	f_s	3 kHz
amplitude of the triangular carrier	V_{tri}	3.05 V

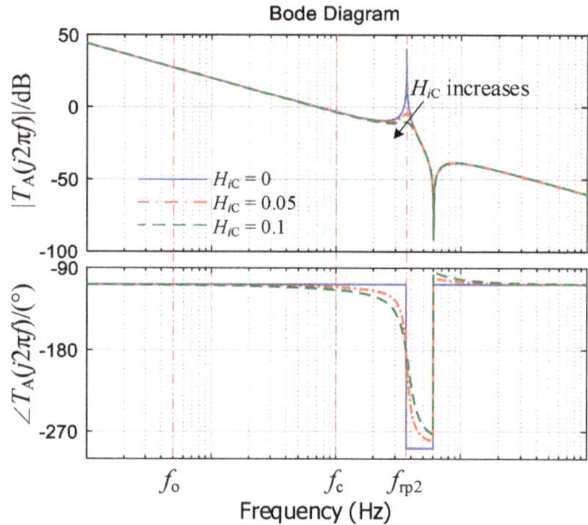

Figure 9. Bode diagrams of the loop gain before compensation when $\alpha + \beta_f = 2$.

The control system parameter design principles for IOLCL-type grid-tied inverters are adopted here to control the FOLLCL-type grid-tied inverter with $\alpha + \beta_f = 2$. The bode diagrams of the loop gain after compensation ($G_i(s) = K_p + K_i/s$) are shown in Figure 10. The frequency characteristics without active damping and compensation (green dotted lines) when $(\alpha, \alpha_f, \beta_f) = (1.2, 1.2, 0.8)$ are also plotted in the same figure for comparison purpose. $H_{iC} = 0.1$, $H_{ig} = 0.15$, $K_p = 0.45$, and $K_i = 2200$ are designed in this case to yield a satisfactory overall system performance. Compared with the original system (green dotted lines), the loop gain at the fundamental frequency (T_{fo}) after compensation (blue solid lines) increases and the high-frequency ($f > f_{rp1}$) magnitude characteristics move down, which can guarantee the fundamental current tracking and high-frequency harmonic attenuation capabilities. It can also be seen from Figure 10 that a lower α can guarantee better performance under the condition of $\alpha + \beta_f = 2$. When $\alpha = 1.1$, the system has a sufficient gain margin ($GM = 5.04$ dB) and an acceptable phase margin ($PM = 38.1°$, while $PM > 45°$ is required for a well-designed system), as well as a reasonable cut-off frequency ($f_c = 948$ Hz) and a sufficient fundamental loop gain ($T_{fo} = 49.5$ dB), while when $\alpha = 1.2$, although there is a slightly higher gain margin ($GM = 5.74$ dB), the PM, f_c, and T_{fo} all decrease. The low phase margin ($PM = 17.1°$) especially threatens the system stability.

Case II ($\alpha + \beta_f \neq 2$, PI control): For $\alpha + \beta_f \neq 2$ (taking $\alpha = 1.1$, $\alpha_f = 1.2$, $\beta_f = 0.8$ as an example), the bode diagrams of the loop gain before compensation ($G_i(s) = 1$) are drawn in Figure 11. $H_{iC} = 0$ is set in this case to avoid active damping. The positive resonance peak is damped effectively by selecting appropriate values for orders α and β_f to make their sum unequal to 2. When $H_{ig} = 0.15$, the same value as in case I, the cut-off frequency is very close to the equivalent switching frequency $2f_s$ (6 kHz), which is not acceptable for a grid-tied inverter. Moreover, when $H_{ig} = 0.1$ or 0.15, the magnitude plot has three cut-off frequencies and the system behaves as a conditionally stable system. For a large H_{ig}, even if the system can be stable after compensation, it is not easy to obtain a sufficient gain margin. If we keep decreasing H_{ig} to 0.05, the magnitude plot has only one cut-off frequency. Therefore, $H_{ig} = 0.05$ is chosen for the compensated system in the next step.

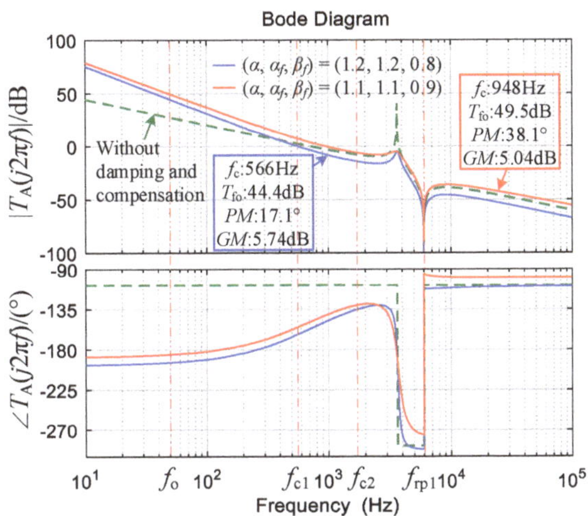

Figure 10. Bode diagrams of the loop gain after compensation when $\alpha + \beta_f = 2$.

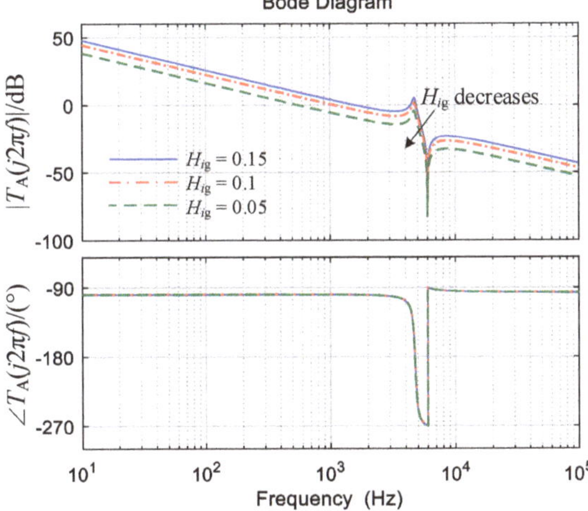

Figure 11. Bode diagrams of the loop gain before compensation when $\alpha + \beta_f \neq 2$.

The bode diagrams of the loop gain after compensation ($G_i(s) = K_p + K_i/s^\lambda$, $\lambda = 1$) are shown in Figure 12. As seen from (46), the decrease of H_{ig} will increase i_{g2}, so K_i should be increased to meet the steady-state error requirement. However, the phase margin when $K_i = 2200$ is only 22.7°, and after increasing K_i from 2200 to 4000, the phase margin decreases to 14.6°, and a sufficient phase margin cannot be guaranteed.

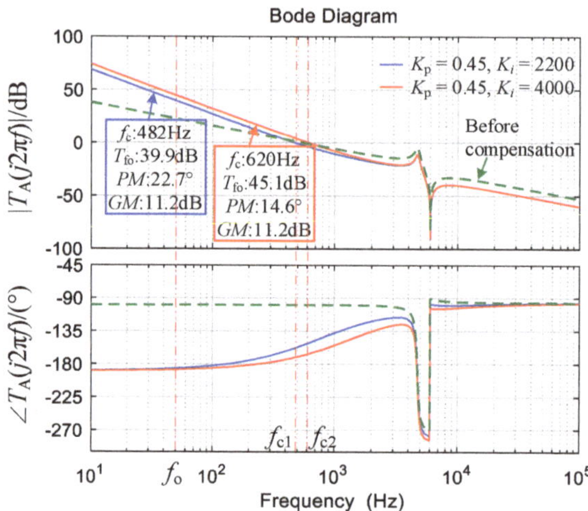

Figure 12. Bode diagrams of the loop gain after compensation with a PI controller when $\alpha + \beta_f \neq 2$ (varying K_i).

Case III ($\alpha + \beta_f \neq 2$, PI$^\lambda$ control): In this case, $(\alpha, \alpha_f, \beta_f) = (1.1, 1.2, 0.8)$, $H_{iC} = 0$, $H_{ig} = 0.05$, and a fractional-order PI$^\lambda$ regulator is used to try to improve the phase margin. When $K_p = 0.45$, $K_i = 2200$, and λ increases from 0.8 to 1.4, the bode diagrams of the loop gain are shown in Figure 13. The phase margin increases with λ, so $\lambda = 1.4$ is selected to leave enough room for K_i adjustment.

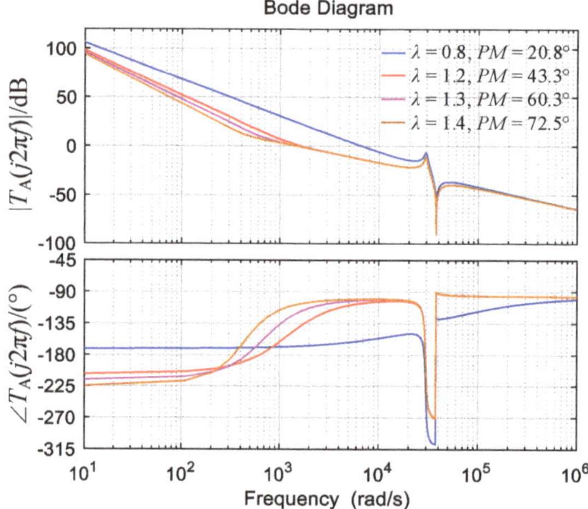

Figure 13. Bode diagrams of the loop gain after compensation with a PI$^\lambda$ controller when $\alpha + \beta_f \neq 2$ (varying λ).

The bode diagrams of the loop gain with varying K_i when $\lambda = 1.4$ are shown in Figure 14. With the increase of K_i, the phase margin decreases, but it is still sufficient even $K_i = 6000$ ($PM = 49.2°$). However, each curve in Figures 13 and 14 has a small T_{fo}, so the steady-state error requirement is still not guaranteed according to (47) and (48). The contradiction between T_{fo} and PM cannot be balanced by a PI$^\lambda$ regulator.

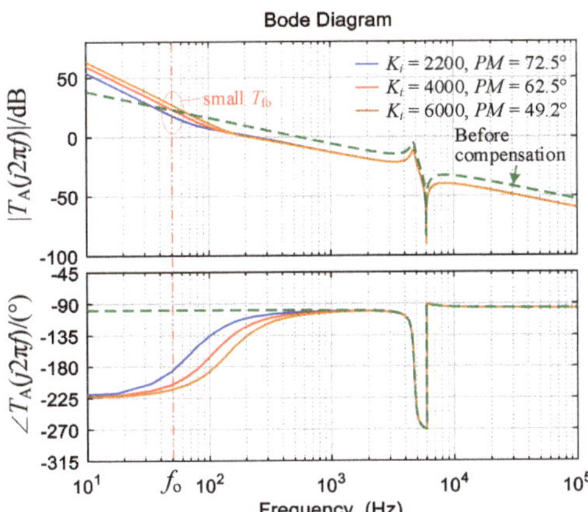

Figure 14. Bode diagrams of the loop gain after compensation with a PI$^\lambda$ controller when $\alpha + \beta_f \neq 2$ ($\lambda = 1.4$ and varying K_i).

Case IV ($\alpha + \beta_f \neq 2$, PR control): In this case, $(\alpha, \alpha_f, \beta_f) = (1.1, 1.2, 0.8)$, $H_{iC} = 0$, $H_{ig} = 0.05$, and a PR regulator is used to control the grid current. The values of the parameters are $K_p = 0.45$, $\omega_o = 2\pi \times 50$ rad/s, and $\omega_i = \pi$ rad/s, and K_r increases from 100 to 300; the bode diagrams of the compensated system are shown in Figure 15. It can be seen that each curve has a large enough T_{fo} to eliminate the steady-state error of i_g. However, PM decreases with the increase of K_r, when $K_r = 100$, $GM = 11.3$ dB, PM is $59.3°$, and the f_c also has a good value, so $K_r = 100$ will be selected in the simulation section.

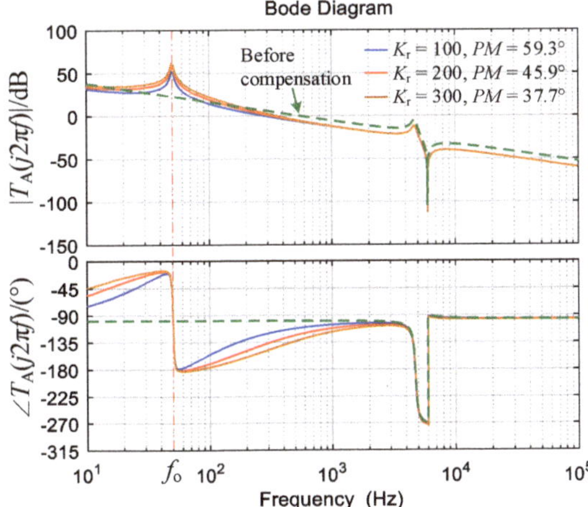

Figure 15. Bode diagrams of the loop gain after compensation with a PR controller when $\alpha + \beta_f \neq 2$ (varying K_r).

Based on the four cases discussed previously, it can be concluded that:

(1) If $\alpha + \beta_f = 2$, the FOLLCL-type grid-tied inverter can be damped by a capacitor current feedback loop. Under PI control, a lower α can achieve a larger PM but a higher f_c.

(2) If $\alpha + \beta_f \neq 2$, the system is stable under the grid current feedback; the capacitor current feedback is avoided. Under PI control, a large K_i should be chosen to reduce the steady-state error of i_g, but the PM decreases significantly.

(3) A PI$^\lambda$ regulator can also make the system stable, but there is a contradiction between T_{fo} and PM.

(4) A PR regulator can simultaneously obtain good T_{fo}, GM, PM, and f_c, which is suitable for controlling the FOLLCL-type grid-tied inverter if $\alpha + \beta_f \neq 2$.

5. Simulations

To verify the characteristics of the FOLLCL-type grid-tied inverter and the effectiveness of the control methods, simulations are conducted with the parameters presented in Tables 1 and 2. In each simulation, the reference grid current i_g^* is set to $50\sin(\omega t)$ A; i_g is magnified three times for observation in the waveform diagram. The fractional-order inductors and capacitor are equivalent to the fractance circuit using the Oustaloup approximation method. In the first simulation, an IOLCL-type grid-tied inverter is studied. As shown in Figure 16, a large amount of harmonics, which is mainly around $2f_s$ (6 kHz), exists in the grid current. The THD of i_g is 14.46%, which is not acceptable in the application. The result indicates that the LCL-type grid-tied inverter has little advantage in low-frequency applications.

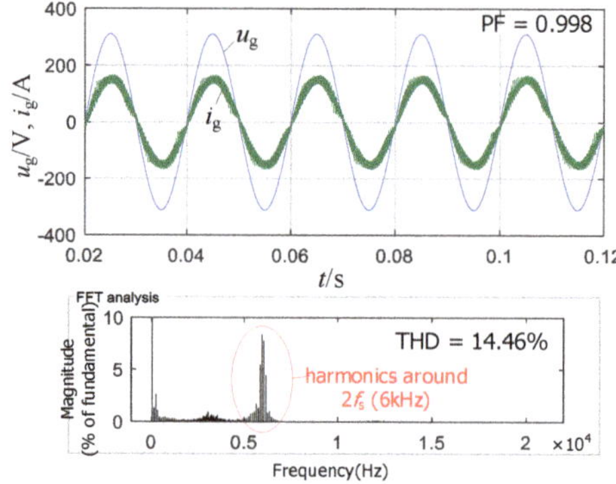

Figure 16. Simulation results of the IOLCL-type grid-tied inverter.

In the second simulation, an FOLLCL-type grid-tied inverter with $\alpha + \beta_f = 2$ and $\alpha_f + \beta_f = 2$ ($\alpha = 1.1$, $\alpha_f = 1.1$, $\beta_f = 0.9$) under PI control is investigated. As shown in Figure 17, when the capacitor current feedback loop is effective before 0.1 s, the system is stable, the grid current has a very low THD (only 0.26%), and the power factor is high (0.998). The FOLLCL-type grid-tied inverter exhibits excellent harmonic suppression ability. However, instability arises after 0.1 s due to the removal of the capacitor current feedback loop, which causes positive resonance.

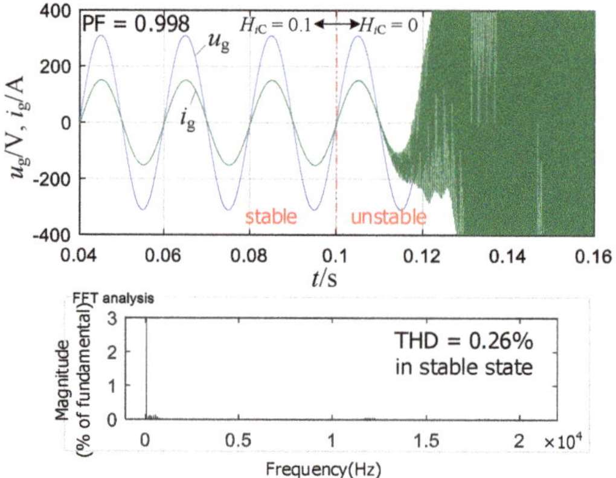

Figure 17. Simulation results of the FOLLCL-type grid-tied inverter with $\alpha + \beta_f = 2$ and a PI regulator.

Moreover, an FOLLCL-type grid-tied inverter with $\alpha + \beta_f \neq 2$ and $\alpha_f + \beta_f = 2$ ($\alpha = 1.1$, $\alpha = 0.2$, $\beta_f = 0.8$) is studied. The inverter is regulated by a PI controller; the control parameters are $K_p = 0.45$ and $K_i = 2200$, respectively. As with the analysis in Section 4 (case II), the capacitor current feedback is eliminated and $H_{ig} = 0.05$. As shown in Figure 18, the system is stable without active damping and the grid current is close to the ideal sine, which has a THD of 0.72%. However, to obtain a sufficient phase margin, K_i cannot be too large, resulting in a certain phase error between u_g and i_g; the power factor is only 0.989.

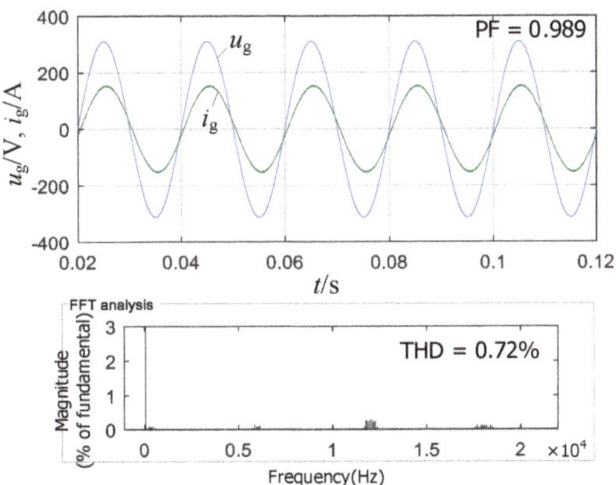

Figure 18. Simulation results of the FOLLCL-type grid-tied inverter with $\alpha + \beta_f \neq 2$ and a PI regulator.

Furthermore, in Figure 19, the simulation results of the FOLLCL-type grid-tied inverter with $\alpha + \beta_f \neq 2$ and $\alpha_f + \beta_f = 2$ ($\alpha = 1.1$, $\alpha_f = 1.2$, $\beta_f = 0.8$) controlled by a PI$^\lambda$ regulator are shown. According to the analysis in Section 4 and case III, when $\lambda = 1.4$ and $K_i = 6000$, although $PM > 45°$, the T_{fo} is very small, as can be seen in Figure 14. Therefore, as shown in Figure 19, both the amplitude error and phase error between i_g^* and i_g are very large, and the power factor is only 0.906.

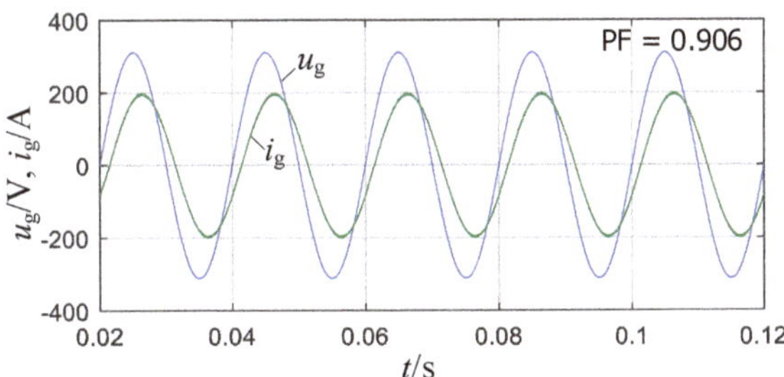

Figure 19. Simulation results of the FOLLCL-type grid-tied inverter with $\alpha + \beta_f \neq 2$ and a PI$^\lambda$ regulator.

Finally, a PR regulator is used to control the FOLLCL-type grid-tied inverter with $\alpha + \beta_f \neq 2$ and $\alpha_f + \beta_f = 2$ ($\alpha = 1.1$, $\alpha_f = 1.2$, $\beta_f = 0.8$). $K_p = 0.45$, $\omega_i = \pi$, and $K_r = 100$ are the parameters of the regulator, and $H_{ig} = 0.05$. The results are shown in Figure 20. The grid current is in strict in-phase with the grid voltage; the power factor is 1. In addition, the amplitude error is close to 0. The results are consistent with the previous analysis in Section 4.

The above simulation results prove that the analyses in previous sections are correct and PR regulator is superior to PI and PI$^\lambda$ regulators to control an FOLLCL-type grid-tied inverter with $\alpha + \beta_f \neq 2$.

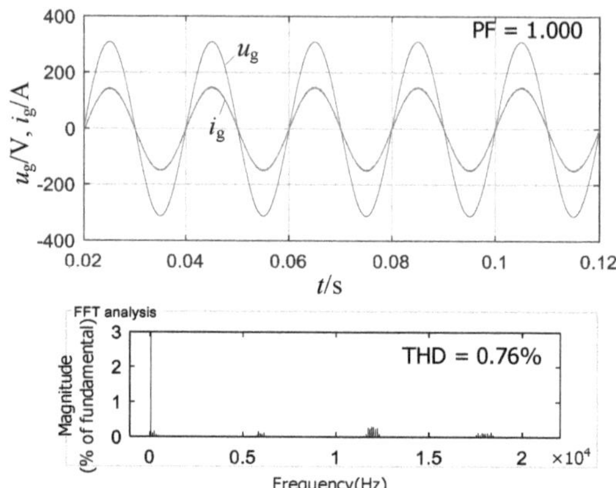

Figure 20. Simulation results of the FOLLCL-type grid-tied inverter with $\alpha + \beta_f \neq 2$ and a PR regulator.

6. Conclusions

In this paper, the fractional-order LLCL filter and the grid-tied inverter based on it are studied. By correctly selecting the orders of the components, the positive resonance can be suppressed and the negative resonance is reserved. Therefore, the passive or active damping can be avoided for the FOLLCL-type grid-tied inverter. Meanwhile, the switching-frequency harmonics in the grid current can be attenuated. For low-frequency applications, it is difficult for the PI controller and fractional-order PI controller to balance all performances simultaneously. PR controllers can guarantee good fundamental frequency loop gain, cut-off frequency, gain margin, and phase margin at the same time. The FOLLCL-type grid-tied inverter without active damping under PR control achieves excellent tracking accuracy and low grid current THD. Simulations are conducted to verify the correctness of the theoretical analyses.

Author Contributions: Methodology, X.W.; Software, R.Z. and J.C.; Validation, R.Z. and J.C.; Investigation, X.W.; Writing—original draft, X.W.; Writing—review & editing, X.W. All authors have read and agreed to the published version of the manuscript.

Funding: This work was supported by the "Guangzhou Science and Technology Plan Project, no. 202102010404".

Institutional Review Board Statement: Not applicable.

Informed Consent Statement: Not applicable.

Data Availability Statement: Not applicable.

Conflicts of Interest: The authors declare no conflict of interest.

References

1. Twining, E.; Holmes, D.G. Grid current regulation of a three-phase voltage source inverter with an LCL input filter. *IEEE Trans. Power Electron.* **2003**, *18*, 888–895. [CrossRef]
2. Pan, D.H.; Ruan, X.B.; Wang, X.H.; Yu, H.; Xing, Z.W. Analysis and design of current control schemes for LCL-type grid-connected inverter based on a general mathematical model. *IEEE Trans. Power Electron.* **2017**, *32*, 4395–4410. [CrossRef]
3. Jayalath, S.; Hanif, M. Generalized LCL-filter design algorithm for grid-connected voltage-source inverter. *IEEE Trans. Ind. Electron.* **2017**, *64*, 1905–1915. [CrossRef]
4. Wu, W.M.; Liu, Y.; He, Y.B.; Chung, H.S.H.; Liserre, M.; Blaabjerg, F. Damping methods for resonances caused by LCL-filter-based current-controlled grid-tied power inverters: An overview. *IEEE Trans. Ind. Electron.* **2017**, *64*, 7402–7413. [CrossRef]
5. Wang, J.G.; Yan, J.D.; Jiang, L.; Zou, J.Y. Delay-dependent stability of single-loop controlled grid-connected inverters with LCL filters. *IEEE Trans. Power Electron.* **2016**, *31*, 743–757. [CrossRef]
6. Wu, W.M.; He, Y.B.; Blaabjerg, F. An LLCL- power filter for single-phase grid-tied inverter. *IEEE Trans. Power Electron.* **2012**, *27*, 782–789. [CrossRef]

7. Wu, W.M.; He, Y.B.; Tang, T.H.; Blaabjerg, F. A new design method for the passive damped LCL and LLCL filter-based single-phase grid-tied inverter. *IEEE Trans. Ind. Electron.* **2013**, *60*, 4339–4350. [CrossRef]
8. Wu, W.M.; Sun, Y.J.; Huang, M.; Wang, X.F.; Wang, H.; Blaabjerg, F.; Liserre, M.; Chung, H.S.H. A robust passive damping method for LLCL-filter-based grid-tied inverters to minimize the effect of grid harmonic voltages. *IEEE Trans. Power Electron.* **2014**, *29*, 3279–3289. [CrossRef]
9. Zhang, Z.H.; Wu, W.M.; Shuai, Z.K.; Wang, X.F.; Luo, A.; Chung, H.S.; Blaabjerg, F. Principle and robust impedance-based design of grid-tied inverter with LLCL-filter under wide variation of grid-reactance. *IEEE Trans. Power Electron.* **2019**, *34*, 4362–4374. [CrossRef]
10. Liu, Y.T.; Jin, D.H.; Jiang, S.Q.; Liang, W.H.; Peng, J.C.; Lai, C.M. An active damping control method for the LLCL filter-based SiC MOSFET grid-connected inverter in vehicle-to-grid application. *IEEE Trans. Veh. Technol.* **2019**, *68*, 3411–3423. [CrossRef]
11. Khan, A.; Gastli, A.; Ben-Brahi, L. Modeling and control for new LLCL filter based grid-tied PV inverters with active power decoupling and active resonance damping capabilities. *Electr. Power Syst. Res.* **2018**, *155*, 307–319. [CrossRef]
12. Attia, H.A.; Freddy, T.K.S.; Che, H.S.; El Khateb, A.H. Design of LLCL filter for single phase inverters with confined band variable switching frequency (CB-VSF) PWM. *J. Power Electron.* **2019**, *19*, 44–57.
13. Alemi, P.; Bae, C.; Lee, D. Resonance suppression based on PR control for single-phase grid-connected inverters with LLCL filters. *IEEE Trans. Emerg. Sel. Topics Power Electron.* **2016**, *4*, 459–467. [CrossRef]
14. Liu, Z.F.; Wu, H.Y.; Liu, Y.; Ji, J.H.; Wu, W.M.; Blaabjerg, F. Modelling of the modified-LLCL-filter-based single-phase grid-tied Aalborg inverter. *IET Power Electron.* **2017**, *10*, 151–1515. [CrossRef]
15. Miao, Z.Y.; Yao, W.X.; Lu, Z.Y. Single-cycle-lag compensator-based active damping for digitally controlled LCL/LLCL-type grid-connected inverters. *IEEE Trans. Ind. Electron.* **2020**, *63*, 1980–1990. [CrossRef]
16. Wang, F.Q.; Ma, X.K. Modeling and analysis of the fractional order Buck converter in DCM operation by using fractional calculus and the circuit-averaging technique. *J. Power Electron.* **2013**, *13*, 1008–1015. [CrossRef]
17. Jia, Z.R.; Liu, C.X. Fractional-order modeling and simulation of magnetic coupled boost converter in continuous conduction mode. *Int. J. Bifurc. Chaos* **2018**, *28*, 1850061. [CrossRef]
18. Yang, R.C.; Liao, X.Z.; Lin, D.; Dong, L. Modeling and analysis of fractional order Buck converter using Caputo–Fabrizio derivative. *Energy Rep.* **2020**, *6*, 440–445. [CrossRef]
19. Wei, Z.H.; Zhang, B.; Jiang, Y.W. Analysis and modeling of fractional-order buck converter based on Riemann-Liouville derivative. *IEEE Access* **2019**, *7*, 162768–162777. [CrossRef]
20. Xie, L.L.; Liu, Z.P.; Zhang, B. A modeling and analysis method for CCM fractional order Buck-boost converter by using R–L fractional definition. *J. Electr. Eng. Technol.* **2020**, *15*, 1651–1661. [CrossRef]
21. Chen, X.; Chen, Y.F.; Zhang, B.; Qiu, D.Y. A modeling and analysis method for fractional-order DC–DC converters. *IEEE Trans. Power Electron.* **2017**, *32*, 7034–7044. [CrossRef]
22. Radwan, A.G.; Emira, A.A.; AbdelAty, A.M.; Azar, A.T. Modeling and analysis of fractional order DC-DC converter. *ISA Trans.* **2018**, *82*, 184–199. [CrossRef] [PubMed]
23. Sharma, M.; Rajpurohit, B.S.; Agnihotri, S.; Rathore, A.K. Development of fractional order modeling of voltage source converters. *IEEE Access* **2020**, *8*, 131750–131759. [CrossRef]
24. Xu, J.H.; Li, X.C.; Liu, H.; Meng, X.R. Fractional-order modeling and analysis of a three-phase voltage source PWM rectifier. *IEEE Access* **2020**, *8*, 13507–13515. [CrossRef]
25. El-Khazali, R. Fractional-order LC$^{\alpha}$L filter-based grid connected PV systems. In *2019 IEEE 62nd International Midwest Symposium On Circuits And Systems (MWSCAS)*; IEEE: New York, NY, USA, 2019; pp. 533–536.
26. Seo, S.W.; Choi, H.H. Digital implementation of fractional order PID-type controller for boost DC–DC converter. *IEEE Access* **2019**, *7*, 142652–142662. [CrossRef]
27. Soriano-Sánchez, A.G.; Rodríguez-Licea, M.A.; Pérez-Pinal, F.J.; Vázquez-López, J.A. Fractional-order approximation and synthesis of a PID controller for a Buck converter. *Energies* **2020**, *13*, 629. [CrossRef]
28. Xie, L.L.; Liu, Z.P.; Ning, K.Z.; Qin, R. Fractional-order adaptive sliding mode control for fractional-order Buck-boost converters. *J. Electr. Eng. Technol.* **2022**, *17*, 1693–1704. [CrossRef]
29. Azghandi, M.A.; Barakati, S.M.; Yazdani, A. Impedance-based stability analysis and design of a fractional-order active damper for grid-connected current-source inverters. *IEEE Trans. Sustain. Energy* **2021**, *12*, 599–611. [CrossRef]
30. Wang, Q.Y.; Ju, B.L.; Zhang, Y.Q.; Zhou, D.; Wang, N.; Wu, G.P. Design and implementation of an LCL grid-connected inverter based on capacitive current fractional proportional–integral feedback strategy. *IET Control Theory Appl.* **2020**, *14*, 2889–2898. [CrossRef]
31. Wang, X.G.; Cai, J.H. Grid-connected inverter based on a resonance-free fractional-order LCL filter. *Fractal Fract.* **2022**, *6*, 374. [CrossRef]

fractal and fractional

Article

Optimal FOPI Error Voltage Control Dead-Time Compensation for PMSM Servo System

Fumin Li [1], Ying Luo [1,*], Xin Luo [1], Pengchong Chen [1] and Yangquan Chen [2]

[1] School of Mechanical Science and Engineering, Huazhong University of Science and Technology, Wuhan 430074, China
[2] School of Engineering, University of California, Merced, CA 95343, USA
* Correspondence: ying.luo@hust.edu.cn

Abstract: This paper proposed a dead-time compensation method with fractional-order proportional integral (FOPI) error voltage control. The disturbance voltages caused by the power devices' dead time and non-ideal switching characteristics are compensated for with the FOPI controller and fed to the reference voltage. In this paper, the actual error voltage is calculated based on the model and actual voltage of the permanent magnet synchronous motor. Considering the parameter error of the permanent magnet synchronous motor and the voltage error caused by the dead-time effect, a FOPI controller is used to calculate the compensation voltage. An improved particle swarm optimization (PSO) algorithm is utilized to design the parameters of the FOPI controller in order to eliminate the dead-time effect, and the optimal fitness function is designed. Compared with other optimization algorithms, the improved PSO algorithm can achieve faster convergence speed in the error voltage controller parameter design. The proposed dead-time compensation method can improve the performance of the current response and eliminate the dead-time effect. This method also eliminates all harmonic disturbances and has a good suppression effect on high-frequency harmonics. The simulation and experimental results show that the dead-time compensation method using optimal FOPI error voltage control makes the current ripple smaller and the response speed faster than that of the traditional optimal integer-order PI control, thus demonstrating the effectiveness and advantages of the proposed method.

Keywords: fractional-order proportional integral (FOPI) controller; dead-time compensation; particle swarm optimization (PSO); voltage source inverter (VSI); permanent magnet synchronous motor (PMSM)

Citation: Li, F.; Luo, Y.; Luo, X.; Chen, P.; Chen, Y. Optimal FOPI Error Voltage Control Dead-Time Compensation for PMSM Servo System. *Fractal Fract.* **2023**, *7*, 274. https://doi.org/10.3390/fractalfract7030274

Academic Editors: Arman Oshnoei and Behnam Mohammadi-Ivatloo

Received: 1 January 2023
Revised: 1 March 2023
Accepted: 10 March 2023
Published: 21 March 2023

1. Introduction

The pulse width modulation (PWM) voltage source inverter (VSI) has been extensively used in permanent magnet synchronous motor drive systems. The dead time should be inserted in switching signals to avoid any shoot-through in the inverter lags of the PWM-VSI system. Because of the dead time of the inverter, there is a voltage drop between the output voltage and the reference voltage. The voltage drop caused by the dead time brings serious problems, such as current distortion and torque pulsation. Especially when the current is nearly zero, the output voltage distortion is more severe. Hence, it is vital to reject the disturbance from the dead time and improve the performance of current trucking.

Practical approaches have been discussed to overcome this problem. These methods can be sorted into three categories: (1) methods based on the modification of the PWM signal; (2) methods based on current harmonics monitoring; (3) methods based on model observation.

In the methods based on the modification of the PWM signal, the pulse width error caused by dead time is compensated for by detecting the current polarity and adjusting the PWM pulse width. In [1], this voltage error is compensated during the next switching period by modification of a reference voltage, the proposed solution can be used to compensate for the voltage error in multilevel multiphase voltage source inverters, but [1] requires additional hardware, which increases the cost and complexity of the system. If the polarity information is not accurate enough, the performance will worsen after compensating. The modified PWM signals are obtained using an off-line calculation in [2], this method relies on the precise detection of current polarity. However, obtaining accurate current polarity at zero-crossing instants is difficult, so it is necessary to determine the current polarity with the help of a more complicated signal processing algorithm. In [3], polarity detection accuracy is improved by using feedback circuits to measure inverter output voltages and Kalman filters to reconstruct fundamental phase currents. These approaches require additional circuits, which increase the end-product cost and hardware complexity.

In the PWM VSI system, the dead-time effect produces sixth-order current harmonics in the synchronous reference frame. According to this theory, the methods based on current harmonic monitoring are proposed [4–11]. The scheme in [4–11] can directly reduce the dead-time current harmonics by generating compensation voltage references while the motor is operating under steady conditions. Although those methods can make the proposed scheme effective in both the steady and transient states, the algorithm converging speed could be improved by limiting computation efforts. In [12], those approaches to the motor design improved the sinusoidal degree of the back electromotive force (EMF) by optimizing the distribution of the stator windings and the structure of the stator slots. Although the approaches can attenuate the harmonics caused by the slot effect and the magnetic saturation, the harmonics generated with the nonlinear characteristics of the inverter still exist [12]. The method based on the proportional-resonant (PR) regulator was also presented. It can track the reference for the positive and negative sequence currents without errors simultaneously [13], although the methods are effective, the interference between the different frequencies needs to be eliminated, and the algorithms need to be simplified. In [4], a dead-time-related harmonic minimization method is proposed based on proportional-integral (PI) controller tuning and under the premise of keeping the same control strategy; however, this method cannot completely eliminate the dead-time effect, which still has a great impact on the control system. This method [4] is compared with the method proposed in this paper. The methods based on current harmonics monitoring cannot take into account the high-frequency harmonic disturbance caused by the dead-time effect; moreover, the algorithm implementation of this method is more complex and requires higher hardware processors.

The methods based on model observation use a disturbance observer to estimate error voltages [14–16], Furthermore, it not only reduces the sixth harmonic as in many other compensation methods but also deals with its multiples concurrently. In [15], this paper introduces a current harmonic elimination method based on a disturbance observer (DOB), Although the DOB was shown to be able to satisfactorily operate under a constant change in the rotational speed of the machine, this method has not been verified for the dynamic response performance during motor startup. In [17], the expression of the inverter input current was derived by considering the deadtime effect. However, these models are not suitable for closed-loop control systems such as VSI-fed field-oriented control (FOC) PMSM servo systems. The control scheme using two extended state observers (ESO) is proposed in [14], which provides a strong ability to suppress dead-time effects, but the effect of this method can be further improved. In the following, the method using ESO [14] is compared with the method proposed in this paper.

In recent years, the number of studies related to the application of fractional controllers has been increasing. Fractional order control achieves better performance than conventional integer-order control, which can provide an opportunity to adjust better the dynamical characteristics of the control system [18–21]. And fractional-order proportional integral (FOPI) control has been widely used for servo systems [20]. Therefore, the FOPI controller is applied in our proposed dead-time compensation method to pursue advanced robustness.

This paper proposes a dead-time compensation method for a PMSM servo system with an optimal FOPI error voltage control. The proposed method based on the model of PMSM can calculate the error between the reference and output voltages, and the FOPI controllers can make the error voltages of the d-axis and q-axis converge to zero quickly. The proposed control strategy uses the PSO algorithm to design the parameters of the FOPI controller, considering the nonlinearity of the dead-time effect. The proposed dead-time compensation strategy can reduce the current disturbance in the VSI system. A PMSM system is used to validate the proposed method. Simulation and experimental results are presented to demonstrate the effectiveness of the proposed method.

The main contributions of this paper are as follows. (1) This paper proposes a dead-time compensation method for PMSM servo systems with optimal FOPI error voltage control. The method proposed in this paper can not only eliminate low-frequency harmonic disturbances but also has a good suppression effect on high-frequency harmonics. A voltage model with dead-time effect and parameter error of permanent magnet synchronous motor is established. The actual error voltage is calculated based on the model and actual voltage of the permanent magnet synchronous motor. Considering the parameter error of the permanent magnet synchronous motor and the voltage error caused by the dead-time effect, the FOPI controller is used to calculate the compensation voltage. (2) An improved particle swarm optimization (PSO) algorithm is utilized to design the parameters of the FOPI controller, and in order to eliminate the dead-time effect, the optimal fitness function is designed. Compared with other optimization algorithms, the improved PSO algorithm can achieve a faster convergence speed in the error voltage controller parameter design. (3) Through theoretical and experimental analysis, it is proven that the method proposed in this paper can have good error voltage control performance and robustness in the case of motor parameter errors.

The rest of this article is organized as follows. The formula derivation process for the error voltage caused by the dead-time effect is presented in Section 2. In Section 3, a dead-time compensation method with optimal FOPI error voltage control is proposed, and the parameter tuning procedure of the FOPI controller is also presented. The simulation and experiments using the proposed dead-time compensation method are presented in Sections 4 and 5, compared with the dead-time compensation method using optimal integer-order proportional integral (IOPI) control, the dead-time compensation method using ESO [14] and the dead-time-related harmonic minimization method [4]. Finally, the conclusion is given in Section 6.

2. Analysis of Dead-Time Effect

The typical three-phase PWM inverter with PMSM load is illustrated in Figure 1.

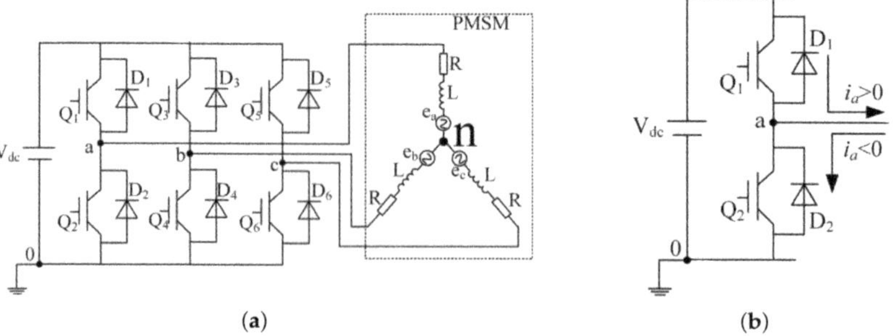

Figure 1. Model component. (**a**) Three-phase PWM drive system. (**b**) One phase leg of the PWM inverter.

In practice, a dead-time, T_d, is inserted in the gating signals to guarantee the safety of the VSI system. The gating signals with the dead time added are shown in Figure 2b. During the dead-time period T_d, the switches in Figure 1 are turned off, and the terminal voltage V_{a0} is determined by the direction of the a-phase current [22]. Figure 2c is the ideal terminal voltage. Considering the dead time and the turn ON/OFF delay, the terminal voltage is illustrated in Figure 2d,e. In this case, the a-phase average terminal voltage error ΔV_{a0} over one PWM period can be given as

$$\Delta V_{a0} = V_{err} * sign(i_a), \quad sign(i_a) = \begin{cases} -1, & i_a < 0 \\ 1, & i_a > 0 \end{cases} \tag{1}$$

where ΔV_{a0} is the a-phase average terminal voltage error over one PWM period, $sign(i_a)$ is the direction of the a-phase current, and V_{err} represents the magnitude of the terminal voltage error, which is defined as

$$V_{err} = \frac{T_d + T_{on} - T_{off}}{T_s} V_{dc} + V_{drop} \tag{2}$$

where V_{dc} is the dc-link voltage, T_d is the dead time, T_s is the PWM carrier period, T_{on} is the turn-on time delay of the switching device, T_{off} is the turn-off time delay of the switching device, and V_{drop} is the forward voltage drop of the switching device and of the diode.

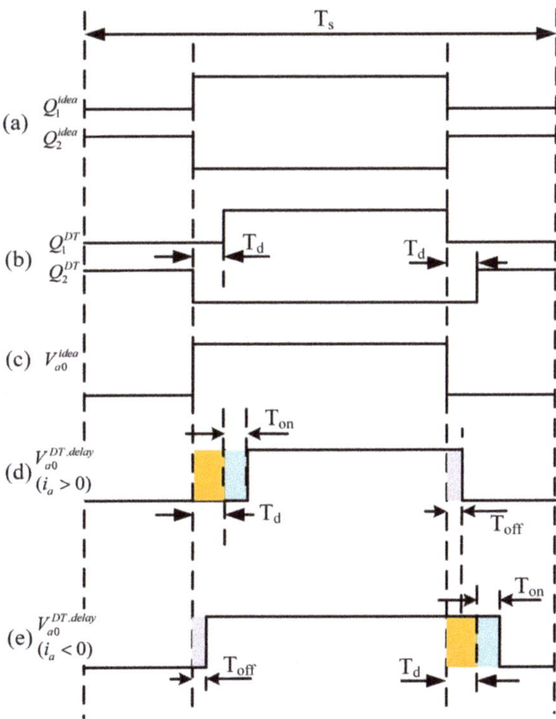

Figure 2. Switching pattern and terminal voltages: (**a**) Ideal gate signals. (**b**) Real gate signals with dead-time; (**c**) Ideal terminal voltage; (**d**) terminal voltage with dead-time and time delay when $i_a > 0$; (**e**) Actual terminal voltage with dead-time and time delay when $i_a < 0$.

In a similar way, the distorted voltage errors of other phases can be obtained. The error voltages in the three-phase stationary reference frame are obtained as

$$
\begin{cases}
\Delta u_{an} = \frac{1}{3}(2\Delta V_{a0} - \Delta V_{b0} - \Delta V_{c0}) \\[2mm]
\Delta u_{bn} = \frac{1}{3}(2\Delta V_{b0} - \Delta V_{a0} - \Delta V_{c0}) \\[2mm]
\Delta u_{cn} = \frac{1}{3}(2\Delta V_{c0} - \Delta V_{a0} - \Delta V_{b0})
\end{cases}
\tag{3}
$$

where ΔV_{a0}, ΔV_{b0}, and ΔV_{c0} are the three-phase terminal voltage errors of VSI and Δu_{an}, Δu_{bn}, and Δu_{cn} are the phase voltage errors of the PMSM in the three-phase stationary reference frame.

The error voltages in the stationary frame can be transformed to the synchronous reference frame as

$$
\begin{bmatrix} \Delta u_{dt} \\ \Delta u_{qt} \end{bmatrix} = T_{2s/2r} \cdot T_{3s/2s} \cdot \begin{bmatrix} \Delta u_{an} \\ \Delta u_{bn} \\ \Delta u_{cn} \end{bmatrix}
\tag{4}
$$

$$
T_{3s/2r} = \frac{2}{3}\begin{bmatrix} 1 & -1/2 & -1/2 \\ 0 & \sqrt{3}/2 & -\sqrt{3}/2 \end{bmatrix}
\tag{5}
$$

$$
T_{2s/2r} = \begin{bmatrix} \cos(\theta_e) & \sin(\theta_e) \\ -\sin(\theta_e) & \cos(\theta_e) \end{bmatrix}
\tag{6}
$$

where Δu_d and Δu_q are the d-axis and q-axis error voltages of the PMSM in the synchronous reference frame caused by dead-time effect, $T_{2s/3r}$ is the Park's transformation, and $T_{2s/3r}$ is Clark's transformation, θ_e is the rotor electrical position.

The analysis mentioned above demonstrates that the dead-time error voltage can be influenced by current polarity, the turn ON/OFF delay, the voltage drop of the switching device, and the diode. It is inaccurate for detecting current polarity because of the clamping of current around the zero-crossing point. Due to the change in operating conditions, such as temperature, many parameters affecting dead-time compensation are difficult to measure. Thus, it takes work to compensate for the effect of dead time directly.

3. Proposed Dead-Time Compensation Strategy

3.1. Error Voltage Calculation Based on PMSM Model

The dynamic model of PMSM in the synchronous reference frame can be represented as (7).

$$
\begin{bmatrix} u_d \\ u_q \end{bmatrix} = \begin{bmatrix} R_s + L_s p & -w_e L_q \\ w_e L_s & R_s + L_q p \end{bmatrix} \begin{bmatrix} i_d \\ i_q \end{bmatrix} + \begin{bmatrix} 0 \\ w_e \psi_f \end{bmatrix}
\tag{7}
$$

where u_d is the d-axis actual voltage, u_q is the q-axis actual voltage, i_d is the d-axis current, i_q is the q-axis current, and L_s, R_s, w_e, and ψ_f represent the stator inductance, resistance, rotor electrical angular velocity, and rotor flux linkage.

Therefore, the dynamic model of PMSM in the discrete-time domain is represented as:

$$
\begin{bmatrix} u_d(k-1) \\ u_q(k-1) \end{bmatrix} = \begin{bmatrix} R_s + \frac{L_s}{T_s} & -w_e L_s \\ w_e L_s & R_s + \frac{L_s}{T_s} \end{bmatrix} \begin{bmatrix} i_d(k) \\ i_q(k) \end{bmatrix} - \frac{L_s}{T_s}\begin{bmatrix} i_d(k-1) \\ i_q(k-1) \end{bmatrix} + \begin{bmatrix} 0 \\ w_e \psi_f \end{bmatrix}
\tag{8}
$$

where k represents the kth PWM period, $i_d(k)$ and $i_q(k)$ are the d-axis current and q-axis current of the kth PWM period, $u_d(k-1)$ and $u_q(k-1)$ are the d-axis voltage and q-axis voltages of the $k-1$ PWM period, and T_s is the sampling period.

Since proper motor parameters cannot be obtained in practice, the equivalent error voltage caused by the motor parameter error can be obtained through (9)

$$
\begin{bmatrix} err_d(k-1) \\ err_q(k-1) \end{bmatrix} = \begin{bmatrix} R_s + \frac{L_s}{T_s} & -w_e L_s \\ w_e L_s & R_s + \frac{L_s}{T_s} \end{bmatrix} \begin{bmatrix} i_d(k) \\ i_q(k) \end{bmatrix} - \frac{L_s}{T_s} \begin{bmatrix} i_d(k-1) \\ i_q(k-1) \end{bmatrix} - \\ \begin{bmatrix} R_c + \frac{L_c}{T_s} & -w_e L_c \\ w_e L_s & R_c + \frac{L_c}{T_s} \end{bmatrix} \begin{bmatrix} i_d(k) \\ i_q(k) \end{bmatrix} + \frac{L_c}{T_s} \begin{bmatrix} i_d(k-1) \\ i_q(k-1) \end{bmatrix}
\tag{9}
$$

where L_c and R_s represent the nominal stator inductance, nominal resistance, and err_d and err_q represent the equivalent error voltage caused by the motor parameter error.

Affected by the dead-time effect of the inverter, there is an error between the reference voltage and the output voltage. According to (8), the dynamic model of PMSM, including the error voltage in the discrete-time domain, can be represented as:

$$
\begin{bmatrix} u_d{}^*(k-1) \\ u_q{}^*(k-1) \end{bmatrix} = \begin{bmatrix} R_c + \frac{L_c}{T_s} & -w_e L_c \\ w_e L_s & R_c + \frac{L_c}{T_s} \end{bmatrix} \begin{bmatrix} i_d(k) \\ i_q(k) \end{bmatrix} - \frac{L_c}{T_s} \begin{bmatrix} i_d(k-1) \\ i_q(k-1) \end{bmatrix} + \begin{bmatrix} err_d(k-1) \\ err_q(k-1) \end{bmatrix} \\ + \begin{bmatrix} \Delta u_{dt}(k-1) \\ \Delta u_{qt}(k-1) \end{bmatrix}
\tag{10}
$$

where $u_d{}^*$ and $u_q{}^*$ represent the d-axis and q-axis reference voltage of the $k-1$ PWM period and $\Delta u_{dt}(k-1)$ and $\Delta u_{qt}(k-1)$ represent d-axis and q-axis error voltages caused by dead-time effect.

Based on (10), the total error voltage caused by the dead-time effect and motor parameter error in the synchronous reference frame can be calculated using (12).

$$
\begin{cases} \Delta u_q(k-1) = \Delta u_{dt}(k-1) + err_d(k-1) \\ \Delta u_d(k-1) = \Delta u_{qt}(k-1) + err_q(k-1) \end{cases}
\tag{11}
$$

$$
\begin{cases} \Delta u_q(k-1) = u_{qc}{}^*(k-1) - u_q(k-1) \\ \Delta u_d(k-1) = u_d{}^*(k-1) - u_{dc}(k-1) \end{cases}
\tag{12}
$$

where u_{dc} and u_q represent the actual nominal voltages of the d-axis and q-axis calculated from the nominal motor parameters, $\Delta u_d(k-1)$ and $\Delta u_q(k-1)$ represent d-axis and q-axis error voltages.

Therefore, to control the error voltage caused by the dead-time effect and the uncertainty error voltage caused by motor parameter error, it is necessary to use a FOPI controller with high control performance.

3.2. Compensation Method with Optimal FOPI Error Voltage Control

Figure 3 shows the system control block diagram. The actual voltage calculation module is shown in Figure 4, which can be represented by (12). When using the compensation method with optimal FOPI error voltage control, the system control block diagram is also shown in Figure 3. The compensation method with optimal FOPI error voltage control uses the FOPI controller as the error voltage controller. The control objective is to make the actual d-axis and q-axis voltages of the motor follow the reference voltages. The error voltage caused by the dead-time effect can be reduced as much as possible.

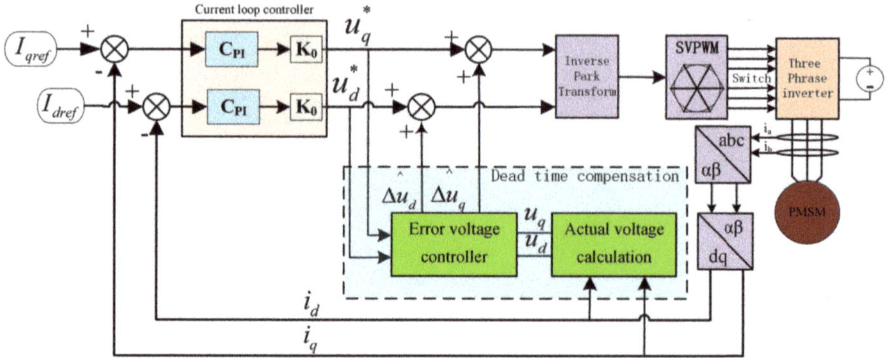

Figure 3. Block diagram of the system control.

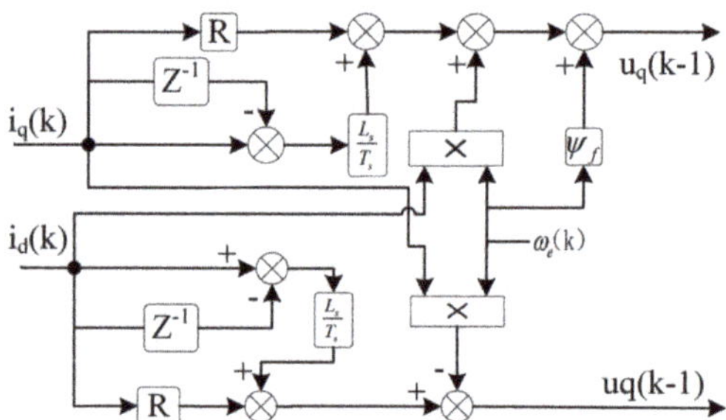

Figure 4. The actual voltage calculation module.

G_{FOPI} is the transfer function of the FOPI controller, which is shown as (13).

$$G_{FOPI} = k_p + k_i/s^\alpha \tag{13}$$

where k_p and k_i are proportional and integral gains and α is a fractional order. In this paper, we consider $\alpha \in (0, 2)$.

According to the impulse response invariance method [23], the higher the order, the higher the accuracy. However, a high approximation order results in more operating time for the microprocessor. Therefore, in practical applications, the approximation order is chosen based on control performance requirements and hardware constraints. The z-domain expression of the fractional operator $1/s^\alpha$ can be obtained with

$$\frac{1}{s^\alpha} = \frac{NUM}{DEN} \tag{14}$$

$$NUM = n_0 + n_1 z^{-1} + n_2 z^{-2} + \ldots + n_5 z^{-5} \tag{15}$$

$$DEN = 1 + d_1 z^{-1} + d_2 z^{-2} + \ldots + d_5 z^{-5} \tag{16}$$

where n_i and $d_i (i = 1 \ldots 5)$ are the discretization coefficients of fractional-order operators.

When using the compensation method with optimal FOPI error voltage control, the error voltage controller is shown in Figure 5 and $\Delta \hat{u}_q(k)$ is the q-axis compensation voltage. The calculation is as in (17).

$$\Delta \hat{u}_q(k) = P_q(k) + I_q^{fo}(k) \tag{17}$$

where $P_q(k)$ and $I_q^{fo}(k)$ represent the proportional and fractional-order integral terms of the q-axis FOPI error voltage controller of the kth PWM period, as in (18) and (19).

$$P_q(k) = kp_q * \Delta u_q(k-1) \tag{18}$$

$$\begin{aligned} I_q^{fo}(k) = ki_q * \big(&n_0 \Delta u_q(k-1) + n_1 \Delta u_q(k-2) + n_2 \Delta u_q(k-3) + n_3 \Delta u_q(k-4) + \\ &n_4 \Delta u_q(k-5) + n_5 \Delta u_q(k-6) - d_1 I_q^{fo}(k-1) - d_2 I_q^{fo}(k-2) - \\ &d_3 I_q^{fo}(k-3) - d_4 I_q^{fo}(k-4) - d_5 I_q^{fo}(k-5) \big) \end{aligned} \tag{19}$$

The $\Delta \hat{u}_d(k)$ is the d-axis compensation voltage. The calculation is as in (20).

$$\Delta \hat{u}_d(k) = P_d(k) + I_d^{fo}(k) \tag{20}$$

where $P_d(k)$ and $I_d^{fo}(k)$ represent the proportional and fractional-order integral terms of the d-axis FOPI error voltage controller of the kth PWM period, as in (21) and (22).

$$P_d(k) = kp_d * \Delta u_d(k-1) \tag{21}$$

$$\begin{aligned} I_d^{fo}(k) = ki_d * \big(&n_0 \Delta u_d(k-1) + n_1 \Delta u_d(k-2) + n_2 \Delta u_d(k-3) + n_3 \Delta u_d(k-4) + \\ &n_4 \Delta u_d(k-5) + n_5 \Delta u_d(k-6) - d_1 I_d^{fo}(k-1) - d_2 I_d^{fo}(k-2) \\ &- d_3 I_d^{fo}(k-3) - d_4 I_d^{fo}(k-4) - d_5 I_d^{fo}(k-5) \big) \end{aligned} \tag{22}$$

where $\Delta \hat{u}_q(k)$ and $\Delta \hat{u}_d(k)$ are the q-axis and d-axis compensation voltages of the kth PWM period, kp_d and ki_d are proportional and integral gains of d-axis FOPI error voltage controller, and kp_q and ki_q are proportional and integral gains of q-axis FOPI error voltage controller.

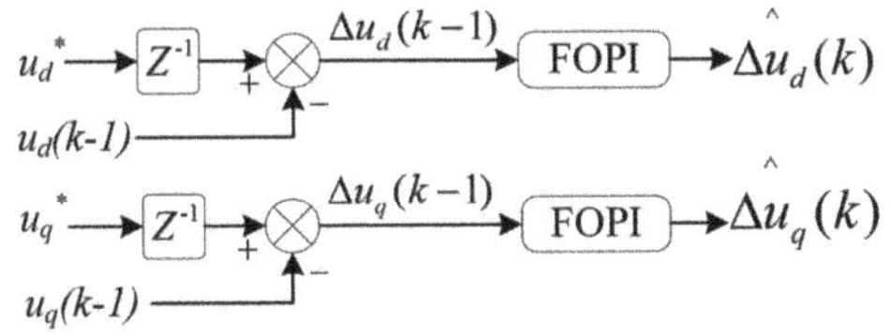

Figure 5. FOPI error voltage controller.

3.3. Compensation Method with Optimal IOPI Error Voltage Control

The FOPI controller can be superior to the IOPI in control performance and robustness. To prove the necessity and superiority of using the FOPI controller, the IOPI controller is used for error voltage control and compared with the FOPI controller. The compensation method with optimal integer order proportional integral (IOPI) error voltage control uses the IOPI controller as the error voltage controller.

G_{IOPI} is the transfer function of the IOPI controller, which is shown as in (23).

$$G_{IOPI} = k_p + k_i/s \tag{23}$$

where k_p and k_i are proportional and integral gains.

When using the compensation method with optimal IOPI error voltage control, the error voltage controller is shown in Figure 6. The compensation voltages are the output of the IOPI controllers. The calculation is as in (24) and (25).

$$\Delta \hat{u}_q(k) = k_{pq} * \Delta u_q(k-1) + k_{iq} * T_s \sum_{i=1}^{k-1} \Delta u_q(i) \tag{24}$$

$$\Delta \hat{u}_d(k) = k_{pd} * \Delta u_d(k-1) + k_{id} * T_s \sum_{i=1}^{k-1} \Delta u_d(i) \tag{25}$$

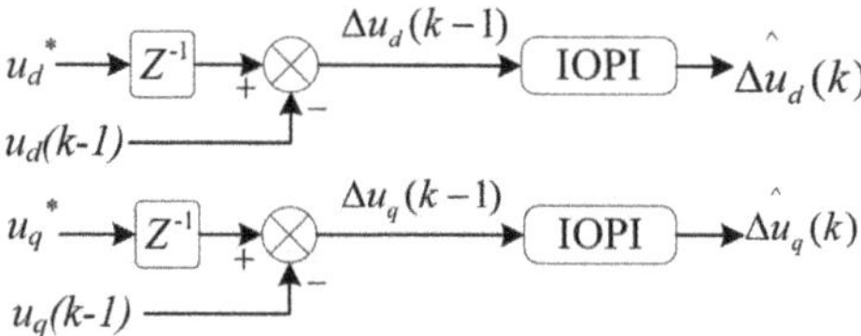

Figure 6. IOPI error voltage controller.

In the above equations (24) and (25), $\Delta \hat{u}_q(k)$ and $\Delta \hat{u}_d(k)$ are the q-axis and d-axis compensation voltages of the kth PWM period, k_{pd} and k_{id} are proportional and integral gains of d-axis IOPI error voltage controller, and k_{pq} and k_{iq} are proportional and integral gains of q-axis IOPI error voltage controller.

3.4. Parameter Design of Error Voltage Controller Based on Improved PSO Algorithm

Due to its simplicity and ease of implementation with only a few parameters, the particle swarm optimization (PSO) algorithm has shown desired optimization performance in continuous optimization problems, and discrete optimization problems [24–26]. Because of the non-linear characteristics of the dead-time effect, it is challenging to use frequency domain analysis to design the parameters of the error voltage controller. In this paper, the improved PSO algorithm is presented to develop the parameters of the error voltage controller, and the process is shown in Figure 7.

In Figure 7, the position of the particle represents the parameters of the error voltage controller, and the fitness function represents the performance index of the compensation effect. When using the method with optimal IOPI error voltage control, the position of the particles corresponds to the parameters of the IOPI controllers. When using the method with optimal FOPI error voltage control, the position of the particles corresponds to the parameters of the FOPI controllers.

The following is the optimization procedure. First, a particle swarm is generated, and the positions of the particles are sequentially assigned to the parameters of the error voltage controller (IOPI controller or FOPI controller). Then the performance index corresponding to the set of parameters can be obtained by the simulation model of the control system. By evaluating the fitness function, the optimal fitness value and particle position can be obtained. Finally, it is judged whether the maximum number of iterations is reached. If not, the operation of updating the particle is performed. If the number of iterations equals the maximum number of iterations, the algorithm obtains the optimal position.

The update operation of particles mainly includes speed update and position update, which are calculated according to (26).

$$\begin{cases} v_{i+1} = \omega_i v_i + c_1 r_1 (P_i - x_i) + c_2 r_2 (G_i - x_i) \\ x_{i+1} = x_i + v_{i+1} \end{cases} \tag{26}$$

where x_i is the particle's position of the ith iteration, i is the number of iteration, v_i is the particle's velocity of the ith iteration, x_{i+1} and v_{i+1} are the position and velocity of the the $i+1$ iteration, and ω_i is the inertia weight at the ith iteration. c_1 and c_2 are the acceleration coefficients, c_1 and c_2 control the relative proportion of cognition and social interaction in the swarm. r_1 and r_2 are random numbers between [0,1], P_i is the optimal position of a single particle until the ith iteration, and G_i is the optimal position of the entire particle swarm until the ith iteration.

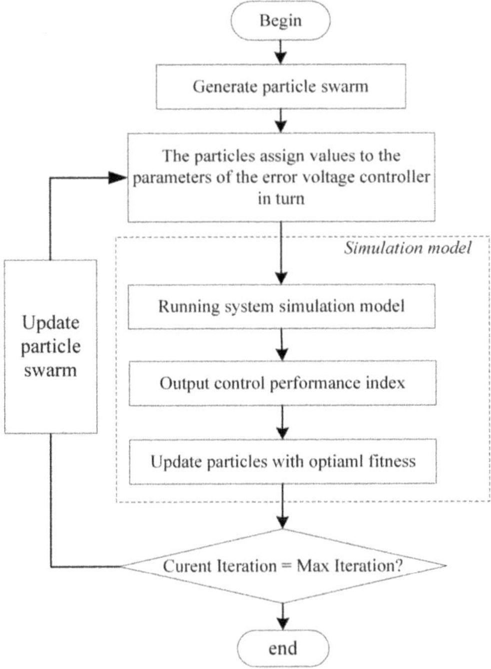

Figure 7. The process of parameter design of error voltage controller.

To dynamically adjust the inertia weight, we consider $\omega_i \in [\omega_{min}, \omega_{max}]$, and the inertia weight at the ith iteration is as in (27).

$$\omega_i = \omega_{max} - \frac{\omega_{max} - \omega_{min}}{\text{Iter}_{max}} \cdot i \tag{27}$$

where ω_{min} and ω_{max} are constants, the $iter_{max}$ is maximum number of iteration.

As in (28), the acceleration coefficient c_1 keeps decreasing over time, and the acceleration coefficient c_2 keeps increasing over time.

$$\begin{cases} c_1 = (c_{1f} - c_{1i}) \frac{i}{\text{Iter}_{max}} + c_{1i} \\ \\ c_2 = (c_{2f} - c_{2i}) \frac{i}{\text{Iter}_{max}} + c_{2i} \end{cases} \tag{28}$$

where c_{1i}, c_{1f}, c_{2i}, and c_{2f} are constants, which are the initial and final values of c_1 and c_2, and the Iter_{max} is the maximum number of iteration.

The purpose of the error voltage controller is to minimize the q-axis voltage error and the d-axis voltage error. The q-axis current ripple directly causes the torque ripple. It greatly influences the system's control performance, so this paper pays more attention to the voltage error control effect of the q-axis. Referring to the evaluation index of ITAE, the fitness function is set to J_{ITAE}, as in (29).

$$J_{ITAE} = 4 * \int_0^\infty t|\Delta u_q|dt + \int_0^\infty t|\Delta u_d|dt \tag{29}$$

where $\Delta u_d(k)$ and $\Delta u_q(k)$ represent the d-axis and q-axis error voltages of the kth PWM period. t is the running time of the system.

When using the compensation method with optimal IOPI error voltage control, the particle position x_i is shown in (30).

$$x_i = \begin{bmatrix} k_{pd} & k_{id} & k_{pq} & k_{iq} \end{bmatrix} \tag{30}$$

where k_{pd} and k_{id} are proportional and integral gains of the d-axis IOPI error voltage controller and k_{pq} and k_{iq} are proportional and integral gains of q-axis IOPI error voltage controller.

For designing the parameters of the IOPI controller, the configuration of the improved PSO algorithms is shown in Table 1.

Based on the configuration in Table 1, the parameters of the IOPI controller are optimized by the PSO algorithm. The convergence curve of the optimal fitness function value with the number of iterations is shown in Figure 8.

Table 1. The configuration of the improved PSO algorithm.

For Optimal IOPI Controler		For Optimal FOPI Controller	
populations	40	populations	40
Iter_{max}	40	Iter_{max}	40
ω_{max}	0.8	ω_{max}	0.8
ω_{min}	0.2	ω_{min}	0.2
c_{1f}	0.3	c_{1f}	0.3
c_{1i}	2.5	c_{1i}	2.5
c_{2i}	0.3	c_{2i}	0.3
c_{2f}	2.5	c_{2f}	2.5
v_{max}	[3,40,3,40]	v_{max}	[3,40,3,40,0.2,0.2]
v_{min}	[−3,−40,−3,−40]	v_{min}	[−3,−40,−3,−40,−0.2,−0.2]
x_{max}	[2,1000,20,1000]	x_{max}	[2,1000,20,1000,0,0]
x_{min}	[0,0,0,0]	x_{min}	[0,0,0,0,−2,−2]

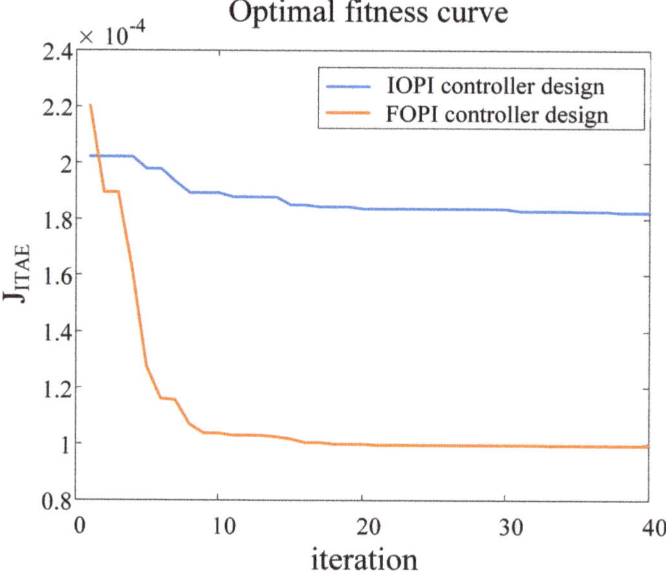

Figure 8. PSO algorithm Optimal fitness function value in FOPI/IOPI controller design.

The parameter iteration result of the IOPI controller based on the improved PSO algorithm is shown in Table 2.

When using the compensation method with optimal FOPI error voltage control, the particle position x_i is shown in (31)

$$x_i = \left[kp_d, ki_d, kp_q, ki_q, \alpha_d, \alpha_q\right] \tag{31}$$

where kp_d and ki_d are proportional and integral gains of the d-axis FOPI error voltage controller, kp_q and ki_q are proportional and integral gains of q-axis FOPI error voltage controller, α_d is the fractional order of the d-axis FOPI error voltage controller, and α_q is fractional order of q-axis FOPI error voltage controller.

For designing the parameters of the FOPI controller, the configuration of the improved PSO algorithms is shown in Table 1.

Based on the configuration in Table 1, the parameters of the FOPI controller are optimized using the PSO algorithm. The convergence curve of the optimal fitness function value with the number of iterations is shown in Figure 8.

Table 2. The parameter iteration result.

Paramters of IOPI Controller		Paramters of FOPI Controller	
k_{pd}	3.05745	kp_d	2.186
k_{id}	431.402	ki_d	491.66
k_{pq}	2.7419	kp_q	1.693
k_{iq}	707.891	ki_q	503.683
		α_d	0.651
		α_q	0.722

The parameter iteration result of the FOPI controller is shown in Table 2.

The Artificial Bee Colony (ABC) Algorithm is an optimization algorithm based on intelligent behavior.

For the parameter design of the IOPI/FOPI controller, the parameter optimization curve of the ABC algorithm is shown in Figure 9 [27]. The comparison of the iterative optimization results of the ABC algorithm and the PSO algorithm is shown in Table 3, it can be seen that the optimal fitness values obtained by the two algorithms are approximately equal, which proves that the controller parameters obtained in Table 2 are the global optimal solution. Compared with the ABC algorithms, the improved PSO algorithm can achieve a faster convergence speed in the error voltage controller parameter design.

Table 3. Comparison of optimal fitness function Values between ABC algorithm and PSO algorithm.

	IOPI Controller	FOPI Controller
Optimal fitness function value of PSO algorithm	1.8350×10^{-4}	9.9584×10^{-5}
Optimal fitness function value of ABC algorithm	1.8310×10^{-4}	9.9618×10^{-5}

The α_d is the fractional order of the d-axis FOPI error voltage controller. According to Table 2, $\alpha_d = 0.651$. The discrete implementation of $1/s^{\alpha_d}$ can be obtained using the impulse response invariance method [23], and the comparison of discretization approximate bode graph and ideal bode graph is shown in Figure 10. The α_q is the fractional order of the q-axis FOPI error voltage controller, and the comparison of discretization approximate bode graph and ideal bode graph of $1/s^{\alpha_q}$ is shown in Figure 11.

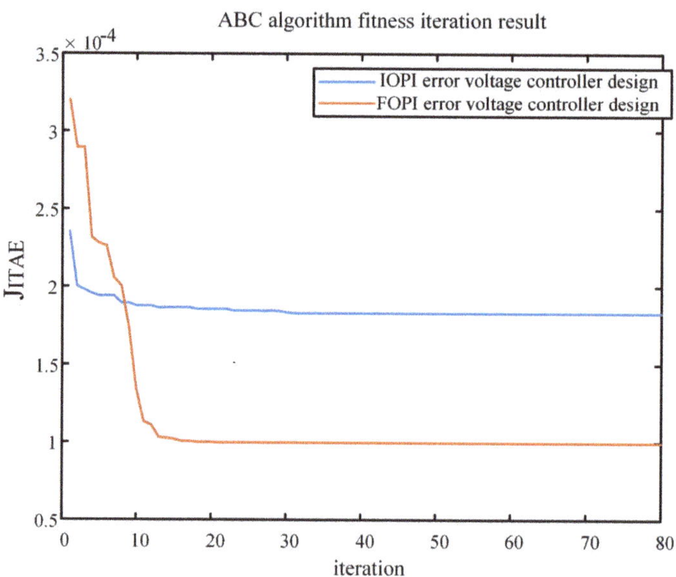

Figure 9. ABC algorithm Optimal fitness function value in FOPI/IOPI controller design.

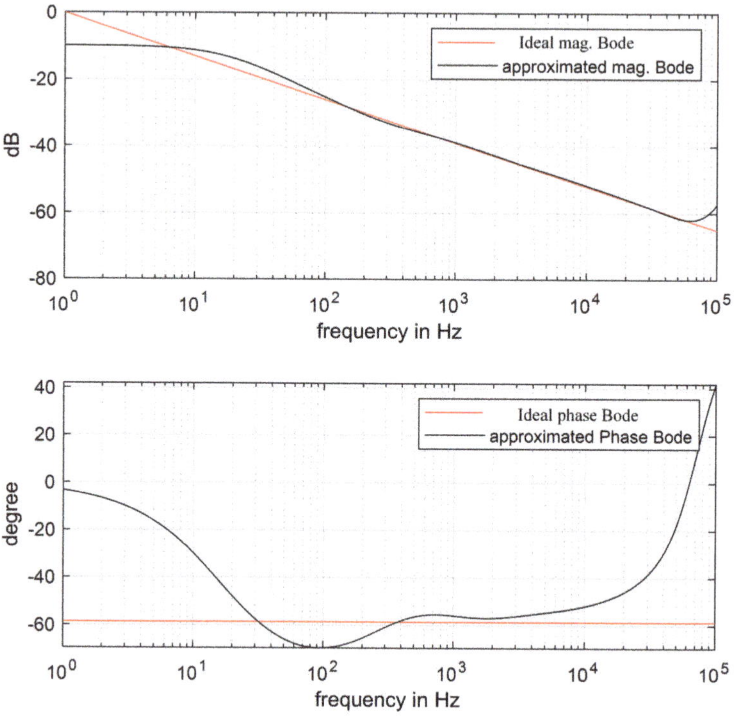

Figure 10. The comparison of discretization approximate bode graph and ideal bode graph of $1/s^{\alpha_d}$.

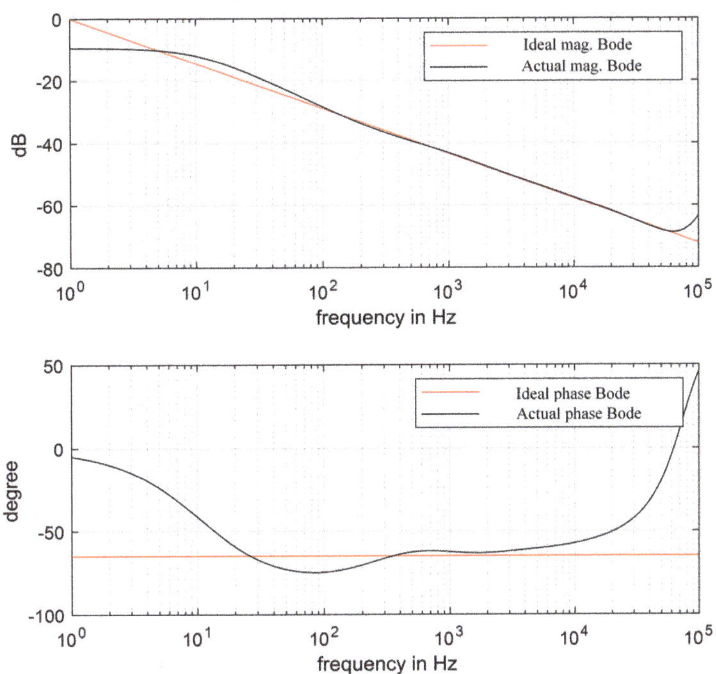

Figure 11. The comparison of discretization approximate bode graph and ideal bode graph of $1/s^{\alpha_q}$.

3.5. Parameters Design of Current Loop Controller and Speed Loop Controller

The PMSM current can be controlled with two PI controllers. Without dead-time compensation, the simplified PMSM current control system is shown in Figure 12.

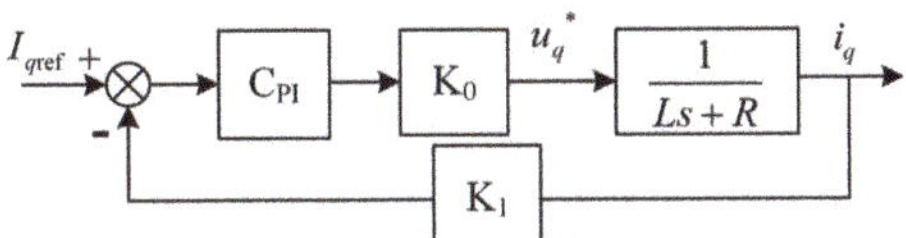

Figure 12. The simplified PMSM current control system.

The current loop controller model C_{PI} is as in (32)

$$C_{PI}(s) = K_{pc} + \frac{K_{ic}}{s} \tag{32}$$

where K_{pc} and K_{ic} are proportional and integral coefficients. The open-loop transfer function of the current loop is as follows:

$$\begin{aligned} G_{co}(s) &= K_0 K_1 C_{PI}(s)\frac{1}{L_s s + R_s} \\ &= K_0 K_1 \left(\frac{K_{pc}s + K_{ic}}{s}\right)\frac{1}{L_s s + R_s}. \end{aligned} \tag{33}$$

where K_0 and K_1 are constants. K_0 represents the voltage conversion coefficient, and K_1 represents the current conversion coefficient. L_s and R_s represent the stator inductance and resistance.

According to the existing system, $K_0 = 179.561$ and $K_1 = 1/19.2$. The current controller is designed by the cancellation method with relationship $K_{pc}/K_{ic} = L/R$ [28]. The gain crossover frequency is set as $\omega_c = 2000$ rad/s,

$$|G_{co}(j\omega_c)| = 1 \tag{34}$$

According to (34), we can get the parameters of current loop controller as follows:

$$K_{pc} = \frac{\omega_c L_s}{K_0 K_1}, \tag{35}$$

$$K_{ic} = \frac{K_{pc} R_s}{L_s} = \frac{\omega_c R_s}{K_0 K_1}. \tag{36}$$

4. Simulation Results

The proposed method is compared with the method without compensation to demonstrate the effectiveness of the dead-time compensation method with optimal FOPI error voltage control. The proposed method is compared with the method using optimal IOPI error voltage control and the compensation with extended state observers (ESO) [14] to indicate higher performance using optimal FOPI control. In order to prove that the FOPI error voltage control method has a good ability to eliminate current harmonics, this dead-time-related harmonic minimization method [4] is compared with the proposed method. The simulation model is built using MATLAB/Simulink, which has the same control scheme and parameters as the drive system. The dead time in the gate of power switches is 2.1 µs. The simulation control block diagram is shown in Figure 3, the Iq_{ref} is set to 1 A, and the Id_{ref} is set to 0 A. According to (35) and (36), The parameters of the current loop controller can be calculated as $K_{pc} = 1.4018$ and $K_{ic} = 121.8973$. The parameters of the error voltage controller are shown in Table 2. Other specifications of the simulation drive system are given in Table 4.

Table 4. Specification of experimental drive system.

Parameters of PMSM		Specification of PWM Inverter	
Pole pairs	5	DC link	310 [V]
Resistance (Rs)	0.38 [Ω]	PWM period	50 [µs]
Inductance (Ls)	4.37 [mH]	Turn-on/off delay	180/320 [µs]
Flux linkage (lam)	0.066 [Wb]	Dead-time	2.1 [µs]
Inertia (J)	0.027 [kg·m²]	IGBT/Diode Ron	36 [mΩ]
viscous daping (B)	0.0502 [N·m·s]	Saturation Volt	1.1 [V]

4.1. Current Closed-Loop Simulation

The simulation result of I_q current is shown in Figure 13 and Table 5. As we can see, the I_q current of the method without compensation is much distorted because of the dead-time effect, and the torque ripple in the steady state is very severe. When using the dead-time compensation method with optimal FOPI error voltage control, the current ripple can be significantly eliminated, and the tracking performance can be improved. Compared with the IOPI error voltage control and the compensation with ESO [14], it can be seen that the method with optimal FOPI error voltage control can further improve tracking performance and reduce the current ripple of the I_q current. Compared with the compensation with ESO [14], the optimal FOPI error voltage control can effectively suppress high-frequency harmonic disturbance.

In the simulation, the comparison of the I_d current response under the three control strategies is shown in Figure 14 and Table 6. The dead-time compensation method with optimal FOPI error voltage control can significantly reduce the current disturbance caused by the dead-time effect, and the method with optimal FOPI error voltage control has a more vital ability to suppress the disturbance of the I_d current than that using optimal IOPI error voltage control and the compensation method with ESO [14]. Figure 15 and Table 7 show the three-phase current response comparison results. The dead-time compensation method with optimal FOPI error voltage control can significantly eliminate the current distortion and reduce the current clamping around the zero-crossing point. The method's performance with optimal FOPI error voltage control is better than that of IOPI error voltage control, which proves that the proposed method has a better dead-time compensation effect. Compared with the compensation with ESO [14], the optimal FOPI error voltage control can effectively suppress high-frequency harmonic disturbance.

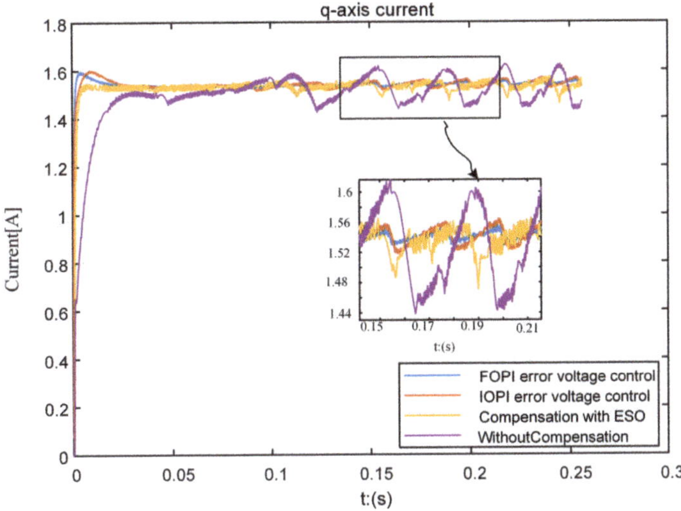

Figure 13. The comparison of I_q step current response of three control strategies.

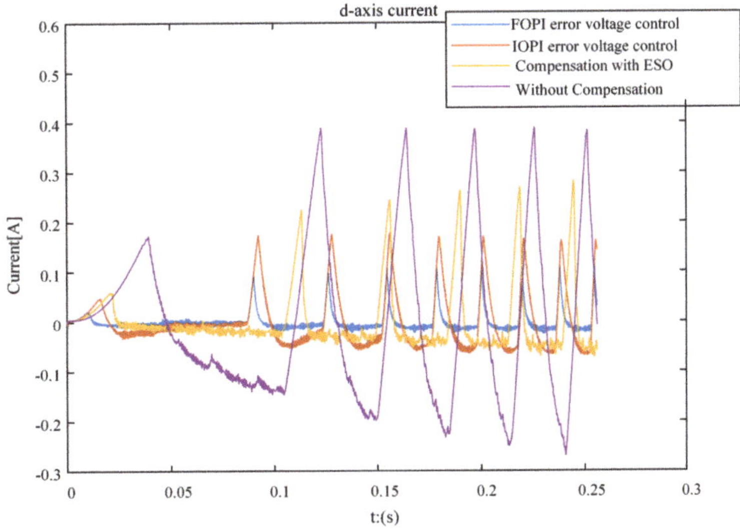

Figure 14. Comparison of I_d current response of three control strategies.

The q-axis error voltage simulation results are shown in Figure 16. The q-axis error voltage varies nonlinearly with increasing speed when using the method without compensation. The dead-time compensation method with optimal FOPI error voltage control can quickly make the error voltage converge to 0. The convergence speed of the proposed method is faster than that of the method using optimal IOPI error voltage control. The method with optimal FOPI error voltage control makes the q-axis error voltage converge to 0 faster than the method with optimal IOPI error voltage control.

Table 5. The performance index comparison of I_q step current response in the simulation.

	without Compensation	IOPI	FOPI	ESO [14]
overshoot (%)	9.05	4.53	4.767	0.0
rise time (s)	0.01825	0.0037	0.0017	0.00986
settling time (s)	\		0.03	0.024	0.00986
current ripple amplitude (A)	0.184	0.0359	0.0175	0.08467

Table 6. The performance index comparison of I_d step current response.

	without Compensation	IOPI	FOPI	ESO [14]
Current ripple amplitude (A)	0.671	0.23	0.097	0.33

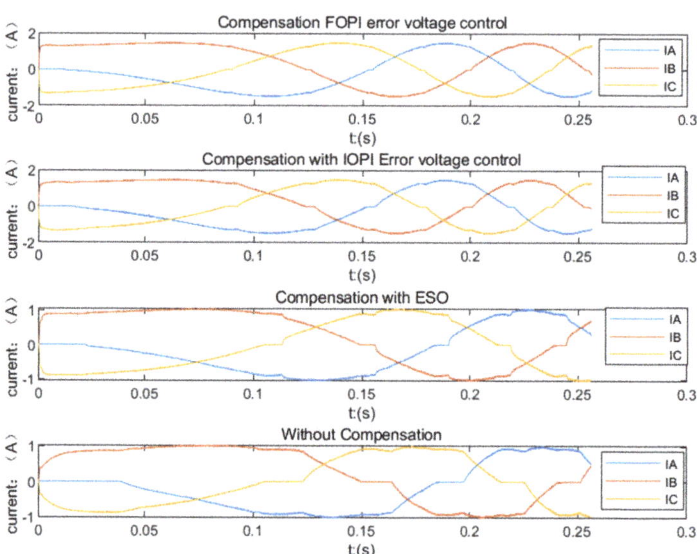

Figure 15. Comparison of three-phase current response of three control strategies.

Table 7. The performance index comparison of three-phrase current step current response.

	without Compensation	IOPI	FOPI	ESO [14]
Current clamping time (s)	0.0102	0.003258	0.001751	0.00482

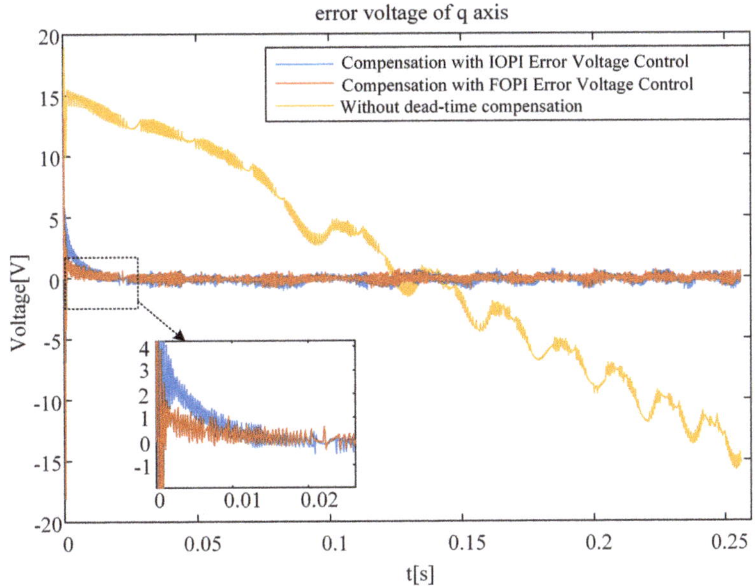

Figure 16. Comparison of the error voltage of q-axis.

The d-axis voltage error simulation results are shown in Figure 17. It can be seen that the d-axis error voltage fluctuates significantly under the influence of the dead-time effect, and the dead-time compensation method with optimal FOPI error voltage control can substantially reduce the d-axis error voltage.

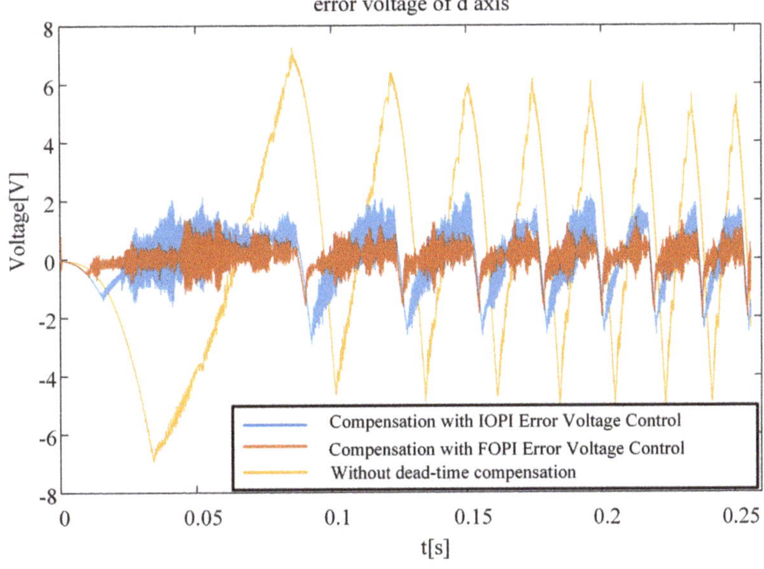

Figure 17. Comparison of the error voltage of d-axis in the simulation.

4.2. Speed Closed-Loop Simulation

In the case of keeping the motor speed constant at 200 r/min, compare the method proposed in this paper with the typical current harmonic elimination method. Observe and compare the three-phase current and its frequency spectrum. In [4], a dead-time-related harmonic minimization method based on proportional-integral (PI) controller tuning is proposed under the premise of keeping the same control strategy. In order to prove that the FOPI error voltage control method has a good ability to eliminate current harmonics, this dead-time-related harmonic minimization method [4] is compared with the proposed method.

The three-phase current simulation results and their spectral analysis for the dead-time-related harmonic minimization method and the method [4] proposed in this paper are shown in Figures 18 and 19. It can be seen from Figure 18 that the proposed method can more effectively eliminate the current clamping phenomenon caused by the dead-time effect. At the point where the current passes through 0, the current clamping time of the proposed method is shorter than the dead-time-related harmonic minimization method [4]. It can be seen from Figure 19 that the proposed method is better than the dead-time-related harmonic minimization method [4] in eliminating the dead-time effect harmonics.

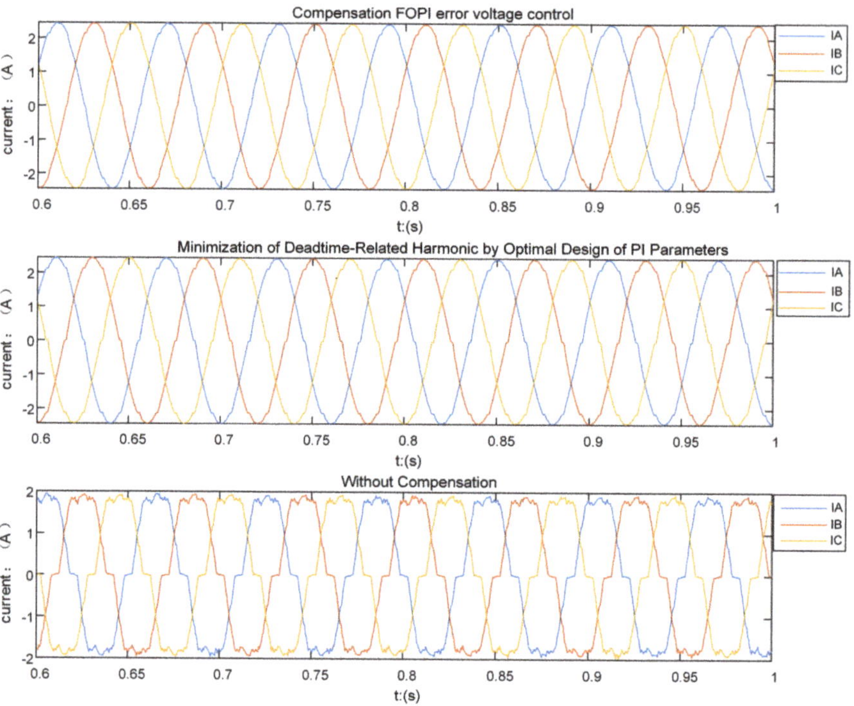

Figure 18. Comparison of three phase current response under 200 r/min speed conditions.

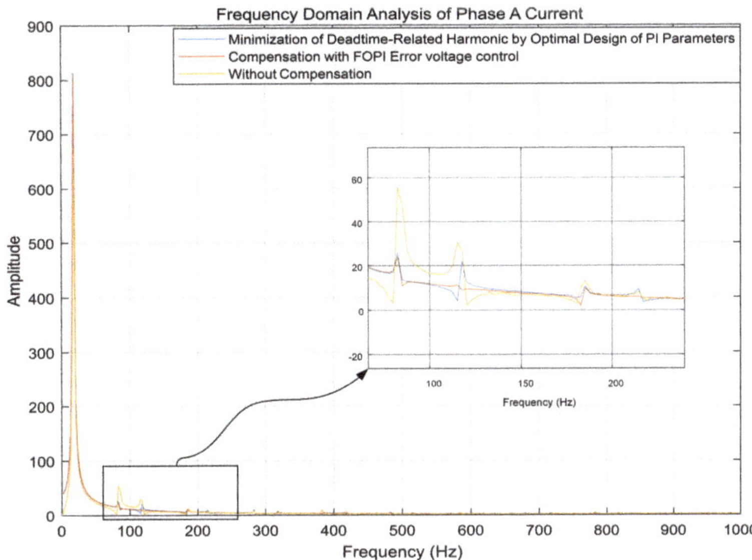

Figure 19. Frequency domain simulation analysis of A phase current under 200 r/min speed conditions.

4.3. Robustness Comparison for Motor Parameter Error

Due to the influence of motor characteristics, motor inductance and other parameters are easy to change during the experiment, so the verification of motor parameter uncertainty is added to the simulation. Three groups of motor parameters are randomly given within a certain range for simulation, and the effectiveness and robustness of the compensation method are verified by comparing the optimal FOPI error voltage control method and the optimal IOPI error voltage control method. The error and nominal motor parameters used in the simulation are shown in Table 8.

Table 8. Three groups of error motor parameters and motor actual parameters.

	Resistance (Ω)	Inductance (mH)	Inertia (kg·m^2)	Viscous Damping (N·m·s)
error parameters (1)	0.514	4.90	0.0396	0.023
error parameters (2)	0.769	2.75	0.033	0.0299
error parameters (3)	0.769	1.96	0.0134	0.0172
actual parameters	0.38	4.37	0.027	0.05027

From Figures 20 and 21, it can be seen that the FOPI error voltage control method still has a good control effect when the motor parameter error occurs. The FOPI error voltage control method can effectively suppress the high-frequency harmonic disturbance caused by the motor parameter error. The control effect of the IOPI controller is poor, and the high-frequency harmonic disturbance cannot be effectively suppressed.

It can be seen from Figures 22 and 23 that in the case of errors in motor parameters, FOPI error voltage control still has good performance in eliminating the current clamping phenomenon and suppressing high-frequency harmonic disturbances. The control effect of FOPI error voltage control is better than that of IOPI error voltage control when there are errors in motor parameters, which proves the necessity of using the FOPI controller.

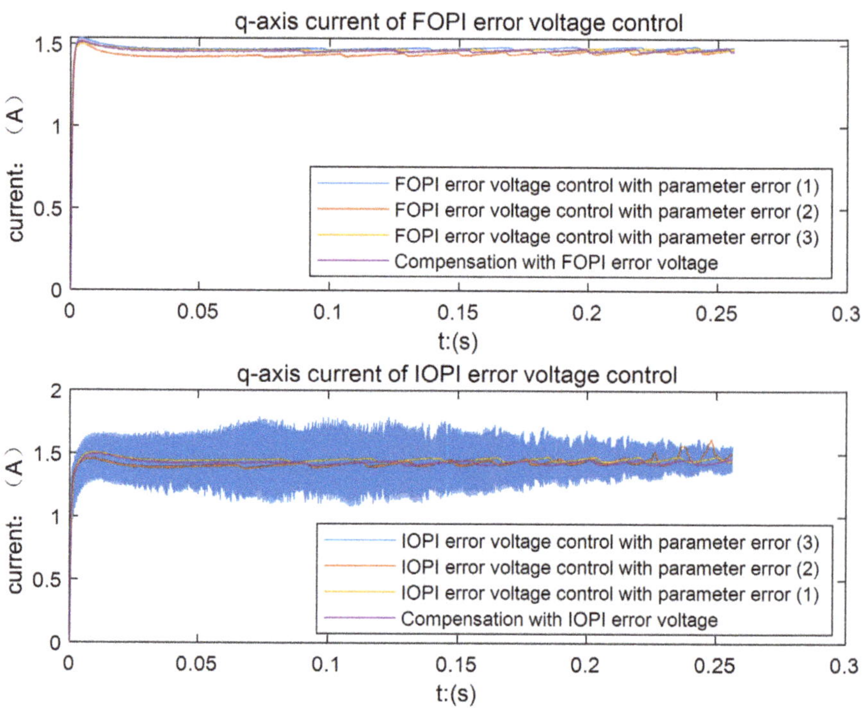

Figure 20. The comparison of I_q step current response in the error motor parameters.

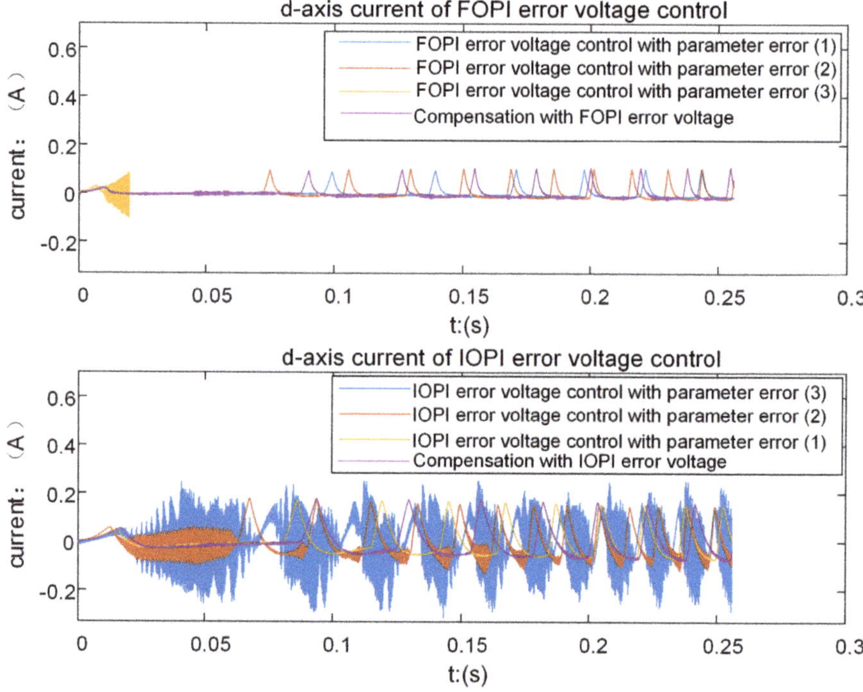

Figure 21. The comparison of I_d step current response in the error motor parameters.

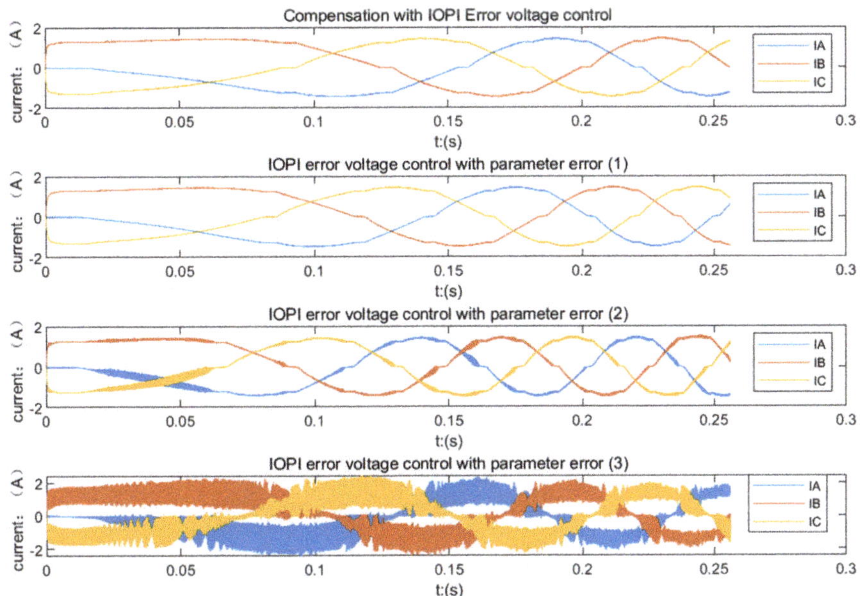

Figure 22. The three-phase current response of the optimal IOPI error voltage control in the error motor parameters.

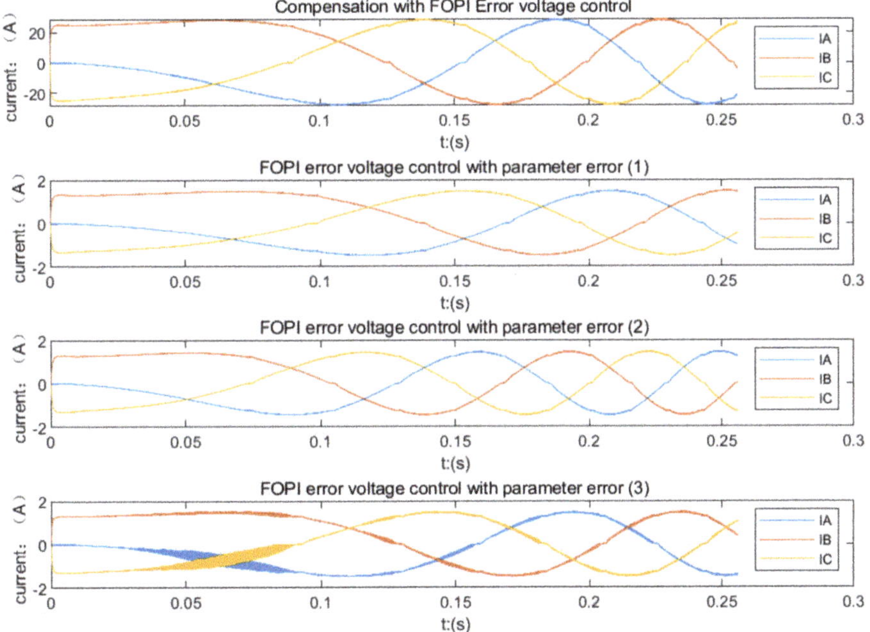

Figure 23. The three-phase current response of the optimal FOPI error voltage control in the error motor parameters.

5. Experimental Results

Experiments are performed under the same operating conditions as the simulation to demonstrate the effectiveness and high performance of the method with optimal FOPI control. A laboratory permanent magnet synchronous motor (PMSM) speed servo platform is shown in Figure 24. The experiments' dead-time setting, controller, and motor parameters are consistent with the simulation. The servo drive is based on the digital signal processor (DSP) TMS320F28335, which is used for AD conversion, encoder sampling, generation of insulated-gate bipolar transistor (IGBT) gate switching signals, and control algorithm implementation.

Figure 24. Experimental platform.

The experimental comparison results of the I_q current step response are shown in Figure 25 and Table 9. The I_q current step response without compensation has the characteristics of a slow rise time and large ripple. Compared with the IOPI error voltage controller and the compensation with ESO [14], it can be seen that the method with optimal FOPI error voltage control can further improve the tracking performance and reduce the current ripple of the I_q current. Compared with the compensation with ESO [14], the optimal FOPI error voltage control can effectively suppress high-frequency harmonic disturbance.

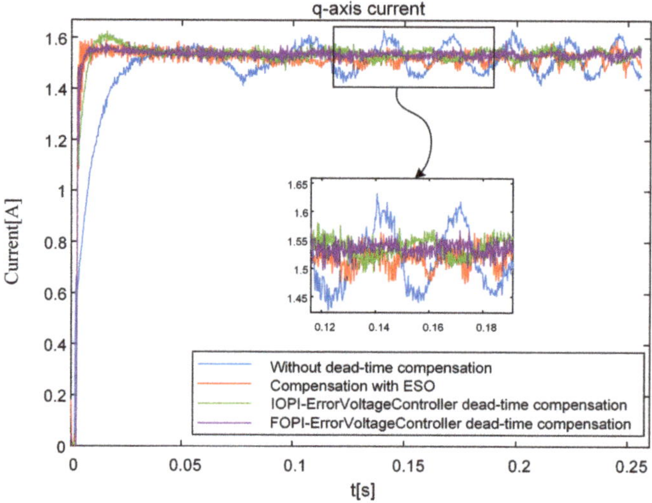

Figure 25. Comparison of I_q current response.

Table 9. The performance index comparison of I_q step current response in the experimental.

	without Compensation	IOPI	FOPI	ESO [14]
overshoot (%)	2.3	3.8	1.7	1.82
rise time (s)	0.036	0.00875	0.00675	0.00375
settling time (s)	\	0.0417	0.007	0.00376
current ripple amplitude (A)	0.1852	0.065	0.0586	0.089

In the experimental results, the I_d current response is compared under the three control strategies in Figure 26 and Table 10. Compared with the method without compensation, the dead-time compensation method with optimal IOPI error voltage control and the compensation method with ESO [14] can significantly reduce the current disturbance caused by the dead-time effect. The method with optimal FOPI error voltage control can further reduce the ripple of the I_d current than the method with optimal IOPI error voltage control and the compensation method with ESO [14].

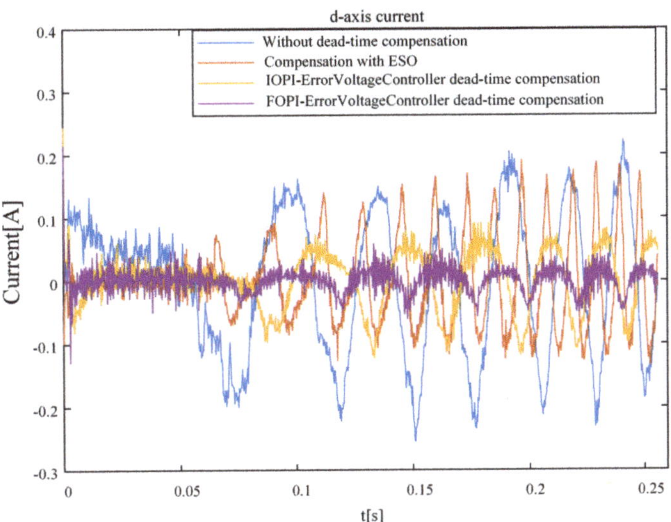

Figure 26. Comparison of I_d current response.

Table 10. The performance index comparison of I_d step current response

	without Compensation	IOPI	FOPI	ESO [14]
Current ripple amplitude (A)	0.424	0.1851	0.0732	0.291

The experimental results of the d-axis voltage error are shown in Figure 27. It can be seen that the d-axis error voltage fluctuates violently under the influence of the dead-time effect. Compared with the method without compensation and optimal IOPI error voltage control, the dead-time compensation method with optimal FOPI error voltage control can significantly reduce the d-axis error voltage.

The q-axis error voltage experimental results are shown in Figure 28. The experimental results of q-axis error voltage are consistent with the simulation results. Without compensation, the q-axis error voltage changes nonlinearly with the speed change. The error voltage can converge to 0 quickly when using the method with optimal FOPI error voltage control.

The convergence speed of the method with optimal FOPI error voltage control is faster than that using optimal IOPI error voltage control.

The above experiment studies the dead-time effect compensation effect of the method proposed in this paper in the case of current closed-loop control. Add a speed closed-loop controller to the previous current closed-loop control system, and the output of the speed closed-loop controller is used as the reference input of the current controller. When the speed is 40 r/min and 200 r/min, the experimental results of the three-phase current of the permanent magnet synchronous motor are shown in Figures 29 and 30.

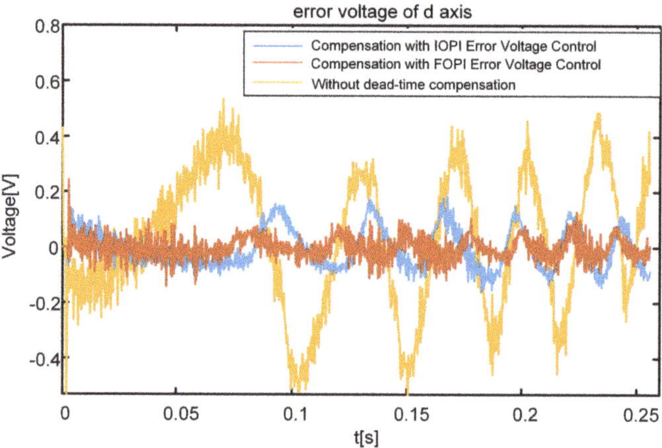

Figure 27. Comparison of the error voltage of d-axis.

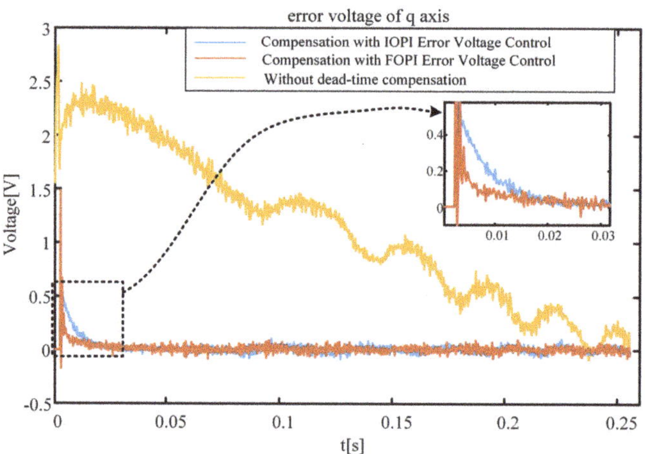

Figure 28. Comparison of the error voltage of q-axis.

The comparison of A-phase current response experimental results under the 40 r/min speed conditions is shown in Figure 29 and Table 11. Compared with the method without compensation and optimal IOPI error voltage control, the dead-time compensation method with optimal FOPI error voltage control can significantly eliminate the current distortion and reduce the clamping of current around the zero-crossing point. The three-phase current clamping time of the method proposed in this paper is less than that of the compensation method using ESO.

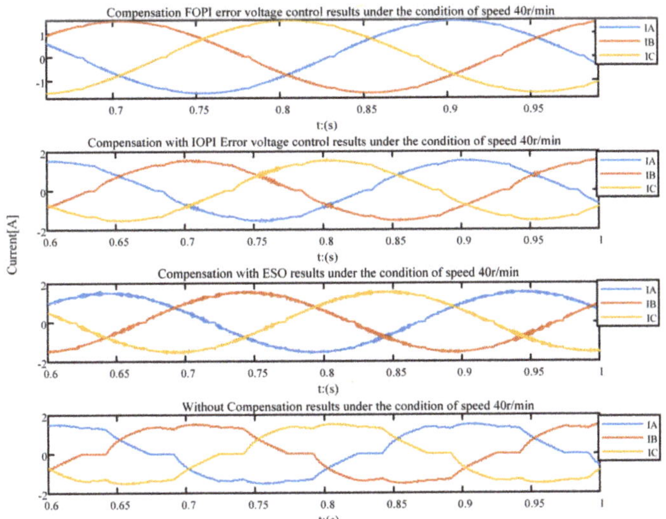

Figure 29. Comparison of Ia current response under 40 r/min speed conditions.

Table 11. The performance index comparison of I_a current step current response under 40 r/min speed conditions.

	without Compensation	IOPI	FOPI	ESO [14]
Current clamping time (s)	0.0112	0.00375	0.00175	0.00195

The comparison of A-phase current response experimental results under 200 r/min speed conditions is shown in Figures 30 and 31.

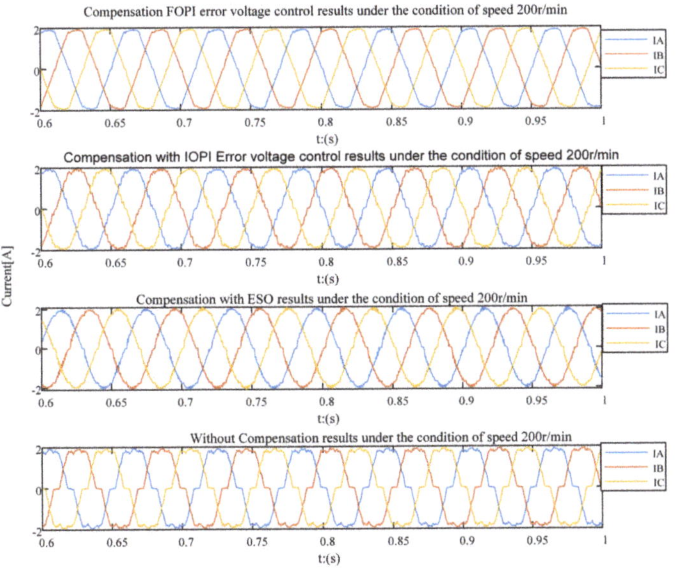

Figure 30. Comparison of three-phase current response under 200 r/min speed conditions.

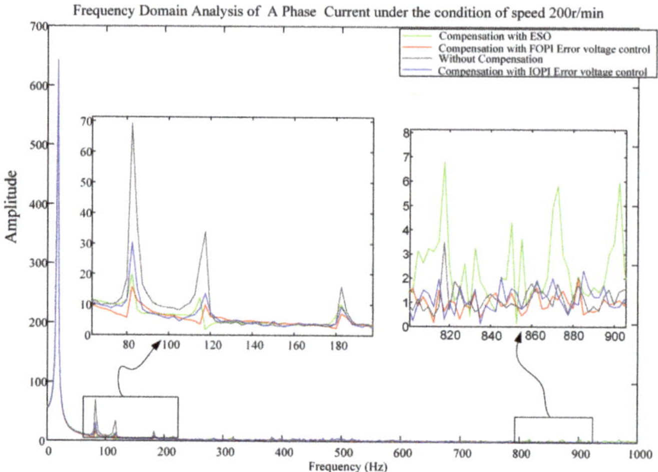

Figure 31. Frequency domain analysis of A phase current under 200 r/min speed conditions (the proposed method, the method with optimal IOPI error voltage control and the method with ESO).

In the case of keeping the motor speed constant at 200 r/min, compare the method proposed in this paper with the typical current harmonic elimination method [4]. Further, observe and compare the three-phase current and its frequency spectrum. The three-phase current experiment results and their spectrum analysis of the dead-time-related harmonic minimization method and the method [4] proposed in this paper are shown in Figures 32 and 33. It can be seen from Figure 18 that the proposed method can more effectively eliminate the current clamping phenomenon caused by the dead-time effect. At the point where the current passes through 0, the current clamping time of the proposed method is shorter than the dead-time-related harmonic minimization method [4]. It can be seen from Figure 19 that the proposed method is better than the dead-time-related harmonic minimization method [4] in eliminating the dead-time effect harmonics.

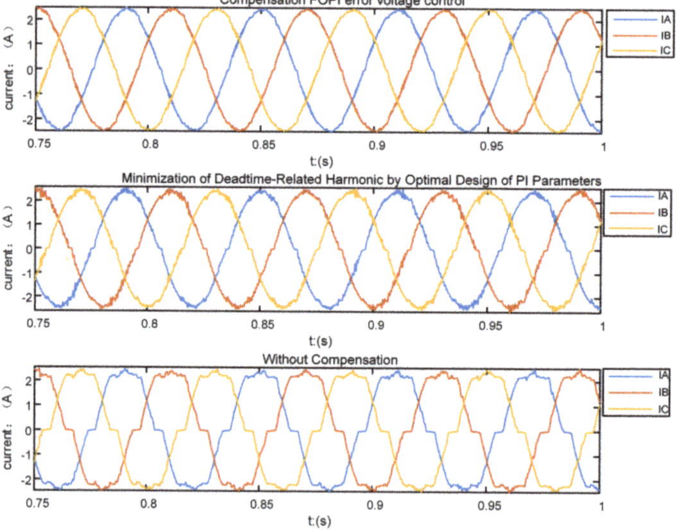

Figure 32. Comparison of three phase current respons under 200 r/min speed conditions.

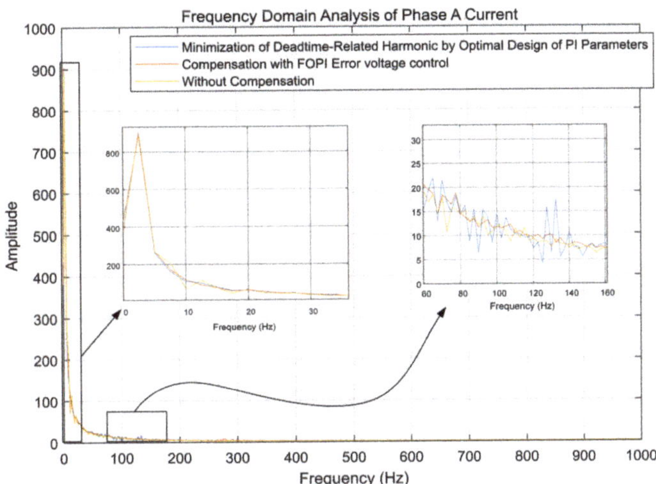

Figure 33. Frequency domain analysis of A phase current under 200 r/min speed conditions (the method with optimal IOPI error voltage control and the minimization method of deadtime-related harmonic).

6. Conclusions

This paper presents a dead-time compensation method for PMSM Servo System with optimal FOPI error voltage control. In this method, the disturbance voltages caused by the power devices' dead time and nonideal switching characteristics are compensated for by the FOPI controller and fed to the reference voltage. With the same parameter design method and parameter design index, the dead-time compensation method with optimal FOPI error voltage control is compared with the compensation method with ESO and the method with optimal IOPI error voltage control. In order to prove that the FOPI error voltage control method has a good ability to eliminate current harmonics, the dead-time-related harmonic minimization method using optimal PI parameters is compared with the proposed method. The simulation and experimental results demonstrate that the dead-time compensation method with optimal FOPI error voltage control can make the current ripple smaller and the response speed faster. Additionally, the proposed method significantly eliminates the current distortion and reduces current clamping around the zero crossing point. Through theoretical and experimental analysis, it is proven that the method proposed in this paper can have good error voltage control performance and robustness in the case of motor parameter errors. It is proven that the proposed dead-time compensation method can improve the performance of the current response by eliminating the dead-time effect.

Author Contributions: Ideas, software, validation, investigation, writing comments and editing: F.L. and Y.L. Review, commentary: X.L., P.C. and Y.C. All authors have read and agreed to the published version of the manuscript.

Funding: This work was supported by the National Natural Science Foundation of China [51975234].

References

1. Lewicki, A. Dead-time effect compensation based on additional phase current measurements. *IEEE Trans. Ind. Electron.* **2015**, *62*, 4078–4085. [CrossRef]
2. Mannen, T.; Fujita, H. Dead-Time Compensation Method Based on Current Ripple Estimation. *IEEE Trans. Power Electron.* **2015**, *30*, 4016–4024. [CrossRef]
3. Dafang, W.; Bowen, Y.; Cheng, Z.; Chuanwei, Z.; Ji, Q. A Feedback-Type Phase Voltage Compensation Strategy Based on Phase Current Reconstruction for ACIM Drives. *IEEE Trans. Power Electron.* **2014**, *29*, 5031–5043. [CrossRef]

4. Wu, Z.; Ding, K.; Yang, Z.; He, G. Analytical Prediction and Minimization of Deadtime-Related Harmonics in Permanent Magnet Synchronous Motor. *IEEE Trans. Ind. Electron.* **2020**, *68*, 7736–7746. [CrossRef]
5. Tang, Z.; Akin, B. A new LMS algorithm based deadtime compensation method for PMSM FOC drives. *IEEE Trans. Ind. Appl.* **2018**, *54*, 6472–6484. [CrossRef]
6. Qiu, T.; Wen, X.; Zhao, F. Adaptive-linear-neuron-based dead-time effects compensation scheme for PMSM drives. *IEEE Trans. Power Electron.* **2015**, *31*, 2530–2538. [CrossRef]
7. Zhu, H.; Chen, Y.; Lei, H.; Chen, D.; Li, Z. A New Dead-time Compensation Method Based on LMS Algorithm for PMSM. *J. Phys. Conf. Ser.* **2021**, *1754*, 012203. [CrossRef]
8. Liu, G.; Wang, D.; Jin, Y.; Wang, M.; Zhang, P. Current -Detection-Independent Dead-Time Compensation Method Based on Terminal Voltage A/D Conversion for PWM VSI. *IEEE Trans. Ind. Electron.* **2017**, *64*, 7689–7699. [CrossRef]
9. Su, J.H.; Hsu, B.C. Application of small-gain theorem in the dead-time compensation of voltage-source-inverter drives. *IEEE Trans. Ind. Electron.* **2005**, *52*, 1456–1458. [CrossRef]
10. Tang, N.; Brown, I.P. Framework and Solution Techniques for Suppressing Electric Machine Winding MMF Space Harmonics by Varying Slot Distribution and Coil Turns. *IEEE Trans. Magn.* **2018**, *54*, 1–12. [CrossRef]
11. Lin, C.; Xing, J.; Zhuang, X. Dead-Time Correction Applied for Extended Flux-Based Sensorless Control of Assisted PMSMs in Electric Vehicles. *Electronics* **2021**, *10*, 220. [CrossRef]
12. Miyama, Y.; Hazeyama, M.; Hanioka, S.; Watanabe, N.; Daikoku, A.; Inoue, M. PWM Carrier Harmonic Iron Loss Reduction Technique of Permanent-Magnet Motors for Electric Vehicles. *IEEE Trans. Ind. Appl.* **2016**, *52*, 2865–2871. [CrossRef]
13. Herman, L.; Papic, I.; Blazic, B. A Proportional-Resonant Current Controller for Selective Harmonic Compensation in a Hybrid Active Power Filter. *IEEE Trans. Power Deliv.* **2014**, *29*, 2055–2065. [CrossRef]
14. Shi, J.; Li, S. Analysis and compensation control of dead-time effect on space vector PWM. *J. Power Electron.* **2015**, *15*, 431–442. [CrossRef]
15. Karttunen, J.; Kallio, S.; Peltoniemi, P.; Silventoinen, P. Current Harmonic Compensation in Dual Three-Phase PMSMs Using a Disturbance Observer. *IEEE Trans. Ind. Electron.* **2016**, *63*, 583–594. [CrossRef]
16. Zhao, Y.; Qiao, W.; Wu, L. Dead-Time Effect Analysis and Compensation for a Sliding-Mode Position Observer-Based Sensorless IPMSM Control System. *IEEE Trans. Ind. Appl.* **2015**, *51*, 2528–2535. [CrossRef]
17. Guha, A.; Narayanan, G. Impact of Dead Time on Inverter Input Current, DC-Link Dynamics, and Light-Load Instability in Rectifier-Inverter-Fed Induction Motor Drives. *IEEE Trans. Ind. Appl.* **2018**, *54*, 1414–1424. [CrossRef]
18. Khurram, A.; Rehman, H.; Mukhopadhyay, S.; Ali, D. Comparative Analysis of Integer-order and Fractional-order Proportional Integral Speed Controllers for Induction Motor Drive Systems. *J. Power Electron.* **2018**, *18*, 723–735.
19. Zheng, W.; Pi, Y. Study of the fractional order proportional integral controller for the permanent magnet synchronous motor based on the differential evolution algorithm. *ISA Trans.* **2016**, *63*, 387–393. [CrossRef] [PubMed]
20. Luo, Y.; Chen, Y. Fractional order [proportional derivative] controller for a class of fractional order systems. *Automatica* **2009**, *45*, 2446–2450. [CrossRef]
21. Luo, Y.; Chen, Y.Q.; Wang, C.Y.; Pi, Y.G. Tuning fractional order proportional integral controllers for fractional order systems. *J. Process. Control.* **2010**, *20*, 823–831. [CrossRef]
22. Choi, J.W.; Sul, S.K. Inverter output voltage synthesis using novel dead time compensation. *IEEE Trans. Power Electron.* **1996**, *11*, 221–227. [CrossRef]
23. Li, Y.; Sheng, H.; Chen, Y. Impulse response invariant discretization of a generalized commensurate fractional order filter. In Proceedings of the 2010 8th World Congress on Intelligent Control and Automation, Jinan, China, 7–9 July 2010; pp. 191–196.
24. Poli, R. Analysis of the Publications on the Applications of Particle Swarm Optimisation. *J. Artif. Evol. Appl.* **2008**, *2008*, 1–10. [CrossRef]
25. Gaing, Z.L. A particle swarm optimization approach for optimum design of PID controller in AVR system. *IEEE Trans. Energy Convers.* **2004**, *19*, 384–391. [CrossRef]
26. Yoshida, H.; Kawata, K.; Fukuyama, Y.; Takayama, S.; Nakanishi, Y. A particle swarm optimization for reactive power and voltage control considering voltage security assessment. *IEEE Trans. Power Syst.* **2000**, *15*, 1232–1239. [CrossRef]
27. Karaboga, D.; Basturk, B. A powerful and efficient algorithm for numerical function optimization: Artificial bee colony (ABC) algorithm. *J. Glob. Optim.* **2007**, *39*, 459–471. [CrossRef]
28. Wang, B.; Wang, S.; Peng, Y.; Pi, Y.; Luo, Y. Design and High-Order Precision Numerical Implementation of Fractional-Order PI Controller for PMSM Speed System Based on FPGA. *Fractal Fract.* **2022**, *6*, 218. [CrossRef]

Article

Disturbance Observer-Based Event-Triggered Adaptive Command Filtered Backstepping Control for Fractional-Order Nonlinear Systems and Its Application

Shuai Song [1], **Xiaona Song** [1,*] **and Inés Tejado** [2,*]

[1] School of Information Engineering, Henan University of Science and Technology, Luoyang 467023, China; shuaisong@haust.edu.cn

[2] Escuela de Ingenierías Industriales, Universidad de Extremadura, 06006 Badajoz, Spain

* Correspondence: xiaona97@haust.edu.cn (X.S.); itejbal@unex.es (I.T.); Tel.: +34-924289300 (ext. 86767) (I.T.)

Abstract: This paper considers the disturbance observer-based event-triggered adaptive fuzzy tracking control issue for a class of fractional-order nonlinear systems (FONSs) with quantized signals and unknown disturbances. To improve the disturbance rejection ability, a fractional-order nonlinear disturbance observer (FONDO) is designed to estimate the unknown composite disturbances. Furthermore, by combining an improved fractional-order command-filtered backstepping control technique and an event-triggered control mechanism, an event-triggered adaptive fuzzy quantized control scheme is established, which guarantees the desired tracking performance can be achieved even in the presence of network constraint. Finally, the validity and superiority of the theoretic results are verified by a fractional-order horizontal platform system.

Keywords: command filtered backstepping control; event-triggered control; fractional-order nonlinear systems; input quantization; nonlinear disturbance observer

Citation: Song, S.; Song, X.; Tejado, I. Disturbance Observer-Based Event-Triggered Adaptive Command Filtered Backstepping Control for Fractional-Order Nonlinear Systems and Its Application. *Fractal Fract.* **2023**, *7*, 810. https://doi.org/10.3390/fractalfract7110810

Academic Editors: António Lopes, Behnam Mohammadi-Ivatloo and Arman Oshnoei

Received: 14 August 2023
Revised: 15 October 2023
Accepted: 7 November 2023
Published: 9 November 2023

1. Introduction

On account of the notable merits of fractional calculus in modeling and characterizing accurate dynamical properties of many real-world systems, fractional-order systems (FOSs) have received wide attention. Therefore, various control methods were extended to investigate the control problem of FOSs [1–8]. In [5], the necessary and sufficient conditions were proposed to guarantee the stability of a class of fractional-order (FO) descriptor systems. In [8], an adaptive neural control design was developed to guarantee the uniform stability of the closed-loop system (CLS) and avoid the violation of the preassigned state constraints. To handle the mismatched uncertainties effectively, the adaptive backstepping control method was widely utilized to achieve the tracking control of fractional-order nonlinear systems (FONSs) due to its structural design and strong robustness to mismatched uncertainties [9–11]. In [10], Liu et al. presented an adaptive fuzzy recursive control algorithm to guarantee the boundedness of the resulting CLS based on direct fractional Lyapunov stability. However, the traditional backstepping design method relies on the repeated differentiation of virtual control laws in the recursive procedure, which undoubtedly will cause the issue of the explosion of complexity and over-parametrization once the dimensionality of the system is overlarge. To overcome this problem, inspired by the integer-order results [12–16], some research works have reported using a modified FODSC technique for FONSs [17–21]. It should be noted that the efforts have barely been made on the composite disturbances consisting of the disturbances and approximation errors in the aforementioned results, where the considered disturbance term is always handled by using the inequality technique or designing a general compensation function. However, it is worth pointing out that the disturbance rejection ability of the existing adaptive control method proposed for the FONSs needs to be further improved to maintain the desired control performance when the investigated system suffers strong changing unknown disturbances.

As we know, unknown disturbances exist in nearly all actual systems; their existence frequently undoubtedly degrades the performance and even destroy the system's stability. To improve the disturbance rejection ability of the system, the disturbance observer (DO)-based control technique has attracted considerable attention and some significant results have been reported for various types of nonlinear systems [22–28]. In [23], Chen et al. proposed a DO-based synchronization control method to handle the robust synchronization issue of two FO chaotic systems. In [26], a super twisting nonlinear disturbance observer was designed to guarantee the finite-time convergence of the estimation errors. Nevertheless, it is noted that most of the available DO-based adaptive control results are concentrated on integer-order nonlinear systems. Thus, developing an NDO-based adaptive backstepping control strategy for a class of FONSs remains an open problem.

On the other hand, it is universally known that real-time data are usually quantized in some practical industrial processing control systems due to the influence of bandwidth limitations. Therefore, one major challenge is how to realize the predefined control goal of the CLSs by the quantized control signals. To this end, substantial attention has been paid to adaptive backstepping quantized control for nonlinear systems [29–32]. In [30], Liu et al. proposed a novel fuzzy quantized recursive control scheme for nonlinear systems. In [32], Sui et al. proposed a finite-time quantized control method for stochastic nonlinear systems, where the traditional power form was not required for determining the control signals. Another effective way of reducing the communication burden is the even-triggered control method. In recent years, a series of event-triggered mechanisms have been developed for nonlinear systems [33,34]. In [33], three different kinds of event-triggered mechanisms were developed to co-design adaptive controllers, which have been widely utilized to address the event-triggered adaptive control problem. Unfortunately, the available adaptive backstepping control methods for FONSs were concentrated on the time-triggered control scheme, as in [17,19,20], where a large number of network communication resources were required since the control signals need to be updated periodically. Therefore, the event-triggered control problem for FONSs needs to be further investigated. Furthermore, a new challenge arises that the complexity of controller design has to be faced when input quantization and the event-triggered control are considered simultaneously. Therefore, how to develop a suitable event-triggered adaptive quantized controller for FONSs also motivates this work.

Inspired by the above observation, we aim to develop a novel FONDO-based event-triggered adaptive fuzzy tracking control method for the FONSs with input quantization in this paper. In comparison with other relevant studies, the main contributions of this paper are as follows:

(1) To the best of our knowledge, this paper first attempts to develop a FONDO-based event-triggered adaptive fuzzy tracking control scheme for unknown FONSs. Compared with the existing adaptive backstepping control results [3,10,11,17–20], the disturbance rejection ability against the mismatched disturbance can be greatly improved by the proposed FONDO.

(2) A co-design consisting of the event-triggered communication mechanism and input quantization is established such that a large amount of communication resources can be saved while fulfilling the preassigned tracking task in comparison to the traditional time-triggered control methods proposed in [3,10,11,17–20].

(3) Distinct from the logarithmic quantizer in [35,36] and hysteretic quantizer in [29,31], an adjustable parameter is introduced to the quantizer for achieving the trade-off between the quantization effects and control performance. Also, the saturation property is adopted to keep the control energy within bounds, making the proposed control method closer to the practical requirements.

The remaining part of this paper consists of the following sections. Section 2 presents the preliminaries and formulates the considered problem. A disturbance observer-based event-triggered adaptive quantized control method is developed for a class of unknown fractional-order nonlinear systems in Section 3. Section 4 provides an application example to verify the feasibility and superiority of the developed method. Finally, the whole work and the potential improvements are concluded in Section 5.

2. Preliminaries and Problem Formulation

2.1. Fractional Calculus

Definition 1 ([37]). *The Caputo fractional derivative of order α of a function $f(t)$ is*

$$D^{\alpha} f(t) = \frac{1}{\Gamma(n-\alpha)} \int_0^t \frac{f^{(n)}(\tau)}{(t-\tau)^{\alpha+1-n}} d\tau \tag{1}$$

where $n-1 < \alpha \leq n$.

Definition 2 ([37]). *The Mittag-Leffler function including two parameters is expressed as:*

$$E_{\alpha_1,\alpha_2}(r) = \sum_{i=0}^{\infty} \frac{r^i}{\Gamma(i\alpha_1 + \alpha_2)}, \tag{2}$$

where $\alpha_1 > 0$, $\alpha_2 > 0$ and r is a complex number. Taking the Laplace transform (LT) for the above equation, one has

$$\mathscr{L}\left\{ t^{\alpha_2-1} E_{\alpha_1,\alpha_2}(-\kappa t^{\alpha_1}) \right\} = \frac{s^{\alpha_1-\alpha_2}}{s^{\alpha_1} + \kappa}. \tag{3}$$

Lemma 1 ([37]). *For $\alpha_1 \in (0,2)$, if there exists a positive constant ℓ such that $\pi\alpha_1/2 < \ell < \min\{\pi, \pi\alpha_1\}$, then the following inequality holds*

$$|E_{\alpha_1,\alpha_2}(r)| \leq \frac{\hbar}{1+|r|}, (\ell \leq |arg(r)| \leq \pi, |r| \geq 0), \tag{4}$$

where β_2 is a real number and $\hbar > 0$.

Lemma 2 ([37]). *For $\alpha_1 \in (0,2)$, if there exist an arbitrary complex number α_2 and a real number ℓ such that*

$$\frac{\pi\alpha_1}{2} < \ell < \min(\pi, \pi\alpha_1), \tag{5}$$

then it can be verified that

$$E_{\alpha_1,\alpha_2}(r) = -\sum_{i=1}^{\infty} \frac{r^{-i}}{\Gamma(\alpha_2 - i\alpha_1)} + o(|r|^{-1-\ell}), |r| \to \infty, \ell \leq |arg(r)| \leq \pi \tag{6}$$

for all integers $\ell \geq 1$.

2.2. Fuzzy Logic Systems

To better achieve the mentioned control objective, fuzzy logic systems (FLSs) are adopted in this paper to handle the unknown nonlinearities. Consider k fuzzy IF–THEN rules with the following form [11]:

$$\mathbb{R}^s\text{: if } x_1 \text{ is } F_1^s \text{ and } \dots \text{ and } x_n \text{ is } F_n^s$$
$$\text{Then, } y \text{ is } G^s, s = 1, \dots, k$$

where $\mathbb{R}^s$ represents the sth rule, $1 \leq s \leq k$, $x_i(i = 1, \dots, n)$, and $y \in \mathbb{R}$ denote the linguistic variables associated with the inputs and outputs of the FLSs, respectively. F_i^s and G^s are the fuzzy set. In this article, the FLSs are described as

$$y(x) = \frac{\sum_{s=1}^{k} w_s \left(\prod_{i=1}^{n} \mu_{F_i^s}(x_i) \right)}{\sum_{s=1}^{k} \left(\prod_{i=1}^{n} \mu_{F_i^s}(x_i) \right)}. \tag{7}$$

Define the weight vector and fuzzy basis function vector as $W = [w_1, \ldots, w_k]^T$ and $\phi(x) = [\triangle_1, \ldots, \triangle_k]^T$, where $\triangle_s = \left[\left(\prod_{i=1}^{n} \mu_{F_i^s}(x_i) \right) / \sum_{s=1}^{k} \left(\prod_{i=1}^{n} \mu_{F_i^s}(x_i) \right) \right]$; then, the above expression can be represented as

$$y(x) = W^T \phi(x). \tag{8}$$

Lemma 3 ([11]). *For any continuous function $f(x)$ defined over a compact set Θ and any given constant o, there exist an FLS and an ideal weight vector W^* such that*

$$\sup_{x \in \Theta} |f(x) - W^{*^T} \phi(x)| \leq \epsilon. \tag{9}$$

Lemma 4 ([38]). *For $z \in \mathbb{R}$ and positive constant ω, the following inequality holds*

$$0 \leq |z| - \frac{z^2}{\sqrt{z^2 + \omega^2}} < \omega. \tag{10}$$

2.3. Problem Formulation

Consider a class of FONSs subject to input quantization and unknown disturbances described by

$$\begin{cases} D^{\alpha} x_i = x_{i+1} + f_i(\bar{x}_i) + d_i(x, t), \, (i = 1, \ldots, n-1), \\ D^{\alpha} x_n = q(u) + f_n(\bar{x}_n) + d_n(x, t) \\ y = x_1, \end{cases} \tag{11}$$

where $x = \bar{x}_n = [x_1, \ldots, x_n]^T \in \mathbb{R}^n$ is the state vector; $y \in \mathbb{R}$ denotes the system output; $f_i(\bar{x}_i)$ represents an unknown but smooth nonlinear function; $d_i(x, t)$ represents the unknown but bounded disturbance terms; and $q(u)$ represents the quantized control signal. To reduce the chattering phenomenon, the following hysteresis quantizer is considered to obtain a quantized control signal:

$$q(u) = \begin{cases} u_i \mathrm{sgn}(u), & \dfrac{u_i}{1+\delta} < |u| \leq u_i, \dot{u} < 0, \text{or} \\[2mm] & u_i < |u| \leq \dfrac{u_i}{1-\delta}, \dot{u} > 0 \\[2mm] u_i(1+\delta)\mathrm{sgn}(u), & u_i < |u| \leq \dfrac{u_i}{1-\delta}, \dot{u} < 0, \text{or} \\[2mm] & \dfrac{u_i}{1-\delta} < |u| \leq \dfrac{u_i(1+\delta)}{1-\delta}, \dot{u} > 0 \\[2mm] 0, & 0 \leq |u| < \dfrac{u_{min}}{1+\delta}, \dot{u} < 0, \text{or} \\[2mm] & \dfrac{u_{min}}{1+\delta} \leq |u| \leq u_{min}, \dot{u} > 0 \\[2mm] q(u(t^-)), & \text{otherwise} \end{cases} \tag{12}$$

where $u_i = \rho^{1-i} u_{min} (i = 1, 2, \ldots)$ with $0 < \rho < 1$ and $\delta = \frac{1-\rho}{1+\rho}$, u_{min} denotes the scope of the dead-zone for $q(u)$ and $q(u(t^-))$ represents the status prior to $q(u(t))$.

Lemma 5 ([30]). *The quantizer $q(u)$ can be expressed as $q(u) = (1-\kappa)u + \kappa\theta$ with the quantizer error ω satisfying*

$$
\begin{cases}
\theta^2 \leq \left(\dfrac{\kappa+\delta}{\kappa}u\right)^2, \forall u \geq |u_{min}|, \\[4mm]
\theta^2 \leq \left(\dfrac{1-\kappa}{\kappa}u_{min}\right)^2, \forall u \leq |u_{min}|,
\end{cases}
\tag{13}
$$

where $0 < \theta < 1$ is a constant.

Remark 1. *The parameter ρ stands for a measure of quantization density. Compared with the logarithmic quantizer, the hysteretic quantizer can reduce the chattering effectively, which can be viewed as a special combination of two asymmetric logarithmic quantizers. In addition, the value of κ, as an adjustable design parameter, could be selected appropriately by the designer to balance the control performance and quantization effects, which also increases the freedom of quantization design.*

Assumption 1. *The reference signal y_r and its FO derivative $D^\alpha y_r$ are available and bounded.*

Assumption 2. *For the external disturbance $d_i(x, t)$, there exist the unknown positive constants $\bar{d}_{i,1}$ and $\bar{d}_{i,2}$ such that the inequalities $|d_i| \leq \bar{d}_{i,1}$ and $|D^\alpha d_i| \leq \bar{d}_{i,2}$ hold.*

The control objective of this work is devoted to presenting a nonlinear disturbance observer-based event-triggered adaptive quantized tracking control scheme for system (11) such that all the signals of the CLS are bounded and the tracking error converges to a small neighborhood of the origin.

3. Main Results

In this section, an NDO-based event-triggered adaptive command-filtered quantized control scheme will be developed for system (11) by integrating with the FOCFB technique and ETC mechanism. Then, the stability analysis of the CLS will be presented based on Mittag-Leffler stability.

NDO-Based Event-Triggered Adaptive Command-Filtered Quantized Control Design

Step 1. At first, we introduce the change of coordinates as:

$$
\begin{cases}
z_1 = x_1 - y_r, \\
z_i = x_i - \vartheta_i^c, (i = 2, \ldots, n),
\end{cases}
\tag{14}
$$

where $z_j (j = 1, \ldots, n)$ denotes surface error and ϑ_i^c is an auxiliary variable used for approximating the virtual control signal, which is produced by

$$
\iota_2 D^\alpha \vartheta_2^c + \vartheta_2^c = \eta_1, \quad \vartheta_2^c(0) = \eta_1(0),
\tag{15}
$$

where $\iota_2 > 0$ is a small time constant and η_1 is the input of the FO filter. Define the filter error $\varepsilon_1 = \vartheta_{2,c} - \eta_1$. To reduce the negative effects caused by filter errors, the compensation signal is constructed as:

$$
D^\alpha \gamma_1 = -a_1 \gamma_1 - \Delta_1 \gamma_1 + \varepsilon_1,
\tag{16}
$$

where $\Delta_1 = \frac{|z_1 \varepsilon_1|}{\gamma_1^2}$ and a_1 is a positive constant.

According to Lemma 3, the FO derivative of z_1 is computed as:

$$
D^\alpha z_1 = z_2 + \eta_1 + \varepsilon_1 + l_1^{-1} W_1^{*T} \phi_1(x_1) + \Lambda_1 - D^\alpha y_r,
\tag{17}
$$

where the FLS is used to approximate the term $F_1(x_1) = l_1 f_1(x_1)$ and $\Lambda_1 = d_1 + l_1^{-1}\epsilon_1$ denotes an unknown composite disturbance satisfying $\Lambda_1 \leq \bar{d}_1^1 + l_1^{-1}\bar{\epsilon}_1^1$.

The first virtual control function α_1 and adaptive law are designed as:

$$\eta_1 = -\frac{z_1\bar{\eta}_1^2}{\sqrt{z_1^2\bar{\eta}_1^2 + \omega_1^2}}, \tag{18}$$

$$\bar{\eta}_1 = (c_1 + \frac{1+b_1}{2})z_1 + l_1^{-1}\hat{W}_1^T\phi_1(x_1) - b_1\gamma_1 - D^\alpha y_r + \hat{\Lambda}_1, \tag{19}$$

$$D^\alpha\hat{W}_1 = l_1^{-1}z_1\phi_1(x_1) - \rho_1\hat{W}_1, \tag{20}$$

where b_1, c_1, ρ_1 and l_1 are all positive constants.

Select the Lyapunov function with the following form:

$$V_1 = \frac{1}{2}z_1^2 + \frac{1}{2}\tilde{W}_1^T\tilde{W}_1 + \frac{1}{2}\gamma_1^2 + \frac{1}{2}\tilde{\Lambda}_1^2 \tag{21}$$

Taking the FO derivative of V_1 yields

$$D^\alpha V_1 \leq z_1\eta_1 + z_1\left(z_2 + \epsilon_1 + l_1^{-1}W_1^{*T}\phi_1(x_1) + \Lambda_1 - D^\alpha y_r\right)$$
$$- \tilde{W}_1^T D^\alpha\hat{W}_1 + \gamma_1 D^\alpha\gamma_1 + \tilde{\Lambda}_1 D^\alpha\tilde{\Lambda}_1. \tag{22}$$

Subsequently, the following FONDO is designed to obtain the estimation of the composite disturbance Λ_1

$$\hat{\Lambda}_1 = l_1 x_1 + \xi_1, \tag{23}$$

with

$$D^\alpha\xi_1 = -l_1\left(\xi_1 + x_2 + l_1^{-1}\hat{W}_1^T\phi_1(x_1) + l_1 x_1\right). \tag{24}$$

Using (23)–(24) yields

$$D^\alpha\hat{\Lambda}_1 = \tilde{W}_1^T\phi_1(x_1) + l_1\tilde{\Lambda}_1. \tag{25}$$

Furthermore, the term $\tilde{\Lambda}_1 D^\alpha\tilde{\Lambda}_1$ in (22) is calculated as:

$$\tilde{\Lambda}_1 D^\alpha\tilde{\Lambda}_1 = \tilde{\Lambda}_1 D^\alpha\Lambda_1 - \tilde{\Lambda}_1\tilde{W}_1^T\phi_1 - l_1\tilde{\Lambda}_1^2, \tag{26}$$

where $D^\alpha\Lambda_1 = D^\alpha d_1 + l_1^{-1}D^\alpha\epsilon_1 \leq \bar{d}_{1,2} + l_1^{-1}\bar{\epsilon}_{1,2} = \bar{\Lambda}_1$.

Using Young's inequality leads to

$$\tilde{\Lambda}_1 D^\alpha\tilde{\Lambda}_1 \leq -(l_1 - 1)\tilde{\Lambda}_1^2 + \frac{1}{2}\bar{\Lambda}_1^2 + \frac{1}{2}\tilde{W}_1^T\tilde{W}_1. \tag{27}$$

On account of Lemma 5, the following relationship holds

$$z_1\eta_1 = -\frac{z_1^2\bar{\eta}_1^2}{\sqrt{z_1^2\bar{\eta}_1^2 + \omega_1^2}} \leq \omega_1 - z_1\bar{\eta}_1. \tag{28}$$

Substituting (18)–(20) and (27)–(28) into (22), one can obtain

$$D^\alpha V_1 \leq -c_1 z_1^2 - \frac{(1+b_1)}{2}z_1^2 + \rho_1\tilde{W}_1^T\hat{W}_1 + z_1\tilde{\Lambda}_1 - a_1\gamma_1^2 + b_1 z_1\gamma_1$$
$$+ \gamma_1\epsilon_1 - (l_1 - 1)\tilde{\Lambda}_1^2 + \frac{1}{2}\bar{\Lambda}_1^2 + \frac{1}{2}\tilde{W}_1^T\tilde{W}_1 + z_1 z_2 + \omega_1. \tag{29}$$

Since $\rho_1 \tilde{W}_1^T \hat{W}_1 \leq -\frac{\rho_1}{2}\tilde{W}_1^T\tilde{W}_1 + \frac{\rho_1}{2}W_1^{*T}W_1^*$ holds, using Young's inequality for (29) yields

$$D^\alpha V_1 \leq -c_1 z_1^2 - \frac{(\rho_1 - 1)}{2}\tilde{W}_1^T\tilde{W}_1 - \left(a_1 - \frac{1+b_1}{2}\right)\gamma_1^2$$

$$- \left(l_1 - \frac{3}{2}\right)\tilde{\Lambda}_1^2 + z_1 z_2 + \Xi_1 \tag{30}$$

where $\Xi_1 = \frac{\rho_1}{2}W_1^{*T}W_1^* + \frac{1}{2}\epsilon_1^2 + \frac{1}{2}\tilde{\Lambda}_1^2 + \omega_1$.

Step i $(i = 2,\dots,n-1)$. Similar to the previous procedure, the FO filter is designed as:

$$l_{i+1}D^\alpha \vartheta_{i+1}^c + \vartheta_{i+1}^c = \eta_i, \quad \vartheta_{i+1}^c(0) = \eta_i(0). \tag{31}$$

Define the filter error $\varepsilon_i = \vartheta_{i+1,c} - \eta_i$ and construct the compensation signal as:

$$D^\alpha \gamma_i = -a_i\gamma_i - \Delta_i\gamma_i + \varepsilon_i, \tag{32}$$

where $\Delta_i = \frac{|z_i \epsilon_i|}{\gamma_i^2}$.

In addition, the FO derivative of z_i is calculated as:

$$D^\alpha z_i = z_{i+1} + \eta_i + \epsilon_i + l_i^{-1}W_i^{*T}\phi_i(\bar{x}_i) + \Lambda_i - D^\alpha \vartheta_i^c, \tag{33}$$

where Λ_i denotes the composite disturbance satisfying $\Lambda_i = d_i + l_i^{-1}\epsilon_i$.

Design the virtual control signal α_i and adaptive law as:

$$\eta_i = -\frac{z_i \bar{\eta}_i^2}{\sqrt{z_i^2 \bar{\eta}_i^2 + \omega_i^2}}, \tag{34}$$

$$\bar{\eta}_i = (c_i + \frac{1+b_i}{2})z_i + z_{i-1} - b_i\gamma_i + l_i^{-1}\hat{W}_i^T\phi_i(\bar{x}_i) - D^\alpha \vartheta_i^c + \hat{\Lambda}_i, \tag{35}$$

$$D^\alpha \hat{W}_i = l_i^{-1}z_i\phi_i(\bar{x}_i) - \rho_i\hat{W}_i, \tag{36}$$

The Lyapunov function is chosen as:

$$V_i = V_{i-1} + \frac{1}{2}z_i^2 + \frac{1}{2}\tilde{W}_i^T\tilde{W}_i + \frac{1}{2}\gamma_i^2 + \frac{1}{2}\tilde{\Lambda}_i^2. \tag{37}$$

Taking the FO derivative of V_i along with (33) leads to

$$D^\alpha V_i \leq -\sum_{j=1}^{i-1}c_j z_j^2 - \sum_{j=1}^{i-1}\frac{(\rho_j - 1)}{2}\tilde{W}_j^T\tilde{W}_j + \Xi_{i-1} - \sum_{j=1}^{i-1}\left(a_j - \frac{1+b_j}{2}\right)\gamma_j^2$$

$$- \sum_{j=1}^{i-1}\left(l_j - \frac{3}{2}\right)\tilde{\Lambda}_j^2 + z_{i-1}z_i + z_i\eta_i + z_i\left(z_{i+1} + \epsilon_i + l_i^{-1}W_i^{*T}\phi_i(\bar{x}_i)\right.$$

$$\left. +\Lambda_i - D^\beta \vartheta_i^c\right) - \tilde{W}_i^T D^\beta \hat{W}_i + \gamma_i D^\beta \gamma_i + \tilde{\Lambda}_i D^\beta \tilde{\Lambda}_i. \tag{38}$$

Similar to (23) in step 1, we construct a FONDO with the following form:

$$\hat{\Lambda}_i = l_i x_i + \xi_i, \tag{39}$$

with

$$D^\alpha \xi_i = -l_i\left(\xi_i + \varphi_i x_{i+1} + l_i^{-1}\hat{W}_i^T\phi_i(\bar{x}_i) + l_i x_i\right). \tag{40}$$

Using (39)–(40) yields

$$D^\alpha \hat{\Lambda}_i = \tilde{W}_i^T\phi_i(\bar{x}_i) + l_i\tilde{\Lambda}_i. \tag{41}$$

It follows from (41) that

$$\tilde{\Lambda}_i D^\alpha \tilde{\Lambda}_i = \tilde{\Lambda}_i D^\alpha \Lambda_i - \tilde{\Lambda}_i \tilde{W}_i^T \phi_i(\bar{x}_i) - l_i \tilde{\Lambda}_i^2. \tag{42}$$

Using Young's inequality obtains

$$\tilde{\Lambda}_i D^\alpha \tilde{\Lambda}_i \leq -(l_i - 1)\tilde{\Lambda}_i^2 + \frac{1}{2}\tilde{\Lambda}_i^2 + \frac{1}{2}\tilde{W}_i^T \tilde{W}_i. \tag{43}$$

Furthermore, one can obtain

$$z_i \eta_i = -\frac{z_i^2 \bar{\eta}_i^2}{\sqrt{z_i^2 \bar{\eta}_i^2 + \omega_i^2}} \leq \omega_i - z_i \bar{\eta}_i. \tag{44}$$

Substituting (34)–(36) and (43)–(44) into (38), one has

$$D^\alpha V_i \leq -\sum_{j=1}^{i} c_j z_j^2 - \sum_{j=1}^{i-1} \frac{(\rho_j - 1)}{2}\tilde{W}_j^T \tilde{W}_j + \Xi_i - \sum_{j=1}^{i-1}\left(a_j - \frac{1+b_j}{2}\right)\gamma_j^2$$
$$- \sum_{j=1}^{i-1}\left(l_j - \frac{3}{2}\right)\tilde{\Lambda}_j^2 + \omega_i - \frac{(1+b_i)}{2}z_i^2 + \rho_i \tilde{W}_i^T \hat{W}_i + z_i \tilde{\Lambda}_i + b_i z_i \gamma_i$$
$$- a_i \gamma_i^2 + \gamma_i \epsilon_i - (l_i - 1)\tilde{\Lambda}_i^2 + \frac{1}{2}\tilde{\Lambda}_i^2 + \frac{1}{2}\tilde{W}_i^T \tilde{W}_i + z_i z_{i+1}. \tag{45}$$

Utilizing Young's inequality, we have

$$D^\alpha V_i \leq -\sum_{j=1}^{i} c_j z_j^2 - \sum_{j=1}^{i} \frac{(\rho_j - 1)}{2}\tilde{W}_j^T \tilde{W}_j - \sum_{j=1}^{i}\left(a_j - \frac{1+b_j}{2}\right)\gamma_j^2$$
$$- \sum_{j=1}^{i}\left(l_j - \frac{3}{2}\right)\tilde{\Lambda}_j^2 + z_i z_{i+1} + \Xi_i, \tag{46}$$

where $\Xi_i = \Xi_{i-1} + \frac{\rho_i}{2}W_i^{*T}W_i^* + \frac{1}{2}\epsilon_i^2 + \frac{1}{2}\tilde{\Lambda}_i^2 + \omega_i$.

Remark 2. *In most existing results, more control energy may be expected for obtaining a desired control performance. However, it is noted that the amplitude of the control signal is usually limited due to the inherent limitations of physical structures. Therefore, making a trade-off between the control performance and input energy is reasonable and significant for practical applications.*

Step n. In this step, the following saturation function is used to bound the actual control signal. Then, one has

$$u(t) = \begin{cases} sign(v)u_{Max}, & |v(t)| \geq u_{Max} \\ v, & |v(t)| < u_{Max} \end{cases} \tag{47}$$

where $u_{Max} > 0$ is the bound of $u(t)$ and v is the input of the saturation nonlinearity $g(v)$ satisfying

$$g(v) = u_{Max} * \frac{e^{\frac{v}{u_{Max}}} - e^{-\frac{v}{u_{Max}}}}{e^{\frac{v}{u_{Max}}} + e^{-\frac{v}{u_{Max}}}}. \tag{48}$$

Then, one has $u(t) = sat(v) = g(v) + h(v)$ with $|h(v)| = |sat(v) - g(v)| \leq u_{Max}(1 - \tanh(1)) = H$.

Utilizing the mean value theorem, the function $g(v)$ can be expressed as:

$$g(v) = g(v^*) + \frac{\partial g(\cdot)}{\partial v}\Big|_{v=v^{\ell_0}}(v - v^*), \tag{49}$$

where $v^{\ell_0} = \ell_0 v + (1 - \ell_0)v^*$ with $0 < \ell_0 < 1$. Letting $v^* = 0$, we can obtain

$$g(v) = \frac{\partial g(\cdot)}{\partial v}\Big|_{v=v^{\ell_0}}(v - v^*)v = g_0(v^{\ell_0})v. \tag{50}$$

Furthermore, one can obtain

$$u(t) = sat(v) = g_0(v^{\ell_0})v + h(v). \tag{51}$$

Since $g(v)$ is a non-decreasing function, there exist two positive constants $\underline{g}_0$ and $\bar{g}_0$ such that $0 < \underline{g}_0 \leq g_0(v^{\ell_0}) \leq \bar{g}_0$.

Design the compensating signal as:

$$D^\alpha \gamma_n = -a_n \gamma_n. \tag{52}$$

Furthermore, the FO derivative of z_n is calculated as:

$$D^\alpha z_n = (1 - \kappa)u(t) + \kappa\theta + l_n^{-1}W_n^{*T}\phi_n(\bar{x}_n) + \Lambda_n - D^\alpha \vartheta_n^c, \tag{53}$$

where $\Lambda_n = d_n + l_n^{-1}\epsilon_n$.

For the purpose of reducing the unnecessary waste of communication resources, the event-triggered mechanism is introduced into controller design. Then, the event-triggered control signal is constructed as:

$$v(t) = \eta(t_k), \forall t \in [t_k, t_{k+1}), \tag{54}$$

with the trigger condition satisfying

$$t_{k+1} = \inf\{t > t_k | |e(t)| \geq \chi|v(t)| + \varsigma\}, \tag{55}$$

where η denotes the transition control signal to be designed and $e(t) = \eta(t) - v(t)$ is the measured error. χ and ς are design parameters satisfying $0 < \chi < 1$ and $\varsigma > 0$.

For the interval $[t_k, t_{k+1})$, it follows from (54)–(55) that

$$\eta(t) = (1 + \mu_1(t)\chi)v(t) + \mu_2(t)\varsigma, \tag{56}$$

in which $\mu_1(t)$ and $\mu_2(t)$ are the time-varying parameters satisfying $|\mu_1(t)| \leq 1$ and $|\mu_2(t)| \leq 1$.

Using Equation (56), one can obtain

$$v(t) = \frac{\eta(t)}{1 + \mu_1(t)\chi} - \frac{\mu_2(t)\varsigma}{1 + \mu_1(t)\chi}. \tag{57}$$

The control signal $\eta(t)$ and parameter update law are given as:

$$\eta(t) = -\frac{(1 + \chi)z_n\bar{\eta}_n^2}{(1 - \kappa)\underline{g}_0\sqrt{z_n^2\bar{\eta}_n^2 + \varpi_n^2}}, \tag{58}$$

$$\bar{\eta}_n = \left(c_n + \frac{1 + b_n}{2}\right)z_n + z_{n-1} - D^\alpha\vartheta_n^c - b_n\gamma_n + \hat{\Lambda}_n$$

$$+ \bar{\varsigma}\tanh\left(\frac{z_n\bar{\varsigma}}{\hbar}\right) + l_n^{-1}\hat{W}_n^T\phi_n(\bar{x}_n) + (1 - \kappa)\mathrm{sgn}(z_n)u_{min} \tag{59}$$

$$D^\alpha\hat{W}_n = l_n^{-1}z_n\phi_n(\bar{x}_n) - \rho_n\hat{W}_n, \tag{60}$$

where $\bar{\varsigma} > \frac{(1-\kappa)\bar{g}_0\varsigma}{1-\chi}$.

Construct the Lyapunov function as:

$$V_n = V_{n-1} + \frac{1}{2}z_n^2 + \frac{1}{2}\tilde{W}_n^T\tilde{W}_n + \frac{1}{2}\gamma_n^2 + \frac{1}{2}\tilde{\Lambda}_n^2. \tag{61}$$

Using (53), the FO derivative of V_n is expressed as:

$$
\begin{aligned}
D^\alpha V_n \leq{} & -\sum_{j=1}^{n-1} c_j z_j^2 - \sum_{j=1}^{n-1} \frac{(\rho_j - 1)}{2}\tilde{W}_j^T\tilde{W}_j - \sum_{j=1}^{n-1}\left(a_j - \frac{1+b_j}{2}\right)\gamma_j^2 - \sum_{j=1}^{n-1}\left(l_j - \frac{3}{2}\right)\tilde{\Lambda}_j^2 \\
& + \frac{z_n(1-\kappa)g_0(v^{\ell_0})\eta(t)}{1+\mu_1(t)\chi} + z_n(1-\kappa)h(v) - \frac{z_n(1-\kappa)g_0(v^{\ell_0})\mu_2(t)\varsigma}{1+\mu_1(t)\chi} \\
& + z_n\left(\kappa\theta + l_n^{-1}W_n^{*T}\phi_n(\bar{x}_n) + \Lambda_n - D^\alpha\vartheta_n^c\right) - \tilde{W}_n^T D^\alpha\hat{W}_n + \gamma_n D^\alpha\gamma_n \\
& + \tilde{\Lambda}_n D^\alpha\tilde{\Lambda}_n + z_{n-1}z_n + \Xi_{n-1}.
\end{aligned}
\tag{62}
$$

Furthermore, design the FONDO as:

$$\hat{\Lambda}_n = l_n x_n + \xi_n, \tag{63}$$

with $D^\alpha\xi_n = -l_n\left(\xi_n + q(u) + l_n^{-1}\hat{W}_n^T\phi_n(\bar{x}_n) + l_n x_n\right)$.

Then, it follows from (63) that

$$D^\alpha\hat{\Lambda}_n = \tilde{W}_n^T\phi_n(\bar{x}_n) + l_n\tilde{\Lambda}_n. \tag{64}$$

According to (64), one yields

$$\tilde{\Lambda}_n D^\alpha\tilde{\Lambda}_n = \tilde{\Lambda}_n D^\alpha\Lambda_n - \tilde{\Lambda}_n\tilde{W}_n^T\phi_n(\bar{x}_n) - l_n\tilde{\Lambda}_n^2. \tag{65}$$

Using Young's inequality obtains

$$\tilde{\Lambda}_n D^\alpha\tilde{\Lambda}_n \leq -(l_n - 1)\tilde{\Lambda}_n^2 + \frac{1}{2}\tilde{\Lambda}_n^2 + \frac{1}{2}\tilde{W}_n^T\tilde{W}_n. \tag{66}$$

It follows from (57) that

$$\frac{z_n(1-\kappa)g_0(v^{\ell_0})\eta(t)}{1+\mu_1(t)\chi} \leq \frac{z_n(1-\kappa)\underline{g}_0\eta(t)}{1+\chi}, \tag{67}$$

$$-\frac{z_n(1-\kappa)g_0(v^{\ell_0})\mu_2(t)\varsigma}{1+\mu_1(t)\chi} \leq \frac{(1-\kappa)\bar{g}_0}{1-\chi}|z_n\varsigma|. \tag{68}$$

Invoking (58)–(60) and (62)–(68), one has

$$
\begin{aligned}
D^\alpha V_n \leq{} & -\sum_{j=1}^{n} c_j z_j^2 - \sum_{j=1}^{n-1}\frac{(\rho_j - 1)}{2}\tilde{W}_j^T\tilde{W}_j - \sum_{j=1}^{n-1}\left(a_j - \frac{1+b_j}{2}\right)\gamma_j^2 - \sum_{j=1}^{n-1}\left(l_j - \frac{3}{2}\right)\tilde{\Lambda}_j^2 \\
& + \Xi_{n-1} + \varpi_n + b_n z_n\gamma_n + 0.2785\hbar + \frac{(1-\kappa)\bar{g}_0}{1-\chi}|z_n\varsigma| - |z_n\bar{\varsigma}| - \left(\frac{2-\kappa+b_n}{2}\right)z_n^2 \\
& + z_n(1-\kappa)h(v) + \rho_n\tilde{W}_n^T\hat{W}_n + z_n\tilde{\Lambda}_n - a_n\gamma_n^2 - (l_n - 1)\tilde{\Lambda}_n^2 + \frac{1}{2}\tilde{\Lambda}_n^2 + \frac{1}{2}\tilde{W}_n^T\tilde{W}_n \\
& - |z_n|(1-\kappa)u_{min} + z_n\kappa\theta.
\end{aligned}
\tag{69}
$$

Using $\rho_n \tilde{W}_n^T \hat{W}_n \leq -\frac{\rho_n}{2} \tilde{W}_n^T \tilde{W}_n + \frac{\rho_n}{2} W_n^{*T} W_n^*$ and $\varsigma > \frac{(1-\kappa)\tilde{g}_0\varsigma}{1-\chi}$, we obtain

$$D^\alpha V_n \leq -\sum_{j=1}^{n} c_j z_j^2 - \sum_{j=1}^{n} \bar{\rho}_j \tilde{W}_j^T \tilde{W}_j - \sum_{j=1}^{n} \bar{a}_j \gamma_j^2 - \sum_{j=1}^{n} \bar{l}_j \tilde{\Lambda}_j^2$$
$$- |z_n|(1-\kappa) u_{min} + z_n \kappa \theta + \Xi_n, \tag{70}$$

where $\Xi_n = \Xi_{n-1} + \varpi_n + \frac{\rho_n}{2} W_n^{*T} W_n^* + 0.2785\hbar + \frac{1}{2} \bar{\Lambda}_n^2 + \frac{1-\kappa}{2} \hbar^2, \bar{a}_s = a_s - \frac{1}{2}, (s = 1, \ldots, n - 1), \bar{a}_n = a_n - \frac{b_n}{2}, \bar{l}_j = l_j - \frac{3}{2}$ and $\bar{\rho}_j = (\rho_j - 1)/2, (j = 1, \ldots, n)$.

By means of the preceding derivations, the following theorem is obtained.

Theorem 1. *For the investigated fractional-order nonlinear plant (11) satisfying Assumptions 1–2, if the recursive control framework consisting of virtual control laws (18)–(19) and (34)–(35), the actual control law (58)–(59), the parameter update laws (20), (36) and (60), and the FONDOs (23), (39) and (63) are adopted, then all the signals of the CLS are bounded and the tracking error converges to a small neighborhood of the origin.*

Proof. Utilizing Lemma 4, the following two cases are considered for inequality (70).

Case (I): $u_{min} \leq u_{Max}$: for this case, the following two sub-cases need to be discussed.

Case (i): $|u(t)| \leq u_{min}$: according to Lemma 4 with $\theta^2 \leq \left(\frac{1-\kappa}{\kappa} u_{min}\right)^2$, we can obtain

$$D^\alpha V_n \leq -\sum_{j=1}^{n} c_j z_j^2 - \sum_{j=1}^{n} \bar{\rho}_j \tilde{W}_j^T \tilde{W}_j - \sum_{j=1}^{n} \bar{a}_j \gamma_j^2 - \sum_{j=1}^{n} \bar{l}_j \tilde{\Lambda}_j^2 + \Xi_n. \tag{71}$$

Case (ii): $|u(t)| \geq u_{min}$, the following inequality holds in accordance with the relationship $\theta^2 \leq \left(\frac{\kappa+\delta}{\kappa} u\right)^2$ in Lemma 4

$$D^\alpha V_n \leq -\sum_{j=1}^{n} c_j z_j^2 - \sum_{j=1}^{n} \bar{\rho}_j \tilde{W}_j^T \tilde{W}_j - \sum_{j=1}^{n} \bar{a}_j \gamma_j^2 - \sum_{j=1}^{n} \bar{l}_j \tilde{\Lambda}_j^2 - |z_n|(1-\kappa) u_{min}$$
$$+ |z_n|(\kappa + \delta)|u| + \Xi_n. \tag{72}$$

Let the quantized parameter satisfy the condition $u_{Max} \leq \frac{1-\kappa}{\kappa+\delta} u_{min}$; then, we have

$$|u| \leq \frac{1-\kappa}{\kappa+\delta} u_{min}. \tag{73}$$

Invoking equalities (72) and (73), one yields

$$D^\alpha V_n \leq -\sum_{j=1}^{n} c_j z_j^2 - \sum_{j=1}^{n} \bar{\rho}_j \tilde{W}_j^T \tilde{W}_j - \sum_{j=1}^{n} \bar{a}_j \gamma_j^2 - \sum_{j=1}^{n} \bar{l}_j \tilde{\Lambda}_j^2 + \Xi_n. \tag{74}$$

Case (II): $u_{min} \geq u_{Max}$: for this case, we have $|u| \leq u_{min}$. Therefore, a similar result can be obtained by referring to Case (i) in Case (I).

Combining with Case (I)–Case (II), by choosing the appropriate parameter c_j, ρ_j, a_j, l_j $(j = 1, \ldots, n)$, we can obtain

$$D^\alpha V_n \leq -\lambda V_n + \Xi_n, \tag{75}$$

where $\lambda = \min\{2c_j, 2\bar{\rho}_j, 2\bar{a}_j, 2\bar{l}_j\}$.

It follows from inequality (75) that

$$D^\beta V_n(t) + \psi(t) = -\lambda V_n(t) + \Xi_n, \tag{76}$$

where $\psi(t) > 0$ is a time-varying parameter.

By taking LT for (76), one has

$$V_n(s) = \frac{s^{\alpha-1}}{s^{\alpha}+\lambda}V_n(0) + \frac{s^{\alpha-(1+\alpha)}\Xi_n}{s^{\alpha}+\lambda} - \frac{\Psi(s)}{s^{\alpha}+\lambda}, \tag{77}$$

where $V_n(s)$ and $\Psi(s)$ stand for the LT of $V_n(t)$ and $\psi(t)$.

Utilizing (2), one has

$$V_n(t) = V_n(0)E_{\alpha,1}(-\lambda t^{\alpha}) + \Xi_n t^{\alpha}E_{\alpha,1+\alpha}(-\lambda t^{\alpha}) - \psi(t) * t^{\alpha-1}E_{\alpha,\alpha}(-\lambda t^{\alpha}). \tag{78}$$

Since both $t^{\alpha-1}$ and $E_{\alpha,\alpha}(-\lambda t^{\alpha})$ are non-negative functions, $\psi(t) * t^{\alpha-1}E_{\alpha,\alpha}(-\lambda t^{\alpha})$ is non-negative. Then, for all $t \geq 0$, we have $\arg(-\lambda t^{\alpha}) = -\pi, |-\lambda t^{\alpha}| \geq 0$. Therefore, it follows from Lemma 1 that there exists a constant $m > 0$ such that $|E_{\alpha,1}(-\lambda t^{\alpha})| \leq \frac{m}{1+\lambda t^{\alpha}}$.

When t tends to ∞, one yields

$$\lim_{t\to\infty} |V_n(0)|E_{\alpha,1}(-\lambda t^{\alpha}) = 0. \tag{79}$$

Moreover, the following relationship holds

$$E_{\alpha,1+\alpha}(-\lambda t^{\alpha}V_n(0)) < \sigma_1 \tag{80}$$

for a time instant $t_1 > 0$ and every $\sigma_1 > 0$.

Using Lemma 2 with $\ell = 1$, we have $E_{\alpha,1+\alpha}(-\kappa t^{\alpha}) = \frac{1}{\Gamma(1)\lambda t^{\alpha}} + o(\frac{1}{|\lambda t^{\alpha}|^2})$. According to the fact that $\Gamma(1) = 1$, the following equality holds

$$t^{\alpha}\Xi_n E_{\alpha,1+\alpha}(-\lambda t^{\alpha}) = \frac{\Xi_n}{\lambda} + t^{\alpha}\Xi_n o(\frac{1}{|\lambda t^{\alpha}|^2}). \tag{81}$$

Moreover, one has

$$t^{\alpha}\Xi_n o(\frac{1}{|\lambda t^{\alpha}|^2}) \leq \sigma_2 \tag{82}$$

for all $t > t_2$ and every $\sigma_2 > 0$. In addition, for every $\sigma_3 > 0$, we have $\frac{\Xi_n}{\lambda} \leq \sigma_3$ by choosing the proper parameters. Therefore, we can obtain

$$t^{\alpha}\bar{\Lambda}_n E_{\alpha,1+\alpha}(-\lambda t^{\alpha}) \leq \sigma_2 + \sigma_3. \tag{83}$$

Since the term $\psi(t) * t^{\alpha-1}E_{\alpha,\alpha}(-\lambda t^{\alpha})$ is non-negative, invoking (78), (80) and (83) obtains

$$V_n(t) \leq \sigma_1 + \sigma_2 + \sigma_3. \tag{84}$$

Using Lemma 3 of [37] and the definition of $V_n(t)$, we can conclude that all resulting signals of the CLS are bounded for $t > \max\{t_1, t_2\}$. Furthermore, it can be obtained that the tracking error z_1 can converge to a small neighborhood of the origin $|z_1| \leq \sqrt{2(\sigma_1 + \sigma_2 + \sigma_3)}$. This completes the proof. $\square$

Moreover, a block diagram, as demonstrated in Figure 1, is presented to clarify the structure of the proposed control approach.

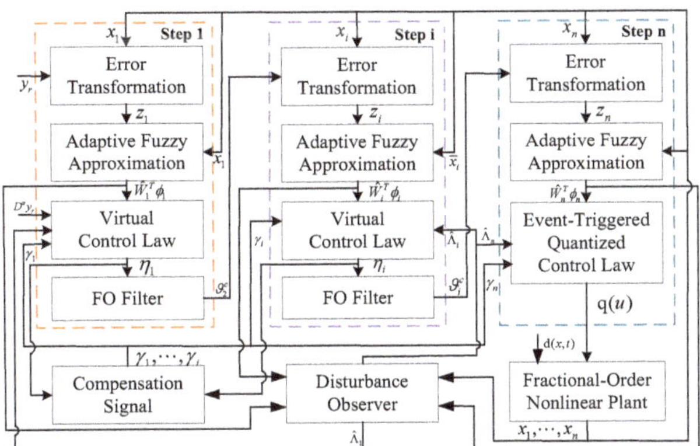

Figure 1. Block diagram of the proposed control scheme.

4. Simulation Verification

In this section, the effectiveness and the practical potential of the presented control approach will be verified through a horizontal platform system (HPS). According to [39], an HPS is mainly composed of two components, i.e., a platform and an accelerometer located on the platform, as shown in Figure 2. When the platform deviates from the horizon, the accelerometer will send an output signal to the torque generator, which generates a torque to invert the rotation of the platform about the rotational axis. The mathematical equation of the HPS is

$$A\ddot{\theta} + D\dot{\theta} + kg\sin\theta - \frac{3g}{R}(B - C)\cos\theta\sin\theta = F\cos\omega t, \tag{85}$$

where θ is the rotation of the platform relative to the earth; $\dot{\theta}$ is the corresponding angular velocity; $F\cos\omega t$ is harmonic torque; A, B and C are the inertia moment of the platform; D is the damping coefficient; k denotes the proportional constant of the accelerometer; and g is the acceleration constant of gravity. The descriptions of relevant parameters are presented in Table 1.

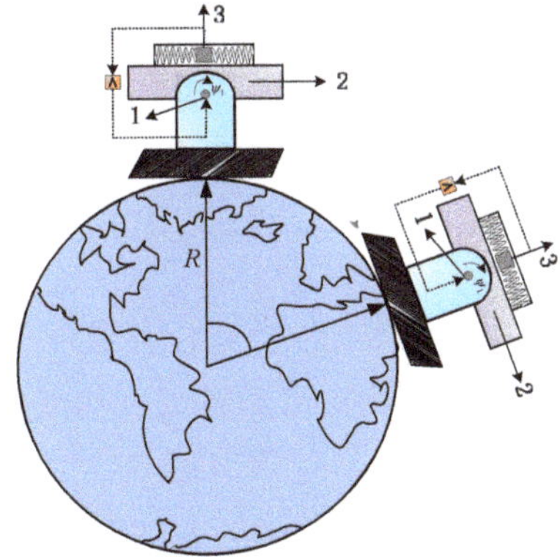

Figure 2. Model of the platform circles along the Earth.

Table 1. Parameter list of the HPS.

Parameters	Nomenclature	Value	Unit
A	Inertia moment of the platform around axis 1	0.3	$\mathrm{kg \cdot m^2}$
B	Inertia moment of the platform around axis 2	0.5	$\mathrm{kg \cdot m^2}$
C	Inertia moment of the platform around axis 3	0.2	$\mathrm{kg \cdot m^2}$
D	Damping coefficient	0.4	$\mathrm{kg \cdot m^2 \cdot s^{-1}}$
F	Amplitude of the harmonic torque	3.4	$\mathrm{N \cdot m}$
g	Acceleration constant of gravity	9.8	$\mathrm{m \cdot s^{-2}}$
k	Proportional constant of the accelerometer	0.11559633	$\mathrm{kg \cdot m \cdot rad}$
R	Radius of the Earth	6,378,000	m
ω	Circular frequency of the harmonic torque	1.8	$\mathrm{rad \cdot s^{-1}}$

We define $x_1 = \theta$ and $x_2 = \dot{\theta}$; then, system (85) can be transformed into the following form:

$$\begin{cases} \dot{x}_1 = x_2 \\ \dot{x}_2 = -ax_2 - b\sin x_1 + l\cos x_1 \sin x_1 + h\cos \omega t \end{cases} \tag{86}$$

where $a = \frac{D}{A}, b = \frac{kg}{A}, l = \frac{3g}{RA}(B\text{-}C)$ and $h = \frac{F}{A}$.

Furthermore, considering that the fractional-order model may provide a more accurate description of physical behavior and the actual system is inevitably influenced by perturbations, the fractional-order model of HPS with input quantization and perturbations can be given by [40]

$$\begin{cases} D^\alpha x_1 = x_2 + d_1(x,t) \\ D^\alpha x_2 = -ax_2 - b\sin x_1 + l\cos x_1 \sin x_1 + h\cos \omega t + q(u) + d_2(x,t) \end{cases} \tag{87}$$

where α denotes fractional order satisfying $\alpha = 0.95$, $d_i(x,t)$ denotes the unknown perturbation and $q(u)$ represents the quantized control signal to be designed. The state response of system (86) without control effort under initial condition $[x_1(0), x_2(0)] = [0.1, -0.1]$ is displayed in Figure 3.

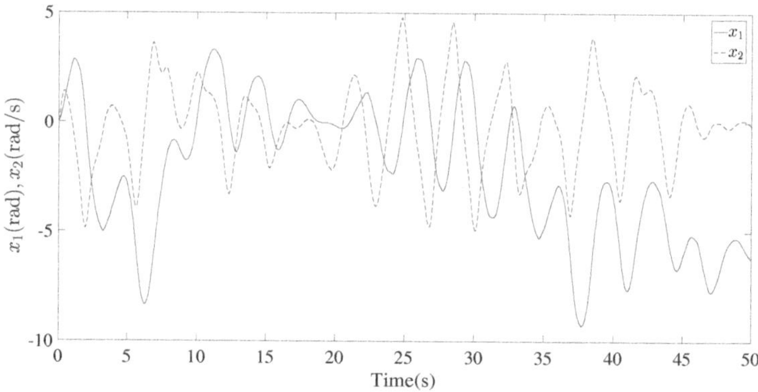

Figure 3. The state response of system (87) without control effort.

The control parameters, initial conditions and disturbances are provided in Table 2. Obviously, the selection of parameters δ, κ, u_{min} and u_{Max} can ensure the quantization condition (73) holds. To achieve an accurate approximation of nonlinear functions, the Gaussian membership functions of FLSs are selected as: $\mu_{F^i}(x) = \exp[-(x - i + 5)^2/4]$ with $i = 1, \ldots, 9$.

Table 2. Selection of simulation parameters.

Design Parameters	Disturbance Terms
$c_1 = c_2 = 50, a_1 = a_2 = \rho_2 = 2, \iota = 0.01,$ $\delta = 0.2, \kappa = 0.15, u_{min} = 5, u_{Max} = 8, l_1 = 10,$ $l_2 = 20, b_1 = b_2 = 1, \chi = 0.5, \alpha = 0.95.$	$d_1(x,t) = 1.5\sin(2t) + 0.5\cos(x_1 x_2),$ $d_2(x,t) = 1.5\cos(2t) + 0.5\sin(x_1 x_2).$

Initial Conditions
$x_1(0) = 0.1, x_2(0) = -0.1, \gamma_1(0) = \gamma_2(0) = 0,$ $\hat{\Lambda}_1(0) = \hat{\Lambda}_2(0) = 0, \hat{W}(0) = [\underbrace{0,0,\dots,0}_{9}]^T.$

Reference Signal
$y_r = 0.5\sin(t) + \sin(0.5t)$

Based on the established control framework as shown in Figure 4, the comparative simulation results under different control schemes are demonstrated in Figures 5–11. The tracking performances are shown in Figures 5 and 6. Furthermore, three kinds of performance indexes, consisting of integral absolute error (IAE), integral time-weighted absolute error (ITAE) and integral square error (ISE), are introduced to quantify the tracking performance under different control methods. It can be concluded from Figures 4 and 5 and Table 3 that better tracking performance can be achieved by using the proposed method in comparison to the FABC method proposed in [10] and the FACFQC method developed in [31]. The responses of composite disturbance Λ_i and its estimation $\hat{\Lambda}_i$ are plotted in Figures 7 and 8. Figures 9 and 10 depict the trajectories of the norm of adaptive parameter $\|\hat{W}\|$ and the quantized control signal $q(u)$. The time interval of each event is demonstrated in Figure 11.

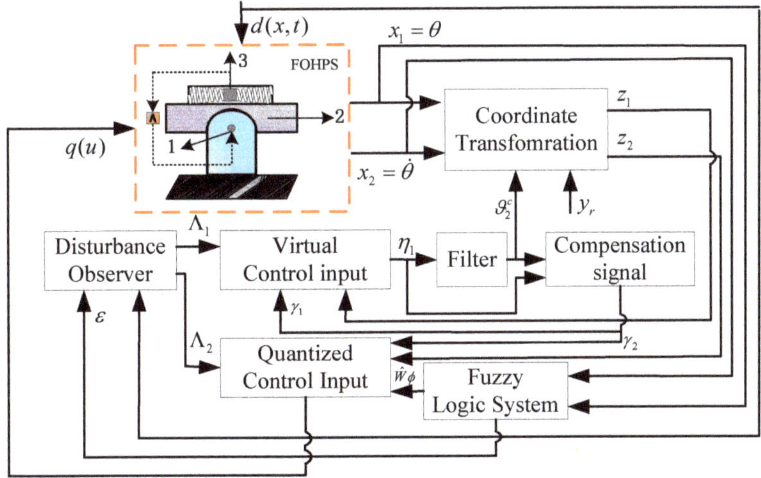

Figure 4. Control framework of the FOHPS.

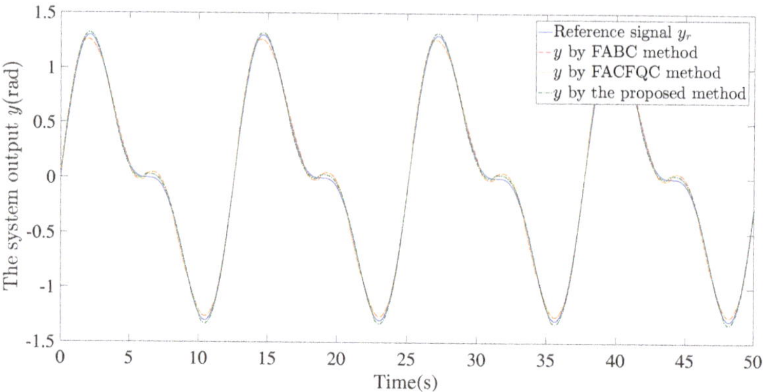

Figure 5. The reference signal y_r and the system output y under different control methods.

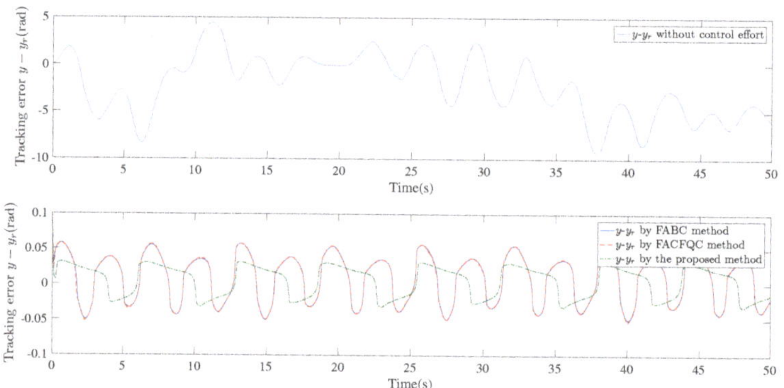

Figure 6. The tracking error y-y_r under different control methods.

Table 3. Performance comparisons under different methods.

Performance Index	Items	FABC in [10]	FACFQC in [31]	Proposed
IAE	$\int_0^T |z_1(t)|dt$	1.75	1.755	**1.153**
ITAE	$\int_0^T t|z_1(t)|dt$	40.62	40.77	**26.09**
ISE	$\int_0^T z_1^2(t)dt$	0.073	0.075	**0.035**

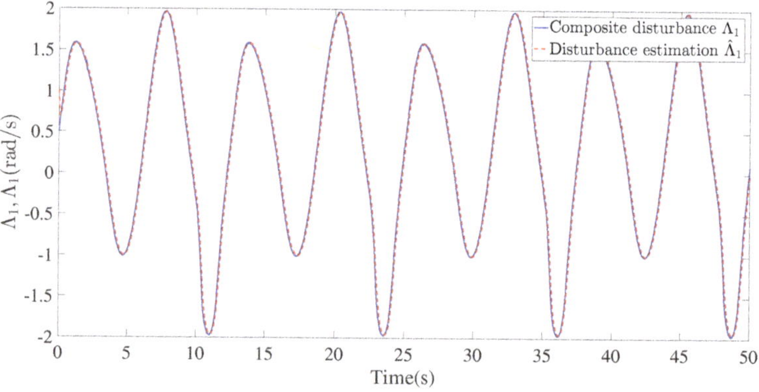

Figure 7. The composite disturbance Λ_1 and its estimation $\hat{\Lambda}_1$.

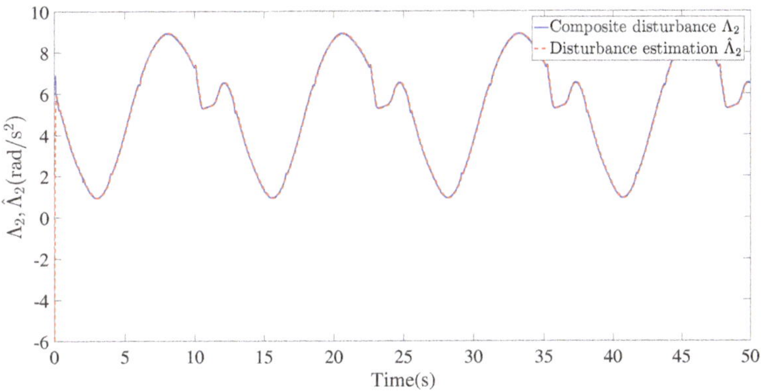

Figure 8. The composite disturbance Λ_2 and its estimation $\hat{\Lambda}_2$.

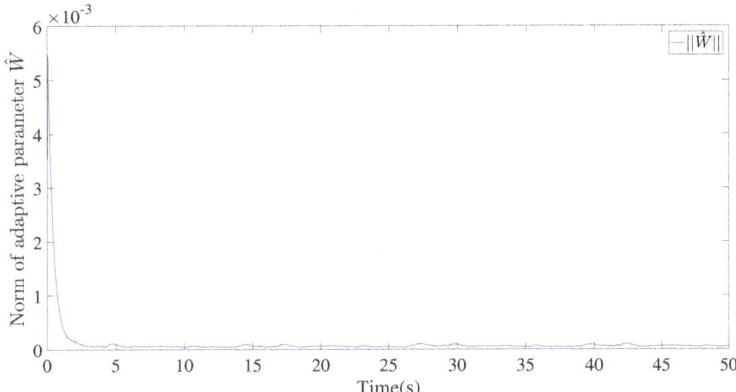

Figure 9. The norm of adaptive parameter $||\hat{W}||$.

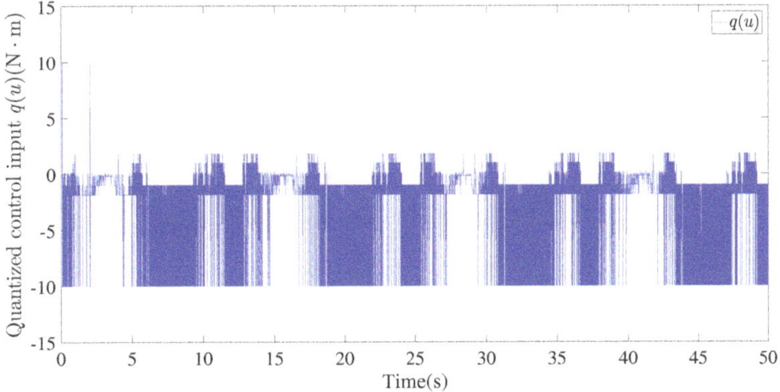

Figure 10. Quantized control signal $q(u)$.

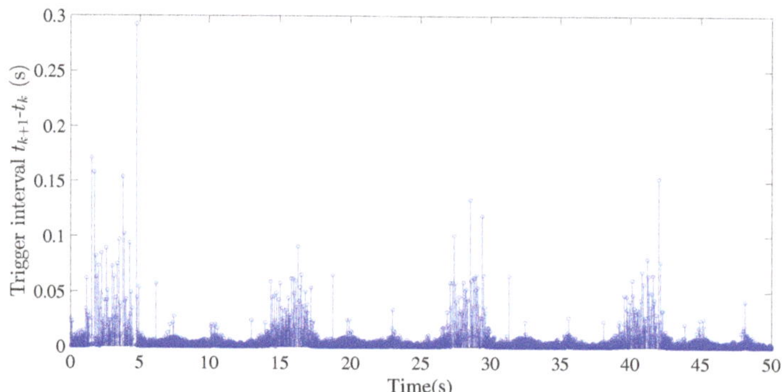

Figure 11. Trigger interval $t_{k+1}-t_k$.

5. Conclusions

In this article, a nonlinear disturbance observer-based event-triggered adaptive command-filtered quantized control approach is developed for fractional-order nonlinear systems with unknown disturbances. By introducing the command-filtered backstepping technique into the recursive design procedure, the potential issue of computational complexity existing in [10] and the negative effect caused by filter error in [17] are successfully avoided. Furthermore, a fractional-order disturbance observer is designed to achieve disturbance estimation, which can improve system robustness against composite disturbances consisting of unknown disturbances and approximation in comparison to the existing recursive control schemes proposed in [10,17,31]. Moreover, differently from the common time-triggered control methods in [10,17,31], the event-triggered control mechanism and input quantization are considered simultaneously, which can save a large amount of communication bandwidth and provide a possible way to make a trade-off between tracking performance and control costs. Finally, the validity and superiority of the proposed method are verified by a fractional-order HPS. However, it should be pointed out that a preassigned transient and steady-state performance cannot be ensured although the design controller can guarantee a relatively satisfactory tracking performance. Therefore, one of our future research works is to propose an adaptive control scheme with assured transient and steady-state performance for fractional-order nonlinear systems.

Author Contributions: Writing—original draft preparation, S.S.; writing—review and editing, S.S. and X.S.; supervision, X.S. and I.T.; funding acquisition, S.S. and X.S. All authors have read and agreed to the published version of the manuscript.

Funding: This work was supported in part by the National Natural Science Foundation of China under Grant 62203153, in part by the Natural Science Fund for Young Scholars of Henan Province under Grant 222300420151, in part by Technology Innovative Teams in University of Henan Province under Grant 23IRTSTHN012 and in part by Top Young Talents in Central Plains under Grant Yuzutong (2021) 44.

Data Availability Statement: Not applicable.

Conflicts of Interest: The authors declare no conflict of interest.

References

1. Tejado, I.; Pérez, E.; Valério, D. Fractional calculus in economic growth modeling of the group of seven. *Fract. Calc. Appl. Anal.* **2019**, *22*, 139–157. [CrossRef]
2. Zhang, X.; Huang, W. Adaptive Neural Network Sliding Mode Control for Nonlinear Singular Fractional Order Systems with Mismatched Uncertainties. *Fractal Fract.* **2020**, *4*, 50. [CrossRef]
3. Zouari, F.; Ibeas, A.; Boulkroune, A.; Cao, J.; Arefi, M.M. Adaptive neural output-feedback control for nonstrict-feedback time-delay fractional-order systems with output constraints and actuator nonlinearities. *Neural Netw.* **2018**, *105*, 256–276. [CrossRef] [PubMed]

4. Bao, H.; Park, J.H.; Cao, J. Adaptive synchronization of fractional-order memristor-based neural networks with time delay. *Nonlinear Dyn.* **2015**, *82*, 1343–1354. [CrossRef]

5. Lin, C.; Chen, B.; Shi, P.; Yu, J. Necessary and sufficient conditions of observer-based stabilization for a class of fractional-order descriptor systems. *Syst. Control. Lett.* **2018**, *112*, 31–35. [CrossRef]

6. Coronel-Escamilla, A.; Gomez-Aguilar, J.F.; Stamova, I.; Santamaria, F. Fractional order controllers increase the robustness of closed-loop deep brain stimulation systems. *Chaos Solut. Fractals* **2020**, *140*, 110149. [CrossRef]

7. Luo, S.; Lewis, F.L.; Song, Y.; Ouakad, H.M.J. Accelerated adaptive fuzzy optimal control of three coupled fractional-order chaotic electromechanical transducers. *IEEE Trans. Fuzzy Syst.* **2021**, *29*, 1701–1714. [CrossRef]

8. Zouari, F.; Ibeas, A.; Boulkroune, A.; Cao, J.; Arefi, M.M. Neural network controller design for fractional-order systems with input nonlinearities and asymmetric time-varying Pseudo-state constraints. *Chaos Solut. Fractals* **2021**, *144*, 110742. [CrossRef]

9. Wei, Y.; Tse, P.W.; Yao, Z.; Wang, Y. Adaptive backstepping output feedback control for a class of nonlinear fractional order systems. *Nonlinear Dyn.* **2016**, *86*, 1047–1056. [CrossRef]

10. Liu, H.; Pan, Y.; Li, S.; Chen, Y. Adaptive fuzzy backstepping control of fractional-order nonlinear systems. *IEEE Trans. Syst. Man Cybern. Syst.* **2017**, *47*, 2209–2217. [CrossRef]

11. Li, Y.; Wang, Q.; Tong, S. Fuzzy adaptive fault-tolerant control of fractional-order nonlinear systems. *IEEE Trans. Syst. Man Cybern. Syst.* **2021**, *51*, 1372–1379. [CrossRef]

12. Yu, J.; Zhao, L.; Yu, H.; Lin, C.; Dong, W. Fuzzy finite-time command filtered control of nonlinear systems with input saturation. *IEEE Trans. Cybern.* **2018**, *48*, 2378–2387. [PubMed]

13. Li, Y.; Li, K.; Tong, S. Finite-time adaptive fuzzy output feedback dynamic surface control for MIMO nonstrict feedback systems. *IEEE Trans. Fuzzy Syst.* **2019**, *27*, 96–110. [CrossRef]

14. Qiu, J.; Sun, K.; Rudas, I.J.; Gao, H. Command filter-based adaptive NN control for MIMO nonlinear systems with full-state constraints and actuator hysteresis. *IEEE Trans. Cybern.* **2020**, *50*, 2905–2915. [CrossRef] [PubMed]

15. Niu, B.; Li, H.; Karimi, H.R. Adaptive NN dynamic surface controller design for nonlinear pure-feedback switched systems with time-delays and quantized input. *IEEE Trans. Syst. Man Cybern. Syst.* **2018**, *48*, 1676–1688. [CrossRef]

16. Liu, Y. Adaptive dynamic surface asymptotic tracking for a class of uncertain nonlinear systems. *Int. J. Robust Nonlinear Control.* **2018**, *28*, 1233–1245. [CrossRef]

17. Ma, Z.; Ma, H. Adaptive fuzzy backstepping dynamic surface control of strict-feedback fractional-order uncertain nonlinear systems. *IEEE Trans. Fuzzy Syst.* **2020**, *28*, 122–133. [CrossRef]

18. Song, S.; Zhang, B.; Song, X.; Zhang, Z. Neuro-fuzzy-based adaptive dynamic surface control for fractional-order nonlinear strict-feedback systems with input constraint. *IEEE Trans. Syst. Man Cybern. Syst.* **2021**, *50*, 3575–3586. [CrossRef]

19. Song, S.; Park, J.H.; Zhang, B.; Song, X. Observer-based adaptive hybrid fuzzy resilient control for fractional-order nonlinear systems with time-varying delays and actuator failures. *IEEE Trans. Fuzzy Syst.* **2021**, *29*, 471–485. [CrossRef]

20. Liu, H.; Pan, Y.; Cao, J. Composite learning adaptive dynamic surface control of fractional-order nonlinear systems. *IEEE Trans. Cybern.* **2020**, *50*, 2557–2567. [CrossRef]

21. Song, X.; Sun, P.; Song, S.; Stojanovic, V. Finite-time adaptive neural resilient DSC for fractional-order nonlinear large-scale systems against sensor-actuator faults. *Nonlinear Dyn.* **2023**, *111*, 12181–12196. [CrossRef]

22. Xu, B.; Sun, F.; Pan, Y.; Chen, B. Disturbance observer-based composite learning fuzzy control of nonlinear systems with unknown dead zones. *IEEE Trans. Syst. Man Cybern. Syst.* **2017**, *47*, 1854–1862. [CrossRef]

23. Chen, M.; Shao, S.; Shi, P. Disturbance-observer-based robust synchronization control for a class of fractional-order chaotic systems. *IEEE Trans. Circuits Syst. II Express Briefs* **2017**, *64*, 417–421. [CrossRef]

24. Min, H.; Xu, S.; Ma, Q.; Zhang, B.; Zhang, Z. Composite-observer-based output-feedback control for nonlinear time-delay systems with input saturation and its application. *IEEE Trans. Ind. Electron.* **2018**, *65*, 5856–5863. [CrossRef]

25. Hou, C.; Liu, X.; Wang, H. Adaptive fault tolerant control for a class of uncertain fractional-order systems based on disturbance observer. *Int. J. Robust Nonlinear Control.* **2020**, *30*, 3436–3450. [CrossRef]

26. Han, S.I. Fuzzy supertwisting dynamic surface control for MIMO strict-feedback nonlinear dynamic systems with supertwisting nonlinear disturbance observer and a new partial tracking error constraint. *IEEE Trans. Fuzzy Syst.* **2019**, *27*, 2101–2114. [CrossRef]

27. Wei, X.; Dong, L.; Zhang, H.; Hu, X.; Han, J. Adaptive disturbance observer-based control for stochastic systems with multiple heterogeneous disturbances. *Int. J. Robust Nonlinear Control.* **2019**, *29*, 5533–5549. [CrossRef]

28. Gao, Z.; Guo, G. Command-filtered fixed-time trajectory tracking control of surface vehicles based on a disturbance observer. *Int. J. Robust Nonlinear Control.* **2019**, *29*, 4348–4365. [CrossRef]

29. Liu, Z.; Wang, F.; Zhang, Y.; Chen, C.L.P. Fuzzy adaptive quantized control for a class of stochastic nonlinear uncertain systems. *IEEE Trans. Cybern.* **2016**, *46*, 524–534. [CrossRef]

30. Liu, W.; Lim, C.C.; Shi, P.; Xu, S. Backstepping fuzzy adaptive control for a class of quantized nonlinear systems. *IEEE Trans. Fuzzy Syst.* **2017**, *25*, 1090–1101. [CrossRef]

31. Song, S.; Park, J.H.; Zhang, B.; Song, X.; Zhang, Z. Adaptive command filtered neuro-fuzzy control design for fractional-order nonlinear systems with unknown control directions and input quantization. *IEEE Trans. Syst. Man Cybern. Syst.* **2021**, *51*, 7238–7249. [CrossRef]

32. Sui, S.; Chen, C.L.P.; Tong, S.; Feng, S. Finite-time adaptive quantized control of stochastic nonlinear systems with input quantization: A broad learning system based identification method. *Int. J. Robust Nonlinear Control.* **2020**, *67*, 8555–8565. [CrossRef]

33. Xing, L.; Wen, C.; Liu, Z.; Su, H.; Cai, J. Event-triggered adaptive control for a class of uncertain nonlinear system. *IEEE Trans. Autom. Control.* **2017**, *62*, 2071–2076. [CrossRef]

34. Song, X.; Wu, C.; Stojanovic, V.; Song, S. 1 bit encoding–decoding-based event-triggered fixed-time adaptive control for unmanned surface vehicle with guaranteed tracking performance. *Control. Eng. Pract.* **2023**, *135*, 105513. [CrossRef]

35. Hayakawaa, T.; Ishii, H.; Tsumurac, K. Adaptive quantized control for nonlinear uncertain systems. *Syst. Control. Lett.* **2009**, *58*, 625–632. [CrossRef]

36. Song, X.; Wang, M.; Ahn, C.K.; Song, S. Finite-time fuzzy bounded control for semilinear PDE systems with quantized measurements and Markov jump actuator failures. *IEEE Trans. Cybern.* **2009**, *52*, 5732–5743. [CrossRef] [PubMed]

37. Li, Y.; Chen, Y.; Podlubny, I. Mittag-Leffler stability of fractional order nonlinear dynamic systems. *Automatica* **2009**, *45*, 1965–1969. [CrossRef]

38. Li, Y.; Yang, G. Adaptive asymptotic tracking control of uncertain nonlinear systems with input quantization and actuator faults. *Automatica* **2016**, *96*, 23–29. [CrossRef]

39. Ge, Z.M.; Yu, T.C.; Chen, Y.S. Chaos synchronization of a horizontal platform system. *J. Sound Vib.* **2003**, *268*, 731–749. [CrossRef]

40. Aghababa, M.P. Chaotic behavior in fractional-order horizontal platform systems and its suppression using a fractional finite-time control strategy. *J. Mech. Sci. Technol.* **2014**, *28*, 1875–1880. [CrossRef]

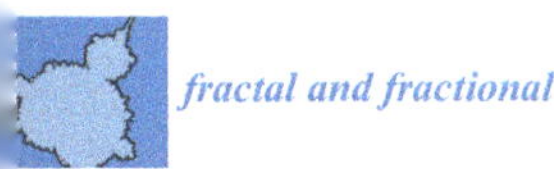
fractal and fractional

Article

Microgrid Frequency Regulation Based on a Fractional Order Cascade Controller

Soroush Oshnoei [1,2], Arman Fathollahi [1], Arman Oshnoei [3,*] and Mohammad Hassan Khooban [1]

[1]   Department of Electrical and Computer Engineering, Aarhus University, 8000 Aarhus, Denmark
[2]   Department of Electrical Engineering, Shahid Beheshti University, Tehran 1983969411, Iran
[3]   Department of Energy (AAU Energy), Aalborg University, 9220 Aalborg, Denmark
*   Correspondence: aros@energy.aau.dk

Abstract: Nowadays, the participation of renewable energy sources (RESs) and the integration of these sources with traditional power plants in microgrids (MGs) for providing demand-side power has rapidly grown. Although the presence of RESs in MGs reduces environmental problems, their high participation significantly affects the system's whole inertia and dynamic stability. This paper focuses on an islanded MG frequency regulation under the high participation of RESs. In this regard, a novel fractional order cascade controller (FOCC) is proposed as the secondary frequency controller. In the proposed FOCC controller structure, a fractional order proportional-integral controller is cascaded with a fractional order tilt-derivative controller. The proposed FOCC controller has a greater degree of freedom and adaptability than integer order controllers and improves the control system's efficiency. The adjustable coefficients of the proposed controller are tuned via the kidney-inspired algorithm. An energy storage system equipped with virtual inertia is also employed to improve the system inertia. The proposed FOCC controller efficiency is compared with proportional-integral-derivative (PID), tilt-integral-derivative (TID), and fractional order proportional-integral-derivative (FOPID) controllers under different disturbances and operating conditions. The results demonstrate that the presented controller provides better frequency responses compared to the other controllers. Moreover, the sensitivity analysis is performed to show the proposed controller robustness versus the parameters' changes in the system.

Keywords: islanded microgrid; frequency regulation; renewable energy sources; fractional order controllers; cascade controller

Citation: Oshnoei, S.; Fathollahi, A.; Oshnoei, A.; Khooban, M.H. Microgrid Frequency Regulation Based on a Fractional Order Cascade Controller. *Fractal Fract.* **2023**, *7*, 343. https://doi.org/10.3390/ fractalfract7040343

Academic Editors: Costas Psychalinos and Da-Yan Liu

Received: 11 December 2022
Revised: 1 April 2023
Accepted: 18 April 2023
Published: 21 April 2023

1. Introduction

Assembling the highest usage of renewable energy sources (RESs) and distributed energy generators (DEGs) has become a worldwide consensus in recent years as the situation surrounding energy security, global warming, and environmental degradation has become more complex [1]. It is now impossible to imagine a future without new power systems that derive a significant portion of their electricity from large-scale RESs such as photovoltaic power plants (PPP) and wind power plants (WPP) [2–4]. In the last decade, the collection and integration of RESs and DEGs in microgrids (MGs) have played an essential role in providing power systems load [5]. Due to information provided by the International Renewable Energy Agency, the generation capacity of RESs across the world reached 2537 GW in recent years. The capacity of hydropower energy sources is estimated to be 1190 GW, assembling them the most extensive share of the global total. This was followed by wind, solar, bioenergy, geothermal, and marine energy resources with the capacity of 623 GW, 586 GW, 124 GW, and 14 GW, respectively. Note that marine energy accounted for the smallest share of the global total, with only 500 MW [6,7].

Traditional generators' frequency modulation capabilities are becoming inadequate as RESs become more integrated into power systems [8,9]. In addition, RESs [10] and energy storage systems (ESSs) [11] are widely used in MGs and power systems to decrease energy consumption and enhance energy utilization efficiency. Although the presence of RESs in electricity grids significantly reduces the concerns related to environmental problems and the lack of fossil fuels, the intermittent nature of these sources affects the stability of grids [12,13]. Hence, ESSs are used in MGs due to the intermittent nature of output power in RESs [14]. Load frequency control (LFC) is a necessary service widely utilized in electricity grids to keep the system frequency in the acceptable range [15]. Since the LFC mechanism has an important impact on the electricity grid's stability, it is essential to study the frequency performance of these grids under the high penetration of RESs.

The MGs operate as part of the upstream utility grid when conditions are standard (grid-connected mode) but can switch to working independently if the utility grid encounters an outage or fault (standalone mode) [16]. One of the significant threats in standalone MGs operating conditions is the output powers of WPPs and PPPs due to the erratic behavior of solar irradiation and wind speed, respectively. These conditions often lead to a discrepancy between supply and demand in MGs [17,18]. As a result, inertia response and frequency regulation may be lost if RESs are widely deployed in MGs. Under these circumstances, even moderate fluctuations in MG frequency may have destructive effects [19]. Therefore, it appears critical to implement appropriate control strategies in isolated MGs to improve the system inertia and frequency regulation.

Numerous control procedures for MGs have been reported in the literature to deal with this challenge. A local controller independently operates local control loops of each distributed generation in an MG at the primary level of hierarchical control [20,21]; the secondary level of control compensates for frequency and voltage changes generated by the primary level of control. Considering its efficiency and comfort of implementation, distributed secondary control has recently gained extensive attention [22]. Moreover, a methodology known as virtual inertia (VI) emulation has been developed to increase the rotational inertia of the MG. Emulating a VI with the help of the derivative technique is one of the more productive manners to go about it [23–25]. In recent years, researchers have been looking into how applying VI in islanded MGs could help improve the MG inertia response and frequency stability [12,26–28].

Nowadays, various control strategies, such as adaptive control [29], fractional order (FO) and integer-order (IO) controllers [16,21,30,31], cascade controllers [32–34], fuzzy-logic techniques [35,36], and H_∞ control theory [37], have been developed in the current works to improve the frequency response of MGs. Robust controller-based linear matrix inequality of an MG with different energy sources is reported in [38]. Ref. [39] proposes a robust control strategy to attain voltage and frequency stability. MG's closed-loop state-space model is elicited after creating a small-signal structure for a single distributed generation unit. The robust stability of the MG is then analyzed using a new Lyapunov–Krasovskii functional. Authors in [40] present a proportional-integral-double derivative controller as the secondary controller of an islanded MG to improve the system frequency regulation. An event-triggered controller regarding input delay and cyber-attack disturbances is proposed in [41] as the secondary frequency controller in an islanded MG. The authors in [42] propose an innovative fuzzy adaptive differential evolution algorithm for modifying a fuzzy logic-based controller to enhance the inertia control in MGs. In [19], the deep deterministic policy gradient is presented as a secondary controller of an island MG. To maintain voltage and frequency stability, the presented controller regulates the power delivered by the storage components. Several analyses employ progressive controlling techniques employing cascaded controllers [43,44] for the improved frequency control process of MGs.

In recent years, the development of science in engineering fields has greatly led to using fractional calculus in control applications. The leading cause for utilizing the IO models was the lack of suitable solution approaches for FO differential equations. FO controllers are dynamic systems modeled by FO differential equations. Numerous physical systems are not modeled with integer-order calculus. This is because their real dynamics include non-integer derivatives. Hence, FO calculus has been introduced to describe such systems precisely. Conventional and IO controllers such as I, PI, and PID are not persistent versus changes in the system coefficients and operating conditions, which causes their performance to be significant. This is a notable flaw of such controllers [16]. In this regard, FO controllers, the FO version of the IO controllers, are introduced to address this problem [31]. The FO controllers have more degrees of freedom and flexibility compared to the IO controllers. In [31], the authors have proposed the hybrid controller—a combination of the tilt-integral-derivative (TID) and FO proportional-integral-derivative (FOPID)—to control the gate-controlled series capacitor installed in the tie-line of a multi-area power grid to enhance the system frequency response. A hybrid controller based on a two-degree-of-freedom design is proposed in [16] as the secondary controller to enhance the frequency stability of a two-area power grid. The authors in [45] have proposed a TID controller for frequency regulation of a power system integrated with WPP. In [46], a secondary PI controller is used to improve the frequency performance of a low-inertia power system in the presence of high penetration of electric vehicles. The authors have not presented any solutions to enhance the system inertia. A FO integral controller has been used in [47] to enhance the frequency stability of an interconnected MG considering the VI equipment. The authors in [48] have proposed a FO controller as the secondary frequency controller of a shipboard MG. In [49], a high-dimensional multiple FO controller has been presented to improve the frequency response of a two-area power system.

The fundamental problem with IO controllers is that constant parameters and no explicit details of the overall procedure and performance are compromised. Additionally, these controllers are linear and exhibit symmetry in special cases. Hence, their performance in nonlinear systems is unstable. Hence, such controllers require to be modified for process control applications. FO derivative is a convolution in FO differential equations. Therefore, it accurately describes the dynamics of inheriting memory and congenital features where IO derivative methods seem to fail. From the control perspective, fractional calculus permits the combination of further degrees of freedom in the control strategies, which can consider more effective limitations when designing the control rules. Fractional calculus can provide appropriate tools when IO calculus fails to perform satisfactorily regarding the design consideration involved. Furthermore, an FO controller is more flexible with a determined complexity growth. Consequently, investigating the control applications according to FO operators can have significant theoretical importance and overall possible significance in addressing the gap between theory and practice.

In addition to changes in the system coefficients and operating conditions, the traditional controllers' effectiveness in today's electricity grids significantly reduces due to the increasing complexities of these grids. To solve this, the cascade controller is introduced as a suitable candidate to improve system control performance. The cascade control concept comes from the control of two sequential procedures, where the inner procedure or output of the first supplies the second or outer procedure in sequence. The rationale behind this configuration is that the fast dynamics of the inner loop enable quick mitigation of perturbation and minimize the possible effects of perturbations before they impact the primary output. Accordingly, cascade controllers are employed in multi-loop control systems to quickly reject disturbance requirements and provide decent set-point tracking [32,33]. In [32], a cascade controller based on the FO and IO controllers is used as the secondary controller of a two-area MG to regulate the system frequency. Ref. [33] has proposed a cascade FO fuzzy PID-IDD controller for effective efficiency of tidal turbines in the LFC task of an island MG. The authors in [50] have proposed a cascade FOPD-PI controller for frequency regulation of an islanded MG without considering VI service. Ref. [51] has studied the LFC issue of a multi-area power system under communication delay and system physical limitations. In this regard, a FOPID controller cascaded with a first-order filter has been presented as the secondary controller.

According to the mentioned explanations, the design of a novel controller based on the FO and cascade controllers (FOCCs) aids in improving the system frequency performance, which is crucial in the hybrid deregulated energy system of MGs. Consequently, the authors are motivated to propose a novel FOCC controller as the secondary controller in an islanded MG. Moreover, a kidney-inspired algorithm (KA) is employed for tuning the optimal coefficients of the suggested controller. Furthermore, to enhance the grid inertia and improve the frequency control problem of the system integrated with RESs, an ESS equipped with VI in coordination with the secondary controller is employed.

The contributions of this study are summarized as: (i) A novel FOCC controller is proposed as the secondary controller in the LFC problem of an islanded MG. In the proposed FOCC controller, FOPI controller is cascaded with a FOTD controller (ii). An ESS equipped with the VI is employed in the studied MG to enhance the total inertia of the system. (iii) The proposed control scheme's dynamic efficiency is compared with PID and TID controllers under various scenarios. (iv) The KA method is employed for optimizing the adjustable coefficients of the presented controllers. (v) The sensitivity analysis is conducted to investigate the presented controller's performance under the changes in the MG system parameters.

The current work is structured as follows: In Section 2, the studied MG modeling is described. Section 3 presents the design of the proposed FOCC controller. In Section 4, the KA optimization method is studied to tune the controller's parameters. The simulation results are illustrated in Section 5. Eventually, the conclusion of this work is provided in Section 6.

2. Microgrid Modeling

This section describes the studied MG modeling. Figure 1 demonstrates the simplified structure of the investigated MG. As illustrated, the MG comprises the different generation units such as a thermal power plant (TPP) with 12 MW, an ESS equipped with the VI with 4 MW, a WPP with 7 MW, and a PPP with 4 MW along with industrial and residential loads with 10 MW and 5 MW, respectively [25]. When a power imbalance occurs between generation power and load, the TPP and ESS receive the frequency deviation signal from the control center and participate in the frequency control problem of the system. Since WPP and PPP do not receive the frequency deviation signal, the output power of these units, together with load changes, are regarded as system disturbances.

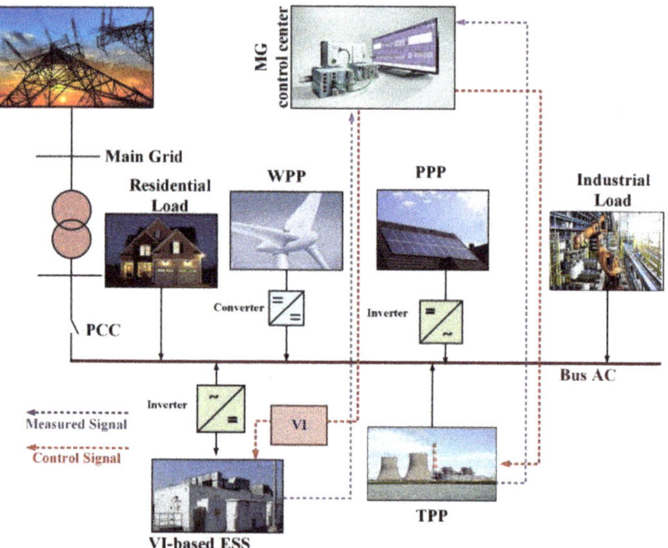

Figure 1. Schematic of the studied MG.

The LFC model of the islanded MG is represented in Figure 2. According to this figure, the TPP unit is composed of a governor, a non-reheat turbine, and a secondary frequency controller. Moreover, the non-linear limitations, such as generation rate constraint (GRC) and governor dead-band (GDB), are also employed in modeling the TPP to attain the real dynamic characteristics of the system. Therefore, the TPP output is restrained by the governor's valve position of 0.002 p.u.MW/s GDB and generator mechanical output of ± 0.5 p.u. MW GRC for both rising and dropping rates. The modeling of the WPP and PPP generation resources is also performed using a first-order transfer function. In this study, a fractional order cascade controller is presented as the secondary frequency controller. The MG frequency deviation considering the generator inertia constant (H) and load damping coefficient (D) is written as follows:

$$\Delta f = \frac{1}{2Hs + D}[\Delta P_{TPP} + \Delta P_{VI} + \Delta P_{WPP} + \Delta P_{PPP} - \Delta P_{Load}] \tag{1}$$

where:

$$\Delta P_{TPP} = \left(\frac{1}{T_t s + 1}\right)\left(\frac{1}{T_g s + 1}\right)\left(\frac{1}{R}\Delta f - \Delta P_{ACE}\right) \tag{2}$$

$$\Delta P_{WPP} = \frac{K_{WPP}}{T_{WPP}s + 1}\Delta P_{wind} \tag{3}$$

$$\Delta P_{PPP} = \frac{K_{PPP}}{T_{PPP}s + 1}\Delta P_{solar} \tag{4}$$

where Δf is the MG frequency deviation, ΔP_{TPP}, ΔP_{VI}, ΔP_{WPP}, and ΔP_{PPP} denote the power changes in TPP, VI-based ESS, WPP, and PPP, respectively; ΔP_{wind} and ΔP_{solar} are the power variations in wind and solar irradiation, respectively; ΔP_{ACE} shows the secondary controller output, and finally, ΔP_{Load} expresses the power changes in the industrial and residential loads.

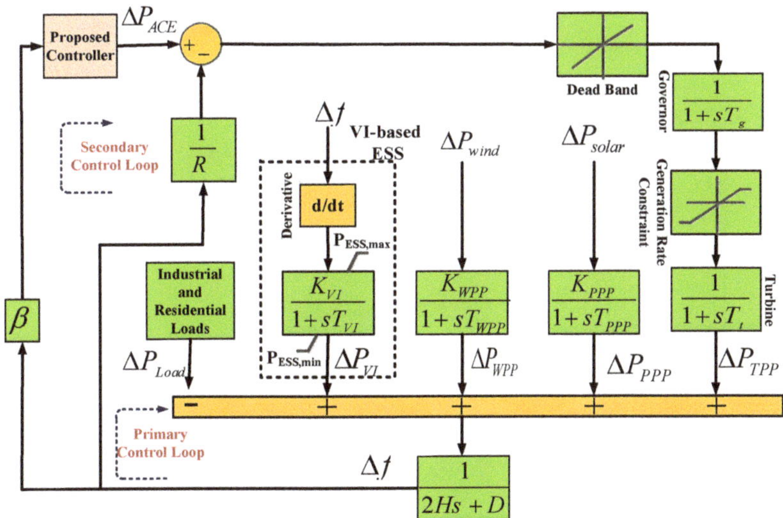

Figure 2. Diagram block of the islanded MG system.

VI is an effective tool to imitate the inertia features of the synchronous generator and enhance the MG frequency stability in the presence of high RESs contribution. The idea of VI is the derivative control that computes the rate-of-change of frequency to regulate the active power to the set point of the MG after the perturbations. In this work, the derivative technique is utilized to emulate the VI concept in the ESS [25]. The VI-based

ESS improves the whole inertia of the system. In addition to the derivative method, a first-order transfer function is also used to imitate the actual dynamic characteristics of the ESS, as demonstrated in Figure 2. Hence, the VI-based ESS can be controlled to provide the necessary active power to the system for enhancing the MG frequency performance. The VI power provided by the ESS ΔP_{VI} is expressed as:

$$\Delta P_{VI} = \frac{K_{ESS}}{T_{ESS}s + 1}\left(\frac{d}{dt}\Delta f\right) \tag{5}$$

Table 1 demonstrates the values and description of the parameters related to the investigated MG [26].

Table 1. Description and values of the MG parameters.

Parameter	Description	Value
H	System inertia (p.u.MW/s)	0.083
D	Load damping coefficient (p.u.MW/Hz)	0.015
T_g	Governor time constant (s)	0.1
T_t	Turbine time constant (s)	0.4
R	Governor droop constant (Hz/p.u.MW)	2.4
B	Frequency bias factor (p.u.MW/Hz)	1
K_{ESS}	ESS gain	0.8
T_{ESS}	ESS time constant (s)	10
K_{WPP}	WPP gain	1
T_{WPP}	WPP time constant (s)	1.5
K_{PPP}	PPP gain	1
T_{PPP}	PPP time constant (s)	1.85

3. Design of the Proposed Fractional Order Cascade Controller

This section describes the design of the presented fractional order cascade controller for the secondary controller to enhance the system's frequency regulation. The idea of the fractional calculus-based cascade controller is known as the FOCC. FO controllers are the FO version of the classical IO controllers [52–54]. These controllers have a greater degree of adaptability and flexibility than IO controllers. They can improve the FO controllers' performance in dealing with oscillations' amplitude and settling time in comparison to the IO controllers. IO controllers are extended to FO controllers utilizing fractional calculus. The generally utilized definitions for fractional derivative and integral are by Riemann–Liouville definitions [30,31] and are expressed by Equations (6) and (7), respectively:

$$_aD_t^\alpha f(t) = \frac{1}{\Gamma(n-\alpha)}\frac{d^n}{dt^n}\int_a^t (t-\tau)^{n-\alpha-1}f(\tau)d\tau \tag{6}$$

$$_aD_t^{-\alpha}f(t) = \frac{1}{\Gamma(\alpha)}\int_a^t (t-\tau)^{\alpha-1}f(\tau)d\tau \tag{7}$$

where $n-1 \leq \alpha < n$, n is an integer, $\Gamma(\cdot)$ shows the Euler's gamma function, and $_aD_t^\alpha f(t)$ denotes the fractional operator. Laplace's transformation of Equation (6) considering zero initial condition is represented as follows:

$$L\{_aD_t^\alpha f(t)\} = s^\alpha F(s) - \sum_{k=0}^{n-1} s^k aD_t^{\alpha-k-1}f(t)|_{t=0} \tag{8}$$

for $n-1 \leq \alpha \leq n$, where $F(s) = L\{f(t)\}$ indicates Laplace's transformation. This paper employs the commande robust d'ordre non-entier (CRONE) approximation, presented by Oustaloup, out of several other approximations [16,21]. CRONE utilizes a recursive distribution of N poles and N zeros, conducting a transfer function during the predefined frequency range $[\omega_l, \omega_h]$ [55–57]. In the simulation process, $\omega_l = 0.01$ rad/s, $\omega_h = 100$ rad/s, and $N = 5$ are presumed.

$$H_f(s) = s^\alpha = K \prod_{n=1}^{N} \frac{1 + (s/\omega_{z,\,n})}{1 + (s/\omega_{p,\,n})} \tag{9}$$

where K shows the tunable gain, $\omega_{z,\,n}$ and $\omega_{p,\,n}$ represent the zeroes and poles of $H_f(s)$, which are computed by (10)–(14).

$$\omega_{z,l} = \omega_l \sqrt{n} \tag{10}$$

$$\omega_{p,n} = \omega_{z,n}\tau, \qquad n = 1,\ldots,l \tag{11}$$

$$\omega_{z,n+1} = \omega_{p,n}\sqrt{n}, \qquad n = 1,\ldots,l-1 \tag{12}$$

$$\tau = (\omega_h/\omega_l)^{\varepsilon/N} \tag{13}$$

$$\sigma = (\omega_n/\omega_l)^{(1-\varepsilon)/N} \tag{14}$$

FOPID and TID controllers are the two well-known FO controllers. Typically, the transfer functions of these controllers are given by (15) and (16), respectively [31].

$$H_{FOPID}(s) = K_P + \frac{K_I}{s^\lambda} + K_D s^\mu \tag{15}$$

$$H_{TID}(s) = K_T s^{(-1/n)} + \frac{K_I}{s} + K_D s \tag{16}$$

where K_I, K_P, K_D, and K_T show the tunable integral, proportional, derivative, and tilt coefficients, respectively. λ and μ are the FO operators of the integral and derivative terms in the FOPID controller, respectively. n indicates the FO operator of the tilt term in the TID controller. λ and μ are adjusted in the range of $(0, 1)$, and n is selected in $(2, 3)$. Similar to controllers FOPID and TID, the transfer functions of the FOPI and FOTD or TD^μ controllers are expressed as:

$$H_{FOPI}(s) = K_P + \frac{K_I}{s^\lambda} \tag{17}$$

$$H_{TD}(s) = K_T s^{(-1/n)} + K_D s^\mu \tag{18}$$

FO controllers are the appropriate candidates to control the system dynamic; accordingly, cascading the FO controllers can significantly improve the system's dynamic efficiency. In this study, cascading the FOPI and TD^μ controllers is proposed as the secondary frequency controller.

The control of two successive processes relates to the idea of cascade control. Cascade control can increase the efficiency of the control system in comparison to single-loop control [33]. Figure 3 demonstrates the structure of the cascade control system. As shown, the cascade control includes two inner and outer control loops, in which the inner loop output provides the second process or input of the outer loop. The inner loop in the cascade control, known as the slave controller, rejects the effects of perturbations in a comparatively quicker procedure before transferring them to different parts of the plant. The inner loop mainly alleviates the impact of deviations related to the internal process coefficients (owing to set-point variations and perturbations) on the control system's efficiency. The outer loop, known as a master controller, handles the final output quality of the process to attain a reference signal $R(s)$ [34].

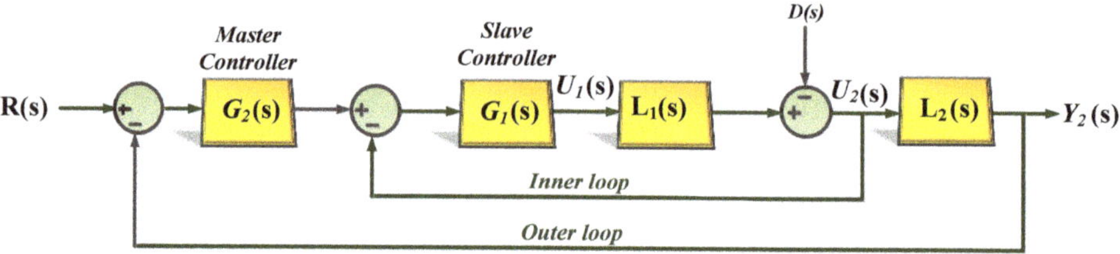

Figure 3. Structure of the cascade controller.

According to Figure 3, $G_1(s)$ and $G_2(s)$ indicate the transfer functions relevant to the slave and master controllers, respectively [58]. Moreover, $L_1(s)$ and $L_2(s)$ are the transfer functions associated with plants of the inner and outer loop, respectively. The system's final output $Y_2(s)$ subjected to load disturbance $D(s)$ is given by (19):

$$Y_2(s) = U_2(s)L_2(s) + D(s) \tag{19}$$

where $U_2(s)$ is the inner loop output or outer loop input. $U_2(s)$ controls $Y(s)$ signal to track $R(s)$. Likewise, the inner loop output $Y_1(s)$ can be obtained as:

$$Y_1(s) = U_2(s) = U_1(s)L_1(s) \tag{20}$$

The proposed control system in this study combines the FOPI&TD$^\mu$ controllers to construct a cascaded system. FOPI controller is cascaded with TD$^\mu$ controller, where FOPI controller forms the master controller $G_2(s)$ and TD$^\mu$ controller the slave one $G_1(s)$.

$$G_2(s) = K_P + \frac{K_I}{s^\lambda} \tag{21}$$

$$G_1(s) = K_T s^{(-1/n)} + K_D s^\mu \tag{22}$$

Consequently, the closed loop transfer function of the cascaded system is represented as follows:

$$Y_2(s) = \left[\frac{L_2(s)L_1(s)G_2(s)G_1(s)}{1 + G_1(s)L_1(s) + L_2(s)L_1(s)G_2(s)G_1(s)}\right]R(s) + \left[\frac{L_2(s)}{1 + G_1(s)L_1(s) + L_2(s)L_1(s)G_2(s)G_1(s)}\right]D(s) \tag{23}$$

Figure 4 illustrates the block diagram of the proposed FOPI and TD$^\mu$ cascade controller. To design the proposed FOCC, the adjustable parameters of the FOPI and TD$^\mu$ controllers are tuned by the KA method.

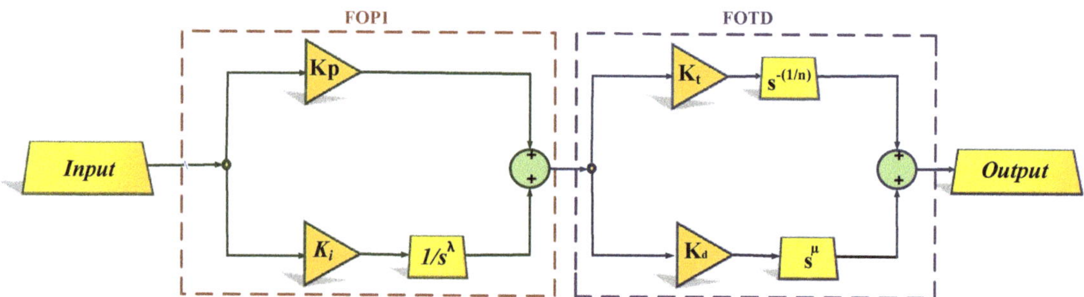

Figure 4. Diagram block of the proposed controller.

4. Optimization Algorithm to Tune the Controller's Coefficients

A new kidney plays a crucial role in the human body by filtering the blood, getting rid of extra fluid and toxins in the form of urine, and regulating the levels of ions in the blood. Since they regulate these processes, they are ultimately in charge of the human body's health. A superior optimization strategy called the KA was invented by exploring and utilizing the features of the human kidney's operating system [59]. In this section, the KA is employed for adjusting the parameters of the suggested FOPI&TD$^\mu$ cascade controller. The four primary components of renal treatments that are mentioned in the imitation are as follows: (1) transferring solutes and water from the blood to the tubules, referred to as filtration; (2) transporting important water and solutes from the tubules to the circulation, referred to as reabsorption; (3) transferring more harmful substances from the circulation to the tubule, known as secretion; and (4) transmitting the toxic substances from the first process stages into the urine, which is called excretion.

A total population of potential solutes is formed based on the real mechanism of kidneys in the KA's first stage. Their goal functions are computed in the same way as in other population-based computational methods. In the biological renal system, each solute can be thought of as water particles and solutes in plasma [60]. By gradually improving upon the best solute found so far, a new solute is generated for all solutes at the end of each cycle. In this algorithm, solute motion is defined as:

$$So_{j+1} = rand\left(So_{best} - So_j\right) + So_j \tag{24}$$

where S denotes a sample solution from the population. At iteration j the answer is So_j. So_{best} is the best solution found by the KA method in prior iterations, while $rand\left(So_{best} - So_j\right)$ is a random number between zero and the provided number.

The filtration operation separates the population's higher-quality solutes into filtered blood (FB), whereas the lower-quality solutes are flushed down the waste (W). A filtering rate is employed in the KA for this purpose and is computed and updated with each iteration. In a way similar to the manner the glomerular filtration rate is determined in a living kidney, the following equations specify the filtration rate (FR) [60,61]:

$$FR = \partial \times \sum_{j}^{n} f\left(y_j\right) / n \tag{25}$$

In this equation, $f\left(y_j\right)$ is the objective function of solution y at iteration j, n is the population size, and ∂ is a constant in the interval (0, 1].

In the KA algorithm, if the solute's quality is better than FR, then it is identified as a member of FB, and if it is lower than FR, then it is taken as a member of W. Solutes that are initially excluded from FB due to the reabsorption operator are given another opportunity to meet the quality criteria for inclusion. This is possible only if, after applying the motion operator in Equation (24) once more, the filtration rate is satisfied. This is analogous to how the kidneys of a living organism recycle healthy molecules back into the bloodstream. If this probability is not satisfied, the solute is extracted from W and substituted by a different solute. Further, the worst solute in FB is secreted (removed) if a solute put to FB is better than the worst in FB after the filtration operation. A solute is secreted if it is not preferable to the worst solute in FB. It is similar to how the kidneys filter dangerous substances out of the blood. The solutes in FB are then ranked so that the best one can be updated [61].

Last but not least, a new population is assembled by merging FB and W, and the filtration rate is consequently modified. This cycle of repetition stops once the termination condition is met. The continuous influx of water and solutes into a biological kidney system's glomerular capillary can be analogized to adding random solutes. The procedure code of the KA algorithm is illustrated in Algorithm 1 [60].

To evaluate the effectiveness of the suggested FOPI&TD$^\mu$ cascade controller, it is essential to define a proper objective function. In this regard, this paper considers the integral of time absolute error (ITAE) as a constrained optimization problem:

Objective Function:

$$Min \quad ITAE = \int_0^{T_{sim}} t.|\Delta f|.dt \tag{26}$$

Decision Variables:

$$K_P,\ K_I,\ K_T,\ K_D,\ \lambda,\ \mu,\ n \tag{27}$$

Subject to:

$$K_{P_{min}} \leq K_P \leq K_{P_{max}},\ K_{I_{min}} \leq K_I \leq K_{I_{max}},\ K_{D_{min}} \leq K_D \leq K_{D_{max}}$$
$$K_{T_{min}} \leq K_T \leq K_{T_{max}},\ 0 \leq \lambda,\ \mu \leq 1,\ 2 \leq n \leq 3 \tag{28}$$

Algorithm 1. The pseudo-code of the KA algorithm

I: set the population
II: evaluate the solute in the papulation
III: set the best solute (So_{best})
IV: set filtration rate (FR, Equation (25))
V: set waste (W)
VI: set filtered blood (FB)
VII: set number of iteration (numofiter)
VIII: **while** (iter < numofiter) **do**
IX: **for** all So_j
X: generate new So_j (Equation (24))
XI: check the So_j using FR
XII: **if** So_j assigned to W
XIV: apply reabsorption and generate So_{new} (Equation (24))
XVI: **if** reabsorption is not satisfied (So_{new} cannot be a part of FB)
XV: remove So_j from W (excretion)
XVII: insert a random S into W to replace So_j
XVIII: **end if**
XIX: So_{new} is reabsorbed
XX: **else**
XXI: **if** it is better than the So_{worst} in FB
XXII: So_{worst} is secreted
XXIII: **else**
XXIV: So_j is secreted
XXV: **end if**
XXIV: **end if**
XXV: **end for**
XXVI: rank the S from FB and update the So_{best}
XXVII: merge W and FB
XXVIII: update filtration rate (Equation (25))
XXIX: **end while**
XXX: return So_{best}

In the considered objective function, the settling time and the fluctuations' amplitude are the criteria that should be improved. The advantages of the ITAE index compared to the integral of absolute error (IAE) and integral of square error (ISE) indices have been presented in [62–64]. Table 2 illustrates the optimal values of the controllers' parameters via the KA method.

Table 2. Optimal values of the controllers' parameters using the KA algorithm.

Controller	K_P	K_I	K_D	K_T	λ	μ	n
PID	0.5	−1.2	0.5	-	-	-	-
TID	-	−1.34	0.67	0.74	-	-	3
FOPID	0.95	−1.75	0.75	-	0.5	0.3	-
Proposed FOCC	−9.5	5.8	0.9	0.5	0.1	0.4	3

5. Simulation Results

This section provides the simulation outcomes of the MG system designed in Section 2 employing the proposed FOCC controller as the secondary controller. The PID, TID, and FOPID controllers are also considered as other comparative control methods to study the effectiveness of the suggested FOCC controller. The presented controller performance is studied under various patterns of solar irradiation, wind speed, and load perturbations. Moreover, the sensitivity analysis is performed to evaluate the presented controller performance against the system's parameter changes. The simulations are accomplished on a system with Intel core 7i, CPU of 2.7 GHz, and 64-bit processor using MATLAB/SIMULINK (R2021b) software.

5.1. Scenario 1

In this scenario, a step industrial load change of 0.15 p.u.MW and a step residential load change of 0.1 p.u.MW are applied to the system at times = 15 and 80 s, respectively, as illustrated in Figure 5a. The patterns of the solar irradiation and wind speed changes are depicted in Figure 5b,c, respectively. The system frequency response related to this scenario is shown in Figure 5d. Concerning this figure, it can be said that the proposed controller presents a better frequency performance from the viewpoints of the lower fluctuations' amplitude and shorter settling time than the other controllers. The performance criteria of the ITAE and ISE related to the considered controllers are demonstrated in Table 3. As indicated, the proposed controller provides the lowest ITAE and ISE values than the other controllers. In addition, Table 4 indicates the mean absolute MG frequency deviation (MAGFD) employing the considered controllers for scenario 1. As depicted, the suggested controller prepares the lowest value compared to other controllers indicating its better efficiency in mitigating the oscillations.

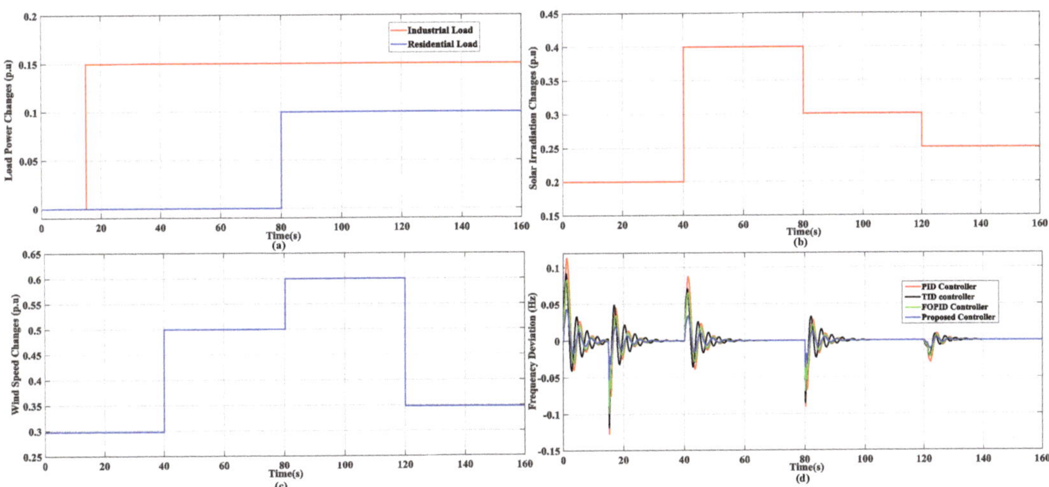

Figure 5. Perturbations and simulation result for scenario 1. (**a**) Load disturbances. (**b**) Wind speed changes. (**c**) Solar irradiation changes. (**d**) Obtained frequency responses.

Table 3. ITAE and ISE performance criteria of the MG frequency deviation employing different controllers.

Index	PID	TID	FOPID	Proposed FOCC
ISE	0.0464	0.0352	0.0285	0.0111
ITAE	0.597	0.472	0.349	0.194

Table 4. Evaluation index of the MG frequency deviation employing different controllers.

	PID	TID	FOPID	Proposed FOCC
Scenario 1	0.0103	0.0097	0.0081	0.005
Scenario 2	0.0541	0.05	0.0425	0.0273
Scenario 3	0.0421	0.038	0.0322	0.0207
Scenario 4	0.0386	0.0348	0.0291	0.0183
Scenario 5	0.0321	0.0283	0.025	0.0152

5.2. Scenario 2

This scenario considers the sequence of step load changes as the load disturbances and random changes for the solar irradiation and wind speed to evaluate the proposed FOCC performance. Figure 6 a–c depicts changes relevant to solar irradiation, wind speed, and load, respectively. The MG frequency response obtained by the presented controllers is shown in Figure 6d. This figure clearly shows that the presented controller provides a better frequency response against the designed disturbances compared to the others. The MAGFD attained by the presented controllers for this scenario is disclosed in Table 4. Clearly, the presented FOCC controller has the lowest value among the controllers and presents a more suitable frequency response than the other controllers.

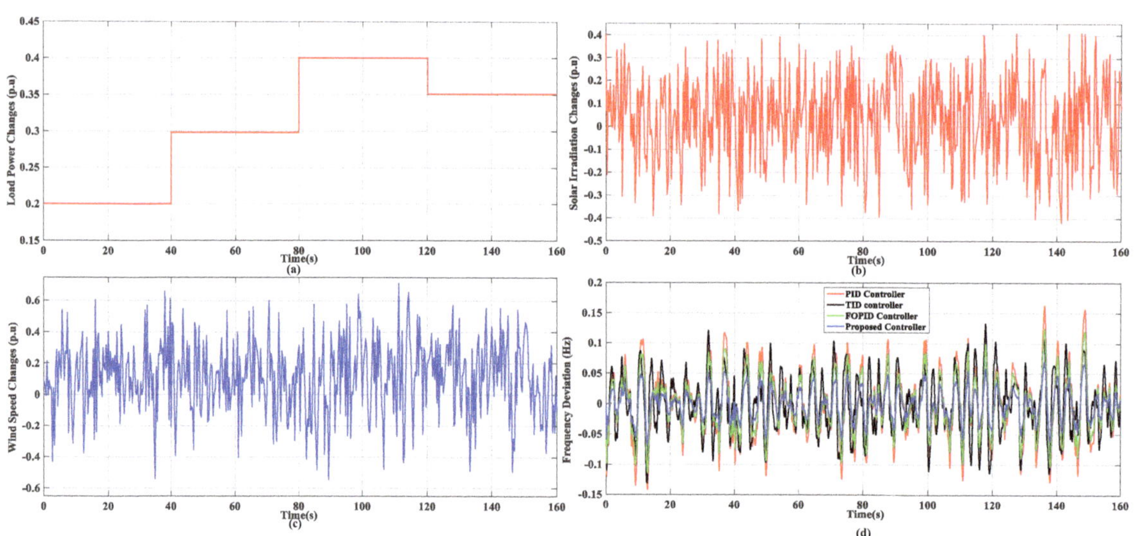

Figure 6. Perturbations and simulation result for scenario 2. (**a**) Load disturbances. (**b**) Wind speed changes. (**c**) Solar irradiation changes. (**d**) Obtained frequency responses.

5.3. Scenario 3

In this scenario, the suggested controller efficiency is investigated under random residential and industrial changes, as shown in Figure 7a. Patterns associated with solar irradiation and wind speed disturbances are presented in Figure 7b,c, respectively. It is evident that the presented FOCC controller provides better handling of the disturbances and improves the system's dynamic performance compared to other controllers. Table 4 indicates the MAGFDs of this scenario. The proposed controller in this scenario has the lowest MAGFD compared to the other controllers, as in the prior scenarios.

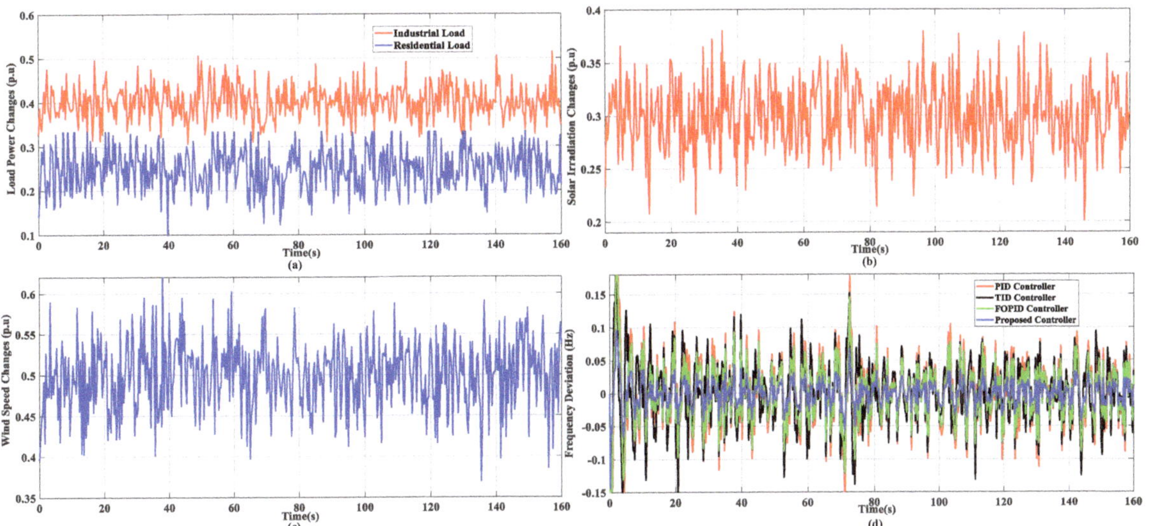

Figure 7. Perturbations and simulation result for scenario 3. (**a**) Load disturbances. (**b**) Wind speed changes. (**c**) Solar irradiation changes. (**d**) Obtained frequency responses.

5.4. Scenario 4

This scenario studies the proposed FOCC efficiency under a 30% reduction in the K_{ESS} value. Figure 8a–c depicts the random perturbations of the loads, solar irradiation, and wind speed, respectively. Figure 8d illustrates the MG frequency response employing the considered controllers. It can be seen that the fluctuations' amplitude is remarkably diminished using the proposed FOCC controller than the other controllers. The MAGFD values related to this scenario are represented in Table 4. It is clear that the proposed controller presents the lowest MAGFD than the other controllers.

5.5. Scenario 5

This scenario evaluates the proposed FOCC performance under a reduction of 50% in the K_{ESS} value to consider more critical conditions. Moreover, severe perturbations, according to Figure 9a–c are also considered as load, solar irradiation, and wind speed, respectively. The system frequency performance for this scenario employing the different controllers is indicated in Figure 9d. As demonstrated, the proposed controller improves the system output from viewpoints of less amplitude oscillations than the other controllers. The MAGFD values of presented controllers related to this scenario are denoted in Table 4. Similar to the prior scenarios, the suggested controller in this scenario also provides the lowest value than the other controllers.

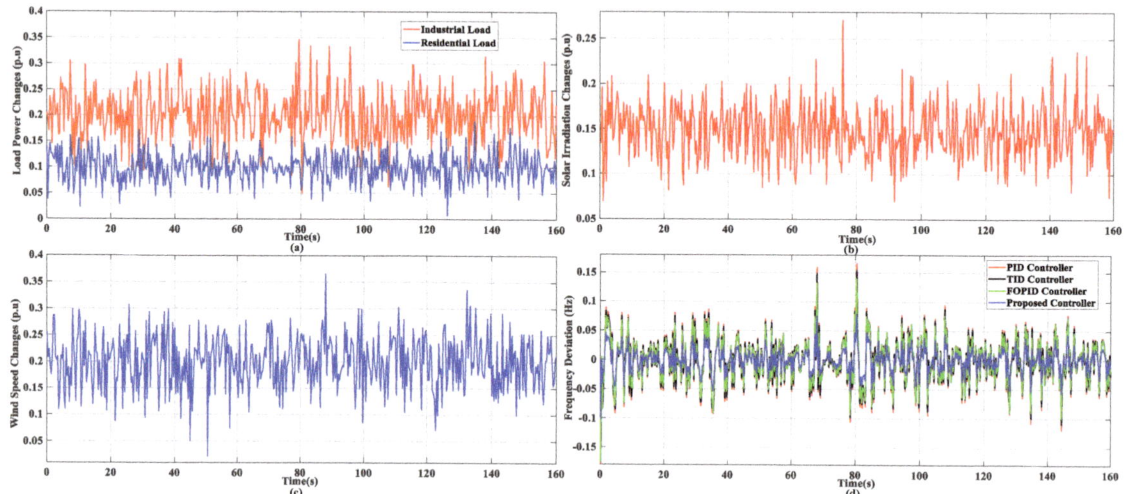

Figure 8. Perturbations and simulation result for scenario 4. (**a**) Load disturbances. (**b**) Wind speed changes. (**c**) Solar irradiation changes. (**d**) Obtained frequency responses.

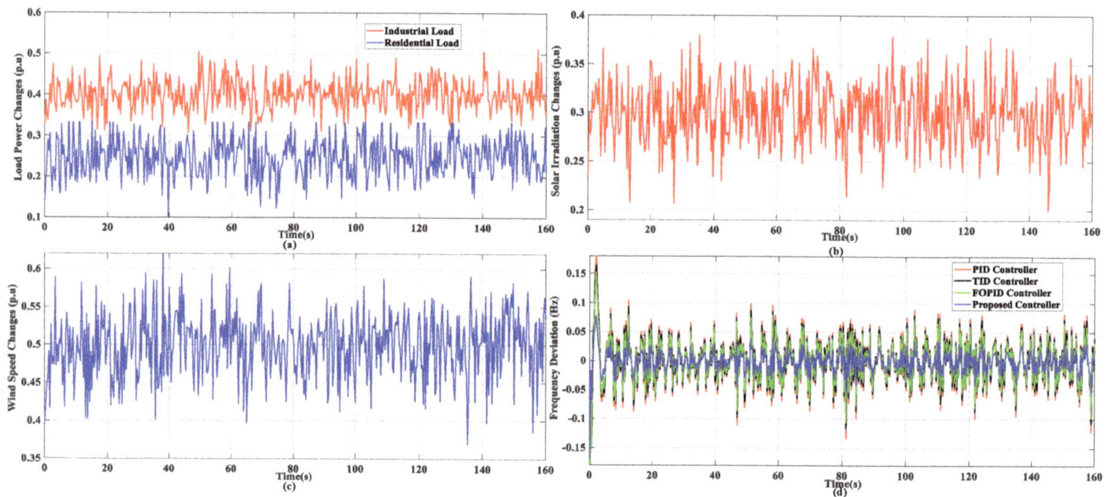

Figure 9. Perturbations and simulation result for scenario 5. (**a**) Load disturbances. (**b**) Wind speed changes. (**c**) Solar irradiation changes. (**d**) Obtained frequency responses.

5.6. Sensitivity Analysis

This subsection analyzes the robustness of the proposed FOCC controller against the system parameters' changes. In this regard, the $\pm30\%$ variations are applied to T_g and T_t parameters under scenario 1 conditions. Figure 10a,b demonstrates the MG frequency responses attained by the proposed FOCC controller under normal conditions and $\pm30\%$ reduction in the T_g and T_t parameters, respectively. According to these figures, it can be noted that the responses under normal conditions and considered changes are almost similar. The MAGFD values for this subsection are presented in Table 5. This table reveals that these changes do not significantly impact the MAGFD values using the proposed FOCC controller. Accordingly, the system stays stable.

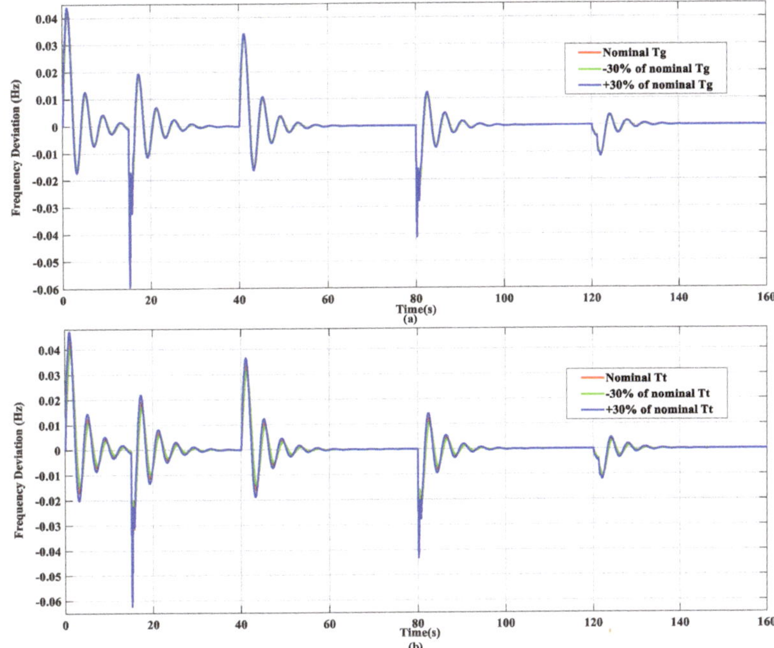

Figure 10. System sensitivity analysis using the presented FOCC controller during ±30% changes in (**a**) T_g (**b**) T_t.

Table 5. Sensitivity analysis for the proposed controller under ±30% changes in Tg and Tt.

Controller	Parameter		MAGFD
Proposed FOCC	T_g	+30%	0.0051
		−30%	0.0045
	T_t	+30%	0.0055
		−30%	0.0043

6. Conclusions

This paper studied the LFC task of an islanded MG in the presence of high participation of RESs. In this regard, an FOCC controller was proposed as the secondary controller to improve the system frequency performance. The proposed FOCC controller has cascaded an FOPI controller with a FOTD controller. An ESS based on the VI control was used to improve the total inertia of the MG. The performance of the suggested FOCC controller was compared with the PID, TID, and FOPID controllers under various perturbations and operating conditions. The tunable parameters of the presented controllers were optimized by the KA method. The results revealed that the presented FOCC controller offers a better frequency response than the other controllers. Eventually, the sensitivity analysis indicated that the suggested FOCC controller is robust versus the coefficients' changes in the system.

Author Contributions: Conceptualization, S.O. and A.F.; methodology, A.O.; software, S.O.; validation, S.O., A.F. and A.O.; formal analysis, M.H.K.; investigation, A.O.; resources, M.H.K.; data curation, A.F.; writing—original draft preparation, S.O.; writing—review and editing, A.F. and A.O.; supervision, M.H.K. All authors have read and agreed to the published version of the manuscript.

Funding: This research received no external funding.

Data Availability Statement: Not applicable.

Conflicts of Interest: The authors declare no conflict of interest.

References

1. Khooban, M.H. An Optimal Non-Integer Model Predictive Virtual Inertia Control in Inverter-Based Modern AC Power Grids-Based V2G Technology. *IEEE Trans. Energy Convers.* **2021**, *36*, 1336–1346. [CrossRef]
2. Zhao, Z.; Guo, J.; Luo, X.; Lai, C.S.; Yang, P.; Lai, L.L.; Li, P.; Guerrero, J.M.; Shahidehpour, M. Distributed Robust Model Predictive Control-Based Energy Management Strategy for Islanded Multi-Microgrids Considering Uncertainty. *IEEE Trans. Smart Grid* **2022**, *13*, 2107–2120. [CrossRef]
3. Bera, A.; Chalamala, B.R.; Byrne, R.H.; Mitra, J. Sizing of Energy Storage for Grid Inertial Support in Presence of Renewable Energy. *IEEE Trans. Power Syst.* **2022**, *37*, 3769–3778. [CrossRef]
4. Fathollahi, A.; Derakhshandeh, S.Y.; Ghiasian, A.; Khooban, M.H. Utilization of dynamic wireless power transfer technology in multi-depot, multi-product delivery supply chain. *Sustain. Energy Grids Netw.* **2022**, *32*, 100836. [CrossRef]
5. Farsizadeh, H.; Gheisarnejad, M.; Mosayebi, M.; Rafiei, M.; Khooban, M.H. An Intelligent and Fast Controller for DC/DC Converter Feeding CPL in a DC Microgrid. *IEEE Trans. Circuits Syst. II Express Briefs* **2020**, *67*, 1104–1108. [CrossRef]
6. Kang, W.; Li, Q.; Gao, M.; Li, X.; Wang, J.; Xu, R.; Chen, M. Distributed secondary control method for islanded microgrids with communication constraints. *IEEE Access* **2017**, *6*, 5812–5821. [CrossRef]
7. Olabi, A.; Abdelkareem, M.A. Renewable energy and climate change. *Renew. Sustain. Energy Rev.* **2022**, *158*, 112111. [CrossRef]
8. Ranjan, M.; Shankar, R. A literature survey on load frequency control considering renewable energy integration in power system: Recent trends and future prospects. *J. Energy Storage* **2022**, *45*, 103717. [CrossRef]
9. Zuo, Y.; Yuan, Z.; Sossan, F.; Zecchino, A.; Cherkaoui, R.; Paolone, M. Performance assessment of grid-forming and grid-following converter-interfaced battery energy storage systems on frequency regulation in low-inertia power grids. *Sustain. Energy Grids Netw.* **2021**, *27*, 100496. [CrossRef]
10. Oshnoei, S.; Aghamohammadi, M.; Oshnoei, S. A Novel Fractional order controller based on Fuzzy Logic for Regulating the Frequency of an Islanded Microgrid. In Proceedings of the 2019 International Power System Conference (PSC), Tehran, Iran, 9–11 December 2019; pp. 320–326.
11. Oshnoei, A.; Kheradmandi, M.; Muyeen, S.M. Robust Control Scheme for Distributed Battery Energy Storage Systems in Load Frequency Control. *IEEE Trans. Power Syst.* **2020**, *35*, 4781–4791. [CrossRef]
12. Kerdphol, T.; Rahman, F.S.; Watanabe, M.; Mitani, Y.; Turschner, D.; Beck, H.-P. Enhanced virtual inertia control based on derivative technique to emulate simultaneous inertia and damping properties for microgrid frequency regulation. *IEEE Access* **2019**, *7*, 14422–14433. [CrossRef]
13. Shahgholian, G.; Mardani, E.; Fattollahi, A. Impact of PSS and STATCOM devices to the dynamic performance of a multi-machine power system. *Eng. Technol. Appl. Sci. Res.* **2017**, *7*, 2113–2117. [CrossRef]
14. Habibi, M.; Oshnoei, A.; Vahidinasab, V.; Oshnoei, S. Allocation and sizing of energy storage system considering wind uncertainty: An approach based on stochastic SCUC. In Proceedings of the 2018 Smart Grid Conference (SGC), Sanandaj, Iran, 28–29 November 2018; pp. 1–6.
15. Khezri, R.; Oshnoei, A.; Oshnoei, S.; Bevrani, H.; Muyeen, S. An intelligent coordinator design for GCSC and AGC in a two-area hybrid power system. *Appl. Soft Comput.* **2019**, *76*, 491–504. [CrossRef]
16. Oshnoei, S.; Oshnoei, A.; Mosallanejad, A.; Haghjoo, F. Novel load frequency control scheme for an interconnected two-area power system including wind turbine generation and redox flow battery. *Int. J. Electr. Power Energy Syst.* **2021**, *130*, 107033. [CrossRef]
17. Fathollahi, A.; Kargar, A.; Derakhshandeh, S.Y. Enhancement of power system transient stability and voltage regulation performance with decentralized synergetic TCSC controller. *Int. J. Electr. Power Energy Syst.* **2022**, *135*, 107533. [CrossRef]
18. Abubakr, H.; Guerrero, J.M.; Vasquez, J.C.; Mohamed, T.H.; Mahmoud, K.; Darwish, M.M.F.; Dahab, Y.A. Adaptive LFC Incorporating Modified Virtual Rotor to Regulate Frequency and Tie-Line Power Flow in Multi-Area Microgrids. *IEEE Access* **2022**, *10*, 33248–33268. [CrossRef]
19. Barbalho, P.I.N.; Lacerda, V.A.; Fernandes, R.A.S.; Coury, D.V. Deep reinforcement learning-based secondary control for microgrids in islanded mode. *Electr. Power Syst. Res.* **2022**, *212*, 108315. [CrossRef]
20. Guerrero, J.M.; Vasquez, J.C.; Matas, J.; De Vicuña, L.G.; Castilla, M. Hierarchical control of droop-controlled AC and DC microgrids—A general approach toward standardization. *IEEE Trans. Ind. Electron.* **2010**, *58*, 158–172. [CrossRef]
21. Oshnoei, S.; Aghamohammadi, M.; Oshnoei, S.; Oshnoei, A.; Mohammadi-Ivatloo, B. Provision of Frequency Stability of an Islanded Microgrid Using a Novel Virtual Inertia Control and a Fractional Order Cascade Controller. *Energies* **2021**, *14*, 4152. [CrossRef]
22. Kanellos, F.D. Real-time control based on multi-agent systems for the operation of large ports as prosumer microgrids. *IEEE Access* **2017**, *5*, 9439–9452. [CrossRef]
23. Fathi, A.; Shafiee, Q.; Bevrani, H. Robust frequency control of microgrids using an extended virtual synchronous generator. *IEEE Trans. Power Syst.* **2018**, *33*, 6289–6297. [CrossRef]

24. Fini, M.H.; Golshan, M.E.H. Determining optimal virtual inertia and frequency control parameters to preserve the frequency stability in islanded microgrids with high penetration of renewables. *Electr. Power Syst. Res.* **2018**, *154*, 13–22. [CrossRef]

25. Kerdphol, T.; Rahman, F.S.; Mitani, Y.; Watanabe, M.; Küfeoğlu, S.K. Robust virtual inertia control of an islanded microgrid considering high penetration of renewable energy. *IEEE Access* **2017**, *6*, 625–636. [CrossRef]

26. Kerdphol, T.; Rahman, F.S.; Watanabe, M.; Mitani, Y. Robust virtual inertia control of a low inertia microgrid considering frequency measurement effects. *IEEE Access* **2019**, *7*, 57550–57560. [CrossRef]

27. Zhang, X.; Zhu, Z.; Fu, Y.; Shen, W. Multi-objective virtual inertia control of renewable power generator for transient stability improvement in interconnected power system. *Int. J. Electr. Power Energy Syst.* **2020**, *117*, 105641. [CrossRef]

28. Ali, H.; Magdy, G.; Li, B.; Shabib, G.; Elbaset, A.A.; Xu, D.; Mitani, Y. A new frequency control strategy in an islanded microgrid using virtual inertia control-based coefficient diagram method. *IEEE Access* **2019**, *7*, 16979–16990. [CrossRef]

29. Abubakr, H.; Vasquez, J.C.; Hassan Mohamed, T.; Guerrero, J.M. The concept of direct adaptive control for improving voltage and frequency regulation loops in several power system applications. *Int. J. Electr. Power Energy Syst.* **2022**, *140*, 108068. [CrossRef]

30. Oshnoei, A.; Khezri, R.; Muyeen, S.; Oshnoei, S.; Blaabjerg, F. Automatic generation control incorporating electric vehicles. *Electr. Power Compon. Syst.* **2019**, *47*, 720–732. [CrossRef]

31. Oshnoei, S.; Oshnoei, A.; Mosallanejad, A.; Haghjoo, F. Contribution of GCSC to regulate the frequency in multi-area power systems considering time delays: A new control outline based on fractional order controllers. *Int. J. Electr. Power Energy Syst.* **2020**, *123*, 106197. [CrossRef]

32. Peddakapu, K.; Srinivasarao, P.; Mohamed, M.R.; Arya, Y.; Krishna Kishore, D.J. Stabilization of frequency in Multi-Microgrid system using barnacle mating Optimizer-based cascade controllers. *Sustain. Energy Technol. Assess.* **2022**, *54*, 102823. [CrossRef]

33. Singh, K.; Arya, Y. Tidal turbine support in microgrid frequency regulation through novel cascade Fuzzy-FOPID droop in de-loaded region. *ISA Trans.* **2022**, *133*, 218–232. [CrossRef] [PubMed]

34. Dash, P.; Saikia, L.C.; Sinha, N. Automatic generation control of multi area thermal system using Bat algorithm optimized PD–PID cascade controller. *Int. J. Electr. Power Energy Syst.* **2015**, *68*, 364–372. [CrossRef]

35. Aziz, S.; Wang, H.; Liu, Y.; Peng, J.; Jiang, H. Variable universe fuzzy logic-based hybrid LFC control with real-time implementation. *IEEE Access* **2019**, *7*, 25535–25546. [CrossRef]

36. Pothiya, S.; Ngamroo, I. Optimal fuzzy logic-based PID controller for load–frequency control including superconducting magnetic energy storage units. *Energy Convers. Manag.* **2008**, *49*, 2833–2838. [CrossRef]

37. Tian, E.; Peng, C. Memory-based event-triggering H∞ load frequency control for power systems under deception attacks. *IEEE Trans. Cybern.* **2020**, *50*, 4610–4618. [CrossRef]

38. Pandey, S.K.; Mohanty, S.R.; Kishor, N.; Catalão, J.P. Frequency regulation in hybrid power systems using particle swarm optimization and linear matrix inequalities based robust controller design. *Int. J. Electr. Power Energy Syst.* **2014**, *63*, 887–900. [CrossRef]

39. Sun, W.; Fang, Z.; Huang, L.; Li, Q.; Li, W.; Xu, X. Distributed robust secondary control of islanded microgrid with stochastic time-varying delays and external disturbances. *Int. J. Electr. Power Energy Syst.* **2022**, *143*, 108448. [CrossRef]

40. Mohanty, B.; Hota, P.K. Comparative performance analysis of fruit fly optimisation algorithm for multi-area multi-source automatic generation control under deregulated environment. *IET Gener. Transm. Distrib.* **2015**, *9*, 1845–1855. [CrossRef]

41. Khalili, J.; Dehkordi, N.M.; Hamzeh, M. Distributed event-triggered secondary frequency control of islanded AC microgrids under cyber attacks with input time delay. *Int. J. Electr. Power Energy Syst.* **2022**, *143*, 108506. [CrossRef]

42. Chamorro, H.R.; Riaño, I.; Gerndt, R.; Zelinka, I.; Gonzalez-Longatt, F.; Sood, V.K. Synthetic inertia control based on fuzzy adaptive differential evolution. *Int. J. Electr. Power Energy Syst.* **2019**, *105*, 803–813. [CrossRef]

43. Veerasamy, V.; Wahab, N.I.A.; Ramachandran, R.; Othman, M.L.; Hizam, H.; Irudayaraj, A.X.R.; Guerrero, J.M.; Kumar, J.S. A Hankel matrix based reduced order model for stability analysis of hybrid power system using PSO-GSA optimized cascade PI-PD controller for automatic load frequency control. *IEEE Access* **2020**, *8*, 71422–71446. [CrossRef]

44. Smith, E.J.; Robinson, D.A.; Agalgaonkar, A.P. A secondary strategy for unbalance consensus in an islanded voltage source converter-based microgrid using cooperative gain control. *Electr. Power Syst. Res.* **2022**, *210*, 108097. [CrossRef]

45. Daraz, A.; Malik, S.A.; Azar, A.T.; Aslam, S.; Alkhalifah, T.; Alturise, F. Optimized Fractional Order Integral-Tilt Derivative Controller for Frequency Regulation of Interconnected Diverse Renewable Energy Resources. *IEEE Access* **2022**, *10*, 43514–43527. [CrossRef]

46. Tripathi, S.; Singh, V.P.; Kishor, N.; Pandey, A. Load frequency control of power system considering electric Vehicles' aggregator with communication delay. *Int. J. Electr. Power Energy Syst.* **2023**, *145*, 108697. [CrossRef]

47. Zaid, S.A.; Bakeer, A.; Magdy, G.; Albalawi, H.; Kassem, A.M.; El-Shimy, M.E.; AbdelMeguid, H.; Manqarah, B. A New Intelligent Fractional-Order Load Frequency Control for Interconnected Modern Power Systems with Virtual Inertia Control. *Fractal Fract.* **2023**, *7*, 62. [CrossRef]

48. Bahrampour, E.; Dehghani, M.; Asemani, M.H.; Abolpour, R. Load frequency fractional-order controller design for shipboard microgrids using direct search alghorithm. *IET Renew. Power Gener.* **2023**, *17*, 894–906. [CrossRef]

49. Yin, L.; Cao, X.; Chen, L. High-dimensional Multiple Fractional Order Controller for Automatic Generation Control and Automatic Voltage Regulation. *Int. J. Control Autom. Syst.* **2022**, *20*, 3979–3995. [CrossRef]

50. Sahoo, G.; Sahu, R.K.; Samal, N.R.; Panda, S. Analysis of type-2 fuzzy fractional-order PD-PI controller for frequency stabilisation of the micro-grid system with real-time simulation. *Int. J. Sustain. Energy* **2022**, *41*, 412–433. [CrossRef]

51. Kumar, A.; Pan, S. Design of fractional order PID controller for load frequency control system with communication delay. *ISA Trans.* **2022**, *129*, 138–149. [CrossRef]
52. Warrier, P.; Shah, P. Fractional order control of power electronic converters in industrial drives and renewable energy systems: A review. *IEEE Access* **2021**, *9*, 58982–59009. [CrossRef]
53. Tepljakov, A.; Alagoz, B.B.; Yeroglu, C.; Gonzalez, E.A.; Hosseinnia, S.H.; Petlenkov, E.; Ates, A.; Cech, M. Towards industrialization of FOPID controllers: A survey on milestones of fractional-order control and pathways for future developments. *IEEE Access* **2021**, *9*, 21016–21042. [CrossRef]
54. Muresan, C.I.; Birs, I.; Ionescu, C.; Dulf, E.H.; De Keyser, R. A review of recent developments in autotuning methods for fractional-order controllers. *Fractal Fract.* **2022**, *6*, 37. [CrossRef]
55. Podlubny, I. *Chapter 10—Survey of Applications of the Fractional Calculus*; Fractional Differential Equations; Elsevier: Amsterdam, The Netherlands, 1999; Volume 198.
56. Aboelela, M.A.; Hennas, R.H.M. Development of a Fractional-Order PID Controller Using Adaptive Weighted PSO and Genetic Algorithms with Applications. In *Fractional Order Systems*; Elsevier: Amsterdam, The Netherlands, 2018; pp. 511–551.
57. Thakar, U.; Joshi, V.; Mehta, U.; Vyawahare, V.A. Fractional-order PI controller for permanent magnet synchronous motor: A design-based comparative study. In *Fractional Order Systems*; Elsevier: Amsterdam, The Netherlands, 2018; pp. 553–578.
58. Shahgholian, G.; Fattollahi, A. Improving power system stability using transfer function: A comparative analysis. *Eng. Technol. Appl. Sci. Res.* **2017**, *7*, 1946–1952. [CrossRef]
59. Hall, J.E. *Pocket Companion to Guyton & Hall Textbook of Medical Physiology E-Book*; Elsevier Health Sciences: Amsterdam, The Netherlands, 2015.
60. Ekinci, S.; Demiroren, A.; Hekimoglu, B. Parameter optimization of power system stabilizers via kidney-inspired algorithm. *Trans. Inst. Meas. Control* **2019**, *41*, 1405–1417. [CrossRef]
61. Jaddi, N.S.; Alvankarian, J.; Abdullah, S. Kidney-inspired algorithm for optimization problems. *Commun. Nonlinear Sci. Numer. Simul.* **2017**, *42*, 358–369. [CrossRef]
62. Idir, A.; Kidouche, M.; Bensafia, Y.; Khettab, K.; Tadjer, S.A. Speed control of DC motor using PID and FOPID controllers based on differential evolution and PSO. *Int. J. Intell. Eng. Syst* **2018**, *20*, 21. [CrossRef]
63. Shabani, H.; Vahidi, B.; Ebrahimpour, M. A robust PID controller based on imperialist competitive algorithm for load-frequency control of power systems. *ISA Trans.* **2013**, *52*, 88–95. [CrossRef]
64. Bruni, R.; Celani, F. Combining global and local strategies to optimize parameters in magnetic spacecraft control via attitude feedback. *J. Optim. Theory Appl.* **2019**, *181*, 997–1014. [CrossRef]

fractal and fractional

Article

First-of-Its-Kind Frequency Enhancement Methodology Based on an Optimized Combination of FLC and TFOIDFF Controllers Evaluated on EVs, SMES, and UPFC-Integrated Smart Grid

Sultan Alghamdi [1], Mohammed Alqarni [2], Muhammad R. Hammad [3] and Kareem M. AboRas [3,*]

[1] Department of Electrical and Computer Engineering, Faculty of Engineering, King Abdulaziz University, Jeddah 21589, Saudi Arabia; smalgamdi1@kau.edu.sa
[2] Department of Electrical Engineering, University of Business and Technology, Ar Rawdah, Jeddah 23435, Saudi Arabia; m.alqarni@ubt.edu.sa
[3] Department of Electrical Power and Machines, Faculty of Engineering, Alexandria University, Alexandria 21544, Egypt; muhammad.ragab@alexu.edu.eg
* Correspondence: kareem.aboras@alexu.edu.eg

Citation: Alghamdi, S.; Alqarni, M.; Hammad, M.R.; AboRas, K.M. First-of-Its-Kind Frequency Enhancement Methodology Based on an Optimized Combination of FLC and TFOIDFF Controllers Evaluated on EVs, SMES, and UPFC-Integrated Smart Grid. *Fractal Fract.* **2023**, *7*, 807. https://doi.org/10.3390/fractalfract7110807

Academic Editors: Arman Oshnoei and Behnam Mohammadi-Ivatloo

Received: 22 October 2023
Revised: 2 November 2023
Accepted: 4 November 2023
Published: 6 November 2023

Abstract: The most recent advancements in renewable energy resources, as well as their broad acceptance in power sectors, have created substantial operational, security, and management concerns. As a result of the continual decrease in power system inertia, it is critical to maintain the normal operating frequency and reduce tie-line power changes. The preceding issues sparked this research, which proposes the Fuzzy Tilted Fractional Order Integral Derivative with Fractional Filter (FTFOIDFF), a unique load frequency controller. The FTFOIDFF controller described here combines the benefits of tilt, fuzzy logic, FOPID, and fractional filter controllers. Furthermore, the prairie dog optimizer (PDO), a newly developed metaheuristic optimization approach, is shown to efficiently tune the suggested controller settings as well as the forms of the fuzzy logic membership functions in the two-area hybrid power grid investigated in this paper. When the PDO results are compared to those of the Seagull Optimization Algorithm, the Runge Kutta optimizer, and the Chaos Game Optimizer for the same hybrid power system, PDO prevails. The system model incorporates physical constraints such as communication time delays and generation rate constraints. In addition, a unified power flow controller (UPFC) is put in the tie-line, and SMES units have been planned in both regions. Furthermore, the contribution of electric vehicles (EVs) is considered in both sections. The proposed PDO-based FTFOIDFF controller outperformed many PDO-based traditional (such as proportional integral derivative (PID), proportional integral derivative acceleration (PIDA), and TFOIDFF) and intelligent (such as Fuzzy PID and Fuzzy PIDA) controllers from the literature. The suggested PDO-based FTFOIDFF controller has excellent performance due to the usage of various load patterns such as step load perturbation, multi-step load perturbation, random load perturbation, random sinusoidal load perturbation, and pulse load perturbation. Furthermore, a variety of scenarios have been implemented to demonstrate the advantageous effects that SMES, UPFC, and EV units have on the overall performance of the system. The sensitivity of a system is ascertained by modifying its parameters from their standard configurations. According to the simulation results, the suggested PDO-based FTFOIDFF controller can improve system stability despite the multiple difficult conditions indicated previously. According to the MATLAB/Simulink data, the proposed method decreased the total fitness function to 0.0875, representing a 97.35% improvement over PID, 95.84% improvement over PIDA, 92.45% improvement over TFOIDFF, 83.43% improvement over Fuzzy PID, and 37.9% improvement over Fuzzy PIDA.

Keywords: prairie dog optimizer; load frequency control; renewable energy sources; fractional-order controllers; fuzzy logic control; sustainable

1. Introduction

1.1. Motivation and Incitement

The provision of a continuous supply of electrical power of a quality that is deemed satisfactory to each and every customer inside a power system should be the primary goal of any power system utility. The electric grid will be in equilibrium when there is a balance between the quantity of electrical power that is needed and the amount that is generated. To obtain a desirable profile of voltage (reactive power balance) and desirable frequency ranges (actual power balance), two basic control strategies are used. The first is known as an automatic voltage regulator (AVR), while the second can either be referred to as an automatic generation control (AGC) or a load frequency controller (LFC) [1]. In a power system that is linked, the objective of LFC is to minimize the transient variations in area frequency and tie-line power exchange while also ensuring that their steady-state errors are zero [2]. Induction motors and transformers are susceptible to experiencing increased magnetizing currents if the frequency experiences a significant dip. The widespread usage of electric clocks and the use of frequency for many other timing applications both necessitate the correct preservation of synchronous time, which must be proportional to frequency and must also incorporate the integral of that variable. The loads are shifted about in a haphazard and fleeting manner by the consumers of electric power. As a direct consequence of this, mismatches between generation and load occur all of a sudden. Due to the mismatch, power is pulled from or fed into the rotor, which results in a change in the generator speed and, as a consequence, the frequency of the system (since frequency is strongly connected to the generator speed). Without proper control, it is difficult to keep the generation and load balances in the correct proportions. Therefore, a control mechanism is needed to mitigate the effects of unpredictable shifts in load and to keep the frequency at the required value. Continuously regulating the active power output of the generator in order to correspond with the randomly variable load is the responsibility of the Load frequency control loop [3].

In a power system that is practically integrated, the generation often consists of a mix of several types of power generation, including thermal, hydro, nuclear, and gas. Nevertheless, due to the tremendous efficiency of nuclear facilities, they are often maintained at a base load that is rather near to their maximum output. The generation of electricity from gas is ideally suited for addressing the varied load demand. Consequently, gas plants are only utilised to satisfy peak demand [4,5]. Therefore, traditional nonrenewable resources were the most common kind of installation in the energy industry. Worries are shifting, however, away from these sources of electricity due to their scarcity and the negative effects they have on the environment [6]. These worries centre on the installation of sources of power that are based on renewable energy. Therefore, a greater emphasis must be placed on sustainable development in order to replace non-renewable sources with renewable energy sources (RESs), such as wind generation, solar photovoltaic (PV) generation, bio-diesel production, and so on. In addition, the improvement of power grids that are based on RES by making use of energy storage devices, the cooperative functioning of electric vehicles (EVs) that have been installed, and other similar activities have garnered significant interest from researchers, industry, and government laws and incentives. They have the potential to contribute to the maintenance of power networks' resilience and dependability [7]. In addition, an improvement in the overall efficiency of power grids is attainable by utilizing modern single- and multi-objective optimisation techniques, such as the robust optimisation methods [8] and the stochastic optimisation methods [9].

1.2. Literature Review

Intermittency, inconstant loading profiles, lower inertia, and other issues are only some of the obstacles that power systems based on RES must contend with. The connectivity of electricity grids that are powered by RES is beneficial in a number of different ways. Nevertheless, renewable energy sources (RES) contribute to the development of fragile power networks that have an unstable reaction to disturbance [10]. When compared to typical grids based on non-renewable sources of energy, renewable-based grids have a far lower inertial response, which is the primary cause of grid instability. Since PV and wind generation are coupled with power interface converters, they are unable to withstand a considerable inertial response, which restricts their capacity for balancing power needs [11]. A low inertial response causes significant instability in power grids, which reduces the controllability of frequency deviations in power grids that are based on RES [12]. This instability may be mitigated by increasing the penetration level of RES.

When the governor system is incapable to cope with frequency changes owing to its slow reaction, active power sources with rapid reactions, such (SMES), are very beneficial in boosting the responsiveness of a system [13,14]. The efficiency of tiny (MES) units, including superconducting and conventional loss varieties, for load frequency regulation is investigated in [15]. There are suggestions on how to make the most of these units' limited energy storage capacity in order to improve the responsiveness of extensive power systems. Since the SMES unit is capable of simultaneously managing both active and reactive powers [16], it is one of the most powerful and critical stabilizers of frequency oscillations. This is due to the fact that frequency oscillations may be quite dangerous. There have been reports in the published literature [17,18] that discuss the practicability of using SMES to improve load frequency performance. Recent developments in power electronics have resulted in the creation of controllers known as flexible alternating current transmission systems (FACTS), which are used in power systems. The performance of a power system can be improved by utilizing FACTS controllers due to their ability to quickly manage the state of the network and their potential to be utilised in this capacity. The FACTS family includes the unified power flow controller (UPFC), which has several characteristics that may be used in a variety of contexts. In addition to managing the UPFC bus voltage and shunt reactive power, UPFC, which comprises of a series and shunt converter coupled by a common dc link capacitor, can also regulate the actual and active power flow in transmission lines [19]. This is made possible by the fact that UPFC is constructed from a series and shunt converter. It has been observed in the literature [20,21] that the effect of various FACTS controllers, such as static synchronous series compensator (SSSC) and thyristor controlled phase shifter (TCPS), working in conjunction with SMES for AGC can have a significant influence. In light of the aforementioned, the LFC analysis presented in this work was carried out while SMES and UPFC were also present.

Power grids that are based on RES have the potential to improve their overall performance if improved and optimized control mechanisms are used [22,23]. LFC has seen widespread adoption as a solution to the frequency variation issues plaguing power networks that are based on RES. The control of generated power is the responsibility of LFCs in order to mitigate loading variations, incompatible parameters, the changing nature of RESs, disruptions, and other similar occurrences [24]. The resistance of power grids to disturbances is directly proportional to the kind of LFC method that is put into practice. In addition, the power grid responsiveness as well as the complexity of the design process are both determined by the correct optimal LFC design technique [25].

For the purpose of load frequency regulation in power systems, a variety of control structures have been discussed and described in the aforementioned literature. First, standard integral order controllers such as PI and PID controllers, which are known for their ease of use, have been adapted for use as load frequency controllers [26]. As a direct consequence of this, soft computing technologies have been included into the process of tuning various controllers. Multiple optimisation strategies, such as particle swarm optimisation [27], whale optimisation algorithm [28], cuckoo search algorithm [29], marine predators algorithm [30], sine cosine adopted dingo optimisation algorithm [31], colliding bodies optimizer [32], cohort intelligence optimization [33], and sea horse optimizer [34] have been used to the design of LFC in order to conduct more research and development on the technology. Using PID controllers, several LFC strategies have been suggested as potential solutions for multi-area linked power systems. These methodologies were important contributors to the preliminary phase of the deregulated LFC operation. Contrarily, it has been noticed that the majority of research have focused their emphasis on LFC difficulties unique to the conventional power system. Furthermore, the significant penetration of RES may bring up a number of challenges, such as voltage instability, poor power quality, frequency variation, and reliability issues.

Regrettably, traditional PID controllers have a number of shortcomings when it comes to coping with system uncertainties such as the fluctuation of RES [35]. As a result, various adjustments have been made in order to enhance the functionality of conventional PID controllers such as the tilt-integral derivative (TID) and the (FOPID) controllers. These improvements have been made in order to increase performance. Calculus using fractions serves as the foundation for these controllers [36]. The efficacy of fractional controllers has been demonstrated by the empirical investigations conducted by researchers [37–39], whereby they have applied mathematical principles to actual scenarios. Several other methods of optimization, including as the differential evolution algorithm [40] and the performance index approach [41], have been utilised in the process of designing the LFC for a multi-area power system utilizing the TID compensator as the primary component. In addition, in comparison to the PID controller, the TID controller possesses the benefits of having a greater disturbance rejection ratio, easier tuning, and fewer impacts of system parameter modification on the system response. All of these advantages make the TID controller the superior choice. In addition, the standard PID controller may be made more efficient by using the FOPID controller since the FOPID controller provides a higher number of degrees of freedom [42]. The genetic algorithm [43], particle swarm optimisation [44], hybrid moth flame optimisation with generalized Hopfield neural network [45], and pollination algorithm [46] have all been utilised in the process of fine-tuning the LFC-based FOPID controller. In comparison to the PID controller, the FOPID has delivered dynamic specifications that are superior, as well as positive results.

Ref. [47] reported on research on two-area hybrid power grids that used a combined-FO mixed structure based on PIDD2 and FOPI control. The Dandelion optimizer (DO) was utilised to fulfil the purpose of optimizing the controller design that was provided. In [48], the butterfly optimisation algorithm (BOA) has been used to present the dual stage LFCs that have been optimized. In the paper [49], the authors offer a hybrid FO LFC approach, which they call FOTID. This method is optimized using manta-ray foraging optimization, or MRFO. There have also been presentations in the research literature of sequences of coupled and cascaded structures. Ref. [50] demonstrates an imperialist competitive-optimizer (ICA)-based cascaded FOPID with FLC. Additional combined fuzzy and FO LFC approaches were described in the paper [51], which used the FLC-FOPI-FOPD, in the paper [52], which used the FL-FOPIDF, and in the paper [53], which used the FLC-PIDF-FOI. Additionally, an enhanced ICA-optimized FPIDN-FOPIDN LFC approach was proposed for usage with two-area grids in [54].

Many intelligent controllers have been developed recently for application in LFC design. These controllers include model predictive control (MPC), fuzzy logic control, artificial neural networks (ANN), and adaptive neuro-fuzzy inference systems (ANFIS). The MPC has reportedly been used to stabilize the system that is integrated with wind turbines, according to [55,56]. The ANFIS has also been used as the LFC for a system that includes many RESs that have been optimized by the ant lion [57]. A solar power plant was part of an integrated system that was regulated by artificial neural networks in [58]. Fuzzy logic controllers (FLCs) are the subject of much study at the present, especially when used in combination with more conventional PID controllers or fractional-order controllers. The fuzzy logic controller allows for more accuracy, which produces better outcomes. Therefore, by selecting the best membership functions for both the inputs and the outputs, the system's overall performance may be improved [59,60]. The system will work more efficiently as a consequence. The PID controller and fuzzy logic controller have been integrated, and both systems have been optimized using a number of techniques, including the marine predators algorithm [61] and the sine-cosine algorithm [62], respectively. Additionally, [63,64] discuss the development of a fuzzy-FOPID controller that makes use of a differential evolution method.

1.3. Contribution and Paper Organization

In light of the aforementioned research and the groundbreaking work presented in [65], in which the authors introduced a state-of-the-art controller for the LFC problem using an approach called tilt fractional-order integral-derivative with fractional-filter (TFOIDFF) and whose parameters are optimized using an artificial hummingbird algorithm (AHA), this article makes a first-of-its-kind attempt to merge the benefits of the fuzzy logic controller and the TFOIDFF controller to provide an outstanding controller. This controller is known as fuzzy TFOIDFF (or FTFOIDFF), and its gains are fine-tuned using prairie dog optimizer (PDO), a recently created nature-inspired metaheuristic optimizer that simulates prairie dog activity in their natural habitat. Additionally, the PDO is utilised successfully to optimize the input scaling factors and pick the optimal membership functions for both the inputs and outputs of the FLC. The following is a list of the principal contributions that can be drawn from this body of work:

- The suggestion of a control structure that combines the benefits of tilt, fuzzy logic, FOPID, and fractional filter regulators in a single controller known as FTFOIDFF that efficiently improves frequency stability in a hybrid two-area linked power system incorporated with severe RES penetrations.
- Utilization of a nature-inspired metaheuristic optimization technique that was recently developed (i.e., the prairie dog optimizer, or PDO) for the purpose of fine-tuning the recommended controller settings as well as the input scaling factors and membership functions for both FLC inputs and outputs in an effective manner.
- Validation of the beneficial influence of the integration of SMES, UPFC, and EVs in enhancing frequency performance during load perturbations and RESs penetrations.
- The robustness and superiority of the proposed PDO-based FTFOIDFF have been demonstrated through a fair performance comparison with other available conventional (for example, PID, PIDA [66], and TFOIDFF [65]) and intelligent (for example, FPID [67] and FPIDA [68]) controllers.

The article then divides into the following sections: The Section 2 provides a detailed description of the investigated hybrid power system that includes a UPFC, an EV, and a SMES; the Section 3 introduces the suggested control methodology based on the PDO approach; the Section 4 provides simulation outcomes; and the Section 5 provides a conclusion with pros and cons.

2. Modelling of the Investigated Hybrid Power System with SMES, UPFC and EVs

2.1. The Power System Structure

This topic of LFC relating to power systems has been discussed in this paper by performing study on two-area linked hybrid power systems. These systems have EVs and SMES units in both areas, in addition to a UPFC unit in the tie-line that connects the two areas together. Different types of traditional power units were included in the power system analysis. These included thermal units, hydroelectric units, and gas units. Each region's capacity on the power grid under study, which incorporates the three conventional units, is 2000 MW of rated power, with the thermal power plant contributing 1087 MW, the hydropower plant contributing 653 MW, and the gas turbine contributing 262 MW [69]. Figure 1 is a simplified schematic representation of the electrical network that was studied. Figure 2 is a block diagram of the transfer function for the hybrid power grid that was analysed, which consists of two interconnected areas. Table 1 displays the transfer functions of the investigated power grid. It is proposed that a combination of fuzzy logic and TFOIDFF controllers be installed in both regions for each generation unit in order to reduce disturbances in the frequencies of both regions and the power exchange between them via the tie-line. The proposed PDO-based FTFOIDFF controller's input signal represents the area control error (ACE), while the output signal represents the secondary control on each generation facility in order to increase active power for network efficiency. A description of the system's parameters, considering their typical values can be found in Appendix A.

According to the equations that are presented in [69], it is possible to calculate the ACEs in both areas as follows:

$$ACE_a = B_a \cdot \Delta F_a + \Delta P_{tie} \tag{1}$$

$$ACE_b = B_b \cdot \Delta F_b - \Delta P_{tie} \tag{2}$$

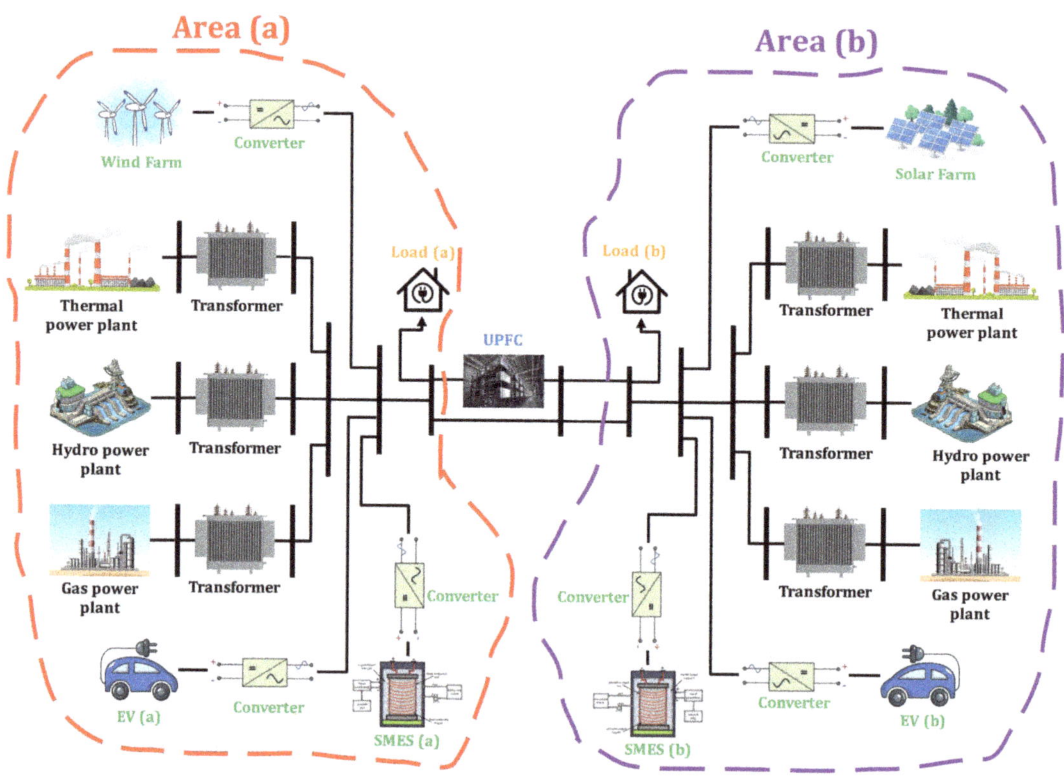

Figure 1. The schematic diagram of investigated power grid.

Table 1. The examined power grid's transfer functions [69].

Power Planet	Model	Transfer Function
Thermal	Governor	$\frac{1}{\tau_{sg}s+1}$
	Reheat	$\frac{k_r\tau_r\,s+1}{\tau_r s+1}$
	Turbine	$\frac{1}{\tau_t s+1}$
Hydraulic	Governor	$\frac{1}{\tau_{gh}s+1}$
	Transient droop compensation	$\frac{\tau_{rs}s+1}{\tau_{rh}s+1}$
	turbine	$\frac{-\tau_w s+1}{0.5\tau_w s+1}$
Gas	Valve positioner	$\frac{1}{b_g s+c_g}$
	Speed governor	$\frac{x_c s+1}{y_c s+1}$
	Fuel system and combustor	$\frac{-\tau_{cr}s+1}{\tau_{fc}s+1}$
	Compressor discharge	$\frac{1}{\tau_{cd}s+1}$

Table 1. *Cont.*

Power Planet	Model	Transfer Function
Others	Power system (a)	$\dfrac{k_{ps1}}{\tau_{ps1}s+1}$
	Power system (b)	$\dfrac{k_{ps2}}{\tau_{ps2}s+1}$
	T-line	$\dfrac{2\pi T_{12}}{s}$
	SMES (a)	$\dfrac{k_{SMES(a)}}{\tau_{SMES(a)}s+1}$
	SMES (b)	$\dfrac{k_{SMES(b)}}{\tau_{SMES(b)}s+1}$
	UPFC	$\dfrac{1}{\tau_{UPFC}s+1}$
	EV (a)	$\dfrac{k_{EV(a)}}{\tau_{EV(a)}s+1}$
	EV (b)	$\dfrac{k_{EV(b)}}{\tau_{EV(b)}s+1}$

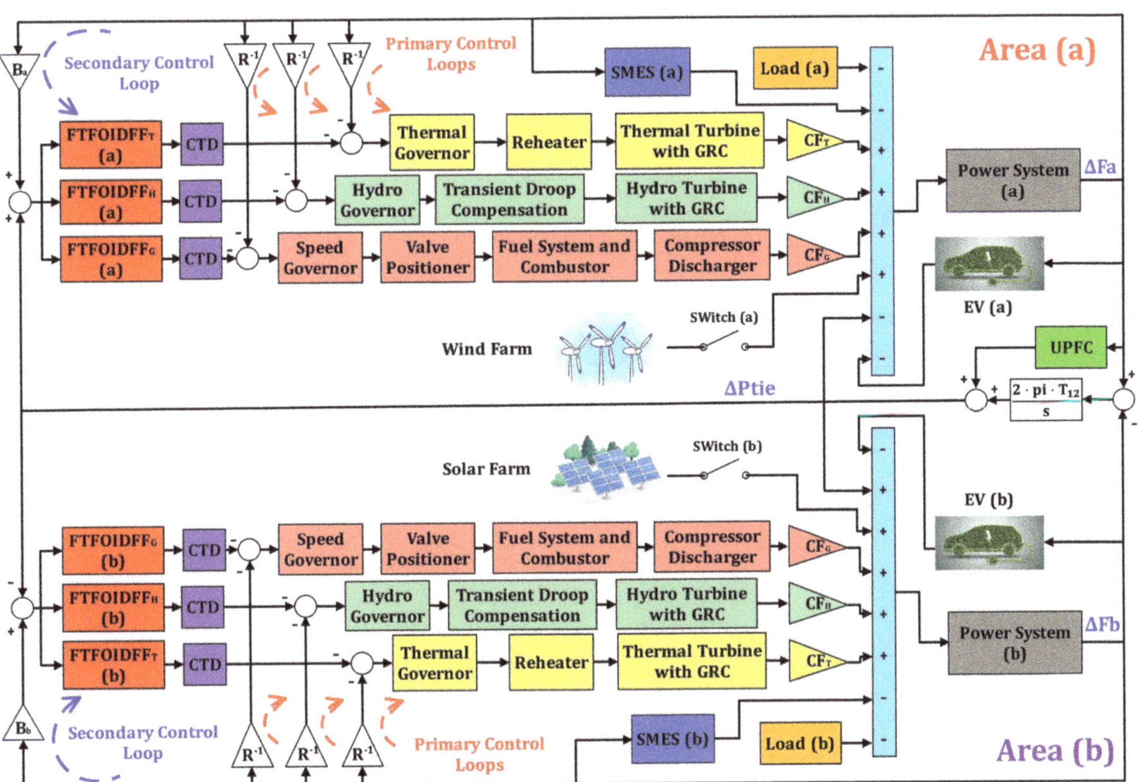

Figure 2. The examined power grid in form of transfer function model.

2.2. Mathematical Representation of UPFC

The ongoing and rapid advancement of power electronics technology over the last decade has made FACTS a viable idea for power system applications. The use of FACTS technology allows for more flexible management of electricity flow along transmission lines. One of the FACTS family's most adaptable devices is the unified power flow controller (UPFC), which has the ability to regulate power flow in the transmission line and provide voltage support as well as enhance transient stability and system oscillation [19,70]. This research takes into account the two-area power system with a UPFC depicted in Figure 3 [71]. A tie-line is linked in series with the UPFC, which reduces oscillations in tie-line power. The series voltage's magnitude and phase angle are shown in Figure 3 as V_{se} and φ_{se}, respectively. The series converter's actual power consumption is balanced by the real component of the current flowing through the shunt branch thanks to the shunt converter's injection of regulated shunt voltage. One may infer from Figure 3 that the complex power at the line's other end can be expressed as follows:

$$P_{\text{real}} - jQ_{\text{reactive}} = \overline{V}_r \times I_{\text{line}} = \overline{V}_r \times \left\{ \frac{(\overline{V}_s + \overline{V}_{se} - \overline{V}_r)}{j(X)} \right\} \tag{3}$$

where

$$\overline{V}_{se} = |V_{se}| \angle (\delta_s - \varphi_{se}) \tag{4}$$

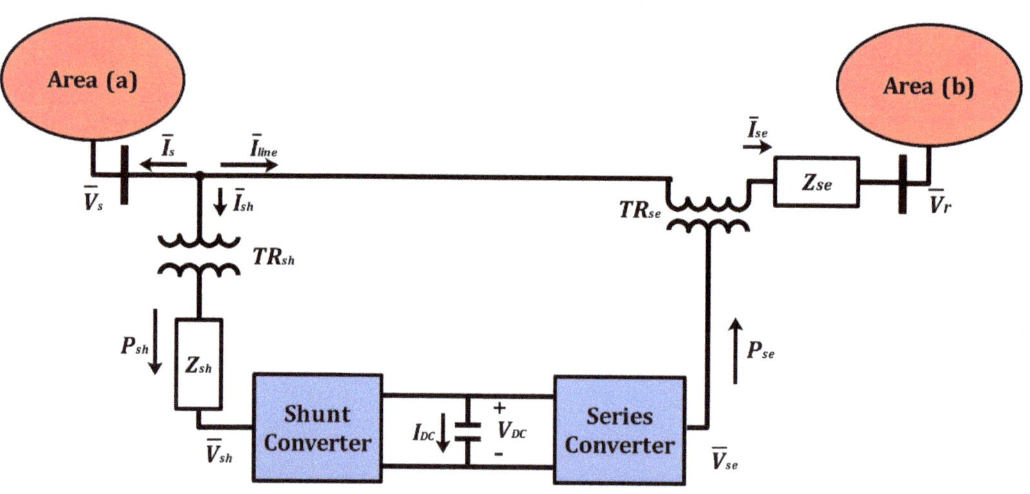

Figure 3. UPFC integration in a dual area linked power system [67].

The real power can be determined by solving Equation (3) as follows:

$$P_{\text{real}} = \frac{|V_s||V_r|}{(X)} \sin(\delta) + \frac{|V_s||V_{se}|}{(X)} \sin(\delta - \varphi_{se}) = P_0(\delta) + P_{se}(\delta, \varphi_{se}) \tag{5}$$

If V_{se} is equal to zero in Equation (5), this indicates that the system's true power is an uncompensated system. In contrast, the UPFC series voltage magnitude may be regulated between zero and $V_{se}(\max)$, and its phase angle (φ_{se}) can be altered between zero and $360°$ degrees at any power angle. A representation of the UPFC-based controller that may be used in LFC can be found in [72].

$$\Delta P_{\text{UPFC}}(s) = \left(\frac{1}{\tau_{\text{UPFC}}s + 1} \right) \cdot \Delta F(s) \tag{6}$$

where τ_{UPFC} denotes the UPFC time constant.

2.3. Mathematical Representation of SMES

Superconducting magnetic energy storage (SMES) is a technology that can store electrical energy from the grid in a coil's magnetic field. The energy loss in the magnetic field of the coil is nearly nonexistent since it is composed of superconducting wire. Small and medium-sized enterprises are capable of storing and regenerating enormous amounts of energy in an almost rapid manner. As a result of this, the power system is able to discharge large amounts of power within a fraction of a cycle in order to prevent a quick decrease in the line power. The SMES is made up of an inductor-conversion unit, an AC/DC converter, a step-down transformer, and a dc superconducting inductor [20]. Due to the fact that all of the SMES unit's components are fixed, it possesses a level of stability that much surpasses that of other types of power storage devices. Diagrammatic representation of a SMES unit found in the power system is shown in Figure 4. The superconducting coil will be charged to a predetermined value from the utility grid when the grid is operating normally. This value is often lower than the maximum charge that the coil is capable of holding. The inverter and rectifier that are part of the power conversion system (PCS) are the components that allow the DC magnetic coil to be linked to the AC grid. After it has been charged, the superconducting coil will conduct current without nearly any losses, which will allow it to support an electromagnetic field. By submerging it in a pool of liquid helium, the coil is maintained at a temperature that is very cold. When there is an abrupt increase in the demand of load, the stored energy is nearly immediately released as AC power through the PCS and into the grid. As the control mechanisms begin to function, the power system is being readjusted to reach a new condition of equilibrium, and the coil's current begins to recharge to its starting value. Whenever there is an abrupt release of loads, the coil quickly becomes charged up to its maximum potential, soaking up part of the surplus energy that is present in the system in the process. During the process of the system returning to its steady state, the excess energy that was absorbed is discharged, and the value of the coil current returns to its typical level. In light of the two SMES described above, units are set up in area (a) and area (b) for the purpose of regulating frequency oscillations, as seen in Figure 2. The input signal to the SMES is the frequency deviation (ΔF), while the output signal is the change in control vector ΔP_{SMES}. The SMES regulator can be formulated as follows [67]:

$$\Delta P_{\text{SMES}}(s) = \left(\frac{k_{\text{SMES}}}{\tau_{\text{SMES}} s + 1} \right) \cdot \Delta F(s) \qquad (7)$$

where k_{SMES} denotes the SMES gain, and τ_{SMES} depicts the SMES time constant.

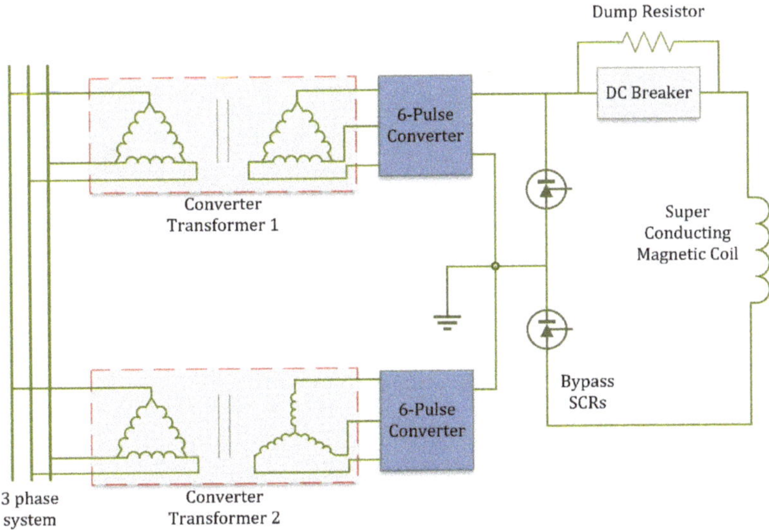

Figure 4. The circuit diagram representation of SMES [67].

2.4. Mathematical Representation of the Wind Farm Unit

The study demonstrates the significant penetration of RES, such as wind power, in the hybrid power system that was evaluated. Wind energy's simplified version has been applied in area (a) of the analyzed power system using the professional software MATLAB/SIMULINK (R2022b). The model is fairly precise since the power produced by the windmill simulation acts exactly in the same way as the energy produced by real wind farms. This is achieved by using a white-noise block, which is used to get a random speed and then multiply it by the wind speed, as shown in Figure 5 [69]. The wind model's collected output power may be expressed mathematically as shown in the following equations [69].

$$P_W = \tfrac{1}{2}\, \rho\, A_T\, V_w^3\, C_P(\lambda,\, \beta) \tag{8}$$

$$C_P(\lambda,\, \beta) = C_1 \cdot \left(\frac{C_2}{\lambda_i} - C_3\beta - C_4\beta^2 - C_5 \right) \cdot e^{\frac{C_6}{\lambda_i}} + C_7\lambda_T \tag{9}$$

$$C_P(\lambda,\, \beta) = C_1 \cdot \left(\frac{C_2}{\lambda_i} - C_3\beta - C_4\beta^2 - C_5 \right) \cdot e^{\frac{C_6}{\lambda_i}} + C_7\lambda_T \tag{10}$$

$$\frac{1}{\lambda_i} = \frac{1}{\lambda_T + 0.08\,\beta} - \frac{0.035}{\beta^3 + 1} \tag{11}$$

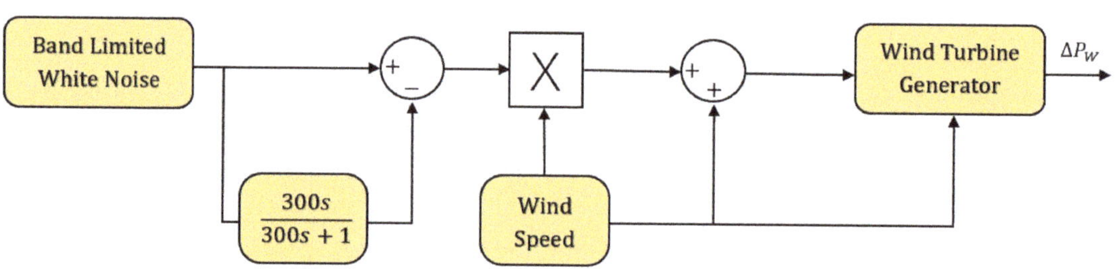

Figure 5. The wind system representation in MATLAB/Simulink program (R2022b) [69].

Ref. [36] contains a listing of all of these aforementioned parameter values for the wind farm that was utilised. Power production from 264 separate wind power units (each generating 0.75 MW) is shown in Figure 6 below. The examined wind farm's power output has a value of around 198 MW.

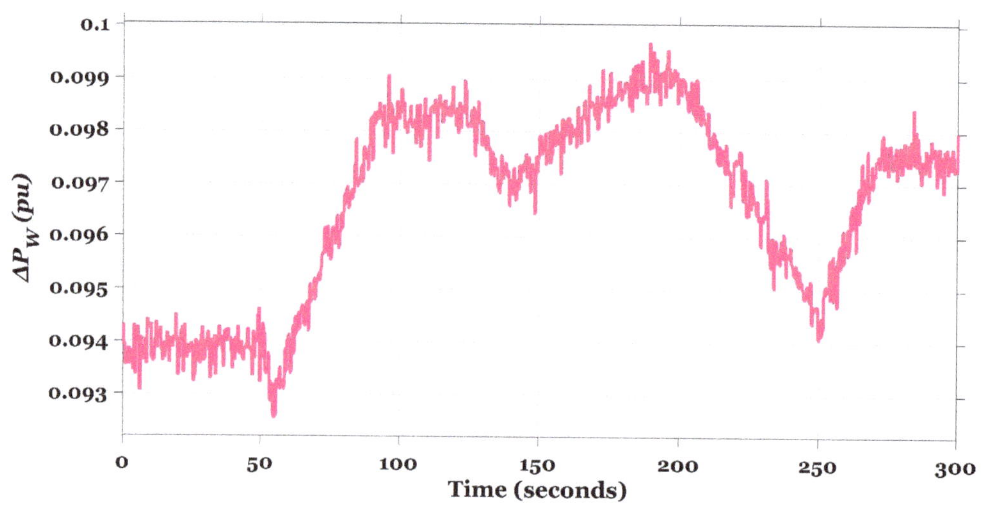

Figure 6. The wind model generated power.

2.5. Mathematical Representation of the PV Unit

Using the professional software MATLAB/SIMULINK (R2022b), the Photovoltaic (PV) model depicted in Figure 7 can be designed. When compared to the output power of an actual PV plant, the model's results are quite close. Additionally, the output energy of the PV model has been penetrating area (b) of the power system that has been under study at around 134 MW. To generate random output oscillations, which are afterwards multiplied by the standard output power offered by a real PV plant, the white-noise block in the MATLAB program (R2022b) is used in this situation. Equation (12) contains the formula for estimating the quantity of power generated by the PV model that was presented [32]. The random output power provided by the PV model is shown and described in Figure 8.

$$\Delta P_{PV} = 0.6 \cdot \sqrt{P_{Solar}} \tag{12}$$

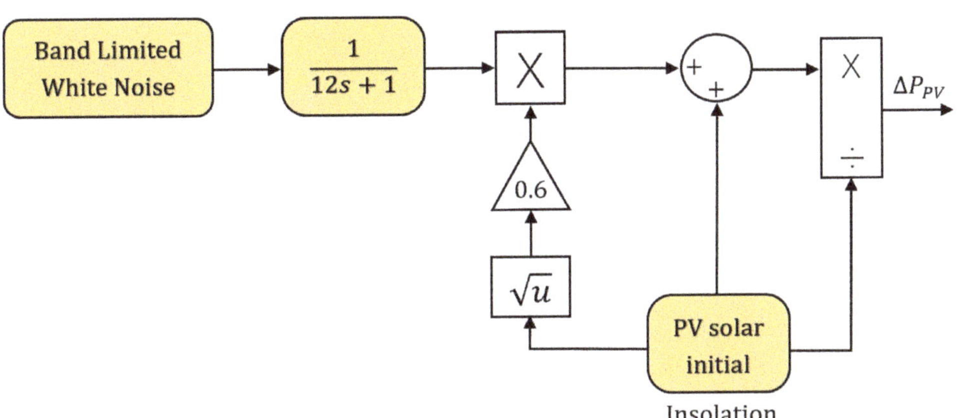

Figure 7. The PV system representation in MATLAB/Simulink program (R2022b) [69].

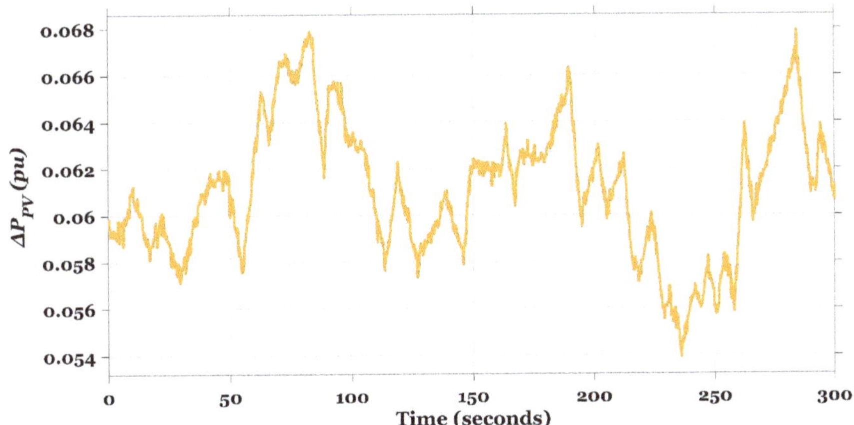

Figure 8. The PV model generated power.

2.6. Mathematical Representation of the EVs Units

Electric vehicles' ability to receive the LFC command and transmit that signal to regulate power consumption while charging and discharging allows them to contribute efficiently to frequency regulation. The presence of a certain number of controlled electric cars (EVs) in the electrical grid under study, as well as the state of charge related to the EVs' capacity, may further limit the LFC signal's ability to respond. In contrast, the EV model is similar to the battery storage system model since it includes batteries that contribute additional energy to the electric network during variations to control frequency deviations. But since EVs are designed for mobility and load, the batteries in them could not be fully charged, which would reduce the amount of additional energy needed to solve the LFC issue. Therefore, it is essential to examine the level of the electric vehicle's charge in order to guarantee more system advancement in spite of the various system fluctuations. The first-order transfer function can be used to estimate the output power of an electric vehicle (EV). This function takes into account the electrical vehicle time constant τ_{EV}, as well as its gain k_{EV}. Equation (13) [69] provides a formulation for the transfer function that is used to represent the EV model. The electric vehicle (EV) model that was constructed in the MATLAB/SIMULINK software (R2022b) can be found described in Figure 9.

$$\Delta P_{EV}(s) = \left(\frac{k_{EV}}{\tau_{EV}s + 1} \right) \cdot \Delta F(s) \tag{13}$$

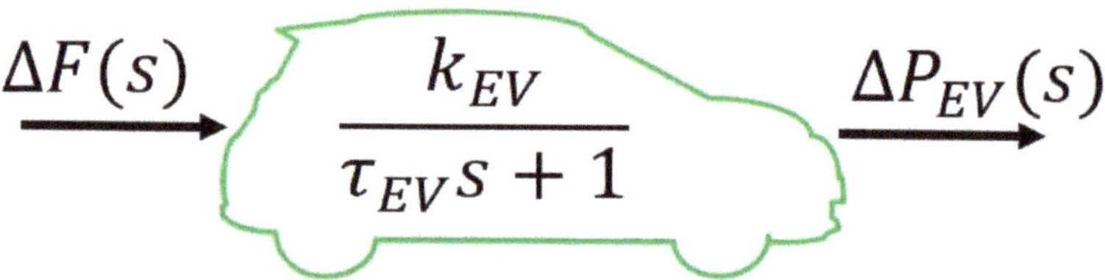

Figure 9. The EV system representation in MATLAB/Simulink program (R2022b).

3. Control Strategy and Problem Presentation

This section covers the design of a hybrid Fuzzy TFOIDFF (FTFOIDFF) controller whose parameters are fine-tuned by PDO algorithm to address the LFC issue, as prior work indicates that classical controllers have limits in handling system uncertainties. The suggested controller improves upon the efficacy and robustness of load frequency management by combining the benefits of fuzzy logic control (FLC) with the recently developed TFOIDFF, which has been reported in [65].

3.1. Prairie Dog Optimizer (PDO)

The PDO technique was initially presented in 2022 by Absalom et al. [73], and it mimics the behaviors of prairie dogs, who are a family of rodents that are herbivorous and mostly lived in the deserts of Northern America. Prairie dogs, which live in one of the wildest regions on Earth, have evolved several survival attributes such as powerful arms, long-nailed toes, and the ability to run quickly over short distances. These traits allow prairie dogs to flee from predators and hide in their connected burrows when they are cornered, allowing them to survive in one of the most wilderness-like environments on the planet.

Prairie dogs are social animals that form large colonies made up of smaller groups called "coteries". These coteries help the colony as a whole by sharing resources and information, such as the foraging call and the alarming sound made when a predator is close by. A prairie dog may be seen in Figure 10 performing its call by rearing up on its hind legs and producing squeaking noises. Although the sound produced by a prairie dog may appear to humans to be nothing more than a simple squeak or yip, the sound really conveys a very specific information to the prairie dog's ear. Exploration and exploitation are the two primary phases that make up the mathematical model of PDO. These phases are motivated by foraging, burrow construction, and their reaction to the source of communications accordingly.

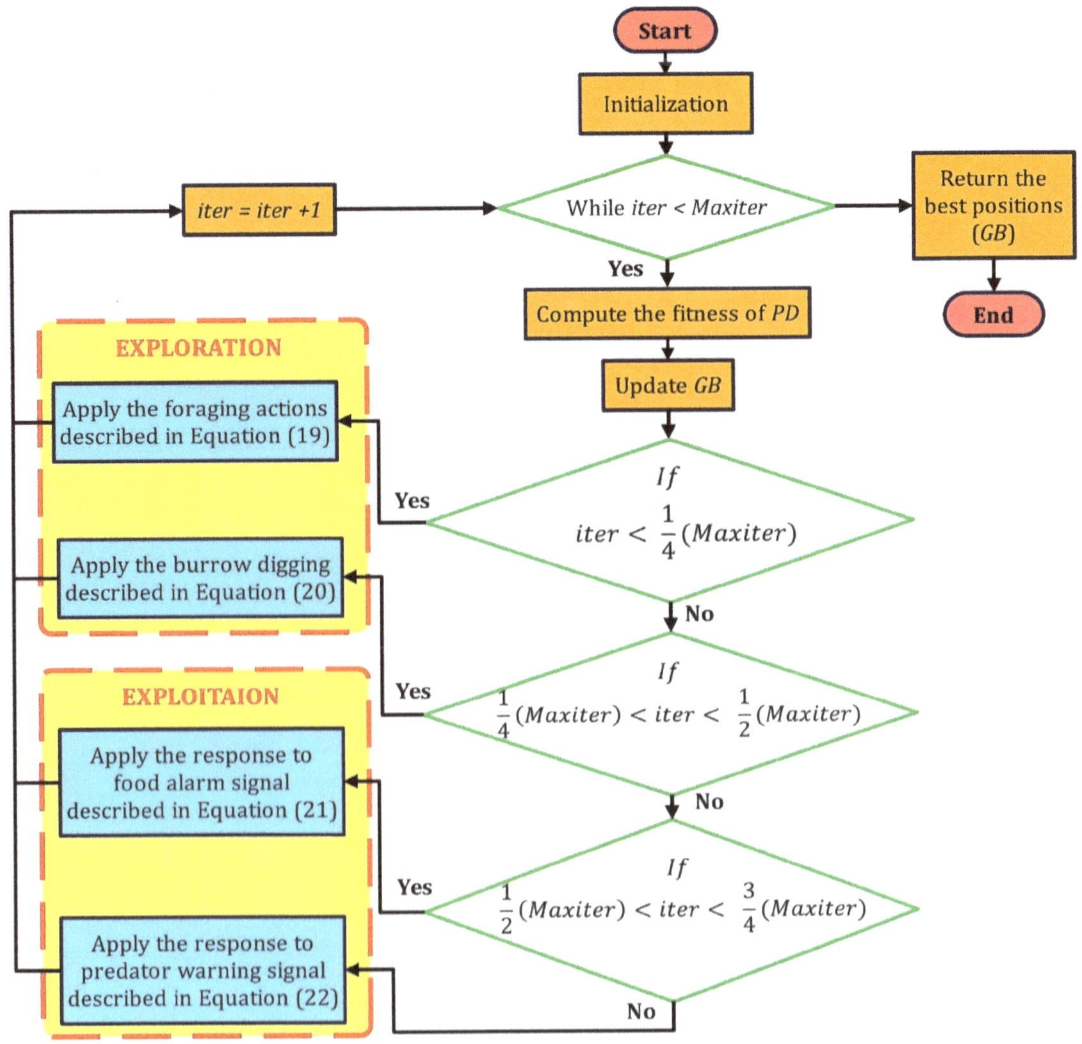

Figure 10. PDO algorithm's flow chart representation.

3.1.1. Initialization

Each prairie dog (PD) is a member of m different coteries, and there are a total of n prairie dogs in a given coterie. Since prairie dogs live and behave as a unit, often known as a coterie, one may use a vector to determine where the ith prairie dog is located inside a given coterie. The matrix that is presented in Equation (14) depicts the locations of all of the coteries (CT) in a colony.

$$CT = \begin{bmatrix} CT_{1,1} & CT_{1,2} & \cdots & CT_{1,d-1} & CT_{1,d} \\ CT_{2,1} & CT_{2,2} & \cdots & CT_{2,d-1} & CT_{2,d} \\ \vdots & \vdots & CT_{i,j} & \vdots & \vdots \\ CT_{m,1} & CT_{m,2} & \cdots & CT_{m,d-1} & CT_{m,d} \end{bmatrix} \tag{14}$$

where $CT_{i,j}$ denotes the jth dimension of the ith coterie inside a colony. The position of each prairie dog in a coterie may be represented by the following equation:

$$PD = \begin{bmatrix} PD_{1,1} & PD_{1,2} & \cdots & PD_{1,d-1} & PD_{1,d} \\ PD_{2,1} & PD_{2,2} & \cdots & PD_{2,d-1} & PD_{2,d} \\ \vdots & \vdots & PD_{i,j} & \vdots & \vdots \\ PD_{n,1} & PD_{n,2} & \cdots & PD_{n,d-1} & PD_{n,d} \end{bmatrix} \tag{15}$$

$PD_{i,j}$ indicates the jth dimension of the ith prairie dog within a coterie, and $n \leq m$. As demonstrated in Equations (16) and (17), each coterie and prairie dog site is assigned using a uniform distribution.

$$CT_{i,j} = rand(0,1) \cdot (UB_j - LB_j) + LB_j \tag{16}$$

$$PD_{i,j} = rand(0,1) \cdot (ub_j - lb_j) + lb_j \tag{17}$$

where UB_j and LB_j are the upper and lower boundaries of the jth dimension of the optimisation challenge, $ub_j = UB_j/m$ and $lb_j = LB_j/m$, and $rand(0,1)$ is a uniformly distributed arbitrary number between zero and one

3.1.2. The Estimation of Objective Function

The value of the objective function is determined at each position of the prairie dog by inputting the solution vector into the objective function that has been constructed. The array described in Equation (18) is used to save the results of the calculation.

$$f(PD) = \begin{bmatrix} f_1([PD_{1,1} & PD_{1,2} & \cdots & PD_{1,d-1} & PD_{1,d}]) \\ f_2([PD_{2,1} & PD_{2,2} & \cdots & PD_{2,d-1} & PD_{2,d}]) \\ \vdots & \vdots & PD_{i,j} & \vdots & \vdots \\ f_n([PD_{n,1} & PD_{n,2} & \cdots & PD_{n,d-1} & PD_{n,d}]) \end{bmatrix} \tag{18}$$

The value of the objective function of each prairie dog is a representation of the quality of the food that can be obtained at a given source, the capacity for digging additional burrows, and the ability to respond appropriately to anti-predator alarms. The array that stores the values of the objective function is prioritized, and the value that corresponds to the lowest possible cost is determined to be the optimal response to the presented optimization challenge. The following two phases are examined, together with the best value for burrow building, due to the role it plays in the animals' ability to hide from predators.

3.1.3. Exploration Phase

During the exploration phase, the mathematical model is created based on the actions of prairie dogs, such as foraging and burrow-digging, in order to start the search in the optimization problem space. In the wild, when a preexisting food supply is no longer able to meet the nutritional needs of the entire colony, prairie dogs will look for a new food source that is likely to provide a greater quantity of food than the one they were using before. After that, a network of tunnels is dug in the area close to the newly discovered food supply in order to produce hiding areas for the several coteries. The most recent location of the food supply that requires foraging is mathematically described as follows in Equation (19), which can be found below:

$$PD_{i+1,j+1} = GB_{i,j} - 0.1 \cdot \left(GB_{i,j} \cdot \Delta + \frac{PD_{i,j} \cdot mean(PD_{n,m})}{GB_{i,j} \cdot (ub_j - lb_j) + \Delta} \right) - \left(1 - \frac{rPD_{i,j}}{GB_{i,j}} \right) \cdot Levy(n) \tag{19}$$

where GB is the best global position found, rPD is a randomly chosen prairie dog position, Δ denotes a tiny number that represents the difference between the prairie dogs, and $Levy(n)$ represents a Levy distribution [74]. When prairie dogs discover a new food source they immediately begin digging tunnels there. The digging ability of prairie dogs is used to determine the target number of burrows, which should decrease as the iteration count rises. Equation (20) provides the current coordinates of the newly dug burrow:

$$PD_{i+1,j+1} = GB_{i,j} \cdot rPD \cdot 1.5 \cdot k \cdot \left(1 - \frac{iter}{Maxiter} \right)^{2\frac{iter}{Maxiter}} \cdot Levy(n) \tag{20}$$

where $iter$ and $Maxiter$ depict the current iteration and the maximum iteration number, respectively, k equals negative unity when the current iteration becomes an odd value and k equals positive unity when the current iteration becomes an even value. The exploration phase will now have a stochastic feature thanks to the addition of the variable k.

3.1.4. Exploitation Phase

The exploitation phase is modelled upon the two distinct reactions exhibited by prairie dogs during communication, one of which is used for food gathering and the other for alerting of potential danger. When information is received regarding the colony's food, the members of the colony are obligated to gather at the site of the food supply. When the message is to alert of the arrival of predators, the nearest member will hide in the burrows while the others await the likely approaching alert to determine whether or not to hide. The exploitation phase of PDO may be represented using Equations (21) and (22) in the following way:

$$PD_{i+1,j+1} = GB_{i,j} - \mu \cdot \left(GB_{i,j} \cdot \Delta + \frac{PD_{i,j} \cdot mean(PD_{n,m})}{GB_{i,j} \cdot (ub_j - lb_j) + \Delta} \right) - \left(1 - \frac{rPD_{i,j}}{GB_{i,j}} \right) \cdot \beta \tag{21}$$

$$PD_{i+1,j+1} = GB_{i,j} \cdot rPD \cdot 1.5 \cdot \left(1 - \frac{iter}{Maxiter} \right)^{2\frac{iter}{Maxiter}} \cdot \beta \tag{22}$$

where μ is a tiny value representing the food source's quality and β is an arbitrary value between zero and one. It should be noted that the exploration and exploitation phases are determined by the number of iterations: For exploration, when $iter \leq \frac{1}{4}(Maxiter)$, forage activities will occur, and when $\frac{1}{4}(Maxiter) \leq iter \leq \frac{1}{2}(Maxiter)$, burrow digging will be undertaken. When $\frac{1}{2}(Maxiter) \leq iter \leq \frac{3}{4}(Maxiter)$, the prairie dogs' reaction to food signal will be carried out, and lastly when $\frac{3}{4}(Maxiter) \leq iter \leq (Maxiter)$, burrow digging will be accomplished. Figure 10 represents the flowchart PDO algorithm.

3.2. The Detailed Configuration of The Proposed FTFOIDFF Regulator

The proposed FTFOIDFF's structure is discussed in this section. The suggested controller is split into two parts: the first is the TFOIDFF controller, which was initially introduced in [65] and has greater efficiency than other traditional and modern controllers. Figure 11 depicts the architecture of the TFOIDFF controller, and Ref. [65] provides further information regarding the design and characteristics of this controller. The mathematical expression of the TFOIDFF controller is given by the following equation [65]:

$$G_C(s) = K_t s^{-\left(\frac{1}{n}\right)} + \frac{K_i}{s^{\lambda_i}} + K_d s^{\mu_d} \frac{N_f}{s^{\lambda_f} + N_f} \tag{23}$$

where, K_d, K_t, K_i represent derivative, tilt, and integral gains. While μ_d, λ_f, λ_i depict the fractional order operators of derivative, fractional filter, and integral terms. Moreover, n denotes the tilt fractional order power, and N_f represents the fractional filter coefficient.

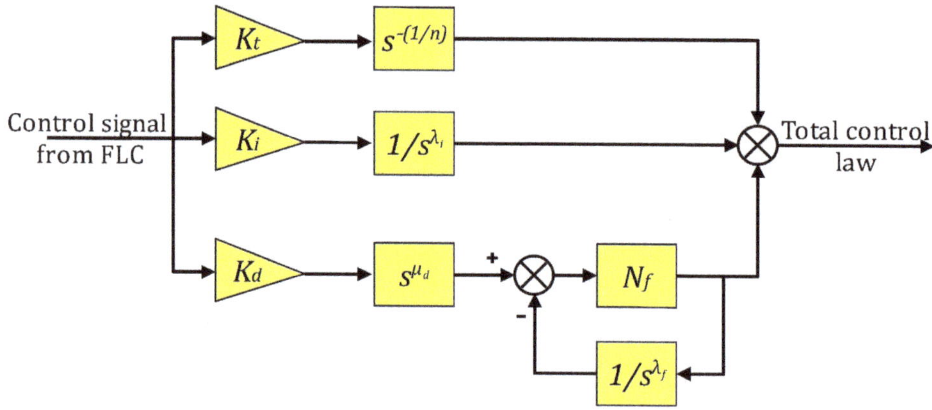

Figure 11. The configuration of TFOIDFF controller [65].

The second part of the proposed FTFOIDFF controller is the fuzzy logic controller (FLC), which was added to the TFOIDFF regulator to improve its performance and functionality. The PDO method has been used to optimize the suggested FLC's MFs in order to get the optimum forms that provide the best results for this research. However, the effectiveness of fuzzy logic regulators is heavily dependent on the membership functions (MFs). This is so due to the fact membership functions (MFs) have a significant impact on the efficacy of fuzzy logic regulators. Additionally, the intricate design of a suitable fuzzy rule base interface system is very important. Figure 12 shows the FTFOIDFF controller architecture used for the LFC study, with the fuzzy controller's inputs being the error (E) and derivative of error (DOE). The gains K_1 and K_2 are the manifestations of the scaling factors. The following is a concise summary of the primary processes involved in the deployment of an FLC [68]:

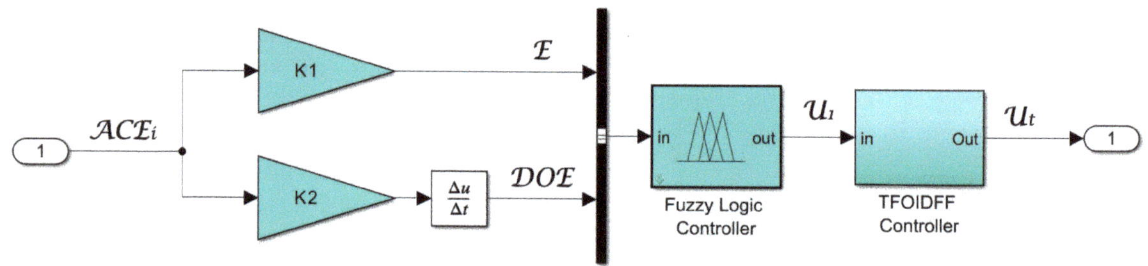

Figure 12. The configuration of the FTFOIDFF controller.

The initial step is referred to as "fuzzification," and at this stage, the FLC is responsible for transforming E and DOE into linguistic variables. In this work, the inputs, and outputs of the FLCs are all triangle membership functions, whose forms have been adjusted by PDO as shown in Figure 13. It is worth noting from Figure 13 that the MFs of the controllers in area (a) differ from those in area (b). In terms of the inputs and outputs, there are five linguistic variables that are utilised. These variables are denoted by the letters NB, NS, Z, PS, and PB, which stand for negative big, negative small, zero, positive small, and positive big, respectively. It is abundantly evident that the membership functions of both the inputs and the outputs are positioned in the interval [−40, 40].

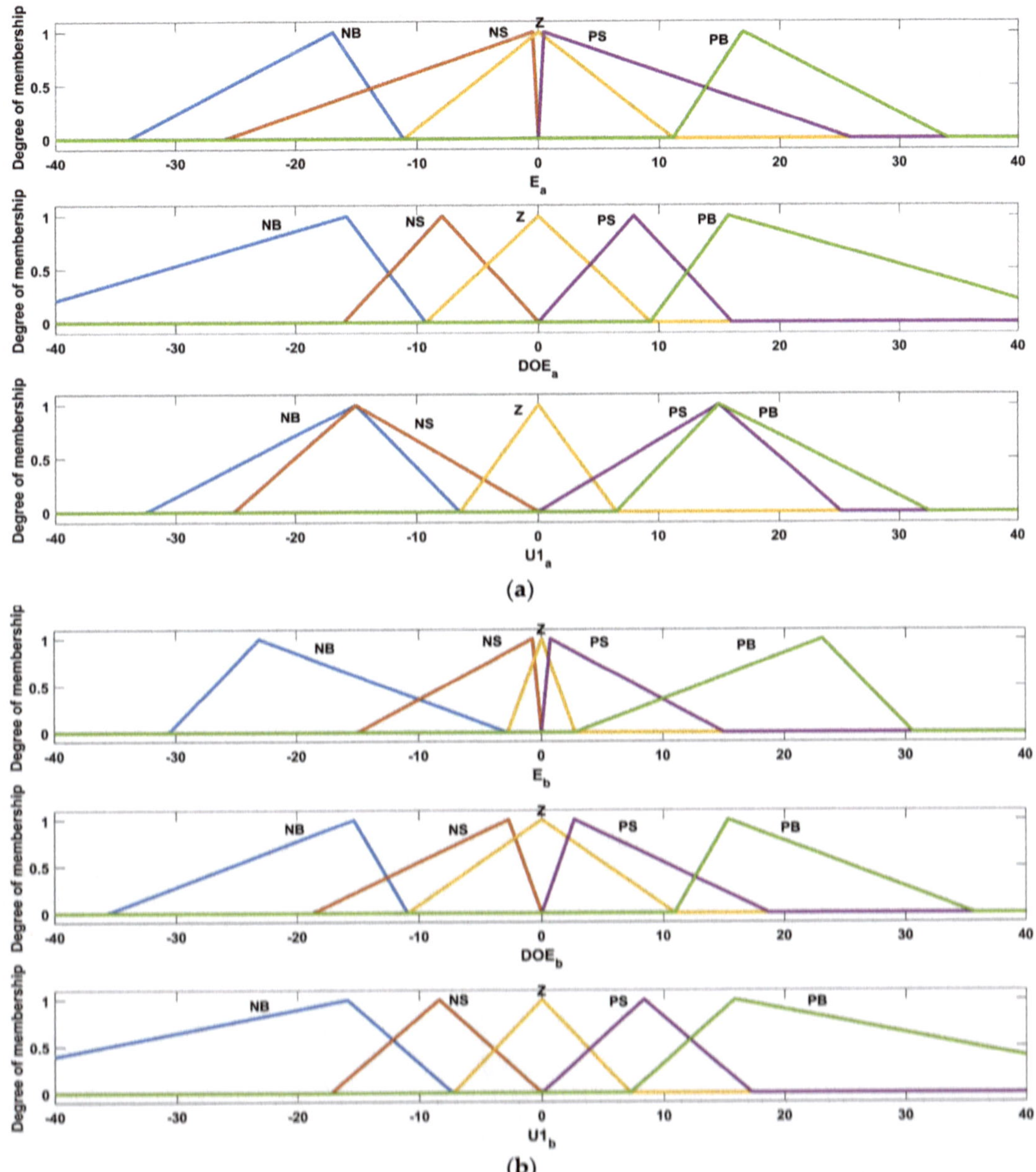

Figure 13. The MFs of the proposed FTFOIDFF for both areas. (**a**) MFs of the controllers in area 1; (**b**) MFs of the controllers in area 2.

The second stage of the procedure involves the implementation of the rule base. The outcomes of the FLC's application of fuzzy rules to the linguistic variables that were produced as a consequence of the fuzzification process are presented in Table 2; the Fuzzy interface system (FIS) that was used in this particular case is Mamdani [68]. The extent and character of the FLC's fundamental rule set are determined by the designer's competence level. Each system uses its own set of rules to provide the best results.

Table 2. FLC rule base.

E	DOE				
	NB	**SN**	**Z**	**SP**	**LP**
NB	NB	NB	NS	NS	Z
NS	NB	NS	NS	Z	PS
Z	NS	NS	Z	PS	PS
PS	NS	Z	PS	PS	PB
PB	Z	PS	PS	PB	PB

We have now arrived at the last phase, which is known as defuzzification. Linguistic variables are used as inputs for the defuzzification operation, and the output of the FIS system is also a linguistic variable. Also, the defuzzification technique converts these variables to crisp variables. The fuzzy output in this study reflects the first control law (U_1) derived utilising the centre of gravity technique of the defuzzification procedure. To obtain the total control law (U_t), which may be expressed as Equation (24), U_1 is sent to the TFOIDFF controller. The main goal of the proposed controller is to reduce system-induced frequency deviations (ΔFa, ΔFb) and tie-line power deviations (ΔPtie) in the LFC loop. Adjusting the parameters of the FTFOIDFF controller (K_t, K_i, K_d, n, λ_i, μ_d, λ_f, N_f, K_1, K_2) in such a way as to achieve this goal is possible.

$$U_t(s) = U_1(s) \cdot G_C(s) \tag{24}$$

In this particular investigation, the integral time absolute error, which is more often referred to as ITAE, was selected to act as the objective function that would be used to evaluate the controller's overall performance. It is believed that ITAE will be the strategy that is most successful in drastically reducing response overshoots, undershoots and settling time in the LFC issue, as shown by Equation (25) [67].

$$FF = ITAE = \int_0^{t_{sim}} t \cdot |\Delta\text{Fa} + \Delta\text{Fb} + \Delta\text{Ptie}| dt \tag{25}$$

In Equation (25), ΔFa and ΔFb represent the system frequency deviations; ΔPtie is the incremental deviation in tie-line power; t_{sim} denotes the simulation time range. The parameters of the suggested FTFOIDFF controller have been restricted according to Equation (26):

$$\begin{cases} K_{t\,min} \leq K_t \leq K_{t\,max} \\ K_{i\,min} \leq K_i \leq K_{i\,max} \\ K_{d\,min} \leq K_d \leq K_{d\,max} \\ n_{min} \leq n \leq n_{max} \\ \lambda_{i\,min} \leq \lambda_i \leq \lambda_{i\,max} \\ \mu_{d\,min} \leq \mu_d \leq \mu_{d\,max} \\ \lambda_{f\,min} \leq \lambda_f \leq \lambda_{f\,max} \\ N_{f\,min} \leq N_f \leq N_{f\,max} \\ K_{1\,min} \leq K_1 \leq K_{1\,max} \\ K_{2\,min} \leq K_2 \leq K_{2\,max} \end{cases} \tag{26}$$

All subsequent Cases will have upper bounds of [20,20,10,10,1,1,1,2,2] and lower bounds of [0,0,0,2,0,0,0,0]. In the next part, we will discuss the outcomes and conclusions from the simulation across a wide range of operational scenarios.

4. Results and Discussion

In this study, the secondary control loop with an extensive integration of RESs is used to restore the examined system frequency to the specified value while accounting for various forms of load variation. The suggested FTFOIDFF controller, which is ideally developed by the PDO algorithm to get the lowest frequency fluctuations for the understudied power grid, is the foundation of the control method that is being presented. The effectiveness of the proposed control approach is also evaluated in comparison to that of existing control techniques, such as PID, PIDA, TFOIDFF, FPID, and FPIDA. MATLAB/SIMULINK® (R2022b) is used to implement all simulation results for the investigated dual-area, multi-unit hybrid power grid in order to verify the suggested controller's efficacy in enhancing the system's performance. The outcomes of the simulation are generated on a computer equipped with an AMD Ryzen 7 3700U-2.30 GHz processor and 20.00 GB of RAM. By computing the value of the optimal objective function, which is represented by the ITAE value across iterations, the effectiveness of the researched power grid may be assessed. Before improving the suggested FTFOIDFF controller using the recommended PDO method, a number of preliminary issues, such as the 30 populations and 100 iterations, must be resolved. Figure 14 depicts a convergence curve that illustrates the performance of the proposed PDO algorithm in comparison to other recent optimization methodologies (i.e., Seagull Optimization Algorithm (SOA), RUNge Kutta optimizer (RUN), and Chaos Game Optimizer (CGO)). The demonstrated convergence curve can be obtained by taking on a 10% SLP at 5 s in area (a) of the investigated hybrid power grid, with no penetration of RESs in both areas. Clearly, the PDO algorithm achieved the lowest objective function value (0.0875) compared to the previously mentioned approaches. Consequently, the convergence curve demonstrates the efficacy of the proposed PDO algorithm.

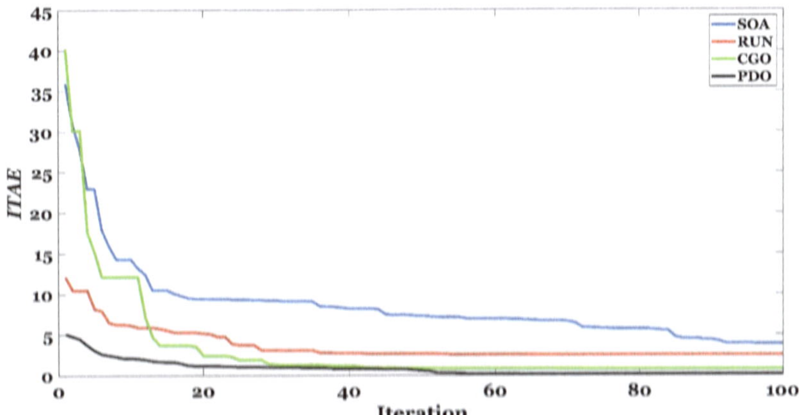

Figure 14. The convergence curve characteristics of SOA, RUN, CGO, and PDO.

4.1. Case I: 10% Step Load Perturbation (SLP) at t = 5 s in Area (a)

To validate the superiority of the suggested PDO based FTFOIDFF regulator over the other traditional (i.e., PID, PIDA [66], TFOIDFF [65]) and intelligent (i.e., FPID [67] and FPIDA [68]) control techniques, which are also fine-tuned by PDO algorithm, a 10% step load perturbation (SLP) is applied in area (a) at t = 5 s. Table 3 displays the PDO-optimized controller parameters for the chosen case study. The convergence curve in Figure 15 shows that the proposed PDO-based FTFOIDFF outperforms the aforementioned control techniques. Figure 16 depicts the frequency and tie-line power responses (ΔFa, ΔFb, and ΔPtie) of the investigated hybrid electrical grid for both area (a) and area (b). According to research, the suggested PDO-based FTFOIDFF controller has better system stability and damping characteristics with less overshoots, undershoots, and settling times than the other controllers. While producing 0.0067 pu variation in tie-line power and frequency deviations of 0.0158 Hz and 0.008 Hz, respectively, in areas (a) and (b). Table 4 contains a comprehensive comparative study of the examined controllers for several parameters such as settling time (ST), maximum overshoot (MOS), and maximum undershoot (MUS).

Table 3. The optimum parameters of the different controllers.

Controller		Thermal		Hydro		Gas
PID	Area (a)	$K_p = 0.024$, $K_i = 0.486$, $K_d = 0.293$, $N_f = 102$	Area (a)	$K_p = 0.025$, $K_i = 3.573$, $K_d = 2.025$, $N_f = 151$	Area (a)	$K_p = 4.993$, $K_i = 4.992$, $K_d = 4.104$, $N_f = 192$
	Area (b)	$K_p = 0.002$, $K_i = 0.011$, $K_d = 0.218$, $N_f = 290$	Area (b)	$K_p = 0.368$, $K_i = 0.176$, $K_d = 0.133$, $N_f = 218$	Area (b)	$K_p = 0.521$, $K_i = 1.209$, $K_d = 4.804$, $N_f = 146$
PIDA	Area (a)	$K_p = 3.365$, $K_i = 0.775$, $K_{d1} = 1.666$, $K_{d2} = 0.001$, $N_{f1} = 265$, $N_{f2} = 174$	Area (a)	$K_p = 0.245$, $K_i = 0.654$, $K_{d1} = 9.895$, $K_{d2} = 0.008$, $N_{f1} = 245$, $N_{f2} = 194$	Area (a)	$K_p = 9.996$, $K_i = 9.999$, $K_{d1} = 0.607$, $K_{d2} = 0.031$, $N_{f1} = 191$, $N_{f2} = 345$
	Area (b)	$K_p = 6.599$, $K_i = 0.487$, $K_{d1} = 1.309$, $K_{d2} = 0.001$, $N_{f1} = 249$, $N_{f2} = 179$	Area (b)	$K_p = 0.714$, $K_i = 0.451$, $K_{d1} = 0.217$, $K_{d2} = 0.005$, $N_{f1} = 214$, $N_{f2} = 201$	Area (b)	$K_p = 5.969$, $K_i = 0.207$, $K_{d1} = 9.246$, $K_{d2} = 0.003$, $N_{f1} = 164$, $N_{f2} = 312$
TFOIDFF	Area (a)	$K_t = 11.173$, $K_i = 0.011$, $K_d = 0.024$, $n = 5.077$, $\lambda_i = 0.16$, $\mu_d = 0.063$, $\lambda_f = 0.011$, $N_f = 166$	Area (a)	$K_t = 0$, $K_i = 4.453$, $K_d = 9.991$, $n = 2.127$, $\lambda_i = 1$, $\mu_d = 0.922$, $\lambda_f = 0.015$, $N_f = 177$	Area (a)	$K_t = 19.988$, $K_i = 6.801$, $K_d = 0.091$, $n = 1.5$, $\lambda_i = 0.004$, $\mu_d = 0.074$, $\lambda_f = 0.694$, $N_f = 113$
	Area (b)	$K_t = 0.408$, $K_i = 0$, $K_d = 8.042$, $n = 5.75$, $\lambda_i = 0.29$, $\mu_d = 0$, $\lambda_f = 0.485$, $N_f = 161$	Area (b)	$K_t = 0.656$, $K_i = 0.242$, $K_d = 7.655$, $n = 3.454$, $\lambda_i = 0.184$, $\mu_d = 0.904$, $\lambda_f = 0.581$, $N_f = 138$	Area (b)	$K_t = 5.632$, $K_i = 11.175$, $K_d = 2.741$, $n = 4.476$, $\lambda_i = 0.003$, $\mu_d = 0.008$, $\lambda_f = 0.095$, $N_f = 396$
FPID	Area (a)	$K_p = 0.078$, $K_i = 0.181$, $K_d = 0.839$, $N_f = 100$, $K_1 = 4.56$, $K_2 = 0.897$	Area (a)	$K_p = 9.889$, $K_i = 0.662$, $K_d = 4.567$, $N_f = 101$, $K_1 = 0.65$, $K_2 = 2.073$	Area (a)	$K_p = 5.426$, $K_i = 9.781$, $K_d = 1.289$, $N_f = 400$, $K_1 = 4.996$, $K_2 = 0.899$
	Area (b)	$K_p = 2.094$, $K_i = 3.213$, $K_d = 0.071$, $N_f = 400$, $K_1 = 4.993$, $K_2 = 3.534$	Area (b)	$K_p = 0$, $K_i = 0.206$, $K_d = 5.046$, $N_f = 364$, $K_1 = 0.001$, $K_2 = 0.448$	Area (b)	$K_p = 0.607$, $K_i = 8.76$, $K_d = 0.055$, $N_f = 156$, $K_1 = 4.793$, $K_2 = 3.077$
FPIDA	Area (a)	$K_p = 4.444$, $K_i = 7.397$, $K_{d1} = 2.01$, $K_{d2} = 0.027$, $N_{f1} = 145$, $N_{f2} = 289$, $K_1 = 0.373$, $K_2 = 4.179$	Area (a)	$K_p = 0.91$, $K_i = 0.558$, $K_{d1} = 3.872$, $K_{d2} = 0.036$, $N_{f1} = 193$, $N_{f2} = 279$, $K_1 = 0.352$, $K_2 = 0.674$	Area (a)	$K_p = 4.151$, $K_i = 4.41$, $K_{d1} = 1.015$, $K_{d2} = 0.05$, $N_{f1} = 279$, $N_{f2} = 241$, $K_1 = 4.94$, $K_2 = 4.045$
	Area (b)	$K_p = 0$, $K_i = 5.903$, $K_{d1} = 9.239$, $K_{d2} = 0.01$, $N_{f1} = 354$, $N_{f2} = 293$, $K_1 = 1.636$, $K_2 = 0.735$	Area (b)	$K_p = 0.852$, $K_i = 1.285$, $K_{d1} = 1.843$, $K_{d2} = 0.028$, $N_{f1} = 172$, $N_{f2} = 128$, $K_1 = 0.006$, $K_2 = 4.879$	Area (b)	$K_p = 0.731$, $K_i = 2.373$, $K_{d1} = 0.387$, $K_{d2} = 0.024$, $N_{f1} = 399$, $N_{f2} = 400$, $K_1 = 2.324$, $K_2 = 3.654$
FTFOIDFF	Area (a)	$K_t = 16.45$, $K_i = 0.025$, $K_d = 0.014$, $n = 6.17$, $\lambda_i = 0.423$, $\mu_d = 0.046$, $\lambda_f = 0.513$, $N_f = 246$, $K_1 = 4.79$, $K_2 = 2.756$	Area (a)	$K_t = 0.459$, $K_i = 2.413$, $K_d = 7.82$, $n = 4.261$, $\lambda_i = 0.876$, $\mu_d = 0.452$, $\lambda_f = 0.094$, $N_f = 284$, $K_1 = 5$, $K_2 = 2.871$	Area (a)	$K_t = 12.762$, $K_i = 4.189$, $K_d = 1.891$, $n = 8.69$, $\lambda_i = 0.02$, $\mu_d = 0.061$, $\lambda_f = 0.815$, $N_f = 189$, $K_1 = 3.62$, $K_2 = 3.112$
	Area (b)	$K_t = 0.783$, $K_i = 0.874$, $K_d = 3.24$, $n = 2.49$, $\lambda_i = 0.481$, $\mu_d = 0.006$, $\lambda_f = 0.147$, $N_f = 188$, $K_1 = 0.782$, $K_2 = 1.023$	Area (b)	$K_t = 8.159$, $K_i = 2.47$, $K_d = 0.489$, $n = 8.421$, $\lambda_i = 0.452$, $\mu_d = 0.394$, $\lambda_f = 0.023$, $N_f = 108$, $K_1 = 0.05$, $K_2 = 0.723$	Area (b)	$K_t = 2.168$, $K_i = 10.631$, $K_d = 1.014$, $n = 6.21$, $\lambda_i = 0.126$, $\mu_d = 0.04$, $\lambda_f = 0.113$, $N_f = 322$, $K_1 = 1.06$, $K_2 = 1.62$

Table 4. The transient response specifications of the studied system for Case I.

Controller	ΔFa (Hz) MOS	MUS	ST	ΔFb (Hz) MOS	MUS	ST	$\Delta Ptie$ (pu) MOS	MUS	ST	ITAE
PID	0.0025	-0.034	30	0	-0.024	13	0.0047	-0.0346	30	3.312
PIDA	0.0026	-0.0424	20	0.001	-0.0475	17	0.0035	-0.0427	28	2.103
TFOIDFF	0	-0.063	13	0.008	-0.0587	13	0.0016	-0.0578	17	1.159
FPID	0	-0.0228	8	0.002	-0.0107	3.3	0	-0.0228	3	0.5282
FPIDA	0	-0.0158	0.4	0	-0.008	1.5	0	-0.0067	1	0.1409
FTFOIDFF	0	-0.0158	0.18	0	-0.0026	1	0	-0.0056	0.7	0.0875

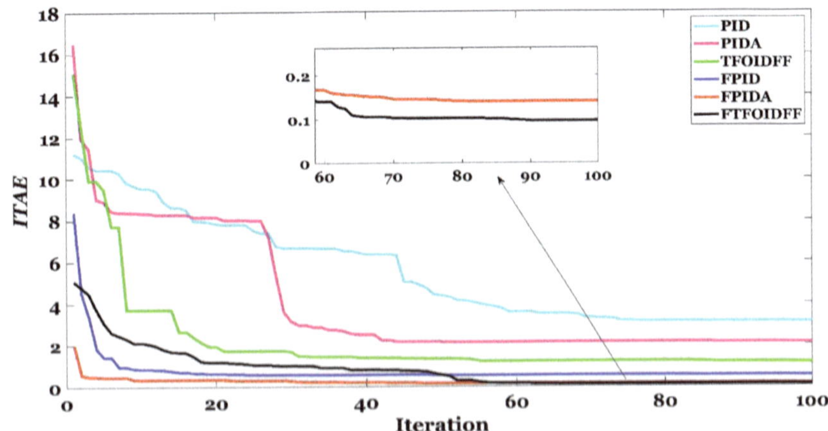

Figure 15. The convergence curve characteristics of the compared controllers.

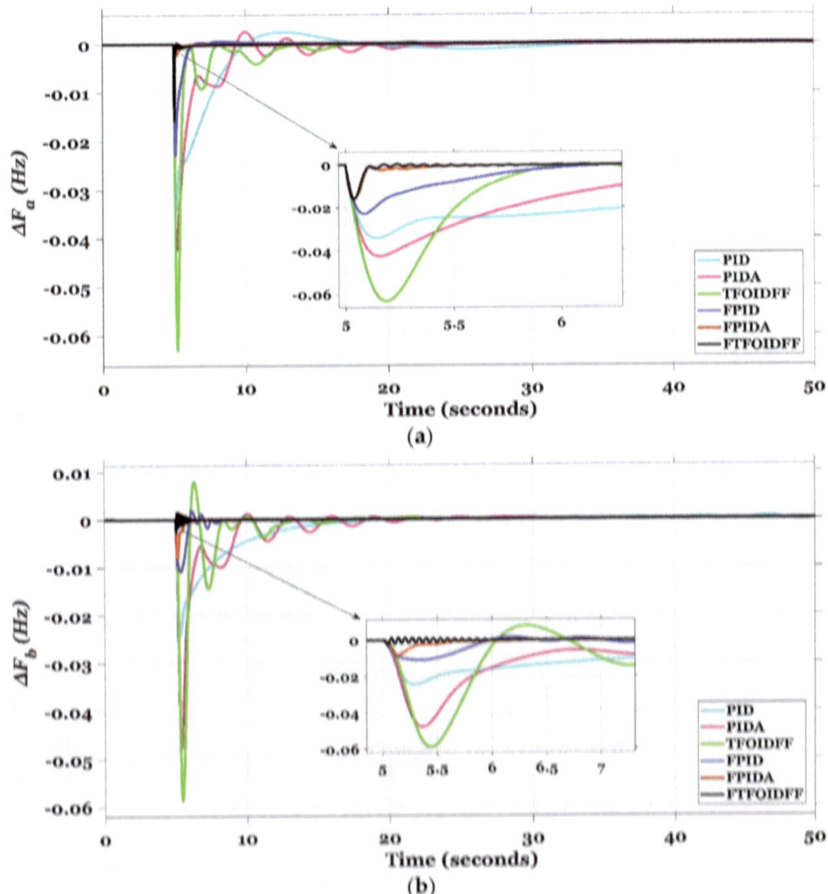

Figure 16. *Cont.*

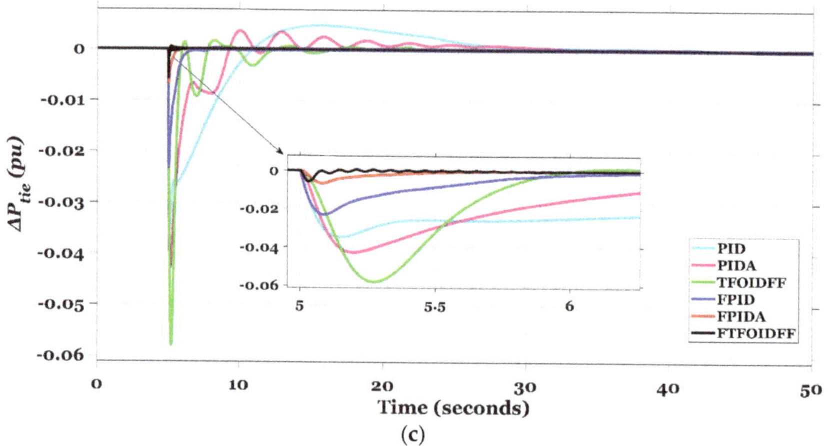

(c)

Figure 16. The system dynamics for Case I. (**a**) ΔFa; (**b**) ΔFb; (**c**) ΔPtie.

4.2. Case II: Multi-Step Load Perturbation (MSLP) in Area (a)

In this specific case, the ability of the suggested PDO-based FTFOIDFF controller is examined and exposed under the influence of applying a harsh multi-step load change pattern in area (a), where the MSLP is depicted in Figure 17. The MSLP is modelled as a simulation of the series change that occurs in actually connected loads. It is possible to say that the MSLP is thought of as a series-forced switch of generators or a series interrupt of the associated loads. Figure 18 depicts the frequency and power deviation waveforms of the system during the impact of the severe multi-step load fluctuation that the proposed regulator was designed to handle. When compared to the dynamic responses of the other control strategies, the suggested control strategy's dynamic responses have faster reactions with a small percentage of deviation in their values. The proposed PDO based FTFOIDFF has effectively maintained the entire ITAE value within 4.605, as shown in Table 5, while having low MOS and MUS, as well as a quick and smooth ST. According to Table 5, for instance, and not as a limitation, the proposed controller has been able to acquire a value of the ITAE that is about 37.11 times less than the PID controller, 19.15 times less than the TFOIDFF controller, and 1.34 times lower than the FPIDA controller.

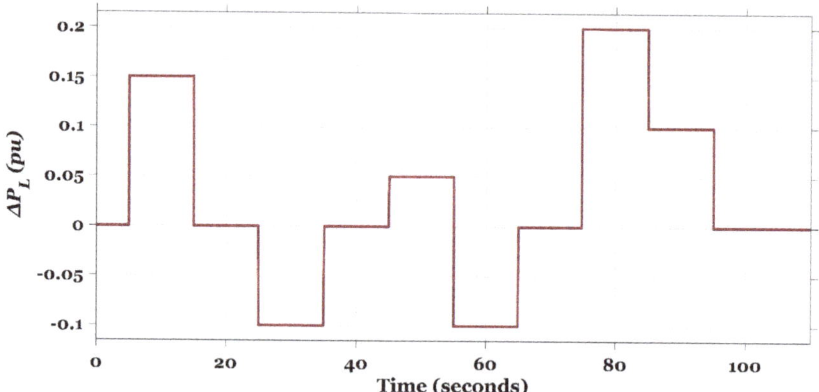

Figure 17. MSLP profile.

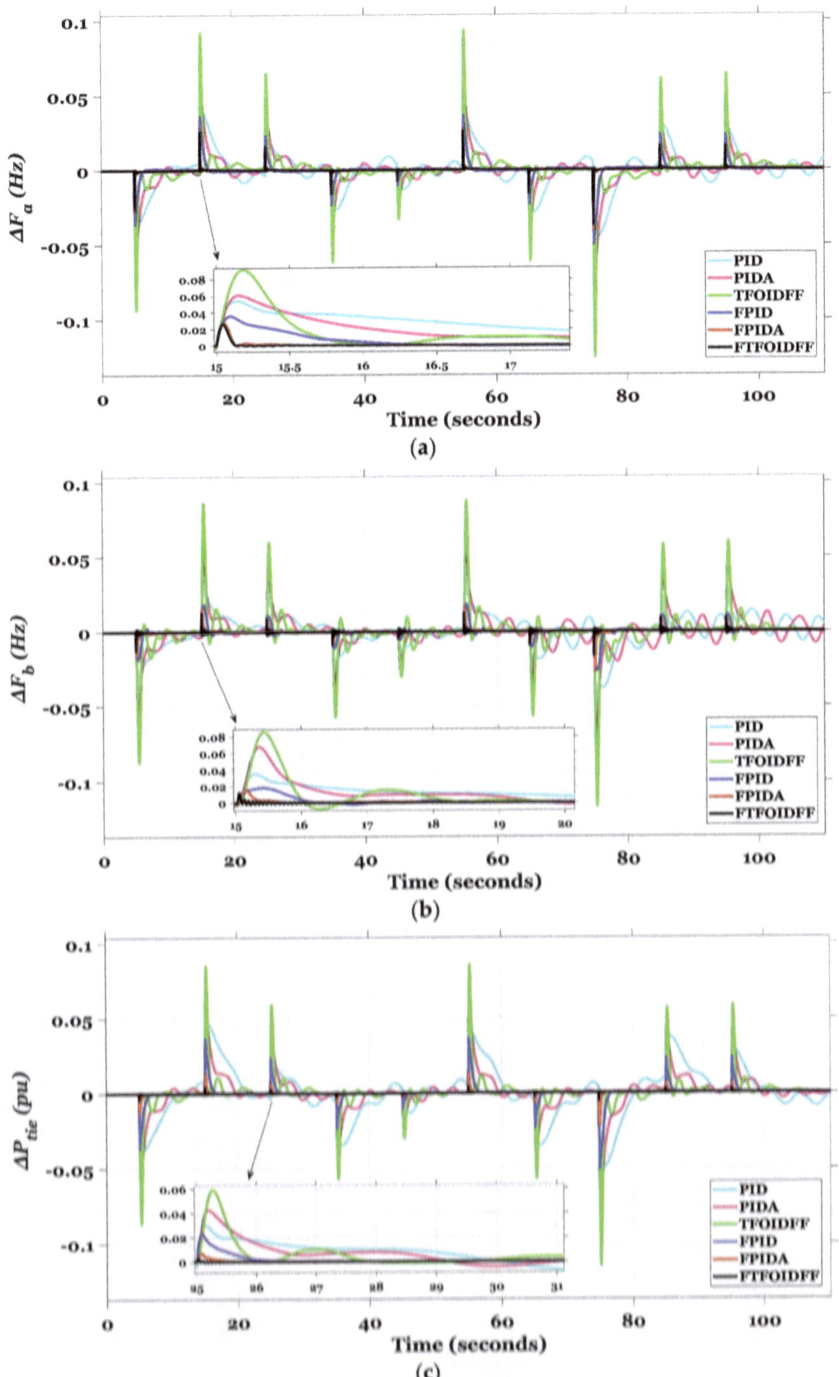

Figure 18. The system dynamics for Case II. (a) ΔFa; (b) ΔFb; (c) ΔPtie.

Table 5. The transient response specifications of the studied system represented as ITAE value using different controllers for Case II.

Controller	ITAE			ITAE$_{tot}$
	ΔFa (Hz)	ΔFb (Hz)	ΔPtie (pu)	
PID	50.32	49.36	71.27	170.9
PIDA	35.35	48.48	39.47	123.3
TFOIDFF	28.73	34.73	24.71	88.17
FPID	8.039	6.663	8.06	22.76
FPIDA	1.804	2.866	1.491	6.162
FTFOIDFF	1.451	2.231	0.923	4.605

4.3. Case III: Random Load Perturbation (RLP) in Area (b)

After determining the effectiveness of the proposed FTFOIDFF controller in the first two cases, we now proceed to a more challenging scenario in which the severe random load perturbation (RLP), depicted in Figure 19, is applied to area (b) of the power system network under investigation. Practically, the RLP may be thought of as a group of series disturbances that might be represented by connected industrial loads. Figure 20 shows a representation of the system's reaction to this scenario utilizing a number of different control techniques (specifically, PIDA, TFOIDFF, FPID, FPIDA, and FTFOIDFF controllers that are based on the PDO). The recommended control strategy's dynamic responses outperform those of the other techniques in terms of speed of response, damping ability, and the values for undershoot, overshoot, and settling time. The system's dynamic performance, as measured by the ITAE index, is summarized in Table 6. The suggested PDO-based FTFOIDFF has successfully kept the ITAE value within 7.976. This number is around 38.29 times lower than the value asserted by the PID controller, 37.24 times smaller than the value claimed by the PIDA controller, 31.07 times lower than the value asserted by the TFOIDFF controller, 5.57 times lower than the value asserted by the FPID controller, and 3.65 times smaller than the value claimed by the FPIDA controller. This demonstrates that the PDO-based FTFOIDFF controller used for LFC is a stable one.

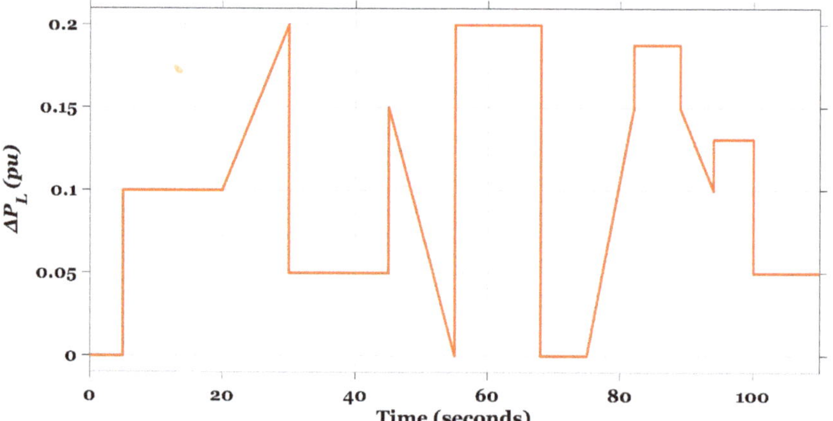

Figure 19. The RLP profile.

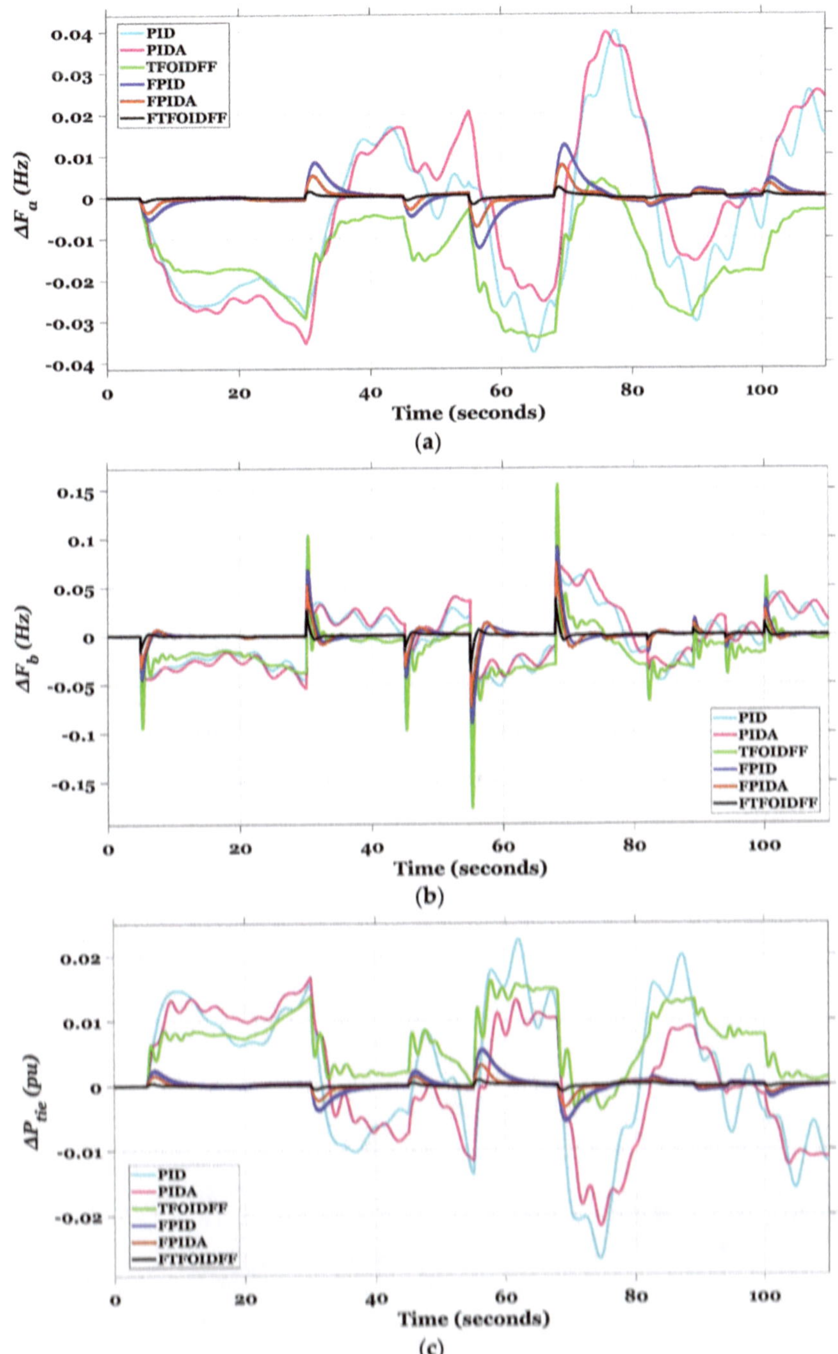

Figure 20. The system dynamics for Case III. (**a**) ΔFa; (**b**) ΔFb; (**c**) ΔPtie.

Table 6. The transient response specifications of the studied system represented as ITAE value using different controllers for Case III.

Controller	ITAE			ITAE$_{tot}$
	ΔFa (Hz)	ΔFb (Hz)	ΔPtie (pu)	
PID	98.32	147.3	59.77	305.4
PIDA	94.31	155.4	47.36	297
TFOIDFF	90.72	117.2	39.85	247.8
FPID	11.62	27.77	5.077	44.46
FPIDA	6.283	20.09	2.724	29.09
FTFOIDFF	1.608	5.676	0.6926	7.976

4.4. Case IV: Pulse Load Perturbation (PLP) in Area (a)

Herein, a severe pulse load perturbation (PLP) with a period of 5 s and high amplitude of 0.5 pu is applied in area (a) twice throughout the simulation duration of 50 s, as shown in Figure 21. Figure 22 presents a visual representation of the area frequencies and tie-line power oscillation reactions that were obtained. When compared with the other control techniques, the results of the simulation reveal that the oscillations are greatly dampened in a short amount of time by the suggested PDO-based FTFOIDFF. Table 7 demonstrates how the suggested controller has the potential to improve the overall performance of the system in terms of the ITAE for each response (i.e., ΔFa, ΔFb, and ΔPtie). According to Table 7, the ITAE index for the ΔFa response is 2.776, the index for Fb is 1.101, and the index for ΔPtie is 0.5113. All of these values are much lower than those of the other controllers that were examined. In addition, as an example and not as a restriction, the proposed controller has been successful in acquiring a value of total ITAE that is about 30.5 times lower than that of the PID controller, 17.7 times lower than that of the PIDA controller, and 13 times lower than that of the FPID controller. This demonstrates the suggested controller's capability to perform robustly in the face of the high-amplitude PLP.

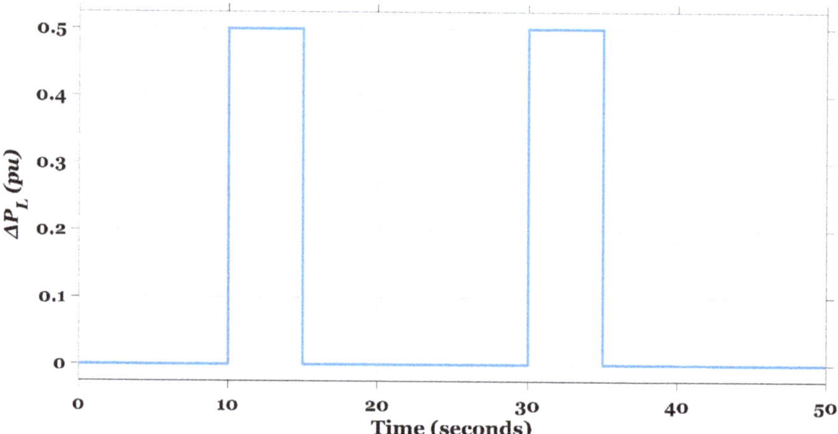

Figure 21. The PLP profile.

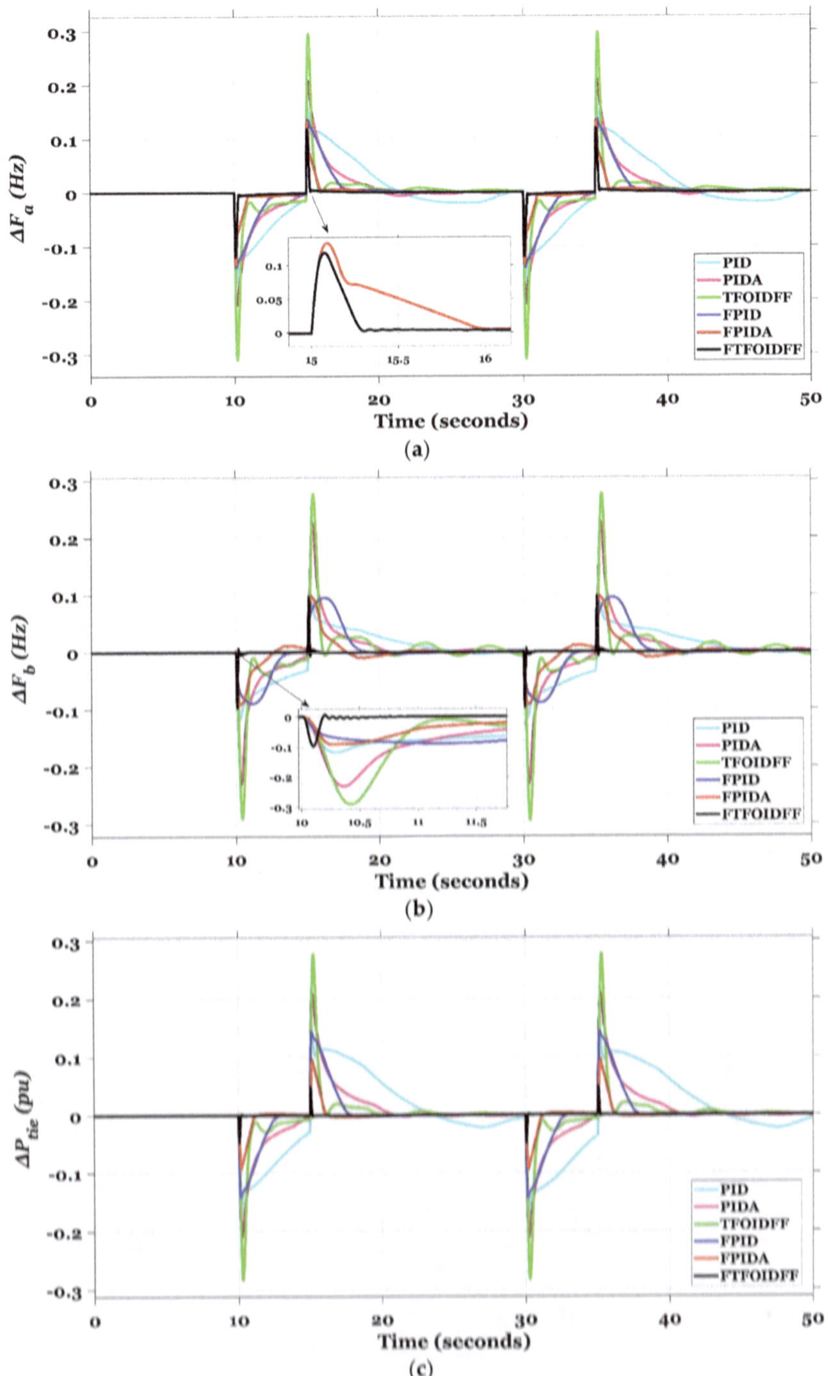

Figure 22. The system dynamics for Case IV. (**a**) ΔFa; (**b**) ΔFb; (**c**) ΔPtie.

Table 7. The transient response specifications of the studied system represented as ITAE value using different controllers for Case IV.

Controller	ITAE			ITAE$_{tot}$
	ΔFa (Hz)	**ΔFb (Hz)**	**ΔPtie (pu)**	
PID	45.48	30.13	58.21	133.8
PIDA	25.53	25.08	27.02	77.64
TFOIDFF	20.16	25.54	17.45	63.15
FPID	17.7	20.84	18.45	56.98
FPIDA	6.826	12.11	5.803	24.74
FTFOIDFF	2.776	1.101	0.5113	4.388

4.5. Case V: Random Sinusoidal Load Perturbation (RSLP) in Area (a)

Within this case, the analysed system is put through rigorous testing by being subjected to a severe random sinusoidal load perturbation (RSLP) profile in area (a), as seen in Figure 23. The formula that describes the RSLP can be expressed as in Equation (27). Following this, a study of the system's performance is provided. The proposed FTFOIDFF is tuned using the PDO algorithm, and its efficacy is measured against the objectives of minimising frequency and tie-line power deviations and maintaining system stability, just as was carried out in the earlier scenarios. Table 8 offers an overview of the ITAE values that were generated by a variety of controllers while taking into account the impact of RSLP. Using the recommended FTFOIDFF controller, one may get the lowest feasible ITAE value. The behaviour of the system under these conditions is shown in Figure 24 as well. The oscillation dampening provided by the proposed combination of fuzzy logic and TFOIDFF is clearly better than that of the other tested controllers. As a result, it is clear that the recommended combination was effective in handling the numerous fluctuations and disturbances.

$$\Delta P_L = 0.15\sin(2.25t) + 0.24\sin(3.45t) - 0.36\sin(4.7t) \tag{27}$$

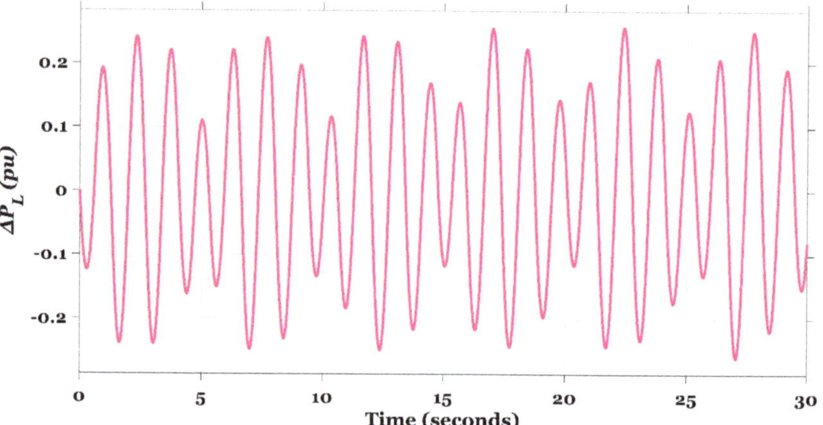

Figure 23. The RSLP profile.

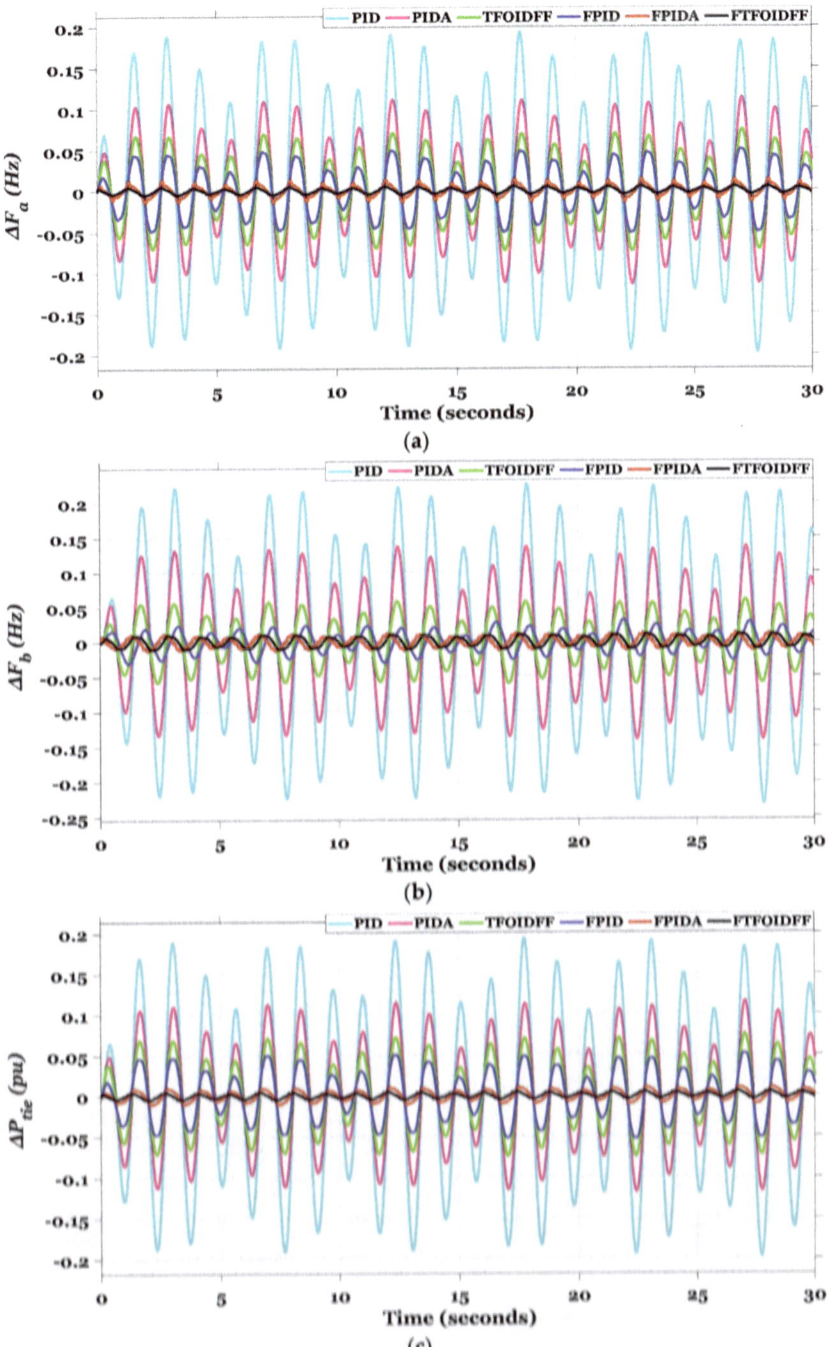

Figure 24. The system dynamics for Case V. (**a**) ΔFa; (**b**) ΔFb; (**c**) ΔPtie.

Table 8. The transient response specifications of the studied system represented as ITAE value using different controllers for Case V.

Controller	ITAE			ITAE$_{tot}$
	ΔFa (Hz)	ΔFb (Hz)	ΔPtie (pu)	
PID	45.53	53.08	45.56	144.2
PIDA	26.19	31.95	26.89	85.03
TFOIDFF	16.82	13.75	17.14	47.71
FPID	11.75	5.644	12.08	29.48
FPIDA	2.485	2.734	2.406	7.626
FTFOIDFF	1.08	2.537	1.076	4.694

4.6. Case VI: MSLP in Area (a) with 0.01 s Communication Time Delay (CTD)

This case study offers the proposal of the CTD challenge that is applied to the controller output with a time delay value of 0.01 s, and it also takes into consideration the application of the MSLP, used in case II, in order to assess the resilience of the recommended FTFOIDFF regulator in terms of system stabilizing. Figure 25 illustrates the various dynamic reactions of the system, which are represented by ΔFa, ΔFb, and ΔPtie. Figure 25 summarizes and explains the performance of the suggested PDO-based FTFOIDFF controller compared to the other controllers in achieving system stability and reliability after evaluating the influence of time delay in the controller action. The PDO-based FTFOIDFF scheme that was developed exhibits good outcomes when it comes to overcoming all of the problems and achieving better system stability. Table 9 shows the dynamic performance of the system as measured by the ITAE value for ΔFa, ΔFb, and ΔPtie. Table 9 also displays the total ITAE value for the system. For the current case, the proposed PDO-based FTFOIDFF has obtained the lowest fitness function with a value of 6.099. This value is almost 28.09 times lower than the PID controller, 20.51 times lower than the PIDA controller, 14.76 times lower than the TFOIDFF controller, 3.78 times lower than the FPID controller, and 2.09 times lower than the FPIDA controller.

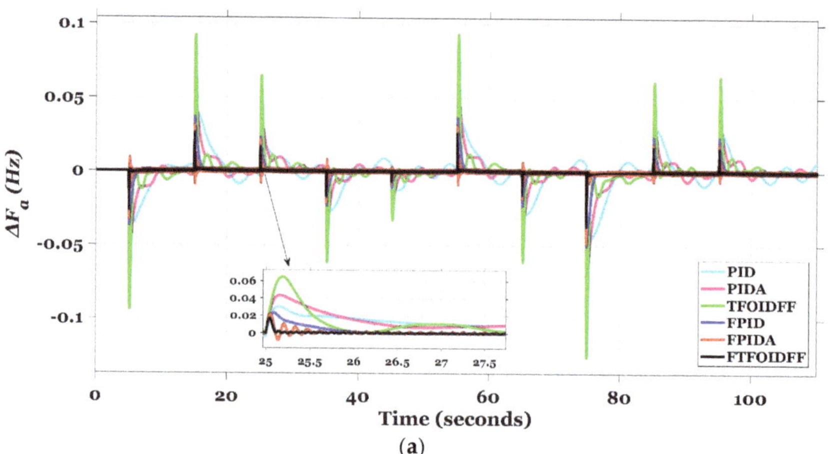

(a)

Figure 25. *Cont.*

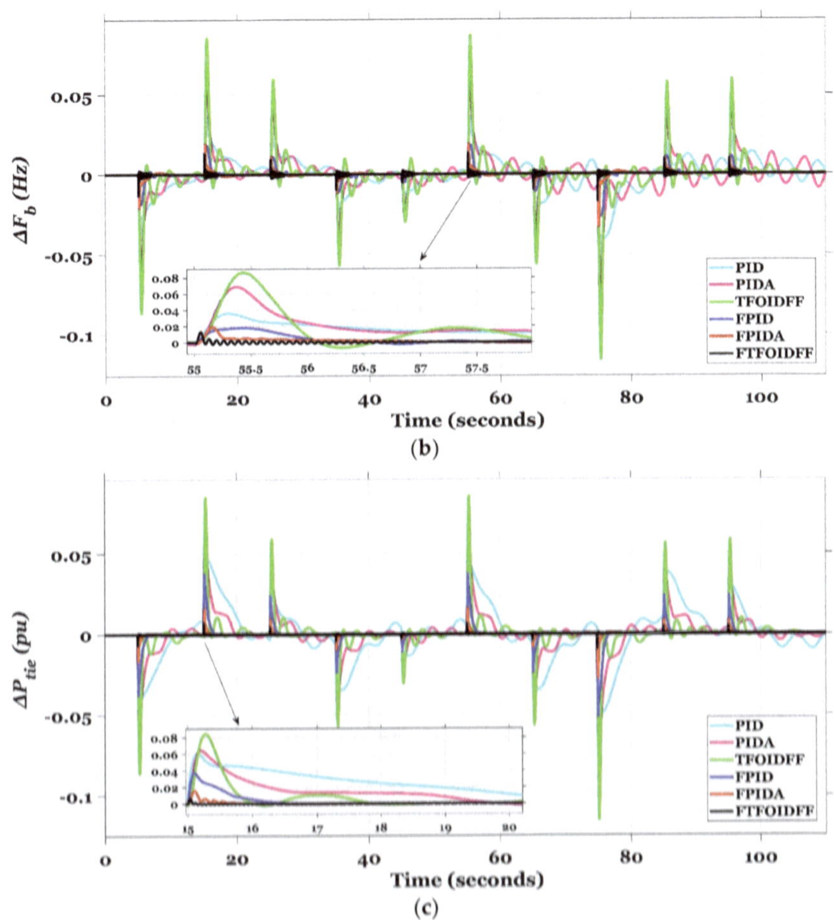

Figure 25. The system dynamics for Case VI. (**a**) ΔFa; (**b**) ΔFb; (**c**) ΔPtie.

Table 9. The transient response specifications of the studied system represented as ITAE value using different controllers for Case VI.

Controller	ITAE ΔFa (Hz)	ΔFb (Hz)	ΔPtie (pu)	ITAE$_{tot}$
PID	50.36	49.53	71.43	171.32
PIDA	35.55	49.72	39.79	125.1
TFOIDFF	28.93	35.9	25.22	90.05
FPID	8.065	6.751	8.225	23.04
FPIDA	6.222	4.129	2.413	12.76
FTFOIDFF	2.132	2.619	1.348	6.099

4.7. Case VII: Applying RESs Fluctuations in Both Areas

This study focuses on high RESs penetration (i.e., the integration of a wind farm unit in area (a) and a PV unit in area (b)) to assess the resilience of the proposed PDO-based FTFOIDFF controller in minimizing the examined system fluctuations. The use of RESs places a strain on the hybrid power grid that was investigated due to the drawbacks associated with these sources (i.e., a lack of inertia in the system). The superiority of the suggested FTFOIDFF has been proved and validated via the use of a variety of control techniques, including PID, PIDA, TFOIDFF, FPID, and FPIDA controllers that are based on the PDO. As shown in Figure 26, the integration of RESs causes severe fluctuations in frequency and flow power in the tie-line. Figure 26 illustrates the system dynamics that assure the reliability and efficacy of the proposed PDO based FTFOIDFF controller in dampening variations in frequency and the flow power in the tie-line and boosting the performance of the investigated power grid. These system dynamics are represented in ΔFa, ΔFb, and $\Delta Ptie$. The dynamic performance of the power system, as evaluated by the ITAE value, is reported in Table 10. With an index of 5.241, the suggested PDO-based FTFOIDFF has proven to have the best overall fitness performance for this scenario. This number is about 101.14 times less than what the PID controller claims, 60.81 times less than what the PIDA controller claims, 65.81 times less than what the TFOIDFF controller claims, 3.47 times less than what the FPID controller claims, and 1.98 times less than what the FPIDA controller claims.

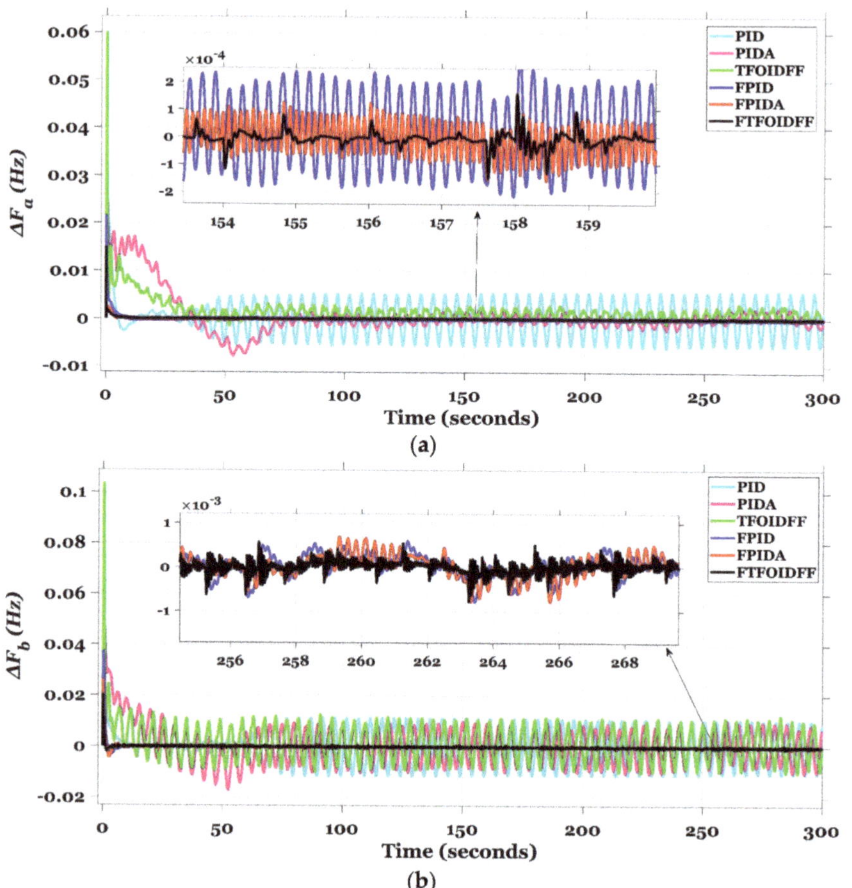

Figure 26. *Cont.*

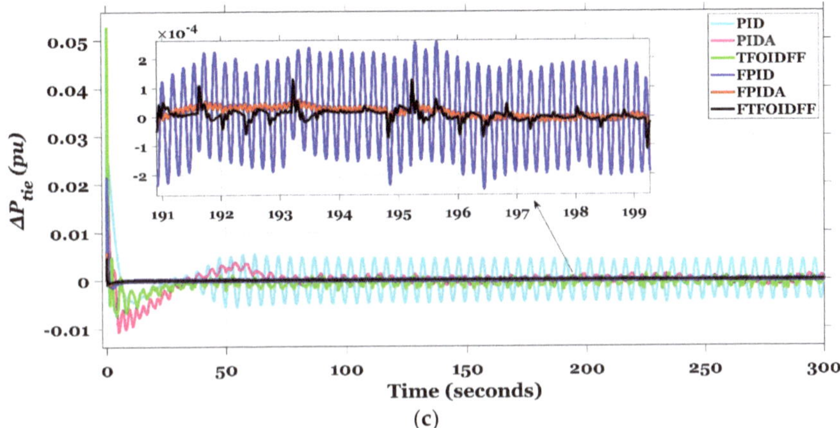

Figure 26. The system dynamics for Case VII. (**a**) ΔFa; (**b**) ΔFb; (**c**) ΔPtie.

Table 10. The transient response specifications of the studied system represented as ITAE value using different controllers for Case VII.

Controller	ITAE			ITAE_tot
	ΔFa (Hz)	ΔFb (Hz)	ΔPtie (pu)	
PID	129.3	230.4	170.5	530.1
PIDA	47.43	245.6	25.69	318.7
TFOIDFF	56.51	258.4	30.04	344.9
FPID	5.713	6.861	5.619	18.19
FPIDA	2.554	7.118	0.7173	10.39
FTFOIDFF	0.777	3.823	0.6413	5.241

4.8. Case VIII: Applying RESs Fluctuations with MSLP in Area (b) and RSLP in Area (a)

In this scenario, the impacts of the three scenarios that came before are taken into consideration simultaneously. As a result, the MSLP profile that is described in Figure 17 is put into effect in area (b), and the RSLP profile that is illustrated in Figure 23 is employed in area (a). Furthermore, the RES penetrations are taken into consideration in both areas (i.e., the integration of the wind farm unit in area (a) and the PV unit in area (b)). The purpose of this is to prove that the recommended PDO-based FTFOIDFF is more superior than existing controllers in preserving system stability under stressful circumstances. The outcomes of this case can be seen in Figure 27 and Table 11. It is possible for the presented controller to obtain a value of the fitness function that is nearly 30.1 times less than the value obtained by the PID controller, 17.9 times lesser than the value obtained by the PIDA controller, 10.22 times fewer than the value obtained by the TFOIDFF controller, 6.21 times lesser than the value obtained by the FPID controller, and 1.63 times smaller than the value obtained by the FPIDA controller. Consequently, it is obvious that the suggested FTFOIDFF controller is an outstanding one that succeeds in effectively managing the many variations and disturbances at the same time.

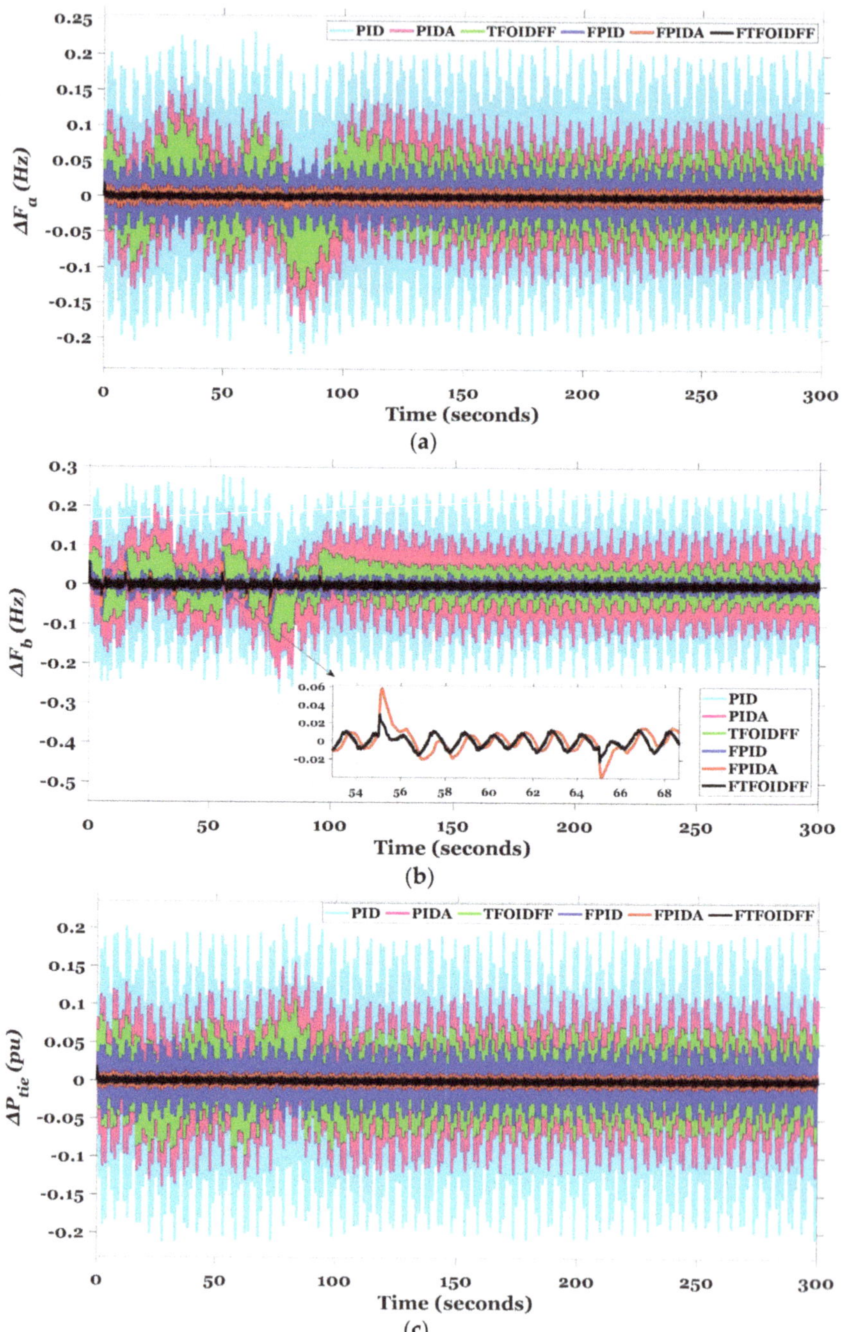

Figure 27. The system dynamics for Case VIII. (**a**) ΔFa; (**b**) ΔFb; (**c**) ΔPtie.

Table 11. The transient response specifications of the studied system represented as ITAE value using different controllers for Case VIII.

Controller	ITAE			ITAE$_{tot}$
	ΔFa (Hz)	ΔFb (Hz)	ΔPtie (pu)	
PID	4517	5261	4501	14,280
PIDA	2626	3214	2671	8511
TFOIDFF	1704	1437	1708	4849
FPID	1164	583.7	1198	2945
FPIDA	249.6	281.5	240.2	771.3
FTFOIDFF	108.6	257.9	108.1	474.6

4.9. Case IX: UPFC and SMES Effect on the Studied System with 30% SLP in Area (a)

In this case, the capability of UPFC and SMES, together with the proposed FTFOIDFF, in improving the dynamic performance of the power system was validated by applying 30% SLP in area (a) while testing system performance for the following cases: without UPFC and SMES units, with UPFC only, with SMES only, and with coordinated application of UPFC and SMES. Figure 28 depicts the dynamics of the system, which indicates very clearly that coordinated application of UPFC and SMES leads to a considerable increase in system performance. This improvement may be measured in terms of least undershoot and overshoot in frequency oscillations, as well as tie-line power exchange. Figure 28 makes it abundantly clear that longer settling times and significant overshoots and undershoots are produced by a system that does not have UPFC and SMES. The undershoot frequency decreased to 0.106 Hz and the settling time decreased to 1.28 s after connecting the SMES units in both regions; the settling time was significantly improved from 1.3 s to 0.5 s when UPFC was applied alone; and the undershoot frequency and settling time were further improved to 0.06 Hz and 0.2 s, respectively, when UPFC and SMES were applied in simultaneously. In addition to that, the fitness function, also known as ITAE, has been improved to 0.201. Table 12 summarizes the whole examination of this case.

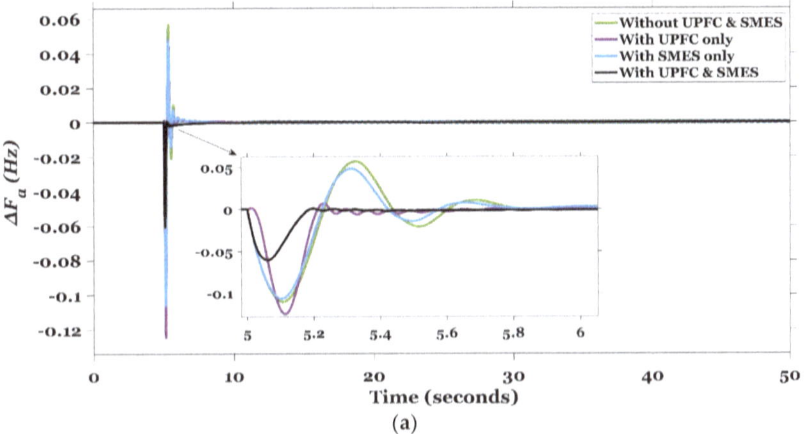

Figure 28. *Cont.*

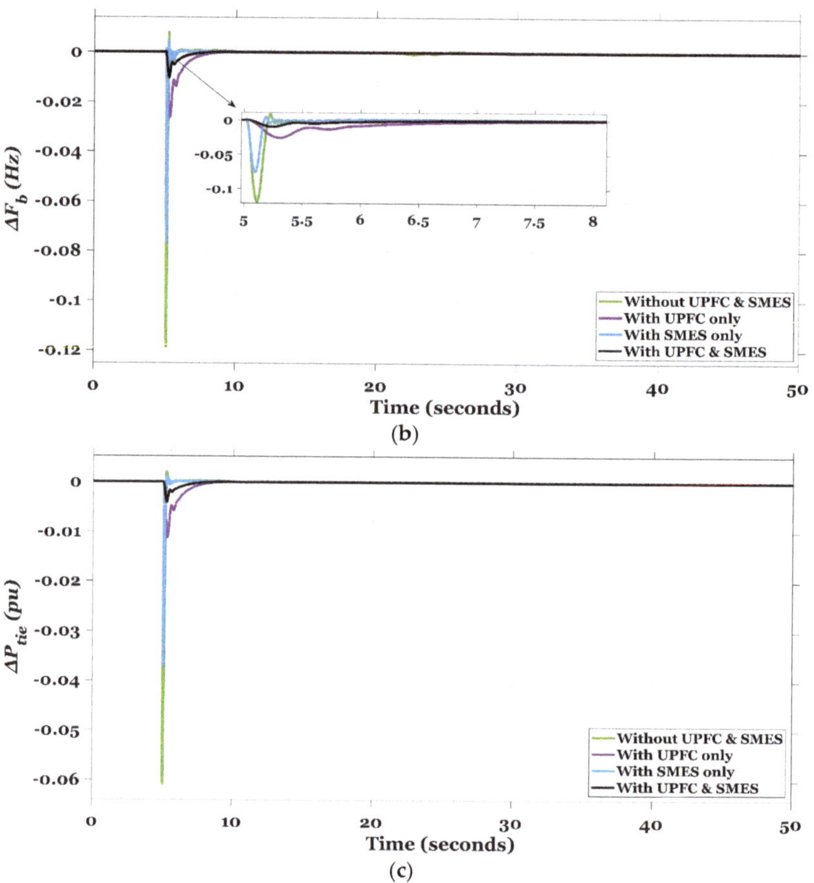

Figure 28. The system dynamics for Case IX. (**a**) ΔFa; (**b**) ΔFb; (**c**) ΔPtie.

Table 12. The transient response specifications of the studied system for Case IX.

Controller	Conditions	ΔFa (Hz)			ΔFb (Hz)			ΔPtie (pu)			ITAE
		MOS	MUS	ST	MOS	MUS	ST	MOS	MUS	ST	
FTFOIDFF	Without UPFC and SMES	0.057	−0.11	1.3	0.008	−0.119	1	0.001	−0.061	0.5	0.614
	With UPFC only	0.007	−0.125	0.5	0	−0.026	2.3	0	−0.011	3	0.377
	With SMES only	0.048	−0.106	1.28	0.005	−0.076	1	0	−0.036	0.5	0.534
	With both UPFC and SMES	0	−0.06	0.2	0	−0.01	2	0	−0.004	2.5	0.201

4.10. Case X: EVs Effect on the Studied System with MSLP in Area (a)

This scenario depicts the integration of EVs into both regions of the power grid that was investigated in order to assess the effectiveness of EVs in managing the studied system frequency and the flow of power between the two areas. Figure 29 provides a description of the different dynamic system responses that are expressed by the parameters ΔFa, ΔFb, and ΔPtie. In Table 13, we can see the ITAE values that correspond to the aforementioned variations in system dynamics as a result of changes in both area frequencies and power flow inside the tie line. If electric vehicles (EVs) are incorporated into the system, the overall ITAE of the system's dynamics falls to 4.605, an improvement of 30.9%. Table 13 demonstrates that the proposed PDO-based FTFOIDFF that takes into account the penetration of EVs in the studied system obtains greater system stability than if these vehicles were not considered. In a nutshell, the incorporation of electric vehicles (EVs) into the power grid that was investigated has the potential to help reduce frequency fluctuations thanks to the energy storage capacity of EVs. This capacity provides the system with additional power under abnormal situations, which helps to ensure that all of the system's dynamic responses remain within acceptable bounds.

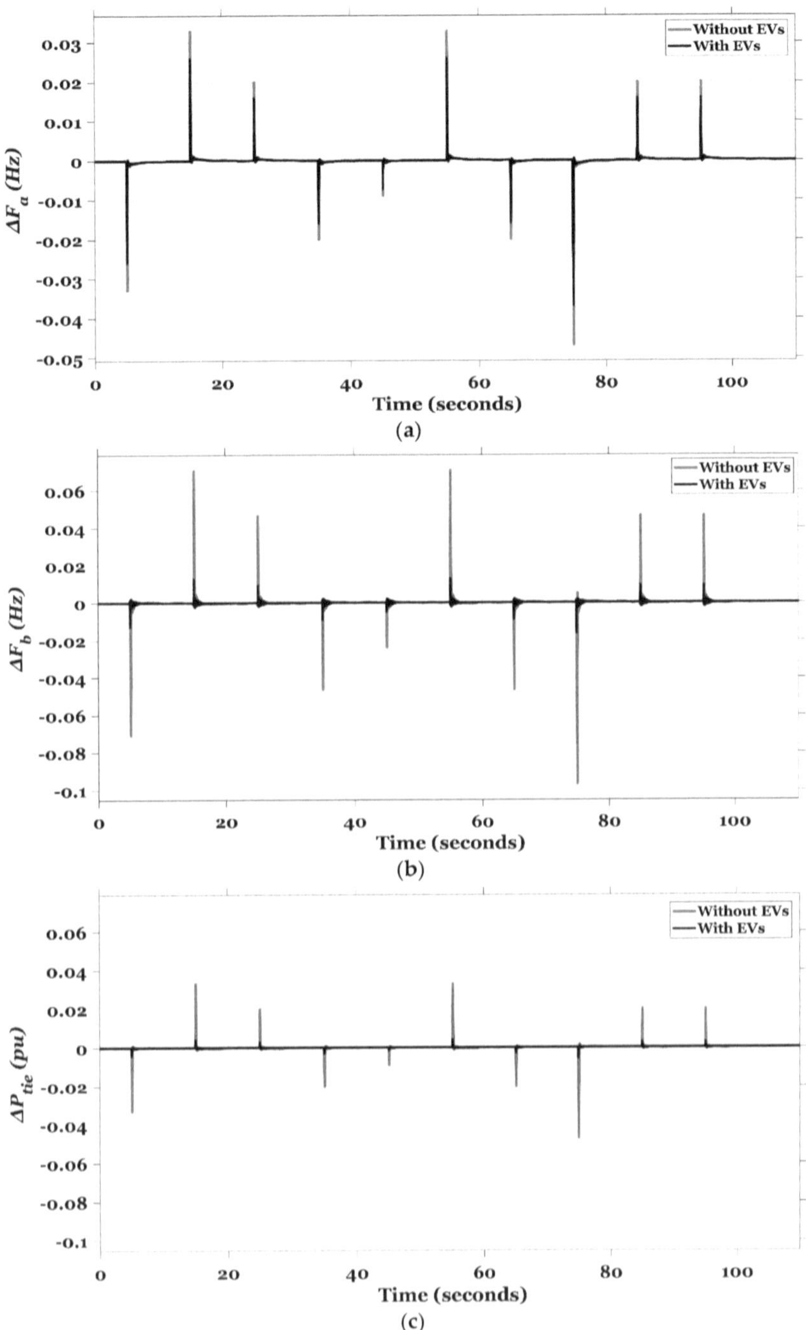

Figure 29. The system dynamics for Case X. (**a**) ΔFa; (**b**) ΔFb; (**c**) $\Delta Ptie$.

Table 13. The transient response specifications of the studied system represented as ITAE value for Case X.

FTFOIDFF Optimized by PDO (Proposed)	ΔFa (Hz)	ITAE ΔFb (Hz)	ΔPtie (pu)	ITAE$_{tot}$
Without EVs	1.946	3.349	1.367	6.662
With EVs	1.451	2.231	0.923	4.605

4.11. Case XI: Sensitivity Analysis

Sensitivity refers to a system's resilience in the face of perturbations to its parameters that fall within a predefined tolerance range. In this part, the resilience of the power system is tested by modifying system parameters such as τ_{gh}, τ_{cd}, y_c, B, k_{EV}, and T_{12} from their nominal values in the range of $+25\%$ to -25% without changing the optimal settings of the proposed FTFOIDFF controller that was provided in case I. The results of the system's performance are presented in Table 14 for a step load change of 10% in area (a) under both nominal and variable conditions. When the above parameters are changed, it is possible to observe that the dynamic responses of ΔFa, ΔFb, and ΔPtie are barely impacted as a result. In addition, the MOS and MUS scarcely vary at all in comparison to the regular operation, however the settling time is somewhat altered in some instances. On the other hand, the dynamic performance of the suggested system is unaffected by any changes in the other system characteristics. As a consequence of this, the PDO-based FTFOIDFF controller that was presented is reliable and demonstrates a high level of effectiveness in preserving system stability even when system parameters are altered.

Table 14. Dynamic response specifications for system parameters change.

Controller	Parameters Variation	% Variation	ΔFa (Hz)			ΔFb (Hz)			ΔPtie (pu)			ITAE
			MOS	MUS	ST	MOS	MUS	ST	MOS	MUS	ST	
FTFOIDFF tuned by PDO (proposed)	Nominal	0	0	−0.0106	0.18	0	−0.003	1	0	−0.0056	0.7	0.0875
	τ_{gh}	+25%	0	−0.0107	0.19	0	−0.003	1.1	0	−0.0057	0.7	0.0878
		−25%	0	−0.0106	0.17	0	−0.003	1	0	−0.0055	0.7	0.0873
	τ_{cd}	+25%	0	−0.0106	0.18	0	−0.0032	1	0	−0.0057	0.7	0.0875
		−25%	0	−0.0106	0.18	0	−0.0029	1	0	−0.0055	0.7	0.0875
	y_c	+25%	0	−0.0107	0.17	0	−0.004	0.9	0	−0.0057	0.5	0.0871
		−25%	0.001	−0.0105	0.2	0.0004	−0.002	1.5	0.0002	−0.0053	1	0.0889
	B	+25%	0	−0.0106	0.17	0	−0.003	0.9	0	−0.0056	0.6	0.0873
		−25%	0	−0.0106	0.2	0	−0.003	1.2	0	−0.0056	0.8	0.0881
	k_{EV}	+25%	0	−0.0105	0.18	0	−0.003	1	0	−0.0055	0.7	0.0874
		−25%	0	−0.0107	0.18	0	−0.003	1	0	−0.0057	0.7	0.0876
	T_{12}	+25%	0	−0.0107	0.18	0	−0.003	1	0	−0.0057	0.7	0.0876
		−25%	0	−0.0105	0.18	0	−0.003	1	0	−0.0055	0.7	0.0874

In closing, the power system's network security has steadily become the primary focus of attention as a result of the ongoing growth of the power system communication network and the subsequent expansion of the coverage area. Since the frequency deviation and the tie-line power signals in the LFC system need to be transferred over a long distance, it is conceivable for an attack to be carried out by the insertion of fake data during the process of signal collecting and transmission. By making unauthorized changes to the system's frequency deviation and tie-line power, the system either incorrectly calculates the value of area control error or compels the load frequency control system to overshoot, which results in frequency oscillation. Moreover, when EVs are integrated into the LFC system, they can be viewed as both the power source and burden for the grid. However, the LFC system that includes EVs is susceptible to covert crimes, and as a result, the power system's dependable operating performance and security will be compromised. Real-time monitoring and detection methodologies have been established that are very prosperous [75,76]. These methodologies were designed so that these security problems could be solved.

5. Conclusions

This study puts out an innovative method for improving load frequency controllers (LFCs) that makes use of a hybrid approach using fuzzy logic control (FLC) and fractional calculus. This article introduces a maiden controller which is called fuzzy tilted fractional-order integral-derivative with fractional-filter (FTFOIDFF) for use in LFC applications. The proposed FTFOIDFF controller combines the best features of tilt, fuzzy logic, FOPID, and fractional filter regulators. Also, a newly developed metaheuristics optimization method called Prairie Dog Optimizer (PDO) is shown to easily tune the recommended regulator parameters. The model of the system takes into account physical restrictions such the Communication Time Delay (CTD), the reheat turbine, and the Generation Rate Constraint (GRC). The results achieved using PDO are compared to those obtained using SOA, RUN, and CGO algorithms, demonstrating PDO's superiority. A UPFC is installed in the tie-line, and SMES units are integrated in both areas so as to test their effect on the performance of the system. Furthermore, EVs contributions are included in both areas. The superior efficacy of the proposed FTFOIDFF controller has been demonstrated by comparing its performance to that of a number of conventional (e.g., PID, PIDA, and TFOIDFF) and intelligent (e.g., FPID and FPIDA) regulators from the literature whose parameters are adjusted using PDO algorithm. It has been shown that the recommended FTFOIDFF controller, which is based on the PDO algorithm, works best when subjected to a wide range of load patterns. In addition, the penetration of renewable energy sources and the latency in communication are considered as potential roadblocks to testing the robustness of the proposed controller and achieving greater system stability. To further illustrate the good impact that SMES, UPFC, and EVs units have on the system as a whole, a variety of scenarios have been created. The sensitivity of the system is evaluated by making alterations to the system's parameters from their baseline values. The simulation results show that, despite the various challenges mentioned above, the proposed FTFOIDFF controller based on the PDO is capable of reaching higher levels of system stability. Furthermore, following research, pros may be summarized as follows:

- The proposed control structure has efficiently improved frequency stability in a multi-area power system with severe RES penetrations.
- Integrating the benefits of tilt, fuzzy logic, FOPID, and fractional filter regulators in a single controller known as FTFOIDFF, which has superior performance over the other recent control structures.
- Application of a nature-inspired metaheuristic optimization technique that was recently developed (i.e., the Prairie Dog Optimizer, or PDO) for the purpose of fine-tuning not only the recommended controller settings but also the MFs of the FLC's inputs and outputs in an effective manner.
- Validation of the positive effect of the integration of SMES, UPFC, and EVs in enhancing frequency performance during several harsh disturbances.

Also, cons can be abridged as following:

- The inclusion of conventional controllers for comparison with intelligent fuzzy-based controllers is unfair, as the incorporation of intelligent controllers, such as fuzzy logic or artificial neural networks, enhances the frequency response performance excessively.
- The use of simple structured models for EV, SMES, and UPFC will not reveal the full impact of these devices or the uncertainties that may be introduced into the systems as a result of their incorporation.

Finally, for future work, the investigated two-area power system can be expanded to three or four areas, and a more complex model for SMES, UPFC, and EVs can be considered.

Author Contributions: The authors confirm the final authorship for this manuscript. All authors have read and agreed to the published version of the manuscript.

Funding: This research work was funded by Institutional Fund Projects under grant no. (IFPIP: 1428-135-1443). The authors gratefully acknowledge technical and financial support provided by the Ministry of Education and King Abdulaziz University, DSR, Jeddah, Saudi Arabia.

Data Availability Statement: The data is available on request from the authors.

Conflicts of Interest: The authors declare no conflict of interest.

Nomenclature

FLC	Fuzzy Logic Control
FTFOIDFF	Fuzzy Tilted Fractional Order Integral Derivative with Fractional Filter
PDO	Prairie Dog Optimizer
SOA	Seagull Optimisation Algorithm
RUN	Runge Kutta optimizer
CGO	Chaos Game Optimizer
LFC	Load Frequency Control
CTD	Communication Time Delay
GRC	Generation Rate Constraint
UPFC	Unified Power Flow Controller
SMES	Superconducting Magnetic Energy Storage
EV	Electric Vehicle
PID	Proportional Integral Derivative
PIDA	Proportional Integral Derivative Acceleration
SLP	Step Load Perturbation
MSLP	Multi-Step Load Perturbation
RLP	Random Load Perturbation
RSLP	Random Sinusoidal Load Perturbation
PLP	Pulse Load Perturbation
AGC	Automatic Generation Control
RES	Renewable Energy Sources
PV	Photovoltaic
FACTS	Flexible Alternating Current Transmission Systems
FO	Fractional Order
PCS	Power Conversion System
PD	Prairie Dog
CT	Coterie
LB	Lower Boundary
UB	Upper Boundary
GB	Global Best
iter	Current Iteration
Maxiter	Maximum Iteration Number
MFs	Membership Functions
E	Error
DOE	Derivative of Error
NB	Negative Big
NS	Negative Small
Z	Zero
PS	Positive Small
PB	Positive Big
FIS	Fuzzy Interface System
U_1	First Control Law
U_t	Total Control Law
G_C	TFOIDFF Controller's Transfer Function
t_{sim}	Simulation Time
K_t	Tilt Gain
K_i	Integral Gain

K_d	Derivative Gain
n	Tilt Fractional Order Power
λ_i	Fractional Order Integral Operator
μ_d	Fractional Order Derivative Operator
λ_f	Fractional Order Filter Operator
N_f	Fractional Filter Coefficient
K_1, K_2	Scaling Factor of the FLC inputs
ACE	Area Control Error
ITAE	Integral Time Absolute Error
MOS	Maximum Overshoot
MUS	Maximum Undershoot
ST	Settling Time
ΔFa	The frequency deviation of Area (a)
ΔFb	The frequency deviation of Area (b)
$\Delta Ptie$	The tie-line power deviation

Appendix A. The Nominal Values of the Power System's Parameters

Parameter	Nominal Value	Parameter Definition
τ_{sg}	0.08 s	Governor time constant
k_r	0.3 s	Gain of reheater steam turbine
τ_r	10.2 s	The time constant of reheater steam turbine
τ_t	0.3 s	Steam turbine time constant
τ_{gh}	0.2 s	Hydroelectric turbine speed governor time constant
τ_{rs}	4.9 s	Hydro turbine speed governor reset time
τ_{rh}	28.749 s	Time constant of the transient droop
τ_w	1.1 s	Average water string time in penstock
b_g	0.049 s	Gas turbine constant of valve positioner
c_g	1	Valves' gas turbine positioner constant
x_c	0.6 s	Gas turbine governor's lead time constant
y_c	1.1 s	Gas turbine governor's lag time constant
τ_{cr}	0.01 s	Combustion response time delay in a gas turbine
τ_{fc}	0.239 s	Gas turbine fuel time constant
τ_{cd}	0.2 s	Volume-time constant for gas turbine compressor discharge
k_{ps1}, k_{ps2}	68.965, 68.965	Power system gains
τ_{ps1}, τ_{ps2}	11.49, 11.49 s	Power system time constants
T_{12}	0.0433 MW	Coefficient of synchronizing
$k_{SMES(a)}$, $k_{SMES(b)}$	1, 1	Gains of SMES
$\tau_{SMES(a)}$, $\tau_{SMES(b)}$	0.07 s	Time constants of SMES units
τ_{UPFC}	0.003 s	Time constant of UPFC unit
$k_{EV(a)}$, $k_{EV(b)}$	1	Gains of EVs
$\tau_{EV(a)}$, $\tau_{EV(b)}$	0.28 s	Time constants of EVs
B_a, B_b	0.431, 0.431 MW/Hz	Frequency bias coefficients
R	2.4 Hz/MW	Governor speed regulation constant for thermal, hydro, and gas units
CF_T, CF_H, CF_G	0.5435, 0.3261, 0.1304	Contribution factors of thermal, hydro, and gas units
GRC with Hydro	--------	(0.045 pu.MW/s) and (0.06 pu.MW/s. For both rising and decreasing rates), respectively
GRC with Thermal	--------	The GRC (generation rate constraint) for the thermal unit is set (0.0017 pu.MW/s) For rising and decreasing rates

References

1. Elgerd, O.I. *Electric Energy Systems Theory—An Introduction*; Tata McGraw Hill: New Delhi, India, 2000.
2. Kundur, P.S.; Malik, O. *Power System Stability and Control*, 2nd ed.; McGraw-Hill Education: Columbus, OH, USA, 2022.
3. Bevrani, H. *Robust Power System Frequency Control*; Springer International Publishing: Cham, Switzerland, 2016.
4. Parmar, K.P.S.; Majhi, S.; Kothari, D.P. Load frequency control of a realistic power system with multi-source power generation. *Int. J. Electr. Power Energ. Syst.* **2012**, *42*, 426–433. [CrossRef]
5. Mohanty, B.; Panda, S.; Hota, P.K. Controller parameters tuning of differential evolution algorithm and its application to load frequency control of multi-source power system. *Int. J. Electr. Power Energ. Syst.* **2014**, *54*, 77–85. [CrossRef]
6. Yildirim, B.; Gheisarnejad, M.; Khooban, M.H. A robust non-integer controller design for load frequency control in modern marine power grids. *IEEE Trans. Emerg. Top. Comput. Intell.* **2022**, *6*, 852–866. [CrossRef]
7. Ahmed, E.M.; Mohamed, E.A.; Elmelegi, A.; Aly, M.; Elbaksawi, O. Optimum Modified Fractional Order Controller for Future Electric Vehicles and Renewable Energy-Based Interconnected Power Systems. *IEEE Access* **2021**, *9*, 29993–30010. [CrossRef]

8.   Sun, P.; Yun, T.; Chen, Z. Multi-objective robust optimization of multi-energy microgrid with waste treatment. *Renew. Energy* **2021**, *178*, 1198–1210. [CrossRef]
9.   Xiao, D.; Chen, H.; Wei, C.; Bai, X. Statistical Measure for Risk-seeking Stochastic Wind Power Offering Strategies in Electricity Markets. *J. Mod. Power Syst. Clean Energy* **2022**, *10*, 1437–1442. [CrossRef]
10. Said, S.M.; Aly, M.; Hartmann, B.; Mohamed, E.A. Coordinated fuzzy logic-based virtual inertia controller and frequency relay scheme for reliable operation of low-inertia power system. *IET Renew. Power Gener.* **2021**, *15*, 1286–1300. [CrossRef]
11. Liu, L.; Hu, Z.; Mujeeb, A. Automatic generation control considering uncertainties of the key parameters in the frequency response model. *IEEE Trans. Power Syst.* **2022**, *37*, 4605–4617. [CrossRef]
12. Magdy, G.; Ali, H.; Xu, D. Effective Control of Smart Hybrid Power Systems: Cooperation of Robust LFC and Virtual Inertia Control Systems. *CSEE J. Power Energy Syst.* **2022**, *8*, 1583–1593. [CrossRef]
13. Tripathy, S.C.; Kalantar, M.; Balasubramanian, R. Dynamics and stability of wind and diesel turbine generators with superconducting magnetic energy storage unit on an isolated power system. *IEEE Trans. Energy Conv.* **1991**, *6*, 579–585. [CrossRef]
14. Abraham, R.J.; Das, D.; Patra, A. Automatic generation control of an interconnected hydrothermal power system considering superconducting magnetic energy storage. *Int. J. Electr. Power Energ. Syst.* **2007**, *29*, 571–579. [CrossRef]
15. Banerjee, S.; Chatterjee, J.K.; Tripathy, S.C. Application of magnetic energy storage unit as load-frequency stabilizer. *IEEE Trans. Energy Conv.* **1990**, *5*, 46–51. [CrossRef]
16. Abraham, R.J.; Das, D.; Patra, A. AGC of a hydrothermal system with SMES unit. In Proceedings of the 2006 IEEE GCC Conference (GCC), Manama, Bahrain, 20–22 March 2006; pp. 1–7.
17. Padhan, S.; Sahu, R.K.; Panda, S. Automatic generation control with thyristor controlled series compensator including superconducting magnetic energy storage units. *Ain Shams Eng. J.* **2014**, *5*, 759–774. [CrossRef]
18. Sudha, K.R.; Vijaya, S.R. Load frequency control of an interconnected reheat thermal system using type-2 fuzzy system including SMES units. *Int. J. Electr. Power Energ. Syst.* **2012**, *43*, 1383–1392. [CrossRef]
19. Hingorani, N.G.; Gyugyi, L. *Understanding FACTS: Concepts and Technology of Flexible AC Transmission System*; IEEE Press: New York, NY, USA, 2000.
20. Praghnesh, B.; Ranjit, R. Load frequency stabilization by coordinated control of Thyristor Controlled Phase Shifters and superconducting magnetic energy storage for three types of interconnected two-area power systems. *Int. J. Electr. Power Energ. Syst.* **2010**, *32*, 1111–1124.
21. Praghnesh, B.; Ranjit, R. Comparative performance evaluation of SMES-SMES, TCPS-SMES and SSSC-SMES controllers in automatic generation control for a two-area hydro-hydro system. *Int. J. Electr. Power Energ. Syst.* **2011**, *32*, 1585–1597.
22. Paliwal, N.; Srivastava, L.; Pandit, M. Application of grey wolf optimization algorithm for load frequency control in multi-source single area power system. *Evol. Intell.* **2022**, *15*, 563–584. [CrossRef]
23. Aryan, P.; Ranjan, M.; Shankar, R. Deregulated LFC scheme using equilibrium optimized Type-2 fuzzy controller. *WEENTECH Proc. Energy* **2021**, *8*, 494–505. [CrossRef]
24. Dreidy, M.; Mokhlis, H.; Mekhilef, S. Inertia response and frequency control techniques for renewable energy sources: A review. *Renew. Sustain. Energy Rev.* **2017**, *69*, 144–155. [CrossRef]
25. Fernández-Guillamón, A.; Gómez-Lázaro, E.; Muljadi, E.; Molina-García, Á. Power systems with high renewable energy sources: A review of inertia and frequency control strategies over time. *Renew. Sustain. Energy Rev.* **2019**, *115*, 109369. [CrossRef]
26. Singh, V.P.; Kishor, N.; Samuel, P. Improved load frequency control of power system using LMI based PID approach. *J. Franklin Inst.* **2017**, *354*, 6805–6830. [CrossRef]
27. Jagatheesan, K.; Anand, B.; Samanta, S.; Dey, N.; Ashour, A.S.; Balas, V.E. Particle swarm optimisation-based parameters optimisation of PID controller for load frequency control of multi-area reheat thermal power systems. *Int. J. Adv. Intell. Paradig.* **2017**, *9*, 464. [CrossRef]
28. Hasanien, H.M. Whale optimisation algorithm for automatic generation control of interconnected modern power systems including renewable energy sources. *IET Gener. Transm. Distrib.* **2018**, *12*, 607–614. [CrossRef]
29. Mehta, P.; Bhatt, P.; Pandya, V. Optimized coordinated control of frequency and voltage for distributed generating system using Cuckoo Search Algorithm. *Ain Shams Eng. J.* **2018**, *9*, 1855–1864. [CrossRef]
30. Sobhy, M.A.; Abdelaziz, A.Y.; Hasanien, H.M.; Ezzat, M. Marine predators algorithm for load frequency control of modern interconnected power systems including renewable energy sources and energy storage units. *Ain Shams Eng. J.* **2021**, *12*, 3843–3857. [CrossRef]
31. Nayak, S.R.; Khadanga, R.K.; Arya, Y.; Panda, S.; Sahu, P.R. Influence of ultra-capacitor on AGC of five-area hybrid power system with multi-type generations utilizing sine cosine adopted dingo optimization algorithm. *Electr. Power Syst. Res.* **2023**, *223*, 109513. [CrossRef]
32. Veerendar, T.; Kumar, D.; Sreeram, V. Maiden application of colliding bodies optimizer for load frequency control of two-area nonreheated thermal and hydrothermal power systems. *Asian J. Control.* **2023**, *25*, 3443–3455. [CrossRef]
33. Murugesan, D.; Jagatheesan, K.; Shah, P.; Sekhar, R. Fractional order PIλDμ controller for microgrid power system using cohort intelligence optimization. *Results Control. Optim.* **2023**, *11*, 100218. [CrossRef]
34. Andic, C.; Ozumcan, S.; Varan, M.; Ozturk, A. A novel Sea Horse Optimizer based load frequency controller for two-area power system with PV and thermal units. *Preprints* **2023**, 2023040368. [CrossRef]

35. Gheisarnejad, M. An effective hybrid harmony search and cuckoo optimization algorithm based fuzzy PID controller for load frequency control. *Appl. Soft Comput.* **2018**, *65*, 121–138. [CrossRef]
36. Ahmed, M.; Magdy, G.; Khamies, M.; Kamel, S. Modified TID controller for load frequency control of a two-area interconnected diverse-unit power system. *Int. J. Electr. Power Energy Syst.* **2022**, *135*, 107528. [CrossRef]
37. Zahariev, A.; Kiskinov, H. Asymptotic Stability of the Solutions of Neutral Linear Fractional System with Nonlinear Perturbation. *Mathematics* **2020**, *8*, 390. [CrossRef]
38. Milev, M.; Zlatev, S. A Note about Stability of Fractional Retarded Linear Systems with Distributed Delays. *Int. J. Pure Appl. Math.* **2017**, *115*, 873–881. [CrossRef]
39. Kiskinov, H.; Milev, M.; Zahariev, A. About the Resolvent Kernel of Neutral Linear Fractional System with Distributed Delays. *Mathematics* **2022**, *10*, 4573. [CrossRef]
40. Kumar Sahu, R.; Panda, S.; Biswal, A.; Chandra Sekhar, G.T. Design and analysis of tilt integral derivative controller with filter for load frequency control of multi-area interconnected power systems. *ISA Trans.* **2016**, *61*, 251–264. [CrossRef] [PubMed]
41. Topno, P.N.; Chanana, S. Load frequency control of a two-area multi-source power system using a tilt integral derivative controller. *J. Vib. Control* **2018**, *24*, 110–125. [CrossRef]
42. Shouran, M.; Anayi, F.; Packianather, M.; Habil, M. Load frequency control based on the Bees Algorithm for the Great Britain power system. *Designs* **2021**, *5*, 50. [CrossRef]
43. Bayati, N.; Dadkhah, A.; Vahidi, B.; Sadeghi, S.H.H. Fopid design for load-frequency control using genetic algorithm. *Sci. Int.* **2015**, *27*, 3089–3094.
44. Pan, I.; Das, S. Fractional order AGC for distributed energy resources using robust optimization. *IEEE Trans. Smart Grid* **2016**, *7*, 2175–2186. [CrossRef]
45. Ramachandran, R.; Satheesh Kumar, J.; Madasamy, B.; Veerasamy, V. A hybrid MFO-GHNN tuned self-adaptive FOPID controller for ALFC of renewable energy integrated hybrid power system. *IET Renew. Power Gener.* **2021**, *15*, 1582–1595. [CrossRef]
46. Sambariya, D.K.; Nagar, O.; Sharma, A.K. Application of FOPID design for LFC using flower pollination algorithm for three-area power system. *Univers. J. Contr. Autom.* **2020**, *8*, 212695735. [CrossRef]
47. Alharbi, M.; Ragab, M.; AboRas, K.M.; Kotb, H.; Dashtdar, M.; Shouran, M.; Elgamli, E. Innovative AVR-LFC design for a multi-area power system using hybrid fractional-order PI and PIDD2 controllers based on dandelion optimizer. *Mathematics* **2023**, *11*, 1387. [CrossRef]
48. Latif, A.; Hussain, S.M.S.; Das, D.C.; Ustun, T.S. Optimum synthesis of a BOA optimized novel dual-stage PI-(1+ID) controller for frequency response of a microgrid. *Energies* **2020**, *13*, 3446. [CrossRef]
49. Mohamed, E.A.; Ahmed, E.M.; Elmelegi, A.; Aly, M.; Elbaksawi, O.; Mohamed, A.-A.A. An optimized hybrid fractional order controller for frequency regulation in multi-area power systems. *IEEE Access* **2020**, *8*, 213899–213915. [CrossRef]
50. Yogendra, A. A novel CFFOPI-FOPID controller for AGC performance enhancement of single and multi-area electric power systems. *ISA Trans.* **2020**, *100*, 126–135. [CrossRef]
51. Yogendra, A. A new optimized fuzzy FOPI-FOPD controller for automatic generation control of electric power systems. *J. Franklin Inst.* **2019**, *356*, 5611–5629. [CrossRef]
52. Gheisarnejad, M.; Khooban, M.H. Design an optimal fuzzy fractional proportional integral derivative controller with derivative filter for load frequency control in power systems. *Trans. Inst. Meas. Control* **2019**, *41*, 2563–2581. [CrossRef]
53. Yogendra, A. Improvement in automatic generation control of two-area electric power systems via a new fuzzy aided optimal PIDN-FOI controller. *ISA Trans.* **2018**, *80*, 475–490. [CrossRef]
54. Arya, Y.; Kumar, N.; Dahiya, P.; Sharma, G.; Çelik, E.; Dhundhara, S.; Sharma, M. Cascade-I λ D μ N controller design for AGC of thermal and hydro-thermal power systems integrated with renewable energy sources. *IET Renew. Power Gener.* **2021**, *15*, 504–520. [CrossRef]
55. Magdy, G.; Shabib, G.; Elbaset, A.A.; Mitani, Y. Frequency stabilization of renewable power systems based on MPC with application to the Egyptian grid. *IFAC-PapersOnLine* **2018**, *51*, 280–285. [CrossRef]
56. Zhang, C.; Wang, S.; Zhao, Q. Distributed economic MPC for LFC of multi-area power system with wind power plants in power market environment. *Int. J. Electr. Power Energy Syst.* **2020**, *126*, 106548. [CrossRef]
57. Fathy, A.; Kassem, A.M. Antlion optimizer-ANFIS load frequency control for multi-interconnected plants comprising photovoltaic and wind turbine. *ISA Trans.* **2018**, *87*, 282–296. [CrossRef]
58. Patowary, M.; Panda, G.; Naidu, B.R.; Deka, B.C. ANN-based adaptive current controller for on-grid DG system to meet frequency deviation and transient load challenges with hardware implementation. *IET Renew. Power Gener.* **2017**, *12*, 61–71. [CrossRef]
59. Dombi, J.; Hussain, A. A new approach to fuzzy control using the distending function. *J. Process. Control.* **2019**, *86*, 16–29. [CrossRef]
60. Valdez, F.; Castillo, O.; Peraza, C. Fuzzy logic in dynamic parameter adaptation of Harmony search optimization for benchmark functions and fuzzy controllers. *Int. J. Fuzzy Syst.* **2020**, *22*, 1198–1211. [CrossRef]
61. Yakout, A.H.; Kotb, H.; Hasanien, H.M.; Aboras, K.M. Optimal fuzzy PIDF load frequency controller for hybrid microgrid system using marine predator algorithm. *IEEE Access* **2021**, *9*, 54220–54232. [CrossRef]
62. Rajesh, K.S.; Dash, S.S. Load frequency control of autonomous power system using adaptive fuzzy based PID controller optimized on improved sine cosine algorithm. *J. Ambient. Intell. Humaniz. Comput.* **2018**, *10*, 2361–2373. [CrossRef]

63. Mishra, D.; Sahu, P.C.; Prusty, R.C.; Panda, S. Fuzzy adaptive Fractional Order-PID controller for frequency control of an Islanded Microgrid under stochastic wind/solar uncertainties. *Int. J. Ambient. Energy* **2021**, *43*, 4602–4611. [CrossRef]
64. Osinski, C.; Leandro, G.V.; Oliveira, G.H.d.C. A new hybrid load frequency control strategy combining fuzzy sets and differential evolution. *J. Control. Autom. Electr. Syst.* **2021**, *32*, 1627–1638. [CrossRef]
65. Mohamed, E.A.; Aly, M.; Watanabe, M. New tilt fractional-order integral derivative with fractional filter (TFOIDFF) controller with artificial hummingbird optimizer for LFC in renewable energy power grids. *Mathematics* **2022**, *10*, 3006. [CrossRef]
66. Yakout, A.H.; Attia, M.A.; Kotb, H. Marine predator algorithm based cascaded PIDA load frequency controller for electric power systems with wave energy conversion systems. *Alex. Eng. J.* **2021**, *60*, 4213–4222. [CrossRef]
67. Pradhan, P.C.; Sahu, R.K.; Panda, S. Firefly algorithm optimized fuzzy PID controller for AGC of multi-area multi-source power systems with UPFC and SMES. *Eng. Sci. Technol. Int. J.* **2016**, *19*, 338–354. [CrossRef]
68. AboRas, K.M.; Ragab, M.; Shouran, M.; Alghamdi, S.; Kotb, H. Voltage and frequency regulation in smart grids via a unique Fuzzy PIDD2 controller optimized by Gradient-Based Optimization algorithm. *Energy Rep.* **2023**, *9*, 1201–1235. [CrossRef]
69. Elkasem, A.H.A.; Khamies, M.; Hassan, M.H.; Agwa, A.M.; Kamel, S. Optimal design of TD-TI controller for LFC considering renewables penetration by an improved chaos game optimizer. *Fractal Fract.* **2022**, *6*, 220. [CrossRef]
70. Gyugyi, L. Unified power-flow control concept for flexible AC transmission systems. *IEE Proc.* **1992**, *139*, 323. [CrossRef]
71. Khadanga, R.K.; Panda, S. Gravitational search algorithm for Unified Power Flow Controller based damping controller design. In Proceedings of the 2011 International Conference on Energy, Automation and Signal, Bhubaneswar, India, 28–30 December 2011.
72. Kazemi, A.; Shadmesgaran, M.R. Extended supplementary Controller of UPFC to improve damping inter-area oscillations considering inertia coefficient. *Int. J. Energy* **2008**, *2*, 25–36.
73. Ezugwu, A.E.; Agushaka, J.O.; Abualigah, L.; Mirjalili, S.; Gandomi, A.H. Prairie dog optimization algorithm. *Neural Comput. Appl.* **2022**, *34*, 20017–20065. [CrossRef]
74. Yang, X.-S.; Deb, S. Cuckoo search via Lévy flights. In Proceedings of the 2009 World Congress on Nature & Biologically Inspired Computing (NaBIC), Coimbatore, India, 9–11 December 2009; pp. 210–214.
75. Zhu, R.; Huang, C.; Deng, S.; Li, Y. Detection of false data injection attacks based on Kalman filter and controller design in power system LFC. *J. Phys. Conf. Ser.* **2021**, *1861*, 012120. [CrossRef]
76. Wu, Z.; Tian, E.; Chen, H. Covert attack detection for LFC systems of electric vehicles: A dual time-varying coding method. *IEEE ASME Trans. Mechatron.* **2023**, *28*, 681–691. [CrossRef]

 *fractal and fractional*

Article

A New Fractional-Order Virtual Inertia Support Based on Battery Energy Storage for Enhancing Microgrid Frequency Stability

Morsy Nour [1,2,*], **Gaber Magdy** [2,3], **Abualkasim Bakeer** [4], **Ahmad A. Telba** [5], **Abderrahmane Beroual** [6], **Usama Khaled** [2] and **Hossam Ali** [2]

[1] Institute for Research in Technology (IIT), ICAI School of Engineering, Comillas Pontifical University, 28015 Madrid, Spain
[2] Department of Electrical Engineering, Faculty of Energy Engineering, Aswan University, Aswan 81528, Egypt; gabermagdy@aswu.edu.eg (G.M.); usamakhaled@energy.aswu.edu.eg (U.K.); hossam_ali@energy.aswu.edu.eg (H.A.)
[3] Faculty of Engineering, King Salman International University, El-Tor 46511, Egypt
[4] Department of Electrical Engineering, Faculty of Engineering, Aswan University, Aswan 81542, Egypt
[5] Electrical Engineering Department, College of Engineering, King Saud University, Riyadh 11421, Saudi Arabia; atelba@ksu.edu.sa
[6] AMPERE Lab UMR CNRS 5005, Ecole Centrale de Lyon, University of Lyon, 36 Avenue Guy de Collongue, 69130 Ecully, France; abderrahmane.beroual@ec-lyon.fr
*   Correspondence: mmohammed@comillas.edu

Citation: Nour, M.; Magdy, G.; Bakeer, A.; Telba, A.A.; Beroual, A.; Khaled, U.; Ali, H. A New Fractional-Order Virtual Inertia Support Based on Battery Energy Storage for Enhancing Microgrid Frequency Stability. *Fractal Fract.* **2023**, 7, 855. https://doi.org/10.3390/fractalfract7120855

Academic Editors: Arman Oshnoei and Behnam Mohammadi-Ivatloo

Received: 4 November 2023
Revised: 26 November 2023
Accepted: 28 November 2023
Published: 30 November 2023

Abstract: Microgrids have a low inertia constant due to the high penetration of renewable energy sources and the limited penetration of conventional generation with rotating mass. This makes microgrids more susceptible to frequency stability challenges. Virtual inertia control (VIC) is one of the most effective approaches to improving microgrid frequency stability. Therefore, this study proposes a new model to precisely mimic inertia power based on an energy storage system (ESS) that supports low-inertia power systems. The developed VIC model considers the effect of both the DC-DC converter and the DC-AC inverter on the power of the ESS used. This allows for more precise and accurate modeling of the VIC compared to conventional models. Moreover, this study proposes a fractional-order derivative control for the proposed VIC model to provide greater flexibility in dealing with different perturbations that occur in the system. Furthermore, the effectiveness of the proposed fractional-order VIC (FOVIC) is verified through an islanded microgrid that includes heterogeneous sources: a small thermal power plant, wind and solar power plants, and ESSs. The simulation results performed using MATLAB software indicate that the proposed VIC scheme provides fast stabilization times and slight deviations in system frequency compared to the conventional VIC schemes. The proposed VIC outperforms the conventional load frequency control by about 80% and the conventional VIC model by about 45% in tackling load/RESs fluctuations and system uncertainty. Additionally, the studied microgrid with the proposed FOVIC scheme is noticeably more stable and responds faster than that designed with integer-order derivative control. Thus, the proposed FOVIC scheme gives better performance for frequency stability of low-inertia power systems compared to conventional VIC schemes used in the literature.

Keywords: fractional-order virtual inertia control; virtual inertia control; virtual synchronous generator; automatic generation control; battery energy storage; frequency regulation

1. Introduction

There is a worldwide interest in generating electricity from renewable energy sources (RESs). Various RESs are being integrated into power systems, such as hydro, solar, wind, geothermal, biomass, ocean energy, etc. Many countries aim to achieve more than 90% electricity generation from RESs by 2050 [1]. Energy security concerns, environmental issues, and economic benefits drive this interest. However, the increase in intermittent RESs penetration results in tedious power system operation and control due to uncertainty and intermittency of RESs production [2].

Synchronous generators (SGs)-led conventional power systems have the inertia capability to maintain voltage and frequency deviations within standard ranges [3]. Though the decreasing inertia makes future power systems driven by power converters more susceptible to system insecurity [4]. Particularly for isolated small power systems and microgrids (MGs), the low-inertia feature adversely affects system stability when distributed generation (DG) penetration increases and loads suddenly change [5]. For instance, the abrupt rise in loads or disconnection of generation units may exacerbate the transient response of MGs, such as the rate of change of frequency (RoCoF) and frequency deviation, leading to system instability [6], and protection devices such as under-frequency load shedding may intervene to prevent a blackout of the whole MG [7]. Many studies investigated the frequency control of microgrids integrated with a high share of RESs [8,9].

Various secondary controllers were developed in previous studies to improve power systems' frequency stability in addition to conventional controllers. The proportional integral derivative derivative (PIDD) controller optimized by the fruit fly algorithm provided better performance than the integral (I), proportional integral (PI), proportional integral derivative (PID), and integral derivative derivative (IDD) controllers for a deregulated power system [10]. The frequency stability of a two-area power system is improved using a new cascaded controller optimized using artificial rabbit optimizers [11]. The authors of [12] enhanced islanded MG frequency stability using a prairie dog optimization-based cascaded controller. Reference [13] used a new cascaded controller optimized by Barnacle Mating Optimizer for improving the frequency stability of two-area interconnected MGs. Another study used a fuzzy cascaded controller to improve frequency control of an islanded MG containing several types of generators and energy storage [14].

In an attempt to cope with the lack of inertia concern, virtual synchronous generators (VSGs) or virtual inertia concepts are proposed [15–18]. With the aid of these techniques, the power electronic devices coupled with RESs or energy storage systems (ESSs) are controlled to emulate the real SGs swing equation and deliver the necessary inertia support to the grid. Consequently, the system frequency stability could be improved, and the frequency fluctuations and RoCoF were reduced following disturbances [19].

Several control techniques have been developed in the literature based on the virtual inertia concept. A conventional PI approach has been applied in [20] to enhance the dynamic security of isolated MGs. Furthermore, a PI-based virtual inertia control (VIC) of a wind turbine has been implemented [21]. In [22], a fuzzy-based virtual inertia strategy is introduced to control frequency deviation in a hybrid power system. Methodologies considering robust control techniques, such as the coefficient diagram method and H_∞ have been proposed in [23,24] for MGs considering high RESs penetration. Other methodologies have adopted model predictive control [25] and adaptive VIC [26,27] to improve system frequency stability.

To attain the best performance from the above-mentioned controllers, they usually contain parameters that need to be optimally fine-tuned. However, selecting the optimal parameters for various virtual inertia frequency controllers in MGs is challenging, as it either employs trial-and-error methods or depends on the designer's experience. To determine the optimal frequency controller parameters, many researchers have implemented a variety of optimization techniques, including particle swarm optimization [28], a modified gray wolf optimization algorithm [29], a chaotic crow search algorithm [30], a Newton-based eigenvalue optimization algorithm [31], a knee point-driven evolutionary algorithm [32], a jellyfish search optimizer [33], and other recently developed optimization algorithms that have been successfully implemented in frequency control applications.

It can be noticed from the literature that the adoption of virtual inertia increases the system order, which causes the output active power to oscillate and adversely affects frequency dynamics. In addition, in the scope of fine-tuning the prior controllers, it is not easy to realize a reasonable trade-off in performance between robustness and control. Hence, it is challenging to guarantee robust stability and efficiency with the aforementioned controlling approaches under a wide range of demand and generation perturbations.

In recent days, the fractional-order (FO) controller has received a lot of attention due to its greater flexibility for adjusting system dynamics, specifically for systems operating in uncertain environments [34–36]. It has been adopted in a deregulated environment for automatic generation control (AGC) [37] and load frequency control (LFC) [38]. The FO controller's superiority is due to its two additional tuning knobs, such as the FO of the differentiator (μ) and the integrator (λ). Because of their inherent flexibility, several researchers have recommended FO controllers over traditional PI and PID controllers for power system stability applications [39,40]. On the other hand, rare work has considered the application of FO derivative in VIC (FOVIC) to improve the conventional VIC (i.e., which uses integer order derivative) response, as in [41,42]. This new concept needs further study on different power systems with different resources and operating conditions. Interestingly, the FOVSG was created in [41,42] with the presumption that the DC link is ideal; nevertheless, the DC link's energy storage has to be further included in the VIC model due to its limitations. In addition, the authors of [41,42], did not study the influence of variation of the virtual inertia constant with different values of FO operators.

Inspired by the above considerations, this work proposes a FOVIC that is applied to an islanded MG, including a small thermal power plant, wind power plant, solar power plants, ESS, and MG loads. The contributions of this research are summarized as follows:

1. Develop a new VIC model separating the DC-DC converter and DC-AC inverter stages. This allows for more precise and accurate modeling of the individual stages compared to the conventional VIC model introduced in [4,20,25,27].
2. Propose a FO derivative control that is applied to the suggested, developed virtual inertia model. The FOVIC has the benefit of reducing system order, which can considerably suppress frequency fluctuation and output power oscillation.
3. Including FOVIC boosts the system's degree of freedom, thus strengthening system stability and further enhancing dynamic performance in the presence of numerous operational circumstances and high RES penetration rates.
4. The considered islanded MG with the proposed FOVIC scheme is examined under different values of the virtual inertia constant and different values of FO operators, highlighting the best operating values for varied case studies.
5. Graphical and numerical outcomes from simulation indicate how competent the suggested control strategy is, where it can significantly reduce frequency deviations compared to the conventional controllers in the literature.

The rest of this paper is structured as follows: Section 2 introduces system configuration. In Section 3, the design of the proposed FOVIC control system is developed. Section 4 verifies the effectiveness of the suggested model and FOVIC through simulation results. The discussion is presented in Section 5. In Section 6, this paper's main conclusions are summarized.

2. System Modelling and Configuration

This study uses an islanded MG system, i.e., a test system, to examine the effectiveness of the proposed VIC model and the impact of the FO control strategy on its performance. The studied MG includes a dispatchable distributed generator (i.e., small thermal power plant), non-dispatchable distributed generators (i.e., wind and solar power plants), ESS, and MG loads [20]. Moreover, the studied system contains a 20 MW thermal power plant, a 4 MW solar power plant, an 8 MW wind power plant, a 1 MW ESS, and 15 MW local loads. The studied island, MG, has a 20 MW power base. A schematic diagram of the investigated system with the proposed control strategy is revealed in Figure 1. Figure 2 shows the frequency response model of the studied system with the proposed developed control strategy, and the studied system's parameters are given in Table 1.

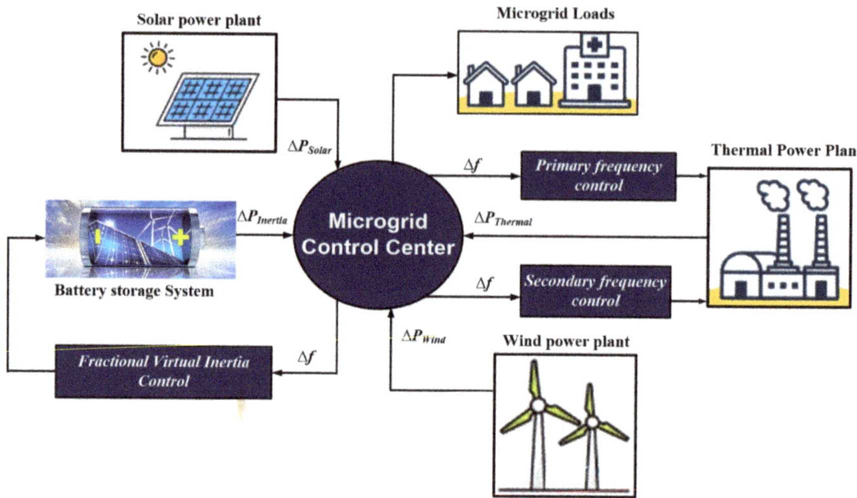

Figure 1. A schematic diagram of the studied islanded MG with the proposed FOVIC strategy.

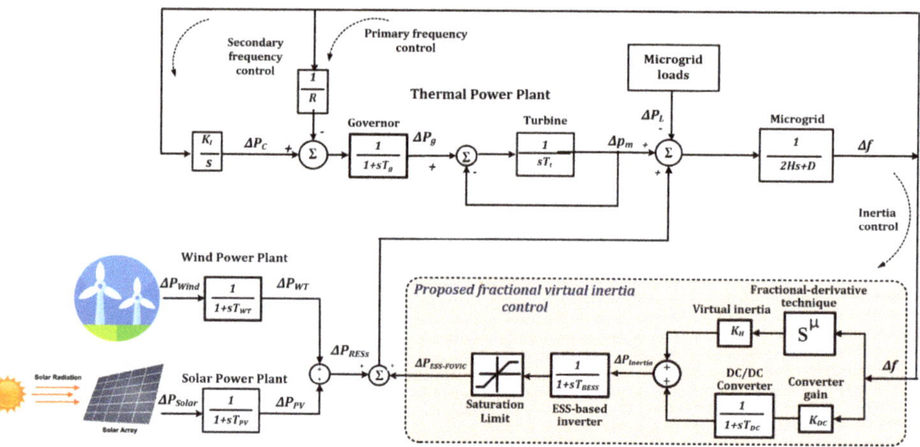

Figure 2. A LFC model of the studied islanded MG with the proposed developed FOVIC.

Table 1. Values of the studied MG parameters [20].

Parameter	Value
Equivalent inertia constant, H (p.u. MWs)	0.082
Microgrid frequency, f (Hz)	50.000
Microgrid damping coefficient, D (p.u.MW/Hz)	0.015
Turbine time constant, T_t (s)	0.400
Governor time constant, T_g (s)	0.100
Speed droop characteristic, R (Hz/p.u.MW)	2.400
Integral control variable gain, K_I	-0.50
Wind turbine time constant, T_{WT} (s)	1.500
Solar system time constant, T_{PV} (s)	1.800
Inverter time constant for conventional VIC, T_{BESS} (s)	0.200
Inverter time constant for proposed VIC, T_{BESS} (s)	0.100
DC/DC converter time constant, T_{DC} (s)	0.100
DC/DC converter gain, K_{DC}	1.000

The fifth-order linearized system for the islanded MG, which considers high penetration of RESs, can be effectively modeled using the state-space approach. The deviation of the examined system's frequency can be obtained by considering the governor action (i.e., the primary control loop) and the LFC (i.e., the secondary control loop), as follows:

$$\dot{\Delta f} = \frac{1}{2H}(\Delta P_m + \Delta P_{WT} + \Delta P_{PV} - \Delta P_L) - \frac{D}{2H} * \Delta f \tag{1}$$

where,

$$\dot{\Delta P_g} = -\frac{1}{T_g}(\Delta P_g) - \frac{1}{R.T_g} * \Delta f + \frac{1}{T_g}(\Delta P_C) \tag{2}$$

$$\dot{\Delta P_m} = -\frac{1}{T_t}(\Delta P_m) + \frac{1}{T_t}(\Delta P_g) \tag{3}$$

$$\dot{\Delta P_{WT}} = \frac{1}{T_{WT}}(\Delta P_{Wind}) - \frac{1}{T_{WT}}(\Delta P_{WT}) \tag{4}$$

$$\dot{\Delta P_{PV}} = \frac{1}{T_{PV}}(\Delta P_{Solar}) - \frac{1}{T_{PV}}(\Delta P_{PV}) \tag{5}$$

In this study, the power changes caused by the wind (ΔP_{Wind}), solar (ΔP_{Solar}), and load (ΔP_L) are taken into account as disturbance signals. Using the state variables from (1) to (5), the linearized state-space model of the investigated MG could be easily developed as follows:

$$\dot{X} = AX + BU + EW \tag{6}$$

$$Y = CX + DU + ZW \tag{7}$$

where,

$$X^T = \begin{bmatrix} \Delta f & \Delta P_g & \Delta P_m & \Delta P_{WT} & \Delta P_{PV} \end{bmatrix}$$

$$W^T = \begin{bmatrix} \Delta P_{Wind} & \Delta P_{Solar} & \Delta P_L \end{bmatrix}$$

$$Y = [\Delta f]$$

where W is the input perturbation vector, U is the control input signal, X is the state vector, and Y is the control output signal. A is the state matrix of the studied system. B and E correspond to the control input signal and the disturbance inputs. The output measurement, or input to the load-frequency controller, is represented by C. D and Z are zero vectors with the same size as the input control signal and disturbance vector, respectively. As a result, the islanded MGs complete state-space representation, taking into account RESs, can be obtained in Equations (8) and (9).

$$\dot{X} = \begin{bmatrix} -\dfrac{D}{2H} & 0 & \dfrac{1}{2H} & \dfrac{1}{2H} & \dfrac{1}{2H} \\ -\dfrac{1}{RT_g} & -\dfrac{1}{T_g} & 0 & 0 & 0 \\ 0 & \dfrac{1}{T_t} & -\dfrac{1}{T_t} & 0 & 0 \\ 0 & 0 & 0 & -\dfrac{1}{T_{WT}} & 0 \\ 0 & 0 & 0 & 0 & -\dfrac{1}{T_{PV}} \end{bmatrix} * \begin{bmatrix} \Delta f \\ \Delta P_g \\ \Delta P_m \\ \Delta P_{WT} \\ \Delta P_{PV} \end{bmatrix} + \begin{bmatrix} 0 \\ \dfrac{1}{T_g} \\ 0 \\ 0 \\ 0 \end{bmatrix} * [\Delta P_c] + \begin{bmatrix} 0 & 0 & -\dfrac{1}{2H} \\ 0 & 0 & 0 \\ 0 & 0 & 0 \\ 1 & 0 & 0 \\ 0 & \dfrac{1}{T_{WT}} & 0 \\ & \dfrac{1}{T_{PV}} & 0 \end{bmatrix} * \begin{bmatrix} \Delta P_{Wind} \\ \Delta P_{Solar} \\ \Delta P_L \end{bmatrix} \tag{8}$$

$$Y = \begin{bmatrix} 1 & 0 & 0 & 0 & 0 \end{bmatrix} X + [0] U + \begin{bmatrix} 0 & 0 & 0 \\ 0 & 0 & 0 \\ 0 & 0 & 0 \\ 0 & 0 & 0 \\ 0 & 0 & 0 \end{bmatrix} W \tag{9}$$

3. Proposed Fractional-Order VIC (FOVIC)

3.1. Description of the Fractional-Order Calculus

Despite being proposed 300 years ago, the utilization of FO has only recently become prevalent due to its complexity [36]. Generic integral and differential notations into any actual number are possible by means of the fractional operators in the controller. The FO differentiator is a mathematical operator that can be viewed as a more generalized form of integral and differential operators, as indicated in the following way:

$$
D_{lb,ub}^{q} = \begin{cases} \frac{d^{d}}{dt^{q}} & q > 0 \\ 1 & q = 0 \\ \int_{lu}^{ub} (d\tau)^{-q} & q < 0 \end{cases}
\tag{10}
$$

where q is the FO operator and ub and lb are the upper and lower bands to calculate operator D.

Two different theories can define the FO principle. The Riemann-Liouville (RL) method is the original one and is utilized to identify the order derivative of a function $f(t)$ [41,42] as follows:

$$
D_{lb,ub}^{q} f(t) = \frac{1}{\Gamma(n-q)} \left(\frac{d}{dt} \right)^{n} \int_{lb}^{ub} \frac{f^{n}(\tau)}{(t-\tau)^{q-n+1}} d\tau
\tag{11}
$$

where $\Gamma(z) = \int_{0}^{\infty} t^{z-1} e^{-t} dt, \Re(z) > 0$ is the function of Gamma, $n \in N$ and the variable q is limited as n-$1 < q < n$.

The previous fractional derivative of RL in (11) can then be transformed using the well-known Laplace method as follows [43]:

$$
\mathcal{L}\left\{ D_{0}^{q} f(t) \right\} = s^{q} F(s) - \sum_{y=0}^{n-1} s^{y} \left(D_{0}^{q-y-1} f(t) \right)|_{t=0}
\tag{12}
$$

where s is the Laplace operator.

The second definition related to the essential notation of FO is Caputo's definition, where the time-domain representation for q order of a function $f(t)$ can be stated as follows:

$$
D_{lb,ub}^{t} f(t) = \begin{cases} \frac{1}{\Gamma(n-q)} \left(\int_{lb}^{ub} \frac{f^{n}(\tau)}{(t-\tau)^{1-n+q}} d\tau \right) & n-1 < q < n \\ \left(\frac{d}{dt} \right)^{n} f(t) & q = n \end{cases}
\tag{13}
$$

Once more, we apply the Laplace transformation to Equation (13) to result in Equation (14), which has an initial condition and represents the integral order with a specific physical significance.

$$
\mathcal{L}\left\{ D_{0}^{q} f(t) \right\} = s^{q} F(s) - \sum_{k=0}^{n-1} s^{q-k-1} f^{(k)}(0)
\tag{14}
$$

The implementation of FO operators in the time domain involves complicated mathematical calculations. The recursive estimation method is typically used to implement the FO definition [44]. The Laplace transformation of the qth derivative is as follows:

$$
s^{q} \approx K \prod_{k=-N}^{N} \frac{s + \omega_{k}'}{s + \omega_{k}}
\tag{15}
$$

where

$$
K = \omega_{h}^{q}
$$

$$
\omega_{k}' = \omega_{b} \left(\frac{\omega_{h}}{\omega_{b}} \right)^{\frac{k+N+(1-q)/2}{2N+1}}
$$

$$\omega_k = \omega_b \left(\frac{\omega_h}{\omega_b} \right)^{\frac{k+N+(1+q)/2}{2N+1}}$$

N is an effective frequency range approximation order $[\omega_b, \omega_h]$ that can be chosen as $[-1000, 1000]$ rad/s.

The fractional calculus expands on the conventional PID controller by allowing more variables to be adjusted. Unlike the traditional PID controller, which only has three variables that can be fine-tuned, the FOPID controller provides five variables for tuning. Consequently, there are two additional variables of integral and differential FOs λ and μ, respectively, as shown in Figure 3. When compared with the conventional PID, these variables have the capability to improve the controller's stability, transient time, and steady-state error. Moreover, it offers the controller greater flexibility and enables them to manage disturbances in the system that occur across a broad range of The transfer function of the FOPID controller is fully represented below:

$$G_c(s) = K_P + K_I \left(\frac{1}{s} \right)^{\lambda} + K_D s^{\mu} \tag{16}$$

where λ and μ lie in the range of 0 and 1, which can decrease the steady-state error and rise time in the system.

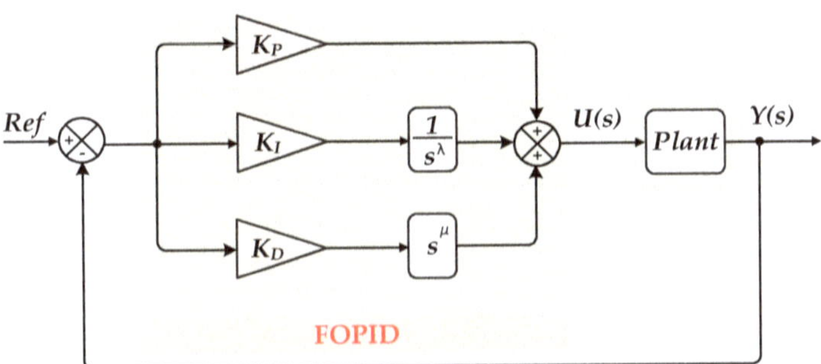

Figure 3. The basic structure of the FOPID controller as an example of applying FO calculus.

3.2. Description of the Proposed Fractional-Order Virtual Inertia Control

The FO controller-based PID has been widely applied for power system stability due to its wide stability and robustness. In contrast, most of the studies are dedicated to LFC [38,45], and AGC [40]. In all prior work in the literature, the notion of FOPID is used as a controller to support the LFC. The synchronous generator, which is regarded as the principal dynamic source of the conventional power system, contributes inertia to the grid via its rotating mass in the traditional power system [46], as in (17).

$$H = \frac{\sum_{x=1}^{G} H_x S_{B,x}}{S_B} \tag{17}$$

where G denotes the whole number of generating units coupled to the grid, S_B is the power system's rated capacity, and H_x and $S_{B,x}$ are the inertia value and rated power of the generation units, respectively. Moreover, the swing equation can be employed to explain the rotating dynamics of actual synchronous machines, as in (18).

$$\Delta P_M(s) - \Delta P_L(s) = (2Hs + D)\Delta f(s) \tag{18}$$

where ΔP_M is the change in mechanical power, ΔP_L is the load power change, and Δf is the frequency deviation.

To enhance the stability of low system inertia in the presence of a significant amount of RESs, it is possible to recreate the synthetic damping attribute and synthetic inertia power in the power system. This action can effectively improve the frequency stability of the power system. Thus, the inverter-based ESS, which is used to inject/absorb active power into/from the power system, is regulated based on the VIC, as in (19).

$$\Delta P_{VIC} = K_H s * \Delta f \tag{19}$$

where K_H is the virtual inertia constant of the VIC.

The typical VIC system on the basis of battery ESS (BESS) is shown in Figure 4, which simulates the real synchronous generator's inertia. When the RESs are penetrated, the VIC-based ESSs dynamic equation that reflects the power system's desired power can be stated as in (20). The amount of power that the inertia of the BESS can generate is constrained by two factors: the maximum capacity of the BESS during the processes of charging and discharging, and the state of charge (SOC) of the BESS. These limitations prevent the inertia output power from exceeding the rated capacity of the BESS and avoid the lifetime degradation of the battery.

$$\Delta P_{ESS-VIC} = \frac{K_H s}{1 + sT_{BESS}}(\Delta f) \tag{20}$$

where T_{BESS} refers to the inverter-based BESS time constant.

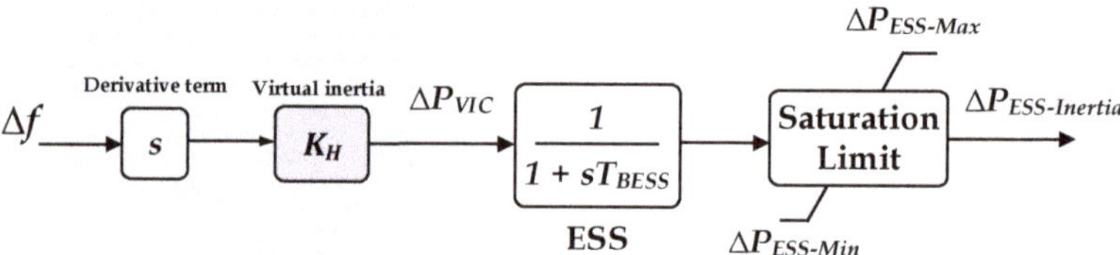

Figure 4. The dynamic structure of conventional VIC-based ESS.

The proposed VIC model separates the DC-DC converter stage and the DC-AC inverter stage, as shown in Figure 5. Each stage is represented by a first-order model with different time constants (T_{DC} for the DC-DC converter and T_{BESS} for the DC-AC inverter) and conversion gains (K_{DC} for the DC-DC converter and unity for the DC-AC inverter). This allows for more precise and accurate modeling of the individual stages. Typically, when the inertia power is not being emulated, the base power of the BESS is directly proportional to the deviation in frequency. This relationship can be represented mathematically as:

$$\Delta P_{ESS} = \Delta f \times \frac{K_{DC}}{1 + sT_{DC}} \times \frac{1}{1 + sT_{BESS}} \tag{21}$$

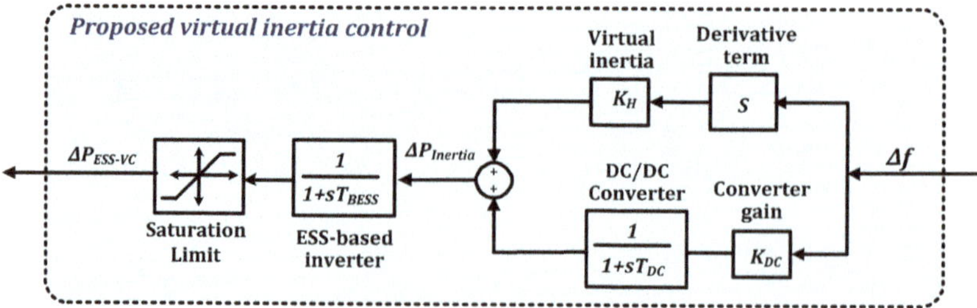

Figure 5. The proposed VIC-based ESS dynamic structure.

The complete transfer function of the proposed VIC can be expressed as follows:

$$\Delta P_{ESS-VIC} = \Delta f \times \left(\frac{K_{DC}}{1 + sT_{DC}} + K_H s \right) \times \frac{1}{1 + sT_{BESS}} \tag{22}$$

where K_H is the virtual inertia constant.

Moreover, this study intends to show how the order of the derivative term of the inertia term can influence the power system's maximum frequency deviation, as shown in Figure 6. After analyzing Figure 6, it can be concluded that the optimal order of the s operator is 0.4. This value is crucial as it ensures the minimum frequency deviation is achieved at various points of K_H. In other words, the s operator plays a significant role in maintaining stability and consistency in the system's frequency response. As we already indicated, the FO controller typically served as the secondary controller in most prior work. The FOVIC has the significant advantage of minimizing the system's order, leading to a substantial decrease in frequency fluctuations and output power oscillations. The complete structure of the proposed FOVIC is shown in Figure 7.

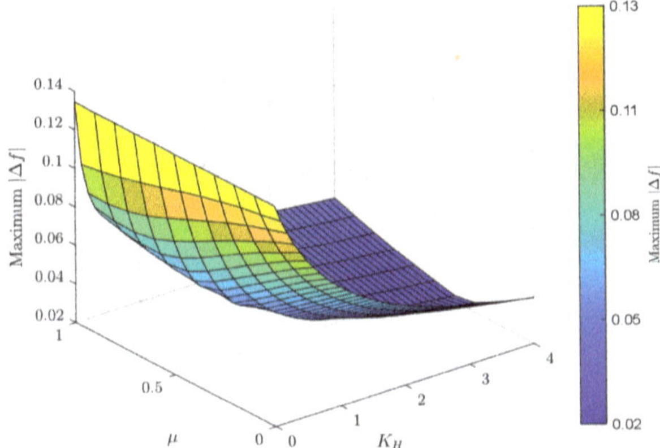

Figure 6. Effect of the virtual inertia constant and FO operator on the system frequency deviation.

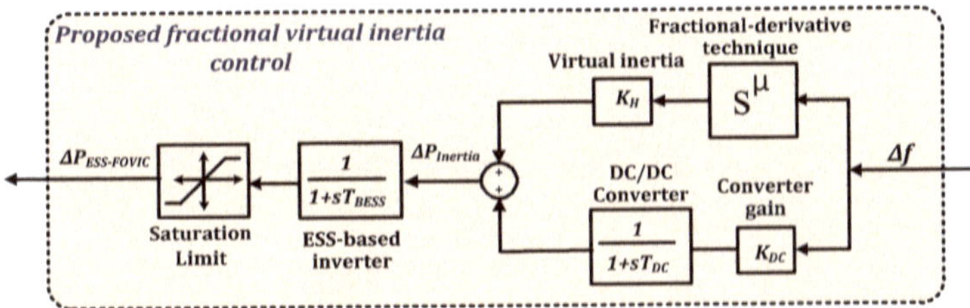

Figure 7. The proposed FOVIC to support the dynamic stability of MG.

The proposed FOVIC will undergo a second modification compared to Figure 4, which involves replacing the entire order of the inertial emulation with a fractional order. To achieve this, the output power and its sign will be combined with the required power from the DC-DC stage. The resulting power will then be set as a command for the inverter to supply power to the power system. The complete transfer function of the proposed FOVIC can be expressed as in (23):

$$\Delta P_{ESS-FOVIC} = \Delta f \times \left(\frac{K_{DC}}{1 + sT_{DC}} + K_H s^\mu \right) \times \frac{1}{1 + sT_{BESS}} \tag{23}$$

where μ is the order of the inertia derivative.

4. Simulation Results

To simulate the researched MG, a MATLAB/Simulink program is employed. With the proposed developed VIC model, which is depicted in Figure 2, and the system's parameters listed in Table 1 [20], the simulation results have been extracted based on two case studies, as follows:

A.   Validate the proposed developed VIC model's superiority compared with the conventional known VIC model, which is presented in many studies, e.g., [2,8,9,12,14,16].

Impact of the FO derivative on the developed structure of the suggested VIC model Both case studies are investigated through different scenarios under various situations of RES, load variation, and system inertia changes, as follows:

A1. Evaluation of the developed VIC model's performance under sudden load changes and different values of the virtual inertia constant

In this case, the suggested developed VIC scheme is compared with both the known conventional VIC scheme and without VIC (i.e., using conventional frequency control only) in the studied islanded MG under different values of virtual inertia constant as well as a 10% step change in load that occurs at time = 0 s. Figure 8 and Table 2 exhibit the frequency deviation of the investigated system with diverse control techniques under the nominal system parameters. From this simulation result, the studied MG with the suggested developed VIC scheme is noticeably more stable and quicker than the conventional VIC scheme. Moreover, the studied system with conventional frequency control (i.e., without the VIC scheme) has high overshoots and slow responses compared to the case of the proposed developed VIC scheme. In contrast, although the frequency deviation of the system with VIC schemes (i.e., the proposed and conventional schemes) is more suppressed when the K_H value is increased, the system with the known conventional VIC needs more time to achieve stability. Thus, by applying the proposed developedVIC scheme, the dynamic stability of the studied MG considering different values of K_H has been preserved.

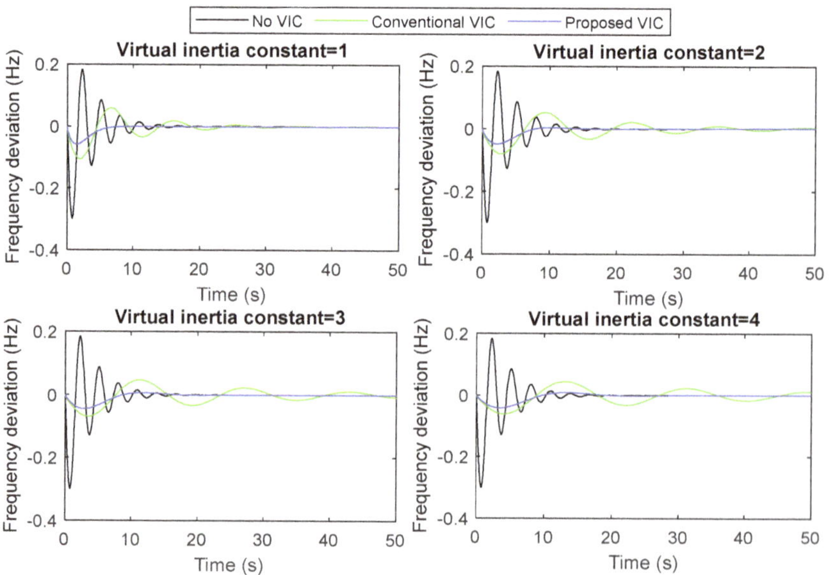

Figure 8. The frequency response of the MG with different values of K_H for case A1.

Table 2. Values of performance indices for case A1.

Control Approach	$\Delta f\,(K_H = 1)$			$\Delta f\,(K_H = 2)$			$\Delta f\,(K_H = 3)$			$\Delta f\,(K_H = 4)$		
	MUS (Hz)	MOS (Hz)	T_S (s)	MUS (Hz)	MOS (Hz)	T_S (s)	MUS (pu)	MOS (pu)	T_S (s)	MUS (pu)	MOS (pu)	T_S (s)
No VIC	2.99×10^{-1}	1.82×10^{-1}	28.72	2.99×10^{-1}	1.82×10^{-1}	28.72	2.99×10^{-1}	1.82×10^{-1}	28.72	2.99×10^{-1}	1.82×10^{-1}	28.72
Con. VIC	1.06×10^{-1}	5.96×10^{-2}	63.74	8.08×10^{-2}	5.24×10^{-2}	108.37	6.84×10^{-2}	4.75×10^{-2}	155.06	6.06×10^{-2}	4.40×10^{-2}	204.26
Pro. VIC	5.85×10^{-2}	1.05×10^{-4}	13.09	4.97×10^{-2}	4.15×10^{-3}	22.41	4.47×10^{-2}	7.06×10^{-3}	33.81	4.13×10^{-2}	8.87×10^{-3}	46.07

A2. Evaluation of the developed VIC model's performance under several load step changes

In this case, a series of sudden load changes are applied in the investigated MG system with varied values of the virtual inertia. The step load change during the period of the simulation is revealed in Figure 9. Figure 10 demonstrates the frequency deviation of the studied MG with different control techniques under this disturbance. The results prove that the suggested developed VIC scheme gives the best performance for the studied islanded MG compared to the other control techniques, and the MG frequency variations are kept within the recommended levels during the load changes. Regarding the use of different values of the virtual inertia constant, the proposed developed VIC scheme showed fast stabilization time and slight deviations in system frequency compared to the conventional VIC scheme, which needs more time to stabilize.

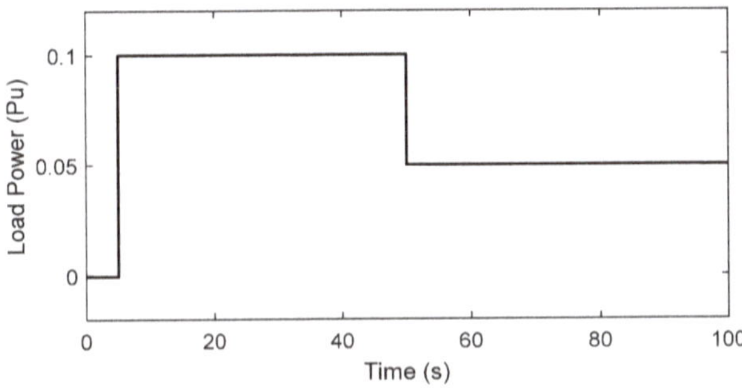

Figure 9. Load step change for case A2.

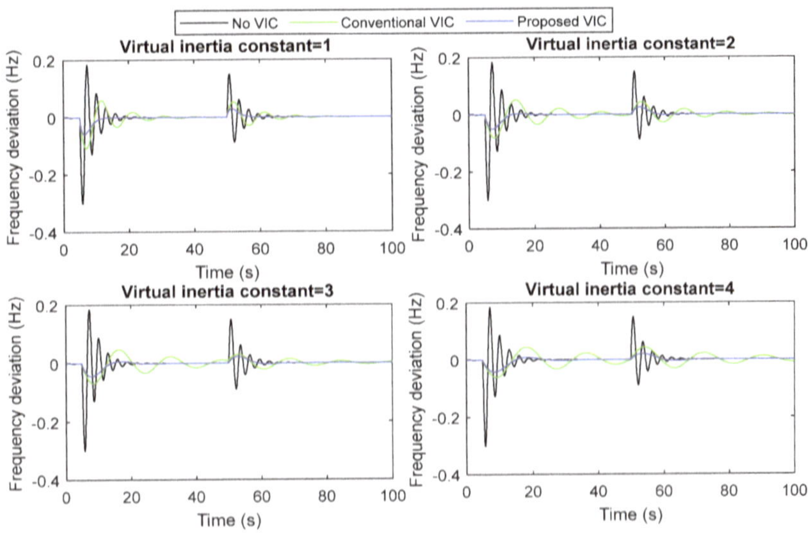

Figure 10. The frequency response of the MG with different values of K_H for case A2.

A3. Evaluation of the developed VIC model's performance under step load change and RES presence

In this case, the MG with the suggested developed VIC scheme is studied under the following operational circumstances: a series of abrupt load changes shown in Figure 11 and power fluctuations of the RESs depicted in Figure 12. Figure 13 shows the frequency response of MG with the different control schemes, considering different values of virtual inertia. Figure 13 shows that the considered islanded MG with the suggested developed VIC scheme is more efficient and dependable. Additionally, in comparison with the conventional VIC scheme, the suggested scheme results in a better frequency response throughout all disturbances and operating conditions of the MG. On the other hand, conventional frequency control (i.e., without the VIC scheme) provided the poorest performance with high values of frequency deviation. Hence, with the proposed developed VIC scheme, the islanded MG system's overall performance is at its finest.

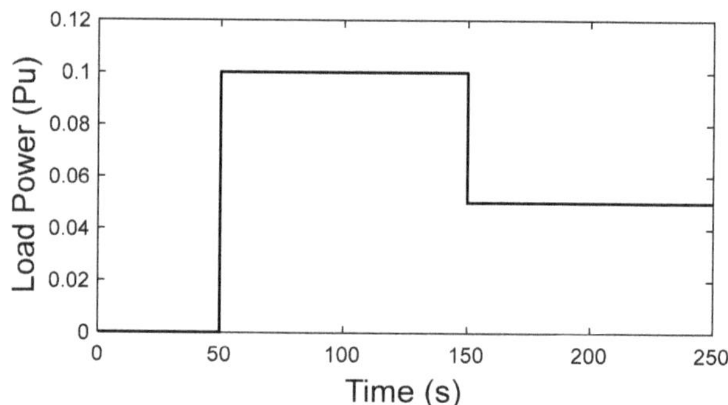

Figure 11. Load step change for case A3.

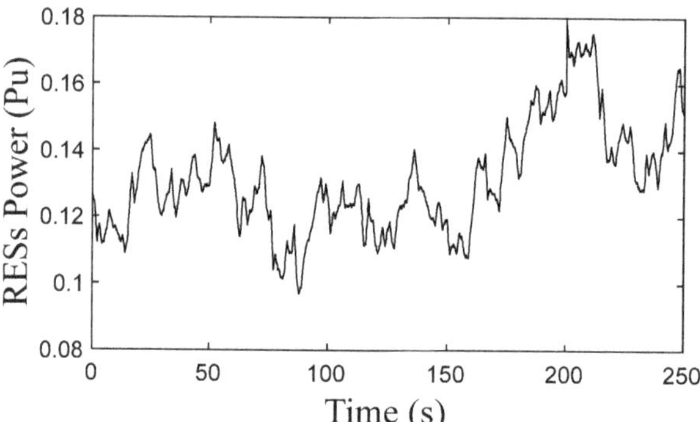

Figure 12. Renewables power generation profile for case A3.

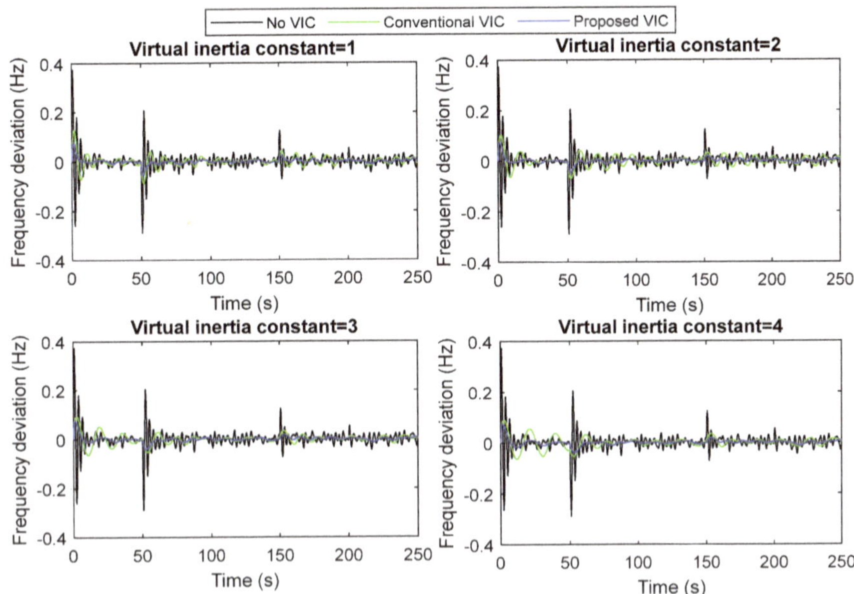

Figure 13. The frequency response of the MG with RES generation and different values of K_H for case A3.

A4. Evaluation of the developed VIC model's performance considering communication delay.

Communication time delays are widely acknowledged in control systems for their potential to diminish system performance and induce instability. Consequently, these delays pose a significant challenge, emerging as a noteworthy uncertainty in analyzing the frequency stability of the power systems, given their ongoing expansion and increasing complexity. The effect of the communications delays with the proposed VIC strategy is tested at different time delays, as shown in Figures 14 and 15. This test considers both the communication time delay affecting the control input (i.e., before the LFC) and the delay in the control action (i.e., after the LFC). The effectiveness of the proposed VIC is evident, as it results in a lower frequency deviation compared to both the conventional VIC strategy and the LFC without VIC implementation. Additionally, with the proposed VIC, the system frequency achieves steady-state more rapidly than with the other two strategies. Notably, as the time delay increases from 0.1 s in Figure 14 to 0.2 s in Figure 15, both the conventional VIC and no VIC strategies exhibit heightened overshoot and settling time in frequency deviations. Consequently, applying the proposed VIC effectively mitigates the impact of communication delays without compromising the system's stability.

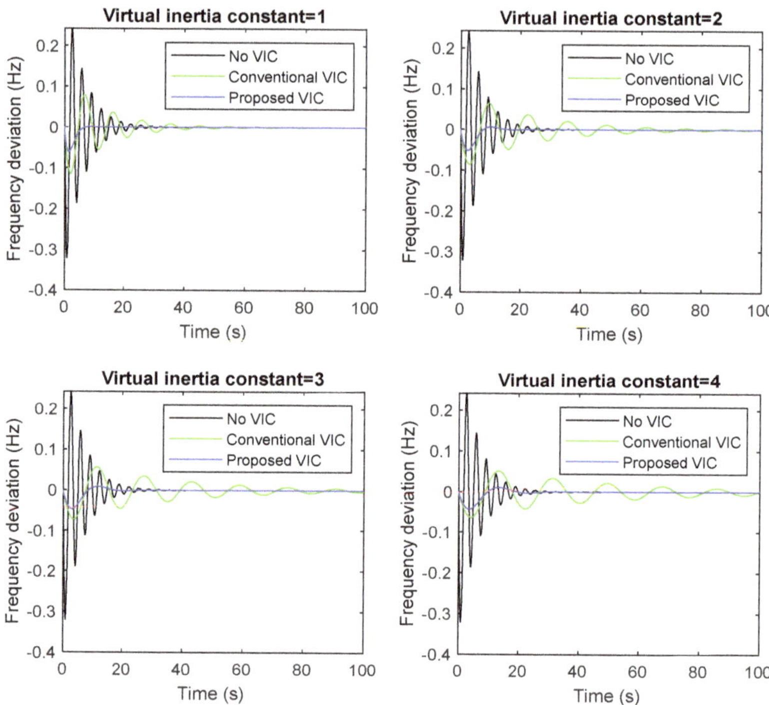

Figure 14. Effect of the 0.1 s communications delay before and after LFC on the MG frequency response for case A4.

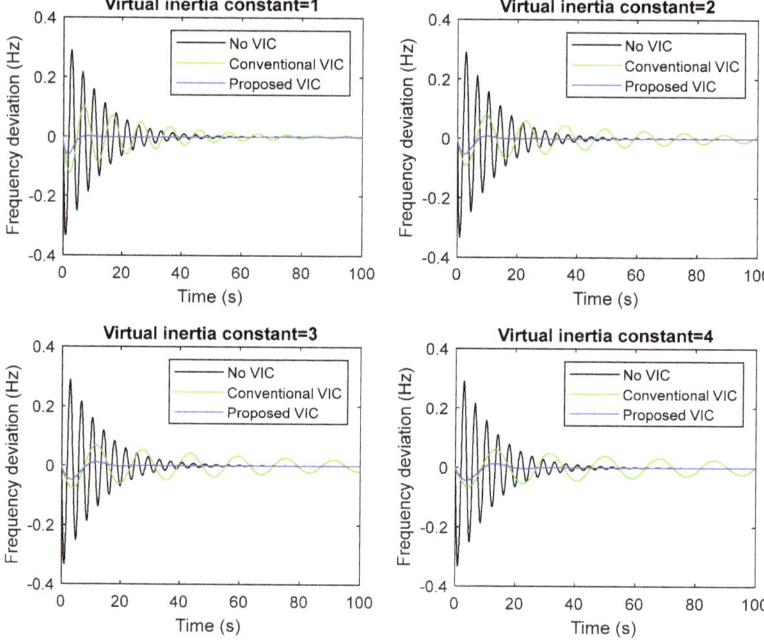

Figure 15. Effect of the 0.2 s communications delay before and after LFC on the MG frequency response for case A4.

B1. Evaluation of the proposed FOVIC model's performance under sudden load change

In this case, the impact of the FO control technique has been considered in the suggested structure of the developed VIC scheme. Therefore, the MG with the suggested FOVIC scheme is examined under different values of virtual inertia constant, different values of FO operators (μ) (i.e., 0.2, 0.4, 0.6, 0.8, and 1), and a 10% step change in load that occurs at time = 0 s. Figure 16 and Table 3 display the frequency deviation of the investigated system with FOVIC under the nominal system parameters. From Figure 16, it is clear that the higher the value of K_H, the more the MG with the proposed FOVIC can dampen the frequency deviations, but it needs more time to achieve stability. On the other side, the role of the FO operator in the suggested FOVIC is shown to change the system's dynamic performance with respect to the overshoot, undershoot, and settling time. Therefore, the results indicate that the MG with the suggested FOVIC scheme at the fractional operator (μ) equals 0.4 is noticeably more stable and quicker than that designed with other fractional operator values. Moreover, it is obvious that this value is consistent with the range of values previously mentioned in the theoretical analysis section.

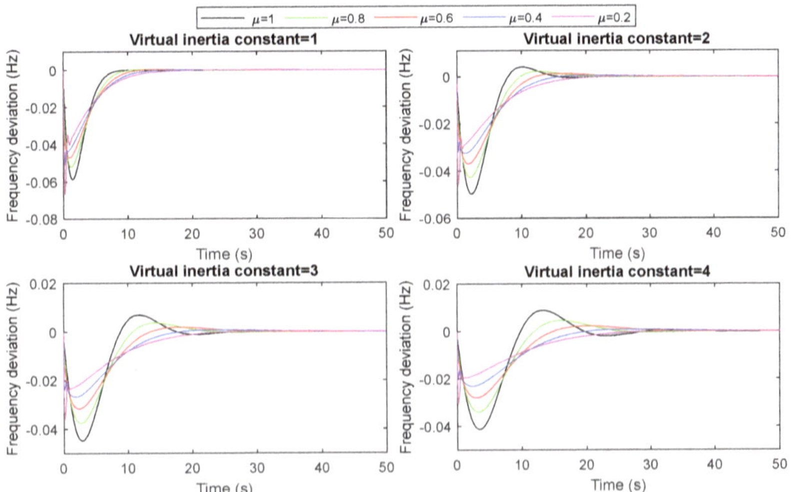

Figure 16. The frequency response of the MG with different values of K_H and fractional operators for case B1.

Table 3. Values of performance indices for case B1.

FO Operator	$\Delta f\,(K_H = 1)$			$\Delta f\,(K_H = 2)$			$\Delta f\,(K_H = 3)$			$\Delta f\,(K_H = 4)$		
	MUS (Hz)	MOS (Hz)	T_S (s)	MUS (Hz)	MOS (Hz)	T_S (s)	MUS (pu)	MOS (pu)	T_S (s)	MUS (pu)	MOS (pu)	T_S (s)
$\mu = 1$	5.85×10^{-2}	1.05×10^{-4}	13.09	4.97×10^{-2}	4.15×10^{-3}	22.41	4.47×10^{-2}	7.06×10^{-3}	33.81	4.13×10^{-2}	8.87×10^{-3}	46.07
$\mu = 0.8$	5.19×10^{-2}	4.78×10^{-4}	22.44	4.26×10^{-2}	2.32×10^{-3}	22.11	3.75×10^{-2}	3.71×10^{-3}	33.87	3.40×10^{-2}	4.58×10^{-3}	46.38
$\mu = 0.6$	4.69×10^{-2}	4.94×10^{-4}	29.27	3.69×10^{-2}	1.35×10^{-3}	32.56	3.16×10^{-2}	1.92×10^{-3}	33.34	2.80×10^{-2}	2.26×10^{-3}	35.51
$\mu = 0.4$	5.10×10^{-2}	3.36×10^{-4}	35.45	3.24×10^{-2}	6.63×10^{-4}	44.32	2.67×10^{-2}	8.35×10^{-4}	49.97	2.31×10^{-2}	9.20×10^{-4}	54.41
$\mu = 0.2$	6.65×10^{-2}	1.31×10^{-4}	38.67	4.59×10^{-2}	1.99×10^{-4}	51.41	3.61×10^{-2}	2.23×10^{-4}	61.93	3.04×10^{-2}	2.28×10^{-4}	70.95

B2. Evaluation of the proposed FOVIC model's performance under system inertia changes

The system inertia (i.e., H) drops when the penetration level of RESs rises by replacing the synchronous generators, causing high-frequency deviations that may cause instability in the power system. In this case, the MG with the suggested FOVIC scheme is examined under different values of virtual inertia constant (K_H), different values of FO operators (μ), and a 10% step change in load that occurs at time = 0 s. Moreover, the performance of the proposed FOVIC in the investigated system is evaluated by considering system inertia changes by $\pm 25\%$ and $\pm 50\%$. Figures 17–20 show the frequency response of the

MG with system uncertainties (i.e., increasing/decreasing the system inertia by ±25% and ±50%, respectively). According to the results, the suggested FOVIC scheme with different values of fractional operators significantly improves the MG frequency performance. It has lower system transients compared to the integer-order VIC. Furthermore, although the frequency variation of the system with the FOVIC scheme is more suppressed when the value of the virtual inertia is increased, the system may need more time to achieve stability. Hence, the proposed FOVIC scheme still maintains the stability of the MG frequency under uncertainties (i.e., system inertia changes).

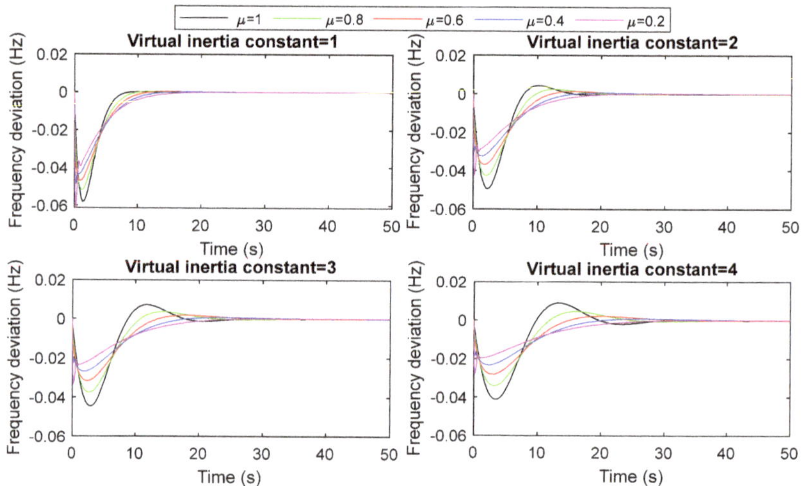

Figure 17. The frequency response of the MG with different values of K_H and fractional operators for case B2 with +25% system inertia (H) variation.

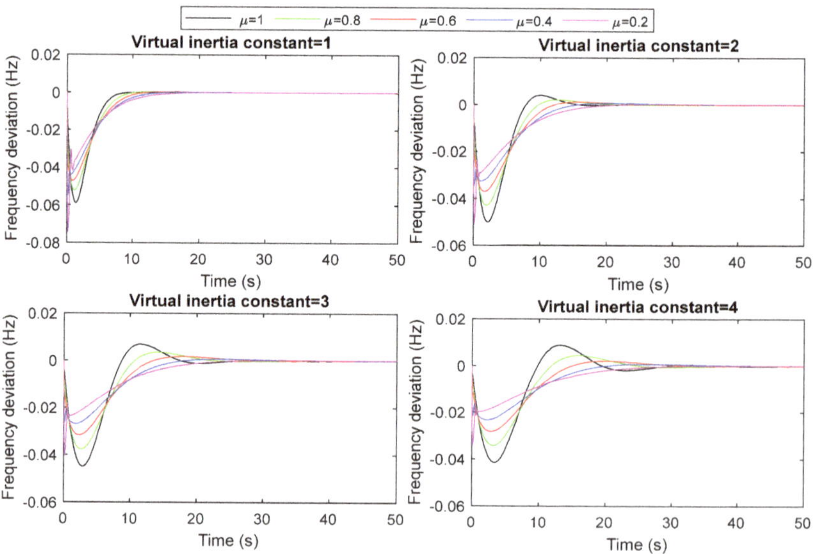

Figure 18. The frequency response of the MG with different values of K_H and fractional operators for case B2 with −25% system inertia (H) variation.

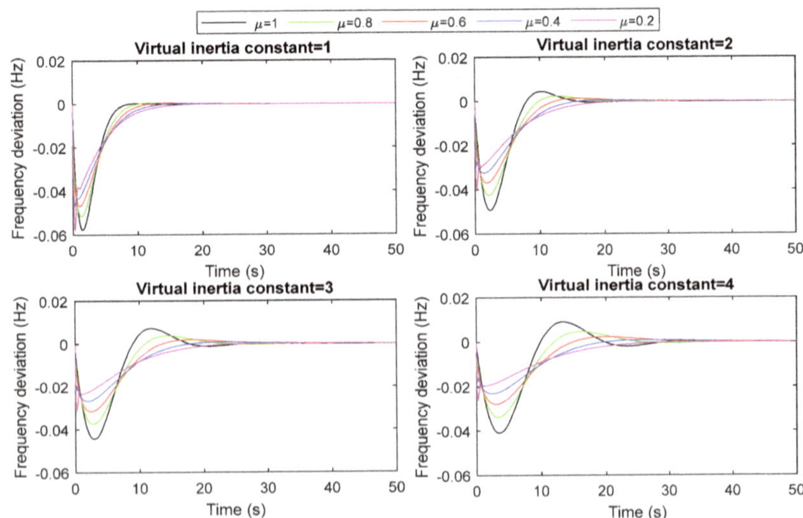

Figure 19. The frequency response of the MG with different values of K_H and fractional operators for case B2 with +50% system inertia (H) variation.

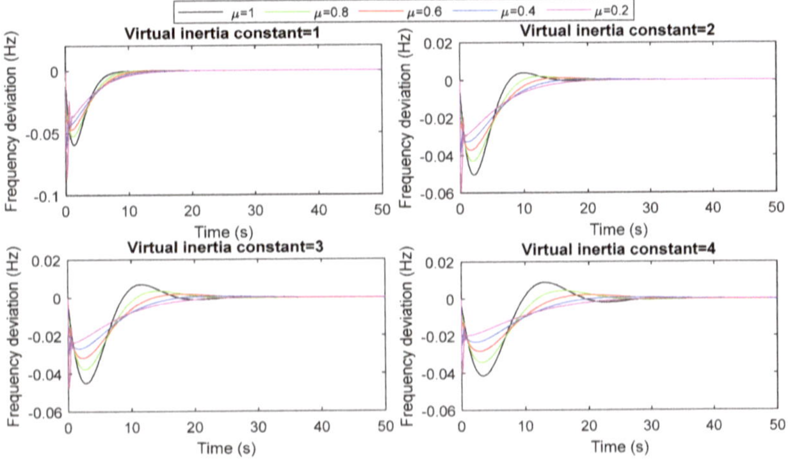

Figure 20. The frequency response of the MG with different values of K_H and fractional operators for case B2 with −50% system inertia (H) variation.

B3. Evaluation of the proposed FOVIC model's performance under step load change and RESs

In this case, the MG with the suggested FOVIC scheme is investigated under different values of virtual inertia constant, different values of FO operators (μ), applying a series of abrupt load changes shown in Figure 21, and power fluctuations of the RESs shown in Figure 22. Figure 23 shows the frequency response of the MG with the suggested FOVIC scheme considering varied values of fractional operators. Out of the simulation results, the considered islanded MG with the proposed FOVIC scheme with different values of fractional operators is more efficient and dependable than the integer-order VIC. Additionally, compared to the proposed integer-order VIC scheme, the proposed FOVIC scheme results in a better frequency response throughout all circumstances of this case study. Although the frequency variation of the MG with the FOVIC scheme is more suppressed when the value of K_H is increased, the system may need more time to achieve stability. Hence, the proposed FOVIC scheme still maintains the stability of the MG frequency under the high penetration of RESs.

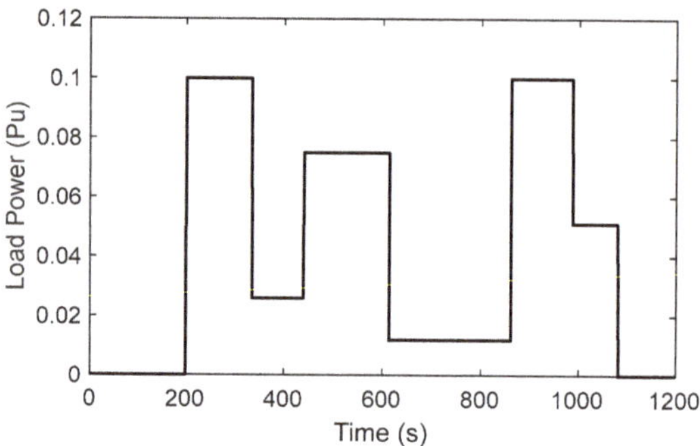

Figure 21. Load step change for case B3.

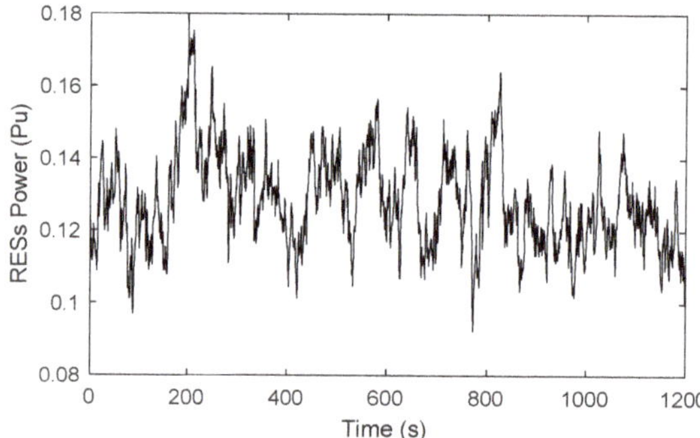

Figure 22. Renewables power generation profile for case B3.

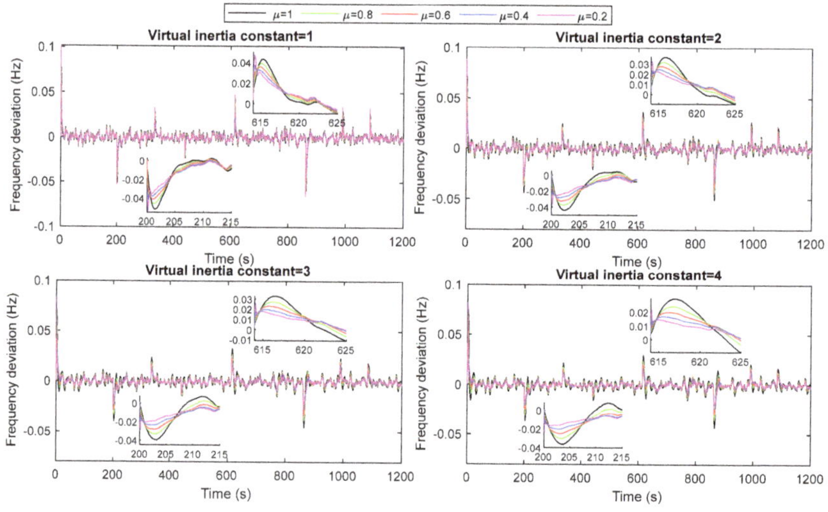

Figure 23. The frequency response of the MG with RES generation, different values of K_H, and fractional operators for case B3.

5. Discussion

The proposed VIC model's developed structure can provide a virtual inertia facility by controlling the battery ESS (BESS) in a way that accurately mimics the behavior of a traditional synchronous generator during the emergency period. Moreover, to provide the proposed VIC model's developed structure with more flexibility in handling various microgrid disturbances, a fractional-order derivative control has been considered. In addition to enhancing frequency response, the proposed fractional-order virtual inertia control based on the BESS permits a greater level of renewable power penetration into the microgrid. The significance of the proposed control system has been supported by the findings obtained from the previous section.

1. The proposed VIC model handles the sudden change in load demand better than the conventional VIC system by about 45% and the secondary frequency control (i.e., without VIC) system by about 80%. As a result, the proposed VIC model's developed structure has been chosen over alternative solutions for microgrid stabilization during sudden/series load changes.

2. The performance of the three control systems used—the proposed VIC model, the conventional VIC model, and the secondary load frequency control, i.e., without VIC—is evaluated under high RESs penetration as well as varying values of the system inertia constant. The system with secondary frequency control (i.e., without the VIC model) gives high oscillations that threaten the system's stability. Nonetheless, the proposed VIC and conventional VIC models are both able to stabilize the system frequency within an acceptable range. Among these two, the proposed VIC system exhibits the best frequency response under different values of the virtual inertia constant, with a maximum frequency deviation of ± 0.059 Hz (in comparison with ± 0.11 Hz by the conventional VIC system).

3. Communication time delay: With the proposed VIC model, the best frequency nadir (about 0.06 Hz) was recorded. In comparison, the conventional VIC model exhibits a frequency drop of about 0.12 Hz, whereas the load frequency control displays about 0.33 Hz. Because of this, the proposed VIC model works best in situations where there is a varying random time delay within the range of (0, 0.2), as in the case of the practical system.

4. The performance of the proposed VIC based on fractional-order derivative control has been evaluated under load/RESs fluctuations, system uncertainties (i.e., increasing/decreasing the system inertia by $\pm 25\%$ and $\pm 50\%$), and different values in the virtual inertia constant and FO operators. The results show that the microgrid with the proposed FOVIC scheme at the fractional operator equal to 0.4 is significantly more stable and faster than that designed with other fractional operator values. Furthermore, it is clear that this value falls within the range of values that were previously mentioned in the section on theoretical analysis.

6. Conclusions

The decreasing inertia makes modern power systems more prone to system instability. The isolated microgrids (MGs) are more susceptible to instability due to the high penetration of non-dispatchable renewable energy sources, low inertia, and small number of dispatchable generation units. Recent studies proved the efficacy of virtual inertia control (VIC) in enhancing the frequency stability of low-inertia power systems. Therefore, this study proposes a new VIC model to accurately mimic inertia power based on an energy storage system (ESS). The new VIC model considers the effect of both the DC-DC converter and the DC-AC inverter on the power of the ESS used. The obtained results prove that the proposed new VIC scheme provides fast stabilization times and slight deviations in system frequency compared to the conventional VIC schemes for all MG operating conditions and disturbances. Moreover, this study proposes a fractional-order VIC to provide greater flexibility in dealing with MG disturbances. The obtained results prove that the proposed VIC outperforms the conventional LFC by about 80% and the conventional VIC model by about 45% in tackling loads/RESs fluctuations and system uncertainty. Hence, the proposed FOVIC scheme is noticeably more stable and faster in response compared to that designed with integer-order VIC for all MG operating conditions and disturbances.

Author Contributions: Conceptualization, M.N., G.M., A.B. (Abualkasim Bakeer) and H.A.; methodology, M.N., G.M., A.B. (Abualkasim Bakeer) and H.A.; software, M.N., G.M. and A.B. (Abualkasim Bakeer); validation, M.N., G.M. and A.B. (Abualkasim Bakeer); formal analysis, M.N., G.M., A.B. (Abualkasim Bakeer) and H.A.; investigation, M.N., G.M., A.B. (Abualkasim Bakeer) and H.A.; resources, M.N., G.M., A.B. (Abualkasim Bakeer) and H.A.; data curation, M.N., G.M., A.B. (Abualkasim Bakeer) and H.A.; writing—original draft preparation, M.N., G.M., A.B. (Abualkasim Bakeer) and H.A.; writing—review and editing, M.N., G.M., A.B. (Abualkasim Bakeer), A.A.T., A.B. (Abderrahmane Beroual), U.K. and H.A.; visualization, M.N., G.M. and A.B. (Abualkasim Bakeer); supervision, A.A.T., A.B. (Abderrahmane Beroual) and U.K.; project administration, A.A.T., A.B. (Abderrahmane Beroual) and U.K.; funding acquisition, A.A.T. and U.K. All authors have read and agreed to the published version of the manuscript.

Funding: This work was supported by the Researchers Supporting Program number (RSPD2023R1108), King Saud University, Riyadh, Saudi Arabia.

Data Availability Statement: Data are available on request from the authors.

Acknowledgments: The authors present their appreciation to King Saud University for funding this research through Researchers Supporting Program Number (RSPD2023R1108), King Saud University, Riyadh, Saudi Arabia.

Conflicts of Interest: The authors declare no conflict of interest.

References

1. International Renewable Energy Agency (IRENA). *Global Energy Transformation: A Roadmap to 2050*; IRENA: Masdar City, United Arab Emirates, 2019.
2. Magdy, G.; Nour, M.; Shabib, G.; Elbaset, A.A.; Mitani, Y. Supplementary Frequency Control in a High-Penetration Real Power System by Renewables Using SMES Application Electrical Systems. *J. Electr. Syst.* **2019**, *15*, 526–538.
3. Dreidy, M.; Mokhlis, H.; Mekhilef, S. Inertia response and frequency control techniques for renewable energy sources: A review. *Renew. Sustain. Energy Rev.* **2017**, *69*, 144–155. [CrossRef]
4. Fathi, A.; Shafiee, Q.; Bevrani, H. Robust Frequency Control of Microgrids Using an Extended Virtual Synchronous Generator. *IEEE Trans. Power Syst.* **2018**, *33*, 6289–6297. [CrossRef]
5. Peng, Q.; Jiang, Q.; Yang, Y.; Liu, T.; Wang, H.; Blaabjerg, F. On the Stability of Power Electronics-Dominated Systems: Challenges and Potential Solutions. *IEEE Trans. Ind. Appl.* **2019**, *55*, 7657–7670. [CrossRef]
6. Ulbig, A.; Borsche, T.S.; Andersson, G. Impact of Low Rotational Inertia on Power System Stability and Operation. *IFAC Proc. Vol.* **2014**, *19*, 7290–7297. [CrossRef]
7. Santurino, P.; Sigrist, L.; Ortega, Á.; Renedo, J.; Lobato, E. Optimal coordinated design of under-frequency load shedding and energy storage systems. *Electr. Power Syst. Res.* **2022**, *211*, 108423. [CrossRef]
8. Ranjan, M.; Shankar, R. A literature survey on load frequency control considering renewable energy integration in power system: Recent trends and future prospects. *J. Energy Storage* **2022**, *45*, 103717. [CrossRef]
9. Makolo, P.; Zamora, R.; Lie, T.-T. The role of inertia for grid flexibility under high penetration of variable renewables—A review of challenges and solutions. *Renew. Sustain. Energy Rev.* **2021**, *147*, 111223. [CrossRef]
10. Mohanty, B.; Hota, P.K. Comparative performance analysis of fruit fly optimisation algorithm for multi-area multi-source automatic generation control under deregulated environment. *IET Gener. Transm. Distrib.* **2015**, *9*, 1845–1855. [CrossRef]
11. El-Sousy, F.F.M.; Aly, M.; Alqahtani, M.H.; Aljumah, A.S.; Almutairi, S.Z.; Mohamed, E.A. New Cascaded 1+PII2D/FOPID Load Frequency Controller for Modern Power Grids including Superconducting Magnetic Energy Storage and Renewable Energy. *Fractal Fract.* **2023**, *7*, 672. [CrossRef]
12. Shayeghi, H.; Rahnama, A.; Bizon, N. TFODn-FOPI multi-stage controller design to maintain an islanded microgrid load-frequency balance considering responsive loads support. *IET Gener. Transm. Distrib.* **2023**, *17*, 3266–3285. [CrossRef]
13. Peddakapu, K.; Srinivasarao, P.; Mohamed, M.; Arya, Y.; Kishore, D.K. Stabilization of frequency in Multi-Microgrid system using barnacle mating Optimizer-based cascade controllers. *Energy Technol. Assess.* **2022**, *54*, 102823. [CrossRef]
14. Sahu, P.C.; Jena, S.; Mohapatra, S.; Debdas, S. Impact of energy storage devices on microgrid frequency performance: A robust DQN based grade-2 fuzzy cascaded controller. *e-Prime* **2023**, *6*, 100288. [CrossRef]
15. Ali, H.; Li, B.; Xu, Z.; Liu, H.; Xu, D. Virtual Synchronous Generator Design Based Modular Multilevel Converter for Microgrid Frequency Regulation. In Proceedings of the 2019 22nd International Conference on Electrical Machines and Systems (ICEMS), Harbin, China, 11–14 August 2019; Volume 1, pp. 1–6.

16. Zhong, Q.-C.; Weiss, G. Synchronverters: Inverters that mimic synchronous generators. *IEEE Trans. Ind. Electron.* **2011**, *58*, 1259–1267. [CrossRef]
17. Fang, J.; Li, H.; Tang, Y.; Blaabjerg, F. Distributed Power System Virtual Inertia Implemented by Grid-Connected Power Converters. *IEEE Trans. Power Electron.* **2018**, *33*, 8488–8499. [CrossRef]
18. Nour, M.; Magdy, G.; Chaves-Ávila, J.P.; Sánchez-Miralles, A.; Jurado, F. A new two-stage controller design for frequency regulation of low-inertia power system with virtual synchronous generator. *J. Energy Storage* **2023**, *62*, 106952. [CrossRef]
19. Mohammadi-ivatloo, B. Provision of Frequency Stability of an Islanded Microgrid Cascade Controller. *Energies* **2021**, *14*, 4152.
20. Magdy, G.; Shabib, G.; Elbaset, A.A.; Mitani, Y. A Novel Coordination Scheme of Virtual Inertia Control and Digital Protection for Microgrid Dynamic Security Considering High Renewable Energy Penetration. *IET Renew. Power Gener.* **2019**, *13*, 462–474. [CrossRef]
21. Gonzalez-Longatt, F.; Chikuni, E.; Rashayi, E. Effects of the Synthetic Inertia from wind power on the total system inertia after a frequency disturbance. In Proceedings of the 2013 IEEE International Conference on Industrial Technology (ICIT), Cape Town, South Africa, 25–28 February 2013; pp. 826–832.
22. Hu, Y.; Wei, W.; Peng, Y.; Lei, J. Fuzzy virtual inertia control for virtual synchronous generator. *Chinese Control Conf. CCC* **2016**, *2016*, 8523–8527.
23. Magnus, D.M.; Scharlau, C.C.; Pfitscher, L.L.; Costa, G.C.; Silva, G.M. A novel approach for robust control design of hidden synthetic inertia for variable speed wind turbines. *Electr. Power Syst. Res.* **2021**, *196*, 107267. [CrossRef]
24. Ali, H.; Magdy, G.; Li, B.; Shabib, G.; Elbaset, A.A.; Xu, D.; Mitani, Y. A New Frequency Control Strategy in an Islanded Microgrid Using Virtual Inertia Control-Based Coefficient Diagram Method. *IEEE Access* **2019**, *7*, 16979–16990. [CrossRef]
25. Sockeel, N.; Gafford, J.; Papari, B.; Mazzola, M. Virtual Inertia Emulator-Based Model Predictive Control for Grid Frequency Regulation Considering High Penetration of Inverter-Based Energy Storage System. *IEEE Trans. Sustain. Energy* **2020**, *11*, 2932–2939. [CrossRef]
26. Li, J.; Wen, B.; Wang, H. Adaptive Virtual Inertia Control Strategy of VSG for Micro-Grid Based on Improved Bang-Bang Control Strategy. *IEEE Access* **2019**, *7*, 39509–39514. [CrossRef]
27. Fawzy, A.; Bakeer, A.; Magdy, G.; Atawi, I.E.; Roshdy, M. Adaptive Virtual Inertia-Damping System Based on Model Predictive Control for Low-Inertia Microgrids. *IEEE Access* **2021**, *9*, 109718–109731. [CrossRef]
28. Alipoor, J.; Miura, Y.; Ise, T. Stability Assessment and Optimization Methods for Microgrid with Multiple VSG Units. *IEEE Trans. Smart Grid* **2016**, *9*, 1462–1471. [CrossRef]
29. Padhy, S.; Panda, S. Simplified grey wolf optimisation algorithm tuned adaptive fuzzy PID controller for frequency regulation of interconnected power systems. *Int. J. Ambient. Energy* **2022**, *43*, 4089–4101. [CrossRef]
30. Ali, H.; Magdy, G.; Xu, D. A new optimal robust controller for frequency stability of interconnected hybrid microgrids considering non-inertia sources and uncertainties. *Int. J. Electr. Power Energy Syst.* **2021**, *128*, 106651. [CrossRef]
31. Liu, J.; Yang, Z.; Yu, J.; Huang, J.; Li, W. Coordinated control parameter setting of DFIG wind farms with virtual inertia control. *Int. J. Electr. Power Energy Syst.* **2020**, *122*, 106167. [CrossRef]
32. Fini, M.H.; Golshan, M.E.H. Determining optimal virtual inertia and frequency control parameters to preserve the frequency stability in islanded microgrids with high penetration of renewables. *Electr. Power Syst. Res.* **2018**, *154*, 13–22. [CrossRef]
33. Mohamed Mostafa Elsaied, H.M.H.; Hameed, W.H.A. Frequency stabilization of a hybrid three-area power system equipped with energy. *IET Renew. Power Gener.* **2022**, *16*, 3267–3286. [CrossRef]
34. Alilou, M.; Azami, H.; Oshnoei, A.; Mohammadi-Ivatloo, B.; Teodorescu, R. Fractional-Order Control Techniques for Renewable Energy and Energy-Storage-Integrated Power Systems: A Review. *Fractal Fract.* **2023**, *7*, 391. [CrossRef]
35. Pan, I.; Das, S. Kriging Based Surrogate Modeling for Fractional Order Control of Microgrids. *IEEE Trans. Smart Grid* **2015**, *6*, 36–44. [CrossRef]
36. Monje, A.; Chen, Y.; Vinagre, B.M.; Xue, D.; Feliu, V. *Fractional Order Systems and Controls*; Springer: Berlin/Heidelberg, Germany, 2010.
37. Debbarma, S.; Saikia, L.C.; Sinha, N. AGC of a multi-area thermal system under deregulated environment using a non-integer controller. *Electr. Power Syst. Res.* **2013**, *95*, 175–183. [CrossRef]
38. Saxena, S. Load frequency control strategy via fractional-order controller and reduced-order modeling. *Int. J. Electr. Power Energy Syst.* **2019**, *104*, 603–614. [CrossRef]
39. Pan, I.; Das, S. Fractional Order AGC for Distributed Energy Resources Using Robust Optimization. *IEEE Trans. Smart Grid* **2015**, *7*, 2175–2186. [CrossRef]
40. Nour, M.; Magdy, G.; Chaves-Avila, J.P.; Sanchez-Miralles, A.; Petlenkov, E. Automatic Generation Control of a Future Multisource Power System Considering High Renewables Penetration and Electric Vehicles: Egyptian Power System in 2035. *IEEE Access* **2022**, *10*, 51662–51681. [CrossRef]
41. Yu, Y.; Guan, Y.; Kang, W.; Chaudhary, S.K.; Vasquez, J.C.; Guerrero, J.M. Fractional-Order Virtual Synchronous Generator. *IEEE Trans. Power Electron.* **2023**, *38*, 6874–6879. [CrossRef]
42. Long, B.; Li, X.; Rodriguez, J.; Guerrero, J.M.; Chong, K.T. Frequency stability enhancement of an islanded microgrid: A fractional-order virtual synchronous generator. *Int. J. Electr. Power Energy Syst.* **2023**, *147*. [CrossRef]
43. Morsali, J.; Zare, K.; Hagh, M.T. Applying fractional order PID to design TCSC-based damping controller in coordination with automatic generation control of interconnected multi-source power system. *Eng. Sci. Technol. Int. J.* **2017**, *20*, 1–17. [CrossRef]

44. Podlubny, I. *Fractional Differential Equations: An Introduction to Fractional Derivatives, Fractional Differential Equations, to Methods of Their Solution and Some of Their Applications*; Elsevier: Amsterdam, The Netherlands, 1998.
45. Moschos, I.; Parisses, C. A novel optimal PIλDND2N2 controller using coyote optimization algorithm for an AVR system. *Eng. Sci. Technol. Int. J.* **2022**, *26*, 100991. [CrossRef]
46. Kerdphol, T.; Rahman, F.S.; Watanabe, M.; Mitani, Y. *Virtual Inertia Synthesis and Control*; Springer: Dordrecht, The Netherlands, 2021.

Article

Optimized Multiloop Fractional-Order Controller for Regulating Frequency in Diverse-Sourced Vehicle-to-Grid Power Systems

Amira Hassan [1], Mohamed M. Aly [1], Mohammed A. Alharbi [2], Ali Selim [1], Basem Alamri [3], Mokhtar Aly [4], Ahmed Elmelegi [1], Mohamed Khamies [5] and Emad A. Mohamed [6,7,*]

1. Department of Electrical Engineering, Faculty of Engineering, Aswan University, Aswan 81542, Egypt; eng_amirahassan@yahoo.com (A.H.); mohamed.aly@eng.aswu.edu.eg (M.M.A.); ali.selim@aswu.edu.eg (A.S.); eng_ahelmelegi2@yahoo.com (A.E.)
2. Electrical Engineering Department, Taibah University, Naif Bin Abdulaziz Road, Al-Madinah Al-Munawarrah 41477, Saudi Arabia; mharbig@taibahu.edu.sa
3. Department of Electrical Engineering, College of Engineering, Taif University, P.O. Box 11099, Taif 21944, Saudi Arabia; b.alamri@tu.edu.sa
4. Facultad de Ingeniería, Arquitectura y Diseño, Universidad San Sebastián, Bellavista 7, Santiago 8420524, Chile; mokhtar.aly@uss.cl
5. Department of Electrical Engineering, Faculty of Engineering, Sohag University, Sohag 82524, Egypt; mohamed.khamies@eng.sohag.edu.eg
6. Department of Electrical Engineering, College of Engineering, Prince Sattam bin Abdulaziz University, Al Kharj 16278, Saudi Arabia
7. Aswan Wireless Communication Research Center (AWCRC), Faculty of Engineering, Aswan University, Aswan 81542, Egypt
* Correspondence: e.younis@psau.edu.sa

Citation: Hassan, A.; Aly, M.M.; Alharbi, M.A.; Selim, A.; Alamri, B.; Aly, M.; Elmelegi, A.; Khamies, M.; Mohamed, E.A. Optimized Multiloop Fractional-Order Controller for Regulating Frequency in Diverse-Sourced Vehicle-to-Grid Power Systems. *Fractal Fract.* **2023**, *7*, 864. https://doi.org/10.3390/fractalfract7120864

Academic Editors: Behnam Mohammadi-Ivatloo and Arman Oshnoei

Received: 16 September 2023
Revised: 18 November 2023
Accepted: 27 November 2023
Published: 5 December 2023

Abstract: A reduced power system's inertia represents a big issue for high penetration levels of renewable generation sources. Recently, load frequency controllers (LFCs) and their design have become crucial factors for stability and supply reliability. Thence, a new optimized multiloop fractional LFC scheme is provided in this paper. The proposed multiloop LFC scheme presents a two-degree-of-freedom (2DOF) structure using the tilt–integral–derivatives with filter (TIDN) in the first stage and the tilt–derivative with filter (TDN) in the second stage. The employment of two different loops achieves better disturbance rejection capability using the proposed 2DOF TIDN-TDN controller. The proposed 2DOF TIDN-TDN method is optimally designed using the recent powerful marine predator optimizer algorithm (MPA). The proposed design method eliminates the need for precise modeling of power systems, complex control design theories, and complex disturbance observers and filter circuits. A multisourced two-area interlinked power grid is employed as a case study in this paper by incorporating renewable generation with multifunctionality electric vehicle (EV) control and contribution within the vehicle-to-grid (V2G) concept. The proposed 2DOF TIDN-TDN LFC is compared with feature-related LFCs from the literature, such as TID, FOTID, and TID-FOPIDN controllers. Better mitigated frequency and tie-line power fluctuations, faster response, lower overshot/undershot values, and shorter settling time are the proven features of the proposed 2DOF TIDN-TDN LFC method.

Keywords: electrical grid; electric vehicle application; fractional order control theory; load frequency controller (LFC); marine predator optimizer (MPA); multisourced power systems; renewable power sources; vehicle to grid (V2G)

MSC: 68N30

1. Introduction

1.1. Overview

Advancements in energy transition have led to the widespread installation of renewable energy sources (RESs) [1]. In addition, wide global plans have been set to continue reducing carbon emissions in the environment. Another key factor for the energy transition is the wide replacement of various conventional fuel-based vehicles by clean electric vehicles (EVs) [2]. In addition, EVs and their charging infrastructures can widely contribute to improving electrical power system performance by controlling their charging/discharging commands, times, and power flows. However, the installed RESs at high levels of penetration and the installed EVs in power grids have led to different characteristics of modern power systems. Additionally, special attention must be paid to designing their associated control systems. The proper design and implementation of control methods have improved performance in several applications, such as unmanned aerial vehicles (UAVs) [3,4], networked mobile robots [5], nonholonomic vehicles [6,7], etc.

The structure of RESs is different from synchronous generator-based conventional electrical generation systems. They are based on power electronics conversion systems (PECSs) for grid integration. However, PECSs-based renewable generation systems lack inertia, and hence, deteriorated power system stability exists in modern electrical power systems [8]. This is due to the low stored mechanical power in power systems' inertia. This mechanical power helps retain system stability and operation during transients of generation, loading, and faults [9]. Thence, high levels of RESs penetration result in stability concerns in power grids. The load frequency controllers (LFCs) have shown great ability to regulate power grids' frequencies and tie-line powers between areas during normal conditions, as well as abnormal operating conditions [10]. Load frequency controllers (LFCs) regulate both the power output from generation sources and the loading demands, hence the enhanced regulation of the frequency signals besides the tie-line powers among areas [11–13].

1.2. Literature Review

The existing LFC methods in the literature can be mainly classified according to the controller type and number of loops. Based on the LFC type, several control methods have been introduced, such as integer-order control (IO) [14], fractional-order control (FO), sliding mode controllers (SMCs) [15], SMC with interval-type-2 fuzzy [16], repetitive control [17], model predictive controls (MPCs) [18,19], robust control [20,21], machine and deep learning (ML and DL) [22,23], etc. Additionally, based on the number of loops, various degrees of freedom (DOF) have been introduced, such as single-DOF (1DOF), two-DOF (2DOF), and three-DOF (3DOF) [24]. In 1DOF, area control error signals (ACEs) are used in the control feedback loop, whereas in 2DOF, another feedforward loop is added based on the frequency deviation signal with the ACE feedback loop. In 3DOF, a third feedforward signal is used based on the tie-line power between areas. Moreover, a variety of metaheuristic-based optimization methods has been introduced for optimally designing various control parameters in a simultaneous way [25,26]. The optimizer algorithms can be coupled with LFC methods to enhance the power system response and performance regarding the frequency with tie-line power regulations.

Several combinations of IO and FO LFC schemes have been extensively provided in the literature in different single and multi-area-based power systems [27,28]. For instance, the IO-based integrator (I), proportional–integrator (PI), and proportional–integrator–derivative (PID) control schemes represent the most widely introduced LFC methods. An optimum PI LFC has been proposed in [29]. Optimization procedures have been achieved using the binary moths–flame optimization (MFO). Another hybrid gravitation search with a firefly algorithm (hGFA)-based optimized PI LFC has been proposed in [30]. Further optimum PI LFC methods have been proposed using Harris Hawks optimizer algorithm (HHO) [31], and gray wolf optimization (GWO) [32]. Additionally, the PID has been provided in the literature for achieving LFC, such as in [33], by using a hybrid improved gravitation search algorithm and binary particle swarm optimizer (IGSA-BPSO). Optimum PID LFC methods were introduced using artificial bees colony (ABC) in [34] and stability boundary locus (SBL) optimizer in [35]. The applications of the I, ID, and PID integer-order LFCs have been presented and compared in [36]. The salp swarm optimization algorithm (SSA), grasshopper optimizer algorithm (GOA), and collective decision optimizer algorithm (CDO) have been presented for the optimum design of compared controllers. Another PI event-triggered LFC has been presented in [37], wherein it achieved lower peak overshoot/undershoot peaks compared with existing controllers. An optimized PID LFC method has been presented in [38] using a hybrid algorithm of sparrow searching and gray wolf optimization (SSAGWO). The above-presented literature is based on 1DOF control structures, and they are shown to be simple, easily designed, and implementable using low-cost processors, etc. However, in modern power grids, nonlinearities and uncertainties due to RESs hinder their superiority.

Additional flexibility has been achieved through using the FO control structures in LFC [39]. The FO terms are combined to form various structures using the FO tilt (T), FO integrator (I^λ), FO derivative (D^μ), and FO-based filter. A comparison between IO and FO LFC has been presented in [40] verifying that FO LFC has more flexibility. The results verify that FOPID has a faster response and better damping than the IO-based PID LFC scheme. For instance, FOPID has been proposed in [41], while its optimization procedures have been made using the movable damped waves optimizer (MDWA). Moreover, the sine–cosine optimizer algorithm (SCA) has been presented for FOPID optimization in [42]. The tilt FO control has also been presented in the literature for LFC [43]. The ABC optimization [44], pathfinder algorithm (PFA) [45], and differential evolutions algorithms (DEs) [43] were provided in the literature for optimizing tilt-based LFC methods. As an extension for FO control, several combinations have been provided and verified in literature work. The combination leads to the added benefits of both control methods. For instance, TID has been combined with FOPID in [46] to form a hybrid TFOID controller. The optimization processes were achieved through using the artificial ecosystem optimizer (AEO) algorithm. An extended hybrid TFOID with FO filter has been presented in [47] using an artificial hummingbird optimization algorithm (AHA).

From another perspective, various high-DOF LFC methods have been proposed in the literature. A 2DOF PID control has been presented in [48] for LFC. The flower pollination optimizer algorithm (FPA) was presented in [49] for optimizing a 2DOF PID controller. In [50], a 2DOF PID was proposed, and it is optimized through the dragonfly optimizer algorithm (DA) through the development of a new integral based on the weighted goals fitness functions (IB-WGFF). In [51], a 2DOF PD-PID controller has been proposed. A 2DOF PI-PID has been provided in [52], wherein the optimization process is achieved through a slap swarm optimization algorithm (SSA). In addition, a 3DOF (1 + PD)-PID LFC method has been provided in [53], in which an African vulture optimizer algorithm (AVOA) has been introduced for optimizing the controller parameters. A 2DOF TIDF LFC method was provided in [54] with the whale optimizer algorithm (WOA) for design optimization. A higher DOF has been proven to offer superior operation and transient performance of LFC compared with 1DOF-based controllers. Therefore, the focus of this paper is to develop a new 2DOF LFC method using the fractional-order control theory. In addition, optimized parameters are designed using the recent powerful marine predator optimizer algorithm (MPA).

1.3. Article Contribution

Existing work in the literature proves that increasing the DOF of the LFC method can lead to better performance and mitigation of expected disturbances in modern power grids. In addition, selecting a proper optimization algorithm jointly with the proposed controller can achieve enhanced selection processes of the optimized parameters. Moreover, future modern power grids are expected to be more volatile and less stable due to the reduced inertia resulting from the extensive use of PECSs. Among the existing optimization algorithms, the MPA-based optimization algorithm has achieved improved performance in several optimum values searching problems [55,56]. Some featured applications include maximum power extraction [57], PV model parameters extraction [58], PV array reconfiguration [59], PID-based LFC [60], and FO-based LFC methods [61,62].

Therefore, the authors of this paper were motivated to present an MPA-based optimization of the newly proposed 2DOF TIDN-TDN LFC method. The main contribution points in this article are as follows:

- A higher degree of freedom FO-based LFC method is proposed. The newly proposed controller is based on developing a multiloop two-degrees-of-freedom (2DOF) fractional-order-based LFC method. The proposed 2DOF LFC method uses the tilt–integral–derivatives with filter (TIDN) in the outer loop and the tilt–derivative with filter (TDN) in the inner loop.
- The proposed TIDN-TDN controller includes high flexibility due to its included FO operators, which help with better optimization of control performance. The proposed TIDN-TDN controller represents a new combination of fractional-order-based LFC compared with existing control structures in the literature.
- The improved performance using the proposed TIDN-TDN controller results from employing a feedback signal in the outer loop using ACE signal to mitigate the low-frequency fluctuations. Furthermore, the proposed 2DOF TIDN-TDN LFC method employs a feedforward loop using the frequency deviation signal to mitigate the high-frequency disturbances. Thence, better disturbance rejection capability is obtained using the proposed 2DOF TIDN-TDN controller. Moreover, the proposed TIDN-TDN LFC method does not require additional components and/or observer design and/or filter elements.
- An effective control and coordination method is proposed to control the participation of installed and future EVs using the TID controller and is coordinated with the proposed 2DOF TIDN-TDN LFC method. Accordingly, the installed EVs in future modern power systems participate in an effective way to dampen existing disturbances by utilizing the inherent EV batteries. This, in turn, leads to better EV utilization in future power systems with the expected continuous replacements of EVs. The coordination process is achieved inherently within the proposed controller and its design optimization method.
- An improved optimized design of the proposed 2DOF TIDN-TDN LFC method and EV TID controller is presented in this paper using the recent powerful marine predator optimizer algorithm (MPA) method. The parameters of the proposed controllers are determined simultaneously in all the studied interconnected power grids. The proposed method eliminates the need for complex control theories and/or mathematical determination processes using classical control methods. Thence, complex control designs and modeling are avoided using the powerful MPA optimizer.
- Further improvements are achieved by the proposed controller by avoiding the common problems of disturbance observer-based control, such as precise model dependency, complex tuning and design requirements, high computational complexity, sensitivity to measurement noise, and limited applicability.

In this work, the practical characteristics of connected renewable energy sources, grid battery contribution, and grid model parameter variations are considered. However, for a more practical extension of this work, the interference between voltage and frequency control designs must be considered. In addition, the effects of existing communication delays in power systems must be modeled and considered, especially with the move toward smarter power grid trends.

The remaining sections in this article are organized as follows: The detailed modeling, mathematical representation, and components of the selected power grid are presented in Section 2. The existing LFCs from the literature are presented in Section 3 with the FO control theory. The proposed 2DOF TIDN-TDN is introduced in Section 4. The design optimization and the MPA optimizer are detailed in Section 5. The obtained simulation results of the 2DOF TIDN-TDN and the selected case study are proven in Section 6 with comparative results. This paper's conclusions are provided in Section 7.

2. Mathematical Models of the System

This study focuses on a hybrid power system consisting of two interlinked areas connected via an AC bus tie-line system, as shown in Figure 1. As detailed in Figure 2, each of the two areas features four types of dynamic energy sources: a reheat-based thermal plant, a hydraulics plant, a gas power unit, and a nuclear plant. The physical power systems' boundaries, such as the generation rate constraints (GRCs) and governors' dead band (GDB), are modeled and considered as system nonlinearity. Additionally, RESs have a wind unit in area 1 and a PV unit in area 2, and both areas are considering the participation of EVs to control frequency, as depicted in Figure 2. Each area in the system being studied is equipped with a frequency controller that regulates the power output from various energy units. To have control of the power injected by EVs and to take part in frequency regulations within studied areas, another controller is added. In the system under study, each area possesses a rated capacity of 2000 MW, and its nominal loading is 1740 MW. The system parameters for the system under study are shown in the Appendix A section of this paper.

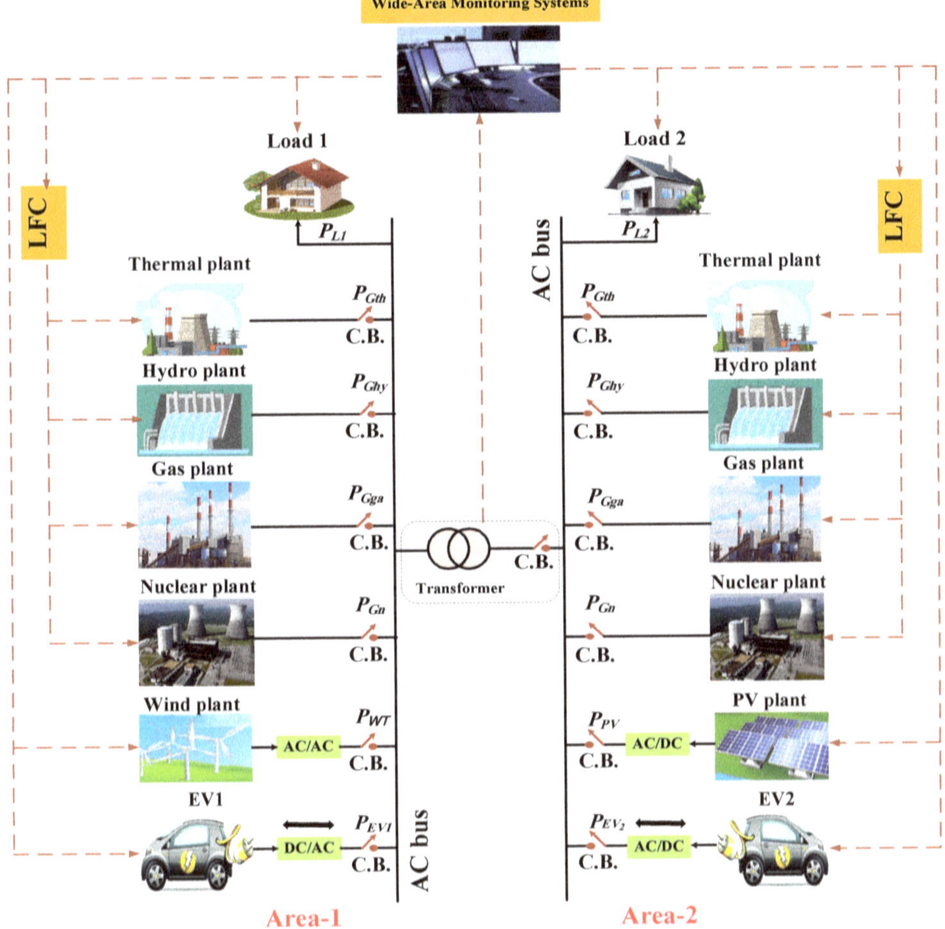

Figure 1. Power system structure of the two studied areas.

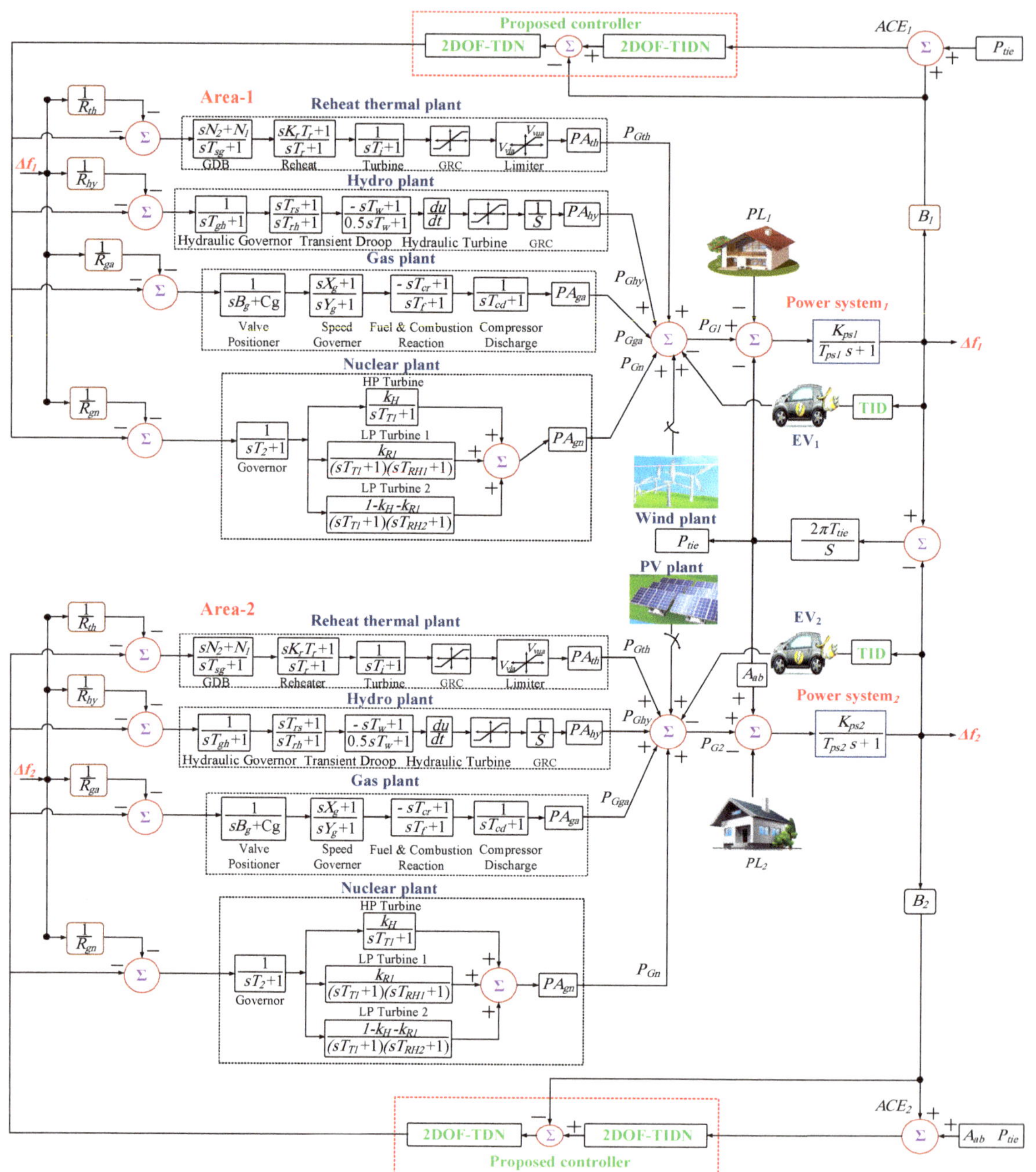

Figure 2. Mathematical model details of the studied diverse-source two-area interlinked power systems with EVs and RESs.

2.1. Modeling Various Generation Sources

2.1.1. Thermal Plant

The first-order transfer function is used to model reheating using the work in [63] as follows:

$$G_{r_{th}}(s) = \frac{sK_rT_r + 1}{sT_r + 1} \tag{1}$$

$$G_{t_{th}}(s) = \frac{1}{sT_i + 1} \tag{2}$$

where K_r is the gain of the steam turbine, and T_r is its time constant, whereas T_i represents the time constant of the reheater transfer function. The symbol s denotes the Laplace transform s-domain. The model includes thermal units governed by generation rate constraints GRC and GDB, with the GRCs set at 10% pu/min for both increasing and decreasing scenarios (0.0017 pu·MW/s). With respect to the change in addition to its rate for speed, a linearized version for modeling GDB can be utilized. The Fourier series was utilized to create the GDB transfer function model with a 0.5 percent for backlash as follows:

$$GDB = \frac{sN_2 + N_1}{sT_{sg} + 1} \tag{3}$$

where N_1 and N_2 denote Fourier coefficients selected as $N_1 = 0.8$ and $N_2 = -0.2/\pi$, as presented in [64].

2.1.2. Hydraulic Plant

The governor, droop compensations, and the penstock turbine make up the Hydraulic turbine's general transfer function. In [46], it can be shown as follows:

$$G_{hy}(s) = \frac{1}{sT_{gh} + 1} \cdot \frac{sT_{rs} + 1}{sT_{rh} + 1} \cdot \frac{-sT_w + 1}{0.5sT_w + 1} \tag{4}$$

The considered GRCs for the hydraulic plant represented as increasing/ decreasing rates are 270% pu/min (0.045 pu·MW/s) and 360% pu/min (0.06 pu·MW/s), respectively.

2.1.3. Gas Plant

The general transfer function of the gas turbine comprises components such as the valve positioner, speed governor, combustion reactions and fuel, and the compressor's discharge. In [63], it can be shown as follows:

$$G_{ga}(s) = \frac{1}{sB_g + C_g} \cdot \frac{sX_g + 1}{sY_g + 1} \cdot \frac{-sT_{cr} + 1}{sT_f + 1} \cdot \frac{1}{sT_{cd} + 1} \tag{5}$$

where, B_g, C_g, X_g, Y_g, T_{cr}, T_f, and T_{cd} denote the valve positioner's time constant, valve position of the gas turbine, the time constant of lead, the time constant of lag for the gas turbine, the time delay of a gas turbine's combustion reactions, the time constant of gas turbine fuel, and the time constant of the compressor's discharge volume, respectively.

2.1.4. Nuclear Plant

The model for the nuclear plant, as detailed in [65], comprises a speed governor, a high-pressure type turbine, and two low-pressure-type turbines. It is represented as follows:

$$G_{gn}(s) = \frac{1}{sT_2 + 1}\left(\frac{K_H}{sT_{T1} + 1} + \frac{K_{R1}}{sT_{T1} + 1} \cdot \frac{1}{sT_{RH1} + 1} + \frac{1 - K_H - K_{R1}}{sT_{T1} + 1} \cdot \frac{1}{sT_{RH2} + 1}\right) \tag{6}$$

where, T_{T2}, T_{T1}, T_{RH1}, and T_{RH2} denote the time constant of the speed governor, the high-pressure (HP)-based turbine, first low-pressure (LP) turbine, and second LP turbine, respectively. Also, K_H and K_{R1} denote the gain of the HP-type turbine and the first LP-type turbine's gain, respectively.

2.2. Modeling Various Renewable Generation Sources

2.2.1. PV Generation

Photovoltaic (PV) generation systems include solar modules, DC/DC converters, DC/AC inverters, and other electrical equipment. Solar modules capture solar energy in DC form, DC/DC converters boost the PV voltage to maximize power extraction, and DC/AC inverters convert the DC voltage into AC for the grid integration process. The PV's transfer function is provided as follows [46]:

$$G_{PV}(s) = \frac{K_{PV}}{T_{PV}s + 1} \tag{7}$$

where K_{PV} and T_{PV} denote the gain and time constant of PV plant transfer function. It is assumed that the MPPT controller operates effectively and that all available power from the sun is injected into the power grid, with variations in the solar irradiance and temperature values.

2.2.2. Wind Generation

The amount of power a wind turbine depends on the wind's velocity at any given moment. The wind turbine system (WTS) determines the pitch angles and creates non-linearities in the system based on wind speed. The WTS transfer function is provided as follows [46]:

$$G_{WT}(s) = \frac{K_{WT}}{T_{WT}s + 1} \tag{8}$$

where K_{WT} and T_{WT} denote the gain and time constant of the WTS plant transfer function. The wind power generator is assumed to have excellent tracking of its maximum power available in the wind, and hence, all its available power is injected into the power grid. Also, the faults in the system in addition to degradation effects are neglected in this work.

2.3. Modeling of EVs

The batteries of current EVs can be used to effectively control the power system's performance. They can be charged or discharged, depending on the power systems' management controller. These batteries have the potential to improve the power system's dynamic response, overall effectiveness, and reliability. Figure 3 shows the utilized dynamic modeling representation for the connected EVs in the power system. It is the equivalent electrical circuit modeling, as shown in [46]. In this study, it is assumed that the distribution of EVs among the two connected areas is equal. The proposed system also makes use of connected EVs to help at lowering frequency fluctuation values. The EV model is expressed as follows [46]:

$$V_{oc}(SOC) = V_{nom} + S\frac{RT}{F}ln\left(\frac{SOC}{C_{nom} - SOC}\right) \tag{9}$$

where $V_{oc}(SOC)$ is the open-circuit voltage V_{oc} for a particular state of charge (SOC), V_{nom} is nominal voltage, and C_{nom} is nominal EV battery capacity (in Ah). Moreover, S is the sensitivity parameter among V_{oc} and SOC. R, F, and T are the gaseous constant, Faraday's constant, and the temperature, respectively. The degradation and its associated problems in EV batteries are not considered in this work. In addition, it is assumed there is a perfect battery management controller between battery cells, and hence, their SOCs are always balanced.

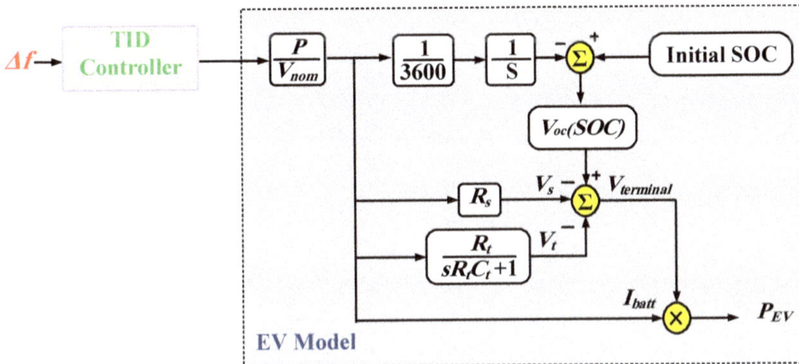

Figure 3. Dynamic modelling of EVs for the intended LFC study.

3. FO Control Theory and Existing LFCs in Literature

3.1. Existing IO LFC Methods

In the literature, IO-based LFC methods have found wide applications. These methods have demonstrated both enhanced performance and simplified control schemes in various industrial applications. As clarified in the literature review, the I, PI, and PID controllers represent the widely studied IO LFC schemes. Furthermore, various metaheuristic optimization techniques are used for determining optimum LFC parameters. The various gain parameters can be optimally tuned for achieving various objectives, such as controlling rise times, transient times, peaks overshoot/undershoot values, stability criteria, steady-state error, etc. The representations of IO-based LFC methods are as follows:

$$C_I(s) = \frac{Y(s)}{E(s)} = \frac{K_i}{s}$$

$$C_{PI}(s) = \frac{Y(s)}{E(s)} = K_p + \frac{K_i}{s}$$

$$C_{PID}(s) = \frac{Y(s)}{E(s)} = K_p + \frac{K_i}{s} + K_d\, s \tag{10}$$

$$C_{PIDF}(s) = \frac{Y(s)}{E(s)} = K_p + \frac{K_i}{s} + K_d\, s\, \frac{N_f}{s + N_f}$$

3.2. FO Control Theory

The FO-based control theories have proven to offer more flexibility and enhanced performance compared with their IO-based counterparts. The FO operators and calculus can be represented using Grunwald–Letnikov, Caputo, and Riemann–Liouville models. The Grundwald–Letnikov scheme represents the α_{th} derivative using the function (f) within the range a to t limits as follows:

$$D^\alpha\big|_a^t = \lim_{h \to 0} \frac{1}{h^\alpha} \sum_{r=0}^{\frac{t-a}{h}} (-1)^r \binom{n}{r} f(t - rh) \tag{11}$$

where h is sample time, and the $[\cdot]$ operator is only integer values in (16). Additionally, n is employed to achieve ($n - 1 < \alpha$ and $\alpha < n$). These binomials coefficients are determined using

$$\binom{n}{r} = \frac{\Gamma(n + 1)}{\Gamma(r + 1)\Gamma(n - r + 1)'} \tag{12}$$

where

$$\Gamma(n+1) = \int_0^\infty t^{x-1} e^{-t}\, dt \tag{13}$$

Riemann and Liouville defined the FO derivative avoiding the use of sums and limits. Instead, the IO derivative is used and represented as follows:

$$D^\alpha \big|_a^t = \frac{1}{\Gamma(n-\alpha)} \left(\frac{d}{dt}\right)^n \int_a^t \frac{f(\tau)}{(t-\tau)^{\alpha-n+1}}\, d\tau \tag{14}$$

Another representation of the FO derivative was made by Caputo, and it is defined as follows:

$$D^\alpha \big|_a^t = \frac{1}{\Gamma(n-\alpha)} \int_a^t \frac{f^{(n)}(\tau)}{(t-\tau)^{\alpha-n+1}}\, d\tau \tag{15}$$

However, $D^\alpha \big|_a^t$ can take different forms, as follows:

$$D^\alpha \big|_a^t = \begin{cases} \alpha > 0 \ \rightarrow\ \frac{d^\alpha}{dt^\alpha} & \text{FO derivative} \\ \alpha < 0 \ \rightarrow\ \int_{t_0}^{t_f} dt^\alpha & \text{FO integral} \\ \alpha = 0 \ \rightarrow\ 1 \end{cases} \tag{16}$$

On the other hand, Oustaloup-based recursive approximations (ORAs) have found wide use in implementing FO derivatives. They are a suitable way for real-time-based digital implementations. They are also a familiar and suitable option for the optimum tuning process of various FO controllers. Thence, ORA-based implementations are employed in this work due to being dominant. The approximation of the mathematical representations of αth FO derivatives (s^α) is expressed as follows:

$$s^\alpha \approx \omega_h^\alpha \prod_{k=-N}^{N} \frac{s+\omega_k^z}{s+\omega_k^p} \tag{17}$$

where ω_k^p and ω_k^z are poles-zeros locations in ω_h, respectively. They are calculated as follows:

$$\omega_k^z = \omega_b \left(\frac{\omega_h}{\omega_b}\right)^{\frac{k+N+\frac{1-\alpha}{2}}{2N+1}} \tag{18}$$

$$\omega_k^p = \omega_b \left(\frac{\omega_h}{\omega_b}\right)^{\frac{k+N+\frac{1+\alpha}{2}}{2N+1}} \tag{19}$$

$$\omega_h^\alpha = \left(\frac{\omega_h}{\omega_b}\right)^{\frac{-\alpha}{2}} \prod_{k=-N}^{N} \frac{\omega_k^p}{\omega_k^z} \tag{20}$$

where the approximate function has $(2N+1)$ poles/zeros. In addition, N determines the ORA filter's order within $(2N+1)$. In this work, ORA representation is utilized with $(M=5)$ inside the range ($\omega \in [\omega_b, \omega_h]$), which is selected between $[10^{-3}, 10^3]$ rad/s.

3.3. Existing FO LFCs Methods

Additionally, IO-based LFCs have been extended by using their FO-based controllers. The FOPI, TID, and FOPID have been applied to LFCs. For instance, the TID represents a simplified version of the FOPID controller. In the TID controller, a tilted proportional gain, denoted as ($K_t\, s^{-\left(\frac{1}{n}\right)}$), replaces the conventional proportional gain (K_p) used in IO-based PID controllers. In this case, the block is referred to as a tilt compensator, and n is called a tilt parameter.

On the other side, the filtering component is added to the derivative component to reduce the chattering noise in the input signal at extreme frequencies. Therefore, the TIDN is used in the literature to improve system stability [46,66]. In the literature, there several studies on integrating one or more IO and/or FO controllers. For instance, FOPID and

TID have been integrated to form the FOTID controller ($TI^\lambda D^\mu$). This, in turn, provides more flexibility and helps reduce the settling time of the system during disturbances [46,67]. The block diagrams of the TID, TIDN, FOTID, and TID-FOPIDN controllers [45,63,64] are shown in Figure 4. The featured FO LFC methods can be represented as follows:

$$C_{FOI}(s) = \frac{Y(s)}{R(s)} = \frac{K_i}{s^\lambda}$$

$$C_{FOPI}(s) = \frac{Y(s)}{R(s)} = K_p + \frac{K_i}{s^\lambda}$$

$$C_{TID}(s) = \frac{Y(s)}{R(s)} = K_t\, s^{-\left(\frac{1}{n}\right)} + \frac{K_i}{s} + K_d\, s$$

$$C_{FOPID}(s) = \frac{Y(s)}{E(s)} = K_p + \frac{K_i}{s^\lambda} + K_d\, s^\mu \tag{21}$$

$$C_{TIDN}(s) = \frac{Y(s)}{R(s)} = K_t\, s^{-\left(\frac{1}{n}\right)} + \frac{K_i}{s} + K_d\, s\frac{N_c s}{N_c + s}$$

$$C_{FOTID}(s) = \frac{Y(s)}{E(s)} = K_t\, s^{-\left(\frac{1}{n}\right)} + \frac{K_i}{s^\lambda} + K_d\, s^\mu$$

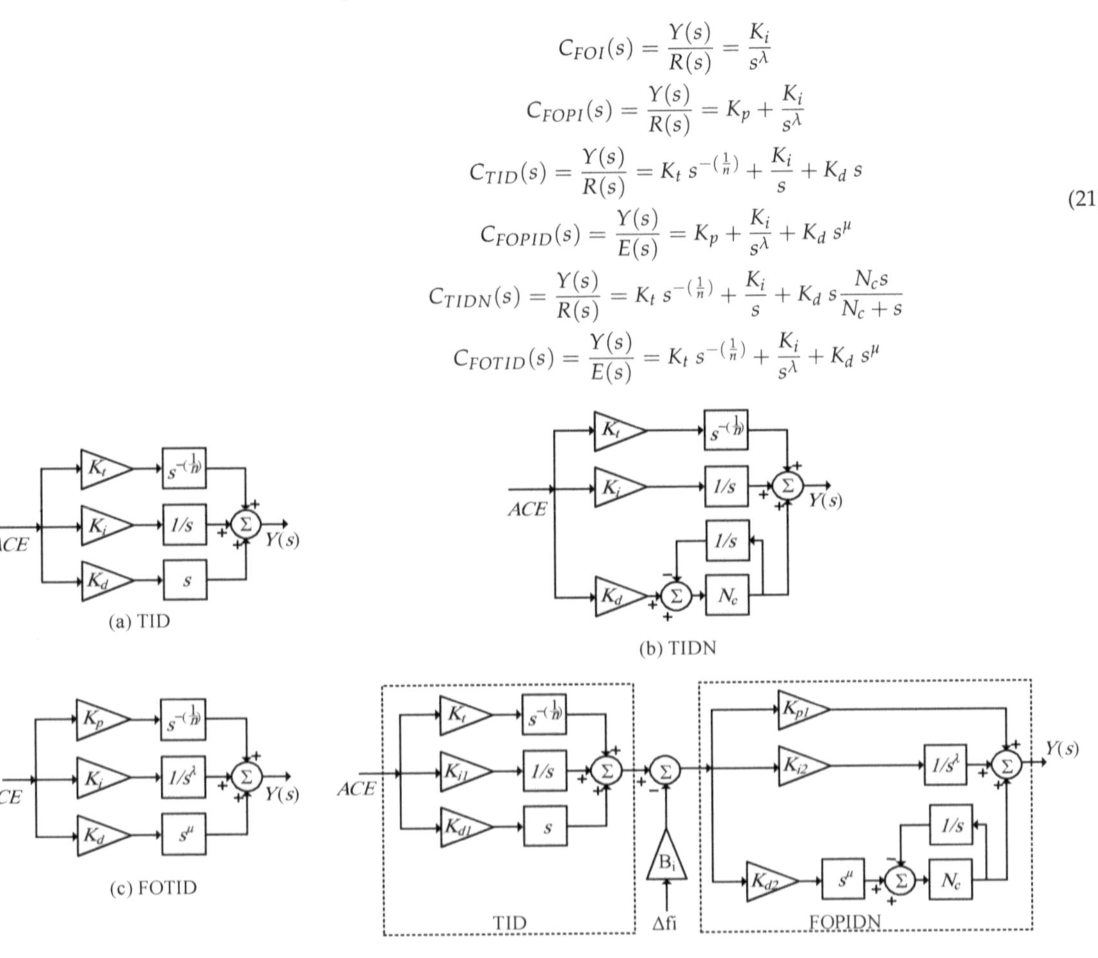

Figure 4. Block diagrams for some selective FO LFC methods.

4. The Proposed 2DOF TIDN-TDN Controller

This study introduces a new configuration for LFC applications, featuring a 2DOF multiloop tilt–integral–derivative filtered controller in cascade with a tilt–derivative filtered controller, termed 2DOF TIDN-TDN. The proposed controller is robust enough to allow the flexible management of an intricately reconstructed system by drastically reducing undershoot/overshoot and improving settling time. This controller also has two degrees of freedom, which improves dynamic response by using both the corresponding area frequency deviation (ΔF_i) and the area control error (ACE_i) signal as inputs. The schematic structure of the proposed cascaded 2DOF TIDN-TDN controller is shown in Figure 5, in which the TIDN controller serves as the outer controller $C_1(s)$ and the TDN controller serves as the inner controller $C_2(s)$. Figure 6 details the elements of the proposed cascaded 2DOF TIDN-TDN controller. Utilizing the frequency deviation signals leads to mitigating the existing high-frequency disturbances, while utilizing the ACE loop leads to mitigating the existing low-frequency disturbances. Hence, the proposed 2DOF TIDN-TDN controller provides better rejection of exiting disturbances due to its high DOF. The proposed multiloop controller is represented mathematically as follows:

$$X(s) = (aK_{t1}\, s^{-\left(\frac{1}{n_1}\right)} + \frac{K_i}{s} + bK_{d1}\,\frac{N_c s}{N_c + s})ACE_i -$$
$$(K_{t1}\, s^{-\left(\frac{1}{n_1}\right)} + \frac{K_i}{s} + K_{d1}\,\frac{N_c s}{N_c + s})B_i\Delta F_i \tag{22}$$

$$U_c(s) = X(s)(cK_{t2}\, s^{-\left(\frac{1}{n_2}\right)} + dK_{d2}\,\frac{N_c s}{N_c + s}) -$$
$$(K_{t2}\, s^{-\left(\frac{1}{n_2}\right)} + K_{d2}\,\frac{N_c s}{N_c + s})B_i\Delta F_i \tag{23}$$

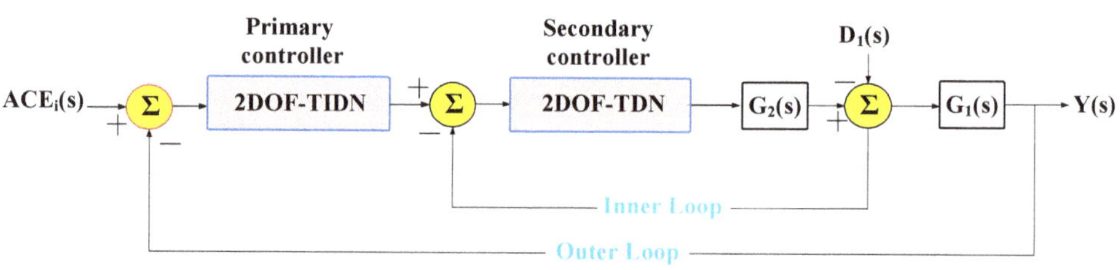

Figure 5. Proposed multiloop cascaded control structure.

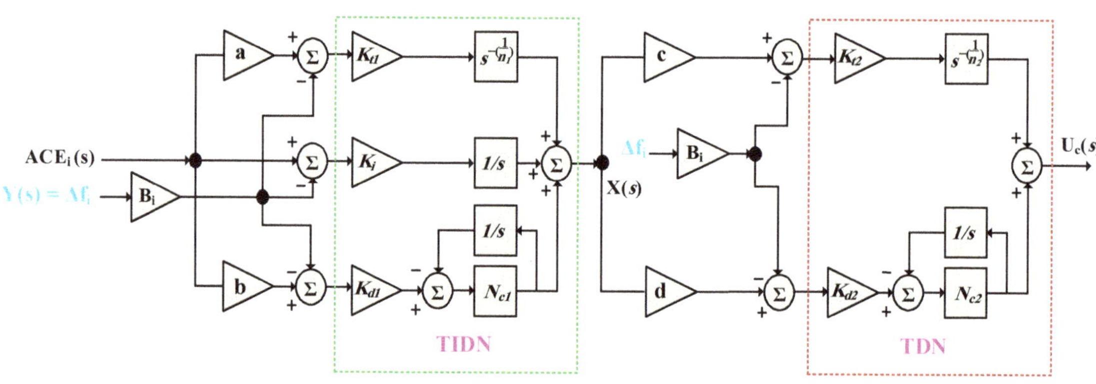

Figure 6. The proposed cascaded 2DOF TIDN-TDN control structure for LFC in each area.

5. The Process for Obtaining Optimized Control Parameters

5.1. Optimization Process

The two main goals of LFCs are to keep tie-line power fluctuations at their minimum value and to cancel out frequency drifts when there are disturbances in the system. To accomplish the goals in the proposed optimization problem targeting LFC applications, an objective function has to be formulated based on these objectives. The objective function, which combines frequency deviations and tie-line power deviations, has been accumulated by utilizing a variety of widely applied error functions. In this work, four different cost functions were used for optimization processes for the proposed 2DOF TIDN-TDN controller. The four objective functions are represented as follows:

$$ISE = \int_0^{t_s} ((\Delta f_1)^2 + (\Delta f_2)^2 + (\Delta P_{tie})^2)\, \mathrm{d}t \tag{24}$$

$$ITSE = \int_{0}^{t_s} \left((\Delta f_1)^2 + (\Delta f_2)^2 + (\Delta P_{tie})^2 \right) t \cdot dt \tag{25}$$

$$IAE = \int_{0}^{t_s} \left(abs(\Delta f_1) + abs(\Delta f_2) + abs(\Delta P_{tie}) \right) dt \tag{26}$$

$$ITAE = \int_{0}^{t_s} \left(abs(\Delta f_1) + abs(\Delta f_2) + abs(\Delta P_{tie}) \right) t \cdot dt \tag{27}$$

The constraints on the problem are the controller tunable parameters' boundary limits for both area 1 and area 2. Figure 7 depicts the main diagram representing the MPA-based process for tuning the controllers' parameters. The design problem can therefore be described as an optimization problem in which the controllers' parameters can be simultaneously determined based on the minimization of the objective function. The constraint for the TID controller can be represented as [64]

$$\begin{aligned}
K_t^{min} &\leq K_t \leq K_t^{max} \\
K_i^{min} &\leq K_i \leq K_i^{max} \\
n^{min} &\leq n \leq n^{max} \\
K_d^{min} &\leq K_d \leq K_d^{max}
\end{aligned} \tag{28}$$

However, they are represented for the FOTID controller as follows [45]:

$$\begin{aligned}
K_t^{min} &\leq K_t \leq K_t^{max} \\
K_i^{min} &\leq K_i \leq K_i^{max} \\
\lambda^{min} &\leq \lambda \leq \lambda^{max} \\
K_d^{min} &\leq K_d \leq K_d^{max} \\
n^{min} &\leq n \leq n^{max} \\
\mu^{min} &\leq \mu \leq \mu^{max}
\end{aligned} \tag{29}$$

The constraints for the TID-FOPIDN controller are represented as follows [63]:

$$\begin{aligned}
K_p^{min} &\leq K_p \leq K_p^{max} \\
K_t^{min} &\leq K_t \leq K_t^{max} \\
K_i^{min} &\leq K_i \leq K_i^{max} \\
K_d^{min} &\leq K_d \leq K_d^{max} \\
n^{min} &\leq n \leq n^{max} \\
\lambda^{min} &\leq \lambda \leq \lambda^{max} \\
\mu^{min} &\leq \mu \leq \mu^{max} \\
N_c^{min} &\leq N_c \leq N_c^{max}
\end{aligned} \tag{30}$$

However, the constraints for the proposed 2DOF TIDN-TDN controller are represented as follows:

$$\begin{aligned}
K_t^{min} &\leq K_t \leq K_t^{max} \\
K_i^{min} &\leq K_i \leq K_i^{max} \\
K_d^{min} &\leq K_d \leq K_d^{max} \\
n^{min} &\leq n \leq n^{max} \\
N_c^{min} &\leq N_c \leq N_c^{max} \\
a^{min} &\leq a \leq a^{max} \\
b^{min} &\leq b \leq b^{max} \\
c^{min} &\leq c \leq c^{max} \\
d^{min} &\leq d \leq d^{max}
\end{aligned} \tag{31}$$

where the constraints for various tuned controllers are limited within the maximum limits as *max* and the minimum values of control parameters as *min*. The minimum values of control gains (K_t^{min}, K_i^{min}, and K_d^{min}) are all zeros, and their maximum (K_t^{max}, K_i^{max}, and K_d^{max}) values are all five. However, (n^{min} and n^{max}) are set to 2 and 5, respectively, (λ^{min} and μ^{min}) are all zeros, and (λ^{max} and μ^{max}) are both set to 1. In addition, (N_c^{min} and N_c^{max}) are set to 5 and 500, respectively. The minimum/maximum tilt/derivative weights (a^{min}, b^{min}, c^{min}, d^{min} and a^{max}, b^{max}, c^{max}, d^{max} are between 0 and 4, respectively. The various optimized parameters obtained based on different optimization algorithms such as MPA, PSO, WOA, and GWO are shown in Tables 1–4, respectively, for the proposed controller, whereas Table 5 summarizes the control parameters of the suggested control methods using the MPA technique.

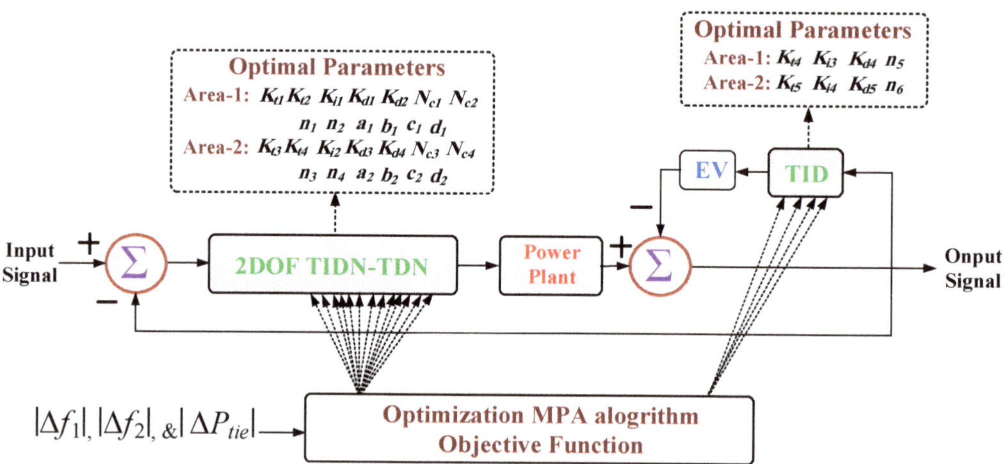

Figure 7. The optimal parameters of optimized controllers using the MPA algorithm.

Table 1. The coefficient parameters of the proposed controller utilizing the MPA algorithm.

Area		Coefficients												
		K_{t1}	K_i	K_{d1}	n_1	Nc_1	a	b	K_{t2}	K_{d2}	n_2	Nc_2	c	d
Area 1	LFC	1.688	1.783	1.444	3.583	313.301	1.789	1.659	1.598	0.989	4.331	202.014	1.416	1.268
	EV	1.943	1.963	1.515	3.076	-	-	-	-	-	-	-	-	-
Area 2	LFC	1.308	0.431	0.136	3.495	499.781	1.433	0.597	0.7058	0.251	4.336	273.051	1.377	1.159
	EV	1.933	0.728	0.611	4.753	-	-	-	-	-	-	-	-	-

Table 2. The coefficient parameters of the proposed controller utilizing the PSO algorithm.

Area		Coefficients												
		K_{t1}	K_i	K_{d1}	n_1	Nc_1	a	b	K_{t2}	K_{d2}	n_2	Nc_2	c	d
Area 1	LFC	1.308	1.684	1.523	2.533	319.511	1.577	1.209	1.251	1.127	4.801	114.156	0.922	1.521
	EV	1.365	1.741	1.048	2.926	-	-	-	-	-	-	-	-	-
Area 2	LFC	1.016	1.009	0.571	3.031	321.091	1.215	1.651	0.943	1.093	3.259	290.375	1.052	1.972
	EV	1.683	1.475	0.991	4.947	-	-	-	-	-	-	-	-	-

Table 3. The coefficient parameters of the proposed controller utilizing the WOA algorithm.

Area		Coefficients												
		K_{t1}	K_i	K_{d1}	n_1	Nc_1	a	b	K_{t2}	K_{d2}	n_2	Nc_2	c	d
Area 1	LFC	0.974	1.292	1.907	4.775	402.015	1.0183	1.033	1.096	0.831	2.364	190.056	1.935	2.057
	EV	0.685	1.495	0.891	3.084	-	-	-	-	-	-	-	-	-
Area 2	LFC	0.884	0.429	1.117	3.157	412.128	0.773	0.945	1.496	1.566	3.047	210.192	0.761	1.0941
	EV	0.551	0.953	0.835	4.045	-	-	-	-	-	-	-	-	-

Table 4. The coefficient parameters of the proposed controller utilizing the GWO algorithm.

Area		Coefficients												
		K_{t1}	K_i	K_{d1}	n_1	Nc_1	a	b	K_{t2}	K_{d2}	n_2	Nc_2	c	d
Area 1	LFC	0.421	0.851	1.006	2.023	399.214	0.475	1.287	1.205	1.0742	2.036	208.109	3.284	1.362
	EV	0.995	0.158	0.931	4.063	-	-	-	-	-	-	-	-	-
Area 2	LFC	0.263	0.273	0.394	2.475	331.074	1.004	1.093	0.929	1.984	3.001	401.001	2.375	0.821
	EV	0.341	1.846	0.894	3.315	-	-	-	-	-	-	-	-	-

Table 5. The coefficient parameters of the different controllers utilizing the MPA algorithm.

Controller	Area		Coefficients										
			K_t	K_{i1}	K_{d1}	K_{p1}/K_{p2}	K_{i2}	K_{d2}	n	Nc	λ	μ
TID	Area 1	LFC	0.1884	0.1238	0.4095	-	-	-	2.1941	-	-	-
		EV	0.1974	0.1436	0.3278	-	-	-	3.0357	-	-	-
	Area 2	LFC	0.2239	0.1131	0.4990	-	-	-	4.3562	-	-	-
		EV	0.2957	0.3537	1.0982	-	-	-	2.4327	-	-	-
FOTID	Area 1	LFC	1.1862	1.4553	2.9561	-	-	-	3.1931	-	0.7882	0.8959
		EV	1.0351	0.2841	1.5951	-	-	-	2.2831	-	-	-
	Area 2	LFC	0.1012	0.1572	1.9997	-	-	-	3.0192	-	0.9813	0.8794
		EV	1.0935	1.2557	0.8324	-	-	-	3.0062	-	-	-
TID-FOPIDN	Area 1	LFC	1.9674	0.6977	1.5785	0.3329	0.9224	0.8098	3.8677	245.04	0.4576	0.5531
		EV	2.5884	1.1238	0.4095	-	-	-	3.0019	-	-	-
	Area 2	LFC	1.9358	0.38442	0.7889	1.3744	0.7179	1.3931	4.6014	300.48	0.4652	0.5592
		EV	1.2239	0.2351	0.5481	-	-	-	3.941	-	-	-

5.2. The Principle of the MPA Optimizer

As clarified in the literature review, the MPA optimizer has been preferred in several optimum parameter determination applications [61]. The MPA principles are based on food search strategies using Levy and the Brownie movement within their surrounding predators. They determine the optimum by modifying the policy using the biological interaction between prey and predators. In [55], more details about the principles with more details about mathematical representations are given. Based on the literature review, the MPA optimizer has proven superior, with several benefits, as follows:

1. High efficiency: The MPA optimizer achieved efficient performance when solving different optimization problems with various properties, particularly where traditional metaheuristic methods fail to converge to optimal solutions.

2. Increased robustness: The optimized MPA showed robust performance against changes within optimization problems and has the ability to adapt itself to the different considered types of problem constraints and/or objectives.
3. More flexibility: The MPA represents a flexible algorithm, which can be modified and/or customized easily to suit the different optimization problems.
4. Scalability: The MPA algorithm was verified and tested on a large variety of optimization problems, wherein it demonstrated promising performance results.

A brief representation of the code and stages is presented in this section. The MPA optimizer has three phases based on the relative speed of the prey and predators. The main stages are shown in the flowchart of the MPA optimizer, shown in Figure 8. Its inherent phases are three phases according to the speed ratio between the prey and predators' speeds. The stages are as follows:

- High Speed Ratio Phase: It corresponds to the first one-third portion of the iteration number. It is related to cases of higher prey speed than predators. The mathematical representation is given as [55]:

$$S_i = R_B \times (Elite_i - R_B \times Z_i), i = 1, 2, \ldots, n \tag{32}$$

$$Z_i = Z_i + P \cdot R \times S_i \tag{33}$$

where R represents the vector from random numbers within the $[0, 1]$ range, whereas P equals 0.5, and R_B is the Brownian motion vector.

- Unity Speed Ratio Phase: This phase is represented by the second one-third portion of the iterations. It is related to the case of equal speed between prey and predators, in which, the predators' movements are represented by Brownian expression and the prey's movements are represented by the Lévy flights model method. Within this phase, the population is divided into two subdivisions. In the first part, (34) and (35) are used, whereas in second part, (36) and (37) are employed for modifying the locations as follows [57]:

$$S_i = R_L \times (Elite_i - R_L \times Z_i), i = 1, 2, \ldots, n/2 \tag{34}$$

$$Z_i = Z_i + P \cdot R \times S_i \tag{35}$$

where R_L is a random variable, and it is generated using Lévy distribution.

$$S_i = R_B \times (R_B \times Elite_i - Z_i), i = 1, 2, \ldots, n/2 \tag{36}$$

$$Z_i = Elite_i + P \cdot CF \times S_i \tag{37}$$

where

$$CF = (1 - \frac{t}{t_{\max}})^{2\frac{t}{t_{\max}}} \tag{38}$$

where t and $t_{\max}$ are the current value and the maximum value for iterations.

- Low Speed Ratio: This phase is formed by the last one-third of the iteration. In this phase, the prey's speed is lower than the predators' speed, in which the location modifications are expressed as follows [55]:

$$S_i = R_L \times (R_L \times Elite_i - Z_i), i = 1, 2, \ldots, n \tag{39}$$

$$Z_i = Elite_i + P \cdot CF \times S_i, CF = (1 - \frac{t}{t_{\max}})^{2\frac{t}{t_{\max}}} \tag{40}$$

In [55], the formation of eddy and fish aggregation devices (FADs) effecting *Fs* is utilized to count for the surrounding environment conditions of prey and predators. The positions of the population are modified based on FADs to avoid the local optimum solution. It is represented as follows:

$$Z_i = \begin{cases} Z_i + CF \cdot [Z_{\min} + R(Z_{\max} - Z_{\min}) \times W , r \leq Fs \\ Z_i + [Fs(1 - r) + r](Z_{r1} - Z_{r2}) , r > Fs \end{cases} \tag{41}$$

where *Fs* is set at 0.2, *W* is a binary number between 0 and 1, and *r* stands for a random number. However, *r*1 and *r*2 are random indices of prey. $Z_{\max}$ and $Z_{\min}$ are the lower and upper bounding vectors.

The tuning process is made offline, and thanks to recently developed powerful computers, the control parameters design process has become possible without consuming more time. Moreover, recent advanced digital signal processors facilitated the fractional-order control implementation and application.

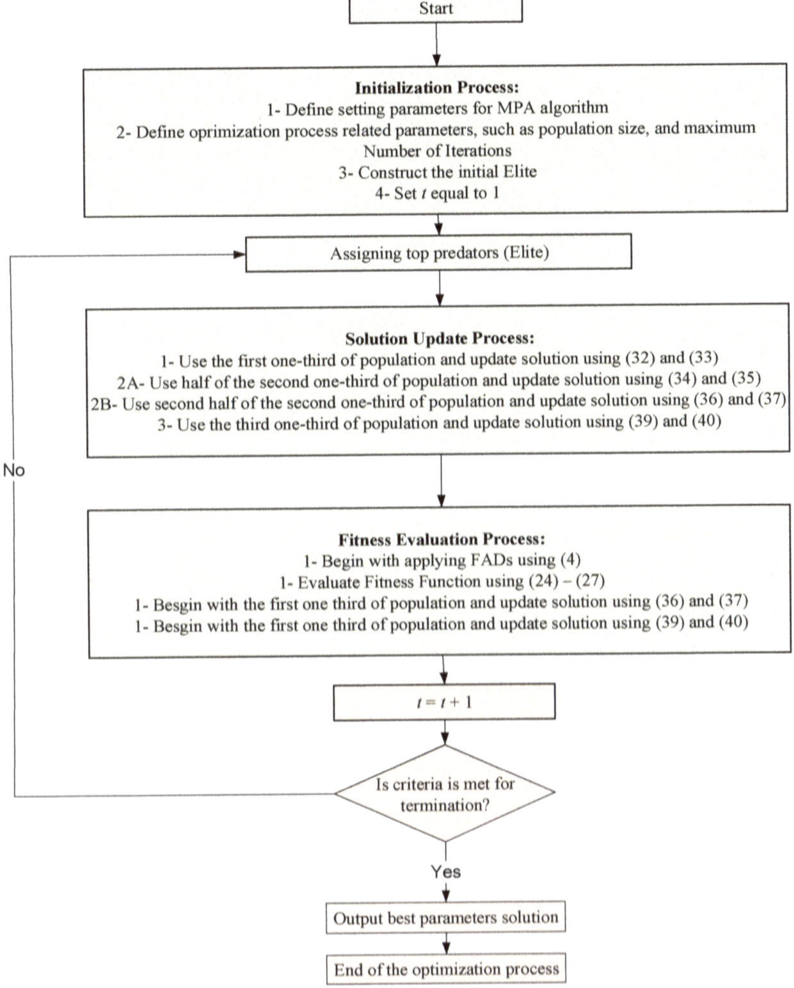

Figure 8. Flowchart of MPA's inherent stages and operation.

6. Simulation Results

In order to improve the LFC of the dual-area MG power systems, this section focuses on the validity and effectiveness of the suggested control technique of the new cascaded 2DOF TIDN-TDN controller coordinated with the EV system. The MPA optimizes the specified control strategy as well as other strategies. The proposed method is checked using the MATLAB/Simulink software 2021a by integrating the two-area system's Simulink with the algorithm code for the MPA to achieve the LFC fitness function. Using a desktop computer with a processor Intel Core i5 CPU clocked at 2.8 GHz, 64-bit version, the entire code of the dual-area microgrid network is implemented. Utilizing the same technique with the EV model based on the MPA method and under the same operating provisions of load change and RESs disturbances of the thoughtful multiarea MG power network, which implicate a decentralized 2DOF TIDN-TDN controller for the AGC and TID for the EV system in each area, the proposed 2DOF TIDN-TDN concept is established by comparing its interpretation with classical and sophisticated control techniques, such as TID, FOTID, and TID-FOPIDN, and the following operational circumstances for the researched multiarea MG system are investigated in terms of the results:

- **Scenario 1:** Action of step load perturbation's effect (SLP).
- **Scenario 2:** Sudden load shedding (SLS).
- **Scenario 3:** Parameters uncertainties of nuclear generation.
- **Scenario 4:** Multiple-load perturbation effects (MLP).
- **Scenario 5:** The action of high RES deployment.
- **Scenario 6:** The effect of low inertia 50% (high RES penetration) with multiple-load perturbation and parameter variations in the nuclear power station.

To judge the efficiency of the proposed MPA-based design, its convergence characteristics are compared with the PSO, GWO, and WOA optimizers. The results are obtained using a computer with a Core-i5 CPU 2.8 GHz and a 64-bit system. The results for IAE, ISE, ITSE, and ITAE are shown in Figure 9. In addition, the calculations of the ISE, ITSE, IAE, and ITAE for the studied optimizers are summarized in Table 6. The results show that the lowest objective function minimization is obtained using the MPA method in the case of ISE and ITAE. The PSO comes in second place in these two objectives. Additionally, MPA and PSO share the best convergence characteristics in the IAE and ITSE cases. However, MPA possesses faster convergence to the optimum values compared with that of the PSO method in those two cases.

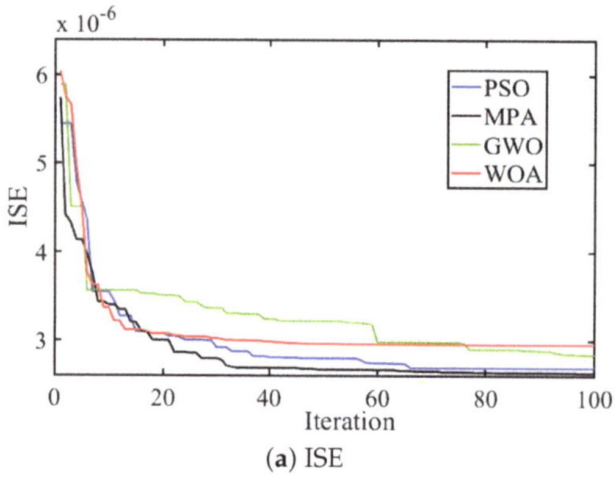

(**a**) ISE

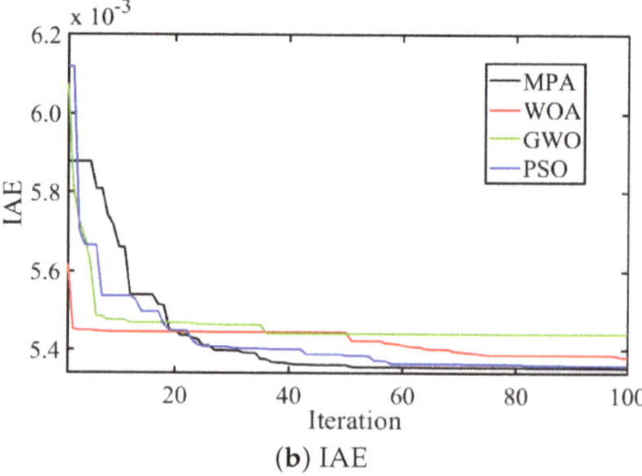

(**b**) IAE

Figure 9. *Cont.*

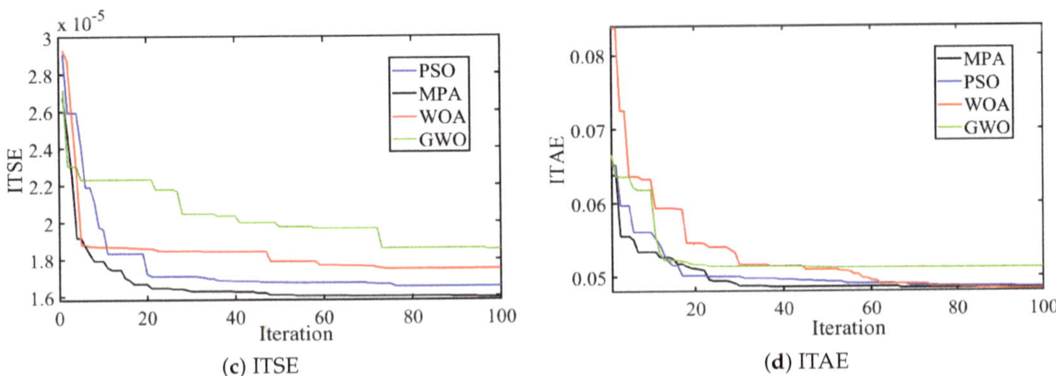

Figure 9. Convergence characteristics of MPA compared to other optimization techniques.

Table 6. Comparative analysis of objective function indices for the study of the different PSO, WOA, and GWO algorithms.

Alogrithm	Proposed Controller	Objective Function			
		ISE	ITSE	IAE	ITAE
PSO	2DOF TIDN-TDN	2.8107×10^{-6}	1.7082×10^{-5}	0.00535	0.0491
WOA	2DOF TIDN-TDN	2.9983×10^{-6}	1.7924×10^{-5}	0.00539	0.0483
GWO	2DOF TIDN-TDN	2.9025×10^{-6}	1.88132×10^{-5}	0.00546	0.0521

6.1. Scenario 1

A 20 MW load is installed in area 1 at an instant of 5 s in this scenario for testing the proposed 2DOF TIDN-TDN controller for LFC and TID controller for an EV system on the studied dual-area MG network based on the MPA method, which is verified by comparing it with PSO, WOA, and GWO in this scenario, as noted in Figure 10. As shown in Figure 11, the suggested approach with SLP is compared with TID, FOTID, and TID-FOPIDN controllers for the suggested dual-area network frequency and power diffraction response. The conventional TID controller has the lowest performance in this graph compared with other approaches, with substantial undershoot values at −0.024 Hz in the first area, −0.019 Hz in the second area, and −0.006 p.u. for tie power. FOTID maintained the frequency variation at −0.014 Hz in area 1 and 0.008 Hz in area 2, with −0.0024 p.u in substitution power between the two areas. However, the cascaded TID-FOPIDN controller provided satisfactory results compared with earlier control methods by dampening the system perversions to acceptable levels, as summarized in Table 7. In contrast to other comparable cascaded controllers, the proposed 2DOF TIDN-TDN controller has the fastest response and the lowest oscillations in regulating frequency and power variations. Further comparisons between the suggested controllers are provided in Table 7. The table displays that the proposed 2DOF TIDN-TDN strategy has the smallest peak overshoots (PO), peak undershoots (PU), and settling time (ST) in terms of frequency errors and exchange tie-line power between the multiarea systems. The obtained results state that the inner loop of the two cascaded TID-FOPIDN and 2DOF TIDN-TDN controllers responded to the varied dynamics originating from the different generation sources in both areas, and the outer loop controller can handle the power system dynamics and the SLPs. Therefore, the performance of the proposed cascaded controllers is more influential than that of the other conventional feedback controllers.

Table 7. Obtained results for the tested scenarios.

Scenario	Controller	Δf_1			Δf_2			ΔP_{tie}		
		PO	PU	ST (s)	PO	PU	ST (s)	PO	PU	ST (s)
No. 1	TID	0.002	0.024	27	0.0023	0.019	33	0.0027	0.0062	42
	FOTID	0.0008	0.014	19	0.0081	0.0077	21	0.0003	0.0024	38
	TID-FOPIDN	0.0006	0.007	17	0.0055	0.0036	18	-	0.0011	27
	Proposed	-	0.003	9	-	0.0014	14	-	0.0006	15
No. 2	TID	0.0238	-	29	0.0188	-	34	0.0058	-	46
	FOTID	0.0141	0.0009	28	0.008	0.0008	33	0.0024	-	37
	TID-FOPIDN	0.00667	0.0007	26	0.0038	0.0005	28	0.0011	0.0001	36
	Proposed	0.00293	-	18	0.00152	-	23	0.00058	-	23
No. 3	TID	0.0016	0.028	28	0.0003	0.022	35	0.0012	0.0068	44
	FOTID	0.0062	0.022	18	0.0021	0.011	28	0.0002	0.0039	37
	TID-FOPIDN	0.0015	0.011	16	0.0011	0.0042	23	0.00008	0.0014	23
	Proposed	0.0005	0.003	10	-	0.0015	11	-	0.0006	17
No. 4	TID	0.369	0.365	36	0.3267	0.3235	41	0.0927	0.0881	48
	FOTID	0.184	0.187	19	0.1228	0.1241	32	0.0368	0.0374	40
	TID-FOPIDN	0.114	0.113	16	0.0745	0.0741	21	0.0214	0.0212	18
	Proposed	0.061	0.059	11	0.0329	0.0327	12	0.0129	0.0126	13
No. 5	TID	0.3852	0.357	OS	0.3465	0.3155	27	0.1114	0.0881	OS
	FOTID	0.1931	0.182	24	0.1325	0.1203	21	0.0406	0.0381	OS
	TID-FOPIDN	0.1066	0.101	15	0.0545	0.0505	17	0.0169	0.0171	19
	Proposed	0.0625	0.058	10	0.0351	0.0315	11	0.0133	0.0128	12
No. 6	TID	0.4299	0.399	OS	0.4299	0.3995	OS	0.1272	0.1054	OS
	FOTID	0.3383	0.322	OS	0.1958	0.1801	28	0.0608	0.0578	OS
	TID-FOPIDN	0.1746	0.169	32	0.0731	0.0695	22	0.0236	0.0232	20
	Proposed	0.0992	0.092	13	0.0441	0.0395	12	0.0144	0.0176	15

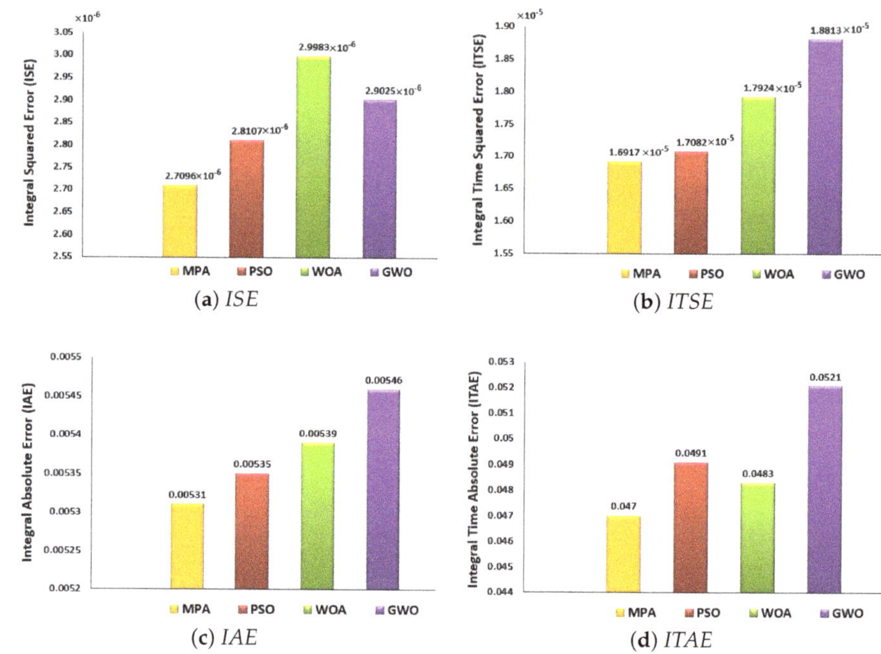

Figure 10. Objective functions comparison in scenario 1: (**a**) ISE, (**b**) ITSE, (**c**) IAE, (**d**) ITAE.

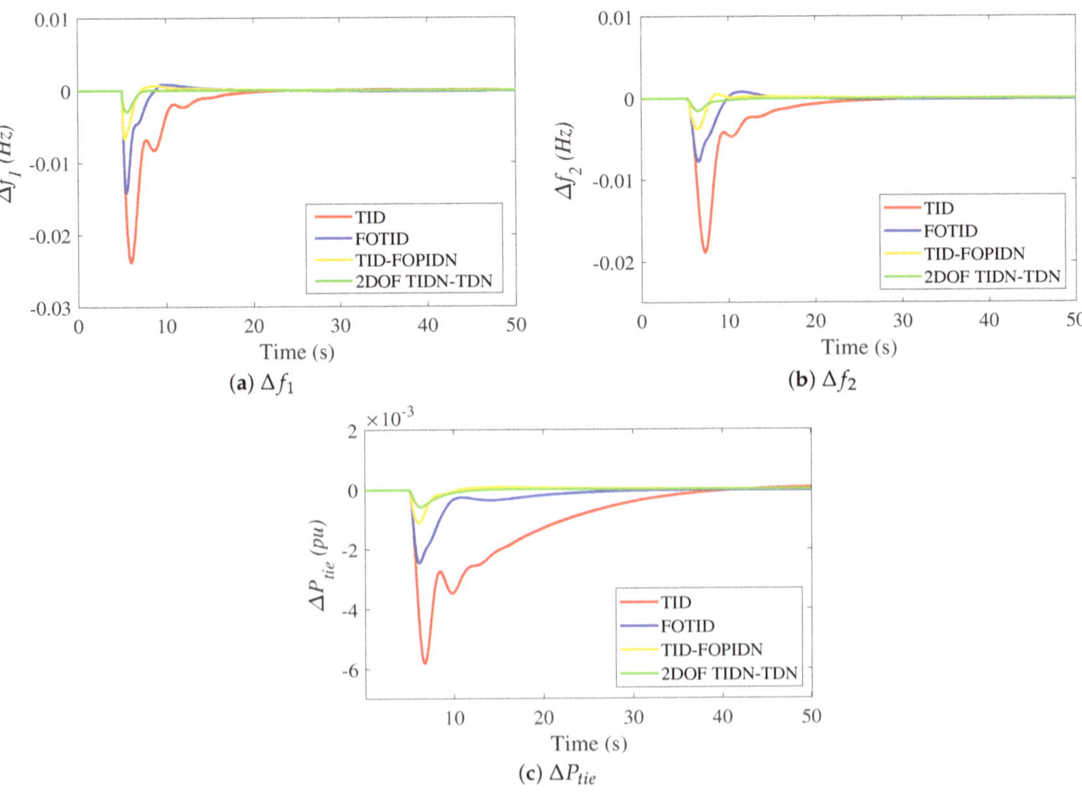

(a) Δf_1

(b) Δf_2

(c) ΔP_{tie}

Figure 11. The frequency response of the dual-MG with step load 1% in scenario 1.

6.2. Scenario 2

The system frequency and power increase dramatically in the case of sudden load failure, which may disturb the generation/demand balancing condition. Therefore, this scenario tested the performance of all the suggested controllers, in this case, by shedding a 200 MW load in a time of 15 s. Figure 12 shows the impact of this critical situation on the tie-line power change and system frequency deviations. It is distinct that the proposed 2DOF TIDN-TDN controller attains the best performance indices among the other LFCs, since the proposed controller achieves the lowest POs, PUs, and STs. Furthermore, the proposed method can damp the frequency oscillations quickly to their steady-state value after 18 s for the frequency error in area 1, whereas the TID, FOTID, and TID-FOPIDN controllers reach this value after 26 s, 28 s, and 29 s, respectively. Overall, these results ensure the superiority of the multiloop cascaded 2DOF TIDN-TDN and TID-FOPIDN controllers in achieving enhanced and fast responses better than the individual controllers during the critical failure case of microgrid loads.

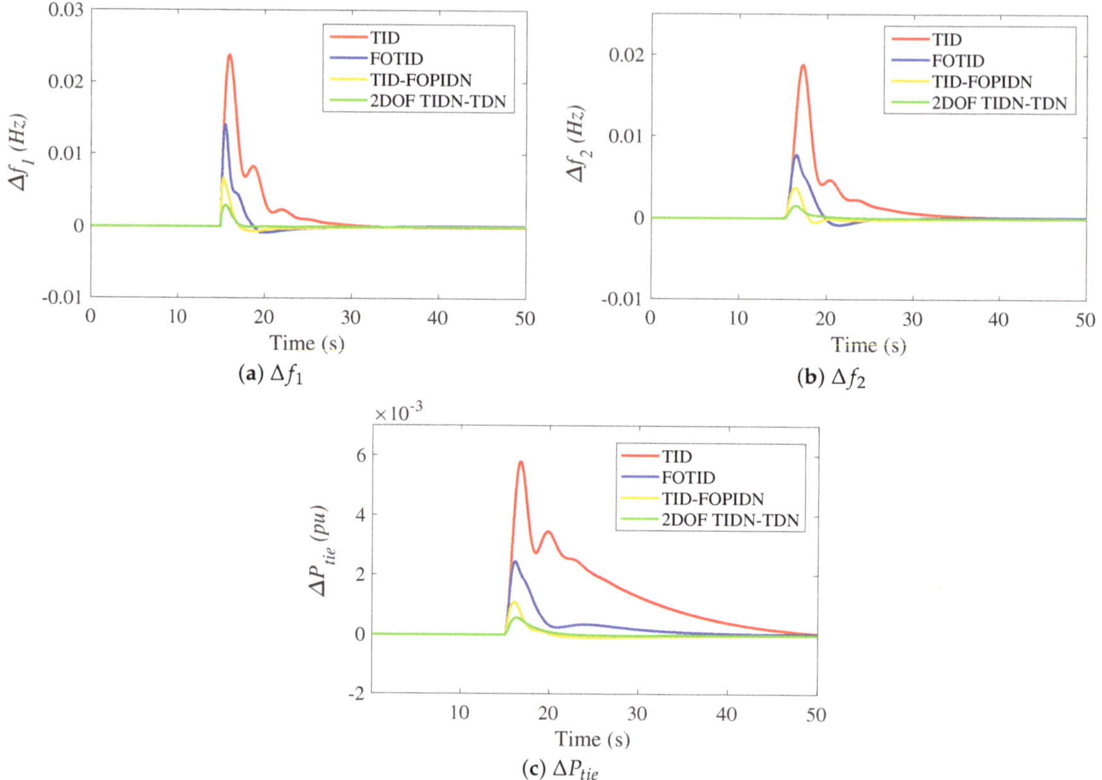

(a) Δf_1

(b) Δf_2

(c) ΔP_{tie}

Figure 12. The frequency response of the dual-MG in scenario 2.

6.3. Scenario 3

This scenario examined the strength of the proposed coordination-based 2DOF TIDN-TDN as an LFC with EVs using the MPA technique under the impact of parameter uncertainties of the nuclear power plant. However, the parameters of the nuclear plant generation are drastically adjusted in accordance with area 1, $[K_H = +40\%, T_{T1} = +70\%, K_{R1} = +50\%, T_{RH1} = +20\%, T_{RH2} = +30\%, T_2 = +60\%]$, and area 2, $[K_H = -40\%, T_{T1} = -70\%, K_{R1} = -50\%, T_{RH1} = -20\%, T_{RH2} = -30\%, T_2 = -60\%]$. The studied dual-area power network system is tested under the same operating condition as the load perturbation of scenario 1. Figure 13 shows the dynamic responses of Δf_1, Δf_2, and ΔP_{tie} of the system, respectively. From the result in this figure, it is evident that when using the TID controller, the frequency deviation is higher compared with previous cases, with undershoot values measuring 0.028 Hz in area 1 and 0.022 Hz in area 2. While the FOTID gave satisfying results with respect to the TID controller, it endured protracted damped oscillations, and it did not have the ability to retrieve the frequency to its steady-state value in a short time duration. However, the cascaded TID-FOPIDN preserved the system frequency at 0.01 Hz in area 1 and at 0.004 Hz in area 2, and the perversion of the power in the tie-line is 0.00494 p.u. On the other hand, the new cascaded 2DOF TIDN-TDN controller is the quickest in curbing the frequency and tie-line power deviations, and it has a lower steady-state error value than that presented by other traditional and cascaded controllers.

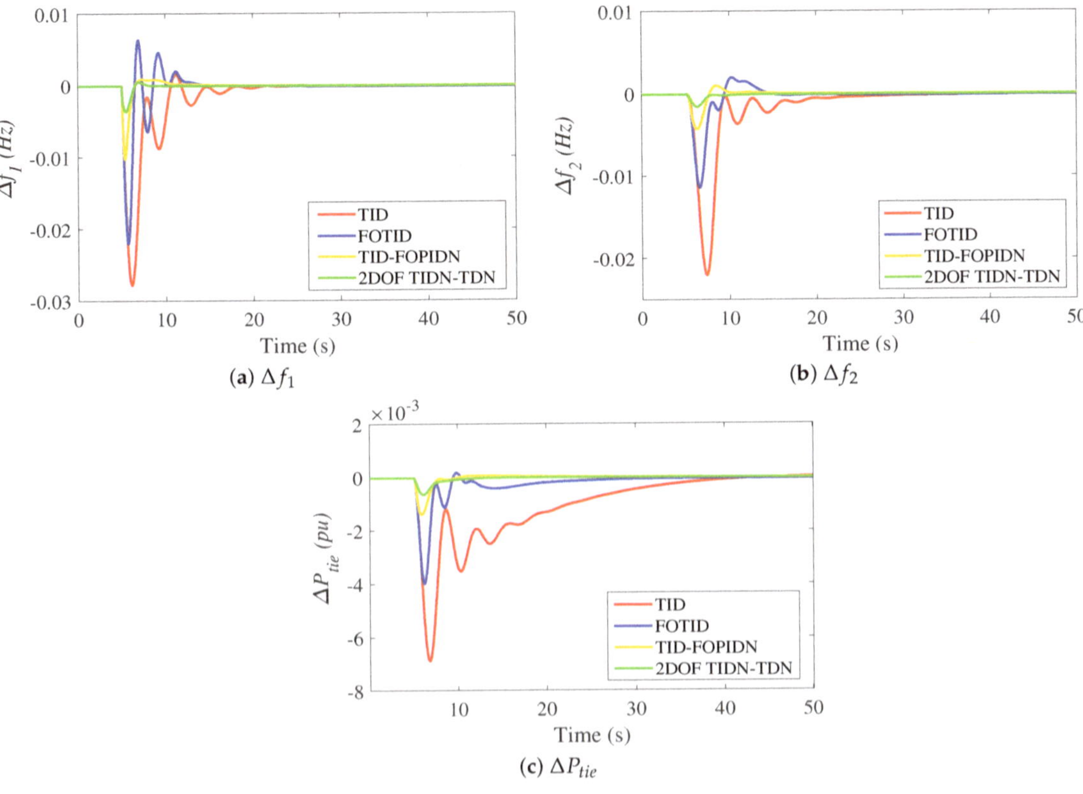

(a) Δf_1

(b) Δf_2

(c) ΔP_{tie}

Figure 13. The dynamic response of the system in scenario 3.

6.4. Scenario 4

The substantial target of this study case is to elucidate the interpretation of the proposed cascaded 2DOF TIDN-TDN control scheme based on the MPA algorithm under the influence of load parameter uncertainties due to multiple-load demands, as depicted in Figure 14 of this scenario. Figure 15 shows the dynamic frequency and energy exchange of the multi-microgrid system due to this significant load change. It is noticed that the proposed 2DOF TIDN-TDN controller has faster action with the slightest deviations compared with other control techniques, especially at the instant of the worst load shift in this scenario ($t = 140$ s). It can minimize the frequency diffractions in the first area 85.92%, 74.1%, and 60% better than the TID, FOTID, and TID-FOPIDN controllers, respectively. Additionally, it can demoralize the frequency vibrations in the second area with percentage gains of 91.9%, 80.85%, and 68.42% compared with the TID, FOTID, and TID-FOPIDN controllers, respectively. Furthermore, Table 7 shows that the proposed cascaded 2DOF TIDN-TDN controller has the lowest tie-line exchange power when compared with the other methods. The calculations of ISE, ITSE, IAE, and ITAE for all the studied scenarios are summarized in Table 8. Moreover, it is clear from this discussion that the percentage coordination of the LFC and EV participation utilizing the new 2DOF TIDN-TDN controller based on the efficient MPA technique yields the best results through the three-step load changes in this scenario. This is due to the function of the inner loop 2DOF-TDN of the proposed controller schematic responding quickly to the variation in the load demand. Furthermore, the results manifest that the other single-loop structures have a phase lag regarding the multiple-load disturbances compared with the suggested cascaded controllers.

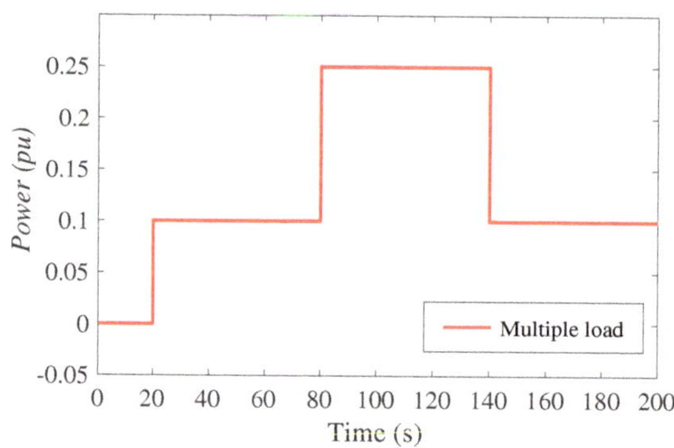

Figure 14. Generation profiles for Scenario 4.

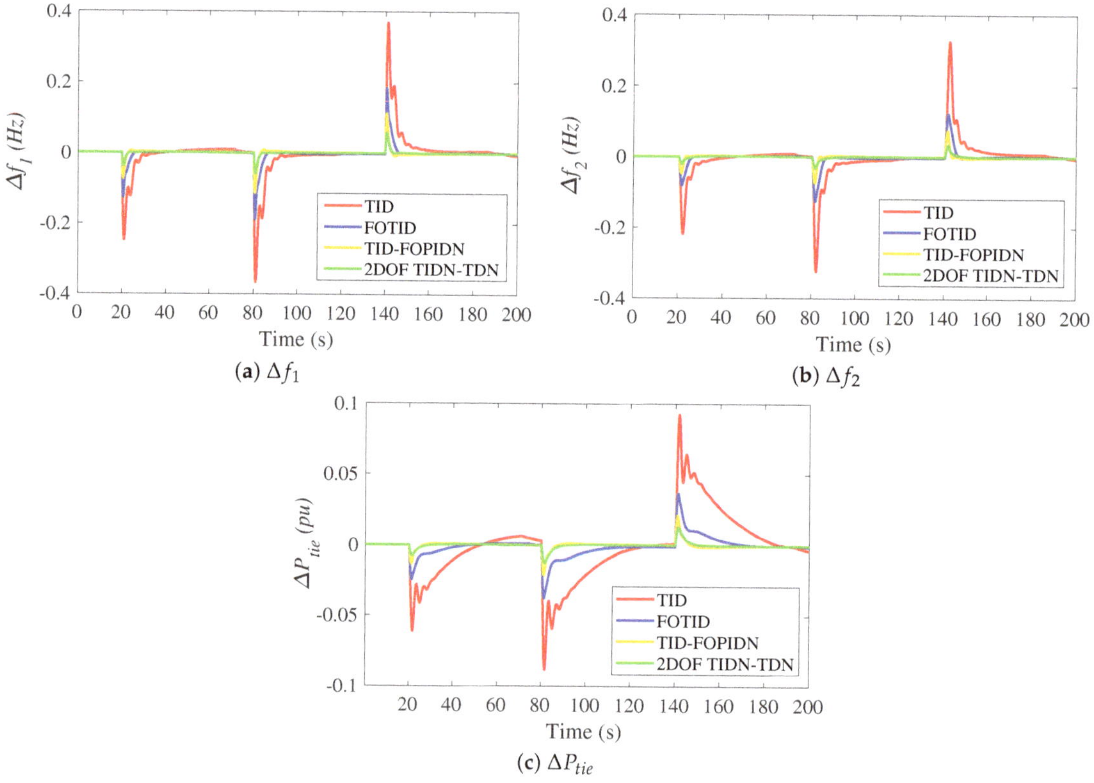

(a) Δf_1

(b) Δf_2

(c) ΔP_{tie}

Figure 15. The frequency response of the dual-MG in scenario 4.

Table 8. Different objective function indices for suggested controllers.

Scenario	Controller Structure	Objective Function			
		ISE	**ITSE**	**IAE**	**ITAE**
No. 1 (SLP 1%)	TID	4.0164×10^{-4}	0.0033	0.1079	1.2661
	FOTID	5.9414×10^{-5}	3.9198×10^{-4}	0.0319	0.3613
	TID-FOPIDN	1.1264×10^{-5}	7.0826×10^{-5}	0.0131	0.1528
	2DOF TIDN-TDN	2.7096×10^{-6}	1.6917×10^{-5}	0.00531	0.0470

Table 8. *Cont.*

Scenario	Controller Structure	Objective Function			
		ISE	**ITSE**	**IAE**	**ITAE**
No. 2 (SLS 1%)	TID	4.0157×10^{-4}	0.0073	0.1072	2.3101
	FOTID	5.9370×10^{-5}	9.8377×10^{-4}	0.0308	0.6193
	TID-FOPIDN	1.1261×10^{-5}	1.8332×10^{-4}	0.0129	0.2691
	2DOF TIDN-TDN	2.7084×10^{-6}	4.3981×10^{-5}	0.0052	0.0955
No. 3 (SLP 1% + change)	TID	4.6412×10^{-4}	0.0037	0.1057	1.2603
	FOTID	1.1785×10^{-4}	7.5945×10^{-4}	0.0398	0.4405
	TID-FOPIDN	1.6056×10^{-5}	9.5954×10^{-5}	0.0132	0.1416
	2DOF TIDN-TDN	2.7891×10^{-6}	1.7453×10^{-5}	0.0057	0.0503
No. 4 (MLP)	TID	0.3264	32.8178	6.5210	659.648
	FOTID	0.0436	4.2085	1.8112	173.6072
	TID-FOPIDN	0.0104	0.9988	0.7770	75.1539
	2DOF TIDN-TDN	0.0034	0.3377	0.4364	43.1104
No. 5 (MLP + RES)	TID	0.5519	52.1715	10.4732	995.6836
	FOTID	0.1317	7.1026	4.3499	290.3492
	TID-FOPIDN	0.0198	0.9122	1.3249	91.8096
	2DOF TIDN-TDN	0.009	0.5497	0.9661	73.1227
No. 6 (MLP + RES + change)	TID	0.6324	58.8358	10.1691	963.1041
	FOTID	0.1738	10.9667	4.6882	331.5865
	TID-FOPIDN	0.0274	1.7433	1.5288	113.6869
	2DOF TIDN-TDN	0.0125	0.8752	1.0647	84.8181

6.5. Scenario 5

This study tests the novel cascaded 2DOF TIDN-TDN controller, augmented with EV participation in the LFC loop, under severe disturbance conditions involving high levels of RES penetration. However, the PV power unit is connected at the initial time, while the wind power is integrated at 110 s, in addition to the effect of multiple-load changes, as shown in Figure 16. Figure 17 illustrates the obtained comparative results for the frequency and tie-line power in this case. It is clear that the fluctuation of Δf_1 and Δf_2 is close to +0.4 Hz, and more than 0.06 p.u. in tie-line power is achieved by using the TID controller. It is followed by the FOTID controller, with a variation in +0.2 Hz in both power system zones. While the two cascaded TID-FOPIDN and 2DOF TIDN-TDN controllers give deviations around +0.1 Hz and 0.08 Hz in area 1 and area 2, respectively, the proposed 2DOF TIDN-TDN controller is efficient and controls the deviations within the lowest time frame with minimum overshoot and undershoot values, especially at the severe load variations instants at 40 s and 80 s. This, in turn, confirms that the new proposed controller acquires superior performance when compared with the other used controllers. In addition, these results prove that the proposed cascaded 2DOF TIDN-TDN controller has a very fast inner control loop, which can respond more quickly to disturbances of load and RES than its outer loop. From another approach, the cascaded (2DOF TIDN-TDN) control signal, which is applied to the conventional power generators and energy storage devices in both area (a) and (b) enables the EVs to have a fast performance in the charge/discharge process and obtain less power from these generators than the other addressed controllers, as shown in Figure 18 and Figure 19, respectively. Therefore, it is clear from this result's explanation that the proposed cascaded 2DOF TIDN-TDN controller based on the MPA technique is the most effective one in this scenario.

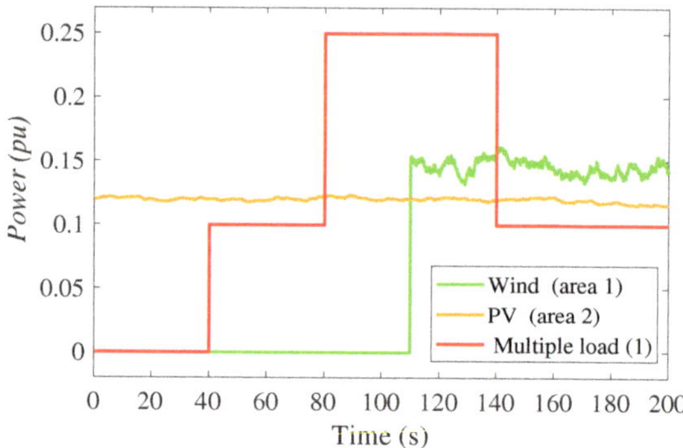

Figure 16. profiles of different generators for Scenario 5.

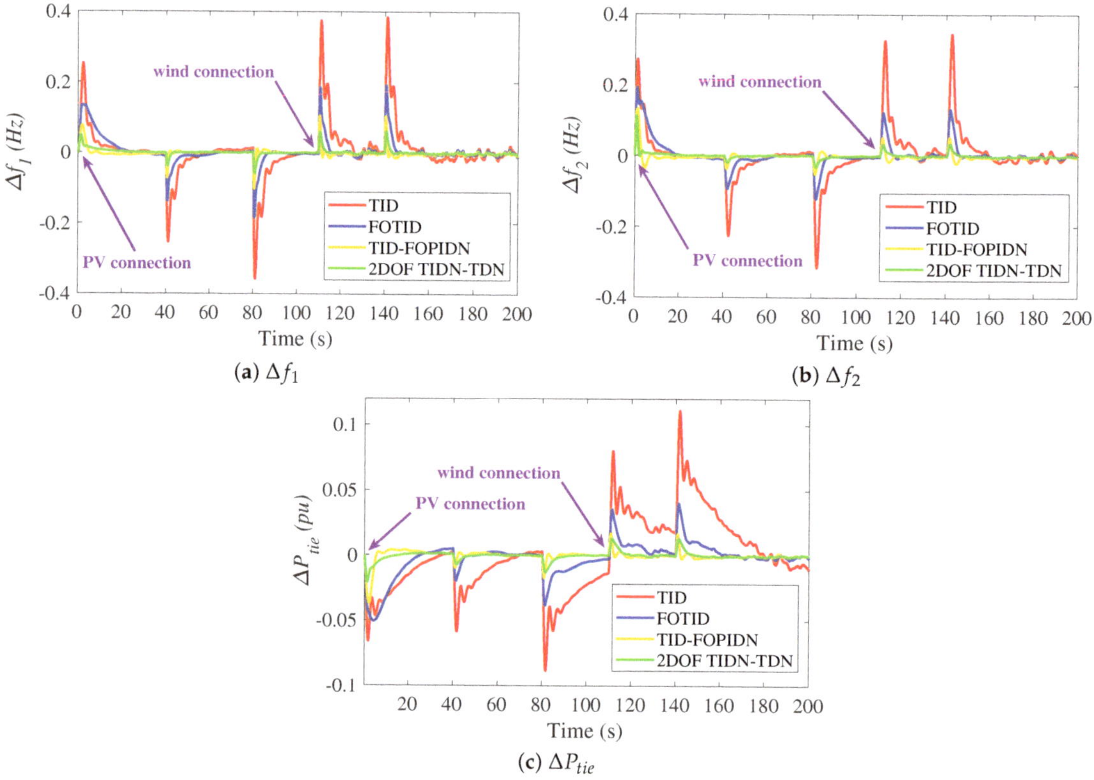

(a) Δf_1

(b) Δf_2

(c) ΔP_{tie}

Figure 17. The frequency response of the dual-MG in scenario 5.

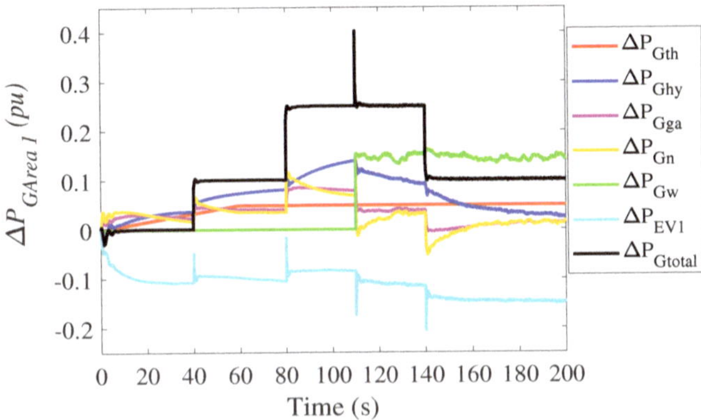

Figure 18. Power generations of area 1 for scenario 5.

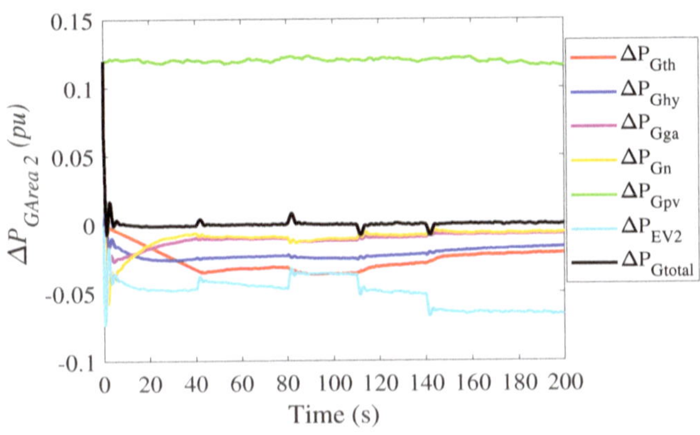

Figure 19. Power generations of area 2 for scenario 5.

6.6. Scenario 6

To scrutinize the interpretation of the proposed 2DOF TIDN-TDN controller, the system is submitted to high perturbations, such as multistep load perturbation, RES power fluctuations, and parameter changes that could cause system instability. In this case, the MG network parameters were changed as follows: low system inertia (i.e., 50% alleviation of its nominal values in two-area) and changes in the parameters of the nuclear plant generation in accordance with area 1, [$K_H = +40\%$, $T_{T1} = +70\%$, $K_{R1} = +50\%$, $T_{RH1} = +20\%$, $T_{RH2} = +30\%$, $T_2 = +60\%$], and area 2, [$K_H = -40\%$, $T_{T1} = -70\%$, $K_{R1} = -50\%$, $T_{RH1} = -20\%$, $T_{RH2} = -30\%$, $T_2 = -60\%$], under the same operating conditions discussed in scenario 4. The obtained results are shown in Figure 20. It is observed that the TID and FOTID controllers have the lowest performance for all perturbation stipulations in this scenario. For instance, the obtained values at 40 s for Δf_1 are 0.28, 0.18, 0.11, and 0.056 for the TIDF, FOTID, TID-FOPIDN, and the 2DOF TIDN-TDN, respectively. It is obvious that the proposed approach acquires the minimum values of the measured PO, PU, and ST in Δf_1, Δf_2, and ΔP_{tie}.

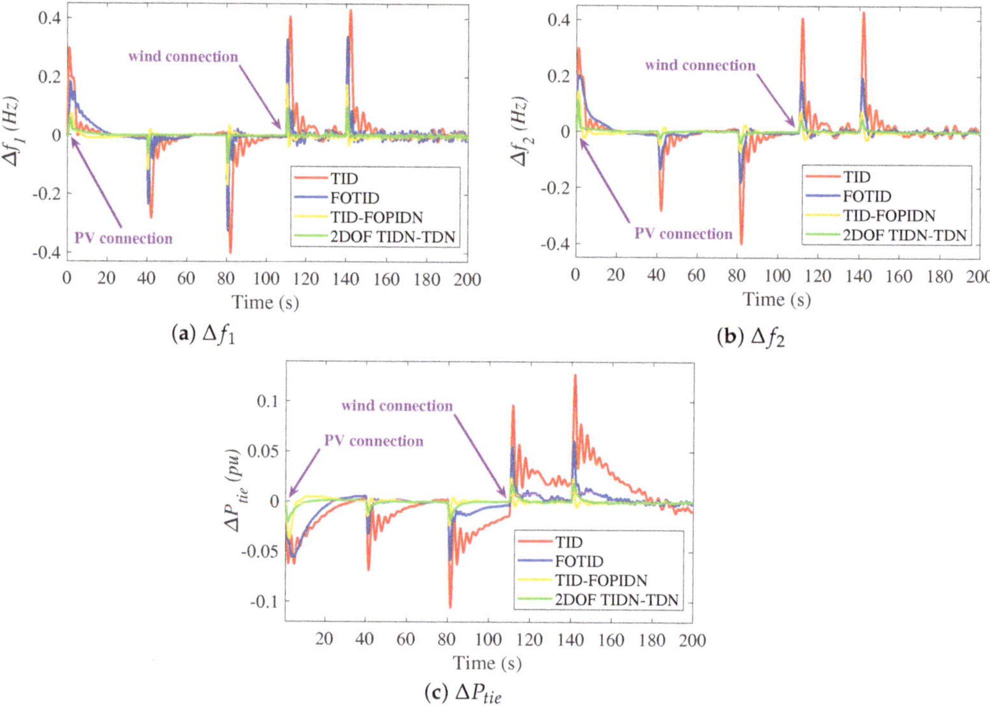

(**a**) Δf_1

(**b**) Δf_2

(**c**) ΔP_{tie}

Figure 20. The frequency response of the dual-MG in scenario 6.

6.7. Stability Analysis of the Closed-Loop System

Based on the studied system modeling and the proposed 2DOF (TIDN-TDN) controller, a stability analysis based on the Bode diagram plot is performed. Figure 21 shows the Bode plot of the examined system loop gains with the proposed controller. The magnitude of the gain margin plot is more steady for all frequencies, according to Figure 21. As a result, the phase margin is infinite, displaying that the proposed controller is able to deal with system uncertainty.

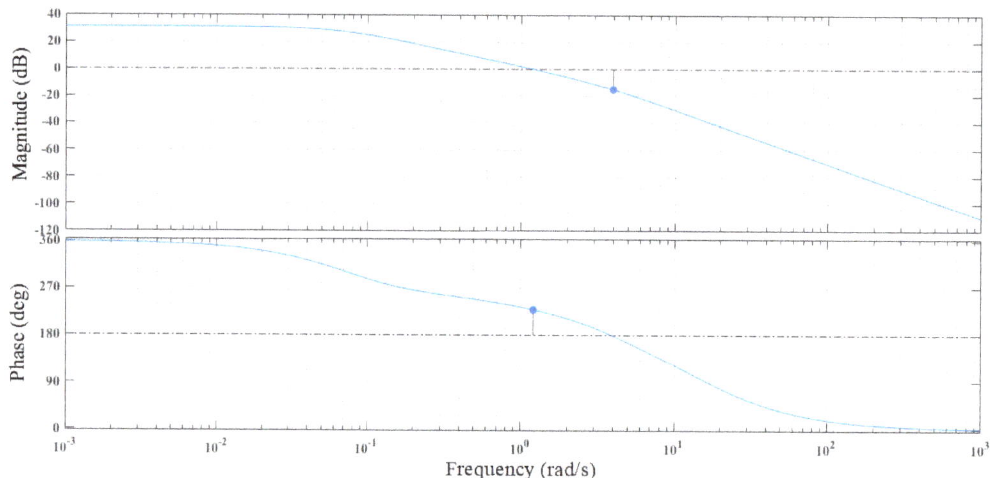

Figure 21. Bode plot of the examined system loop gains with the proposed controller.

7. Conclusions

An optimized 2DOF TIDN-TDN controller was proposed in this paper for the LFC of multisourced interconnected power systems. The proposed controller uses the frequency signals in the inner loops, which enables the mitigation of high-frequency disturbances. Moreover, it uses the ACE in the outer loop, which results in mitigating the low-frequency disturbances. Additionally, the powerful recent marine predator optimizer algorithm (MPA) is proposed for the simultaneous optimization of the LFC and EV controllers' parameters in different areas. Therefore, improved optimum performance is achieved using the proposed MPA-based optimized 2DOF TIDN-TDN controller. Moreover, coordination of EV control is proposed to contribute to mitigating existing disturbances in the power systems (vehicle-to-grid V2G concept). The proposed controller and optimized design were tested and compared using the RES highly penetrated dual-area power systems. The acquired results confirm the superior performance measurements of the proposed 2DOF TIDN-TDN controller over the existing TID, FOTID, cascaded TID-FOPIDN controller. For instance, at step load change, the maximum undershoot in the first area frequency is 0.003 p.u. using the proposed controller compared with 0.007, 0.015, and 0.024 with TID-FOPIDN, FOTID, and TID, respectively. The estimated ISEs in this case were 2.7096×10^{-6}, 1.1264×10^{-5}, 5.9414×10^{-5}, and 4.0164×10^{-4} using the proposed 2DOF TIDN-TDN, TID-FOPIDN, FOTID, and TID, respectively. This signifies that the proposed 2DOF TIDN-TDN LFC has ISE values of 24.06%, 4.56%, and 0.67% compared with TID-FOPIDN, FOTID, and TID, respectively. Future work includes frequency-domain stability analysis and comparison of the fractional-order cascaded controllers. In addition, the consideration of existing communication delays can be presented and investigated from the control design and stability analysis side.

Author Contributions: Conceptualization, A.S., M.A. and E.A.M.; Methodology, A.H.; Software, B.A., M.A., A.E., M.K. and E.A.M.; Validation, A.H., M.A.A. and A.E.; Formal analysis, M.A., M.K. and E.A.M.; Investigation, M.M.A., M.A.A. and A.S.; Resources, A.S. and M.A.; Data curation, M.M.A. and A.E.; Writing—original draft, A.H.; Writing—review & editing, M.A. and E.A.M.; Visualization, B.A.; Supervision, M.M.A. All authors have read and agreed to the published version of the manuscript.

Funding: The authors extend their appreciation to the Deputyship for Research & Innovation, Ministry of Education in Saudi Arabia for funding this research work through the project number (445-9-510).

Data Availability Statement: No new data were created or analyzed in this study. Data sharing is not applicable to this article.

Conflicts of Interest: The authors declare no conflict of interest.

Appendix A

System's basic parameters [63–65,68]:

LFC: $R_{th}, R_{hy}, R_{ga} = 2.4\,\mathrm{Hz/MW}$. $B_1, B_2 = 0.4312\,\mathrm{MW/Hz}$. Power system: $K_{ps1}, K_{ps2} = 68.9655$, $T_{ps1}, T_{ps2} = 11.49\,\mathrm{s}$, $T_{tie} = 0.0433$, $A_{ab} = -1$. Thermal plant $= 846\,\mathrm{MW}$: $T_{sg} = 0.08\,\mathrm{s}$, $T_t = 0.3\,\mathrm{s}$, $K_r = 0.3$, $T_r = 10\,\mathrm{s}$, $PA_{th} = 0.486207$. Hydro plant $= 467\,\mathrm{MW}$: $T_{gh} = 0.2\,\mathrm{s}$, $T_{rh} = 28.749\,\mathrm{s}$, $T_{rs} = 5\,\mathrm{s}$, $T_w = 1\,\mathrm{s}$, $PA_{hy} = 0.268391$. Gas plant $= 227\,\mathrm{MW}$: $B_g = 0.049\,\mathrm{s}$, $C_g = 1$, $X_g = 0.6\,\mathrm{s}$, $Y_g = 1.1\,\mathrm{s}$, $T_{cr} = 0.01\,\mathrm{s}$, $T_f = 0.239\,\mathrm{s}$, $T_{cd} = 0.2\,\mathrm{s}$, $PA_{ga} = 0.130459$. Nuclear plant $= 200\,\mathrm{MW}$: $K_H = 2$, $T_{T1} = 0.5\,\mathrm{s}$, $K_{R1} = 0.3$, $T_{RH1} = 7\,\mathrm{s}$, $T_{RH2} = 9\,\mathrm{s}$, $T_2 = 0.08\,\mathrm{s}$, $PA_{gn} = 0.114943$. PV plant: $T_{PV} = 1.3\,\mathrm{s}$, $K_{PV} = 1$. Wind plant: $T_{WT} = 1.5\,\mathrm{s}$, $K_{WT} = 1$. EV: Penetration Level $= 5$–10%, $V_{nom} = 364.8\,\mathrm{V}$, $C_{nom} = 66.2\,\mathrm{Ah}$, $R_s = 0.074\,\mathrm{ohms}$, $R_t = 0.047\,\mathrm{ohms}$, $C_t = 703.6\,\mathrm{farad}$, $RT/F = 0.02612$, Maximum $SOC = 95\%$, $C_{batt} = 24.15\,\mathrm{kWh}$.

References

1. Ali, M.; Kotb, H.; AboRas, M.K.; Abbasy, H.N. Frequency regulation of hybrid multi-area power system using wild horse optimizer based new combined Fuzzy Fractional-Order PI and TID controllers. *Alex. Eng. J.* **2022**, *61*, 12187–12210. [CrossRef]
2. Ramshanker, A.; Chakraborty, S.; Elangovan, D.; Kotb, H.; Aboras, K.M.; Giri, N.C.; Agyekum, E.B. CO_2 Emission Analysis for Different Types of Electric Vehicles When Charged from Floating Solar Photovoltaic Systems. *Appl. Sci.* **2022**, *12*, 12552. [CrossRef]

3. Li, S.; Shao, X.; Zhang, W.; Zhang, Q. Distributed Multicircular Circumnavigation Control for UAVs with Desired Angular Spacing. *Def. Technol.* **2023** . [CrossRef]

4. Zhang, F.; Shao, X.; Xia, Y.; Zhang, W. Elliptical encirclement control capable of reinforcing performances for UAVs around a dynamic target. *Def. Technol.* **2023**. [CrossRef]

5. Zhang, J.; Shao, X.; Zhang, W.; Na, J. Path-Following Control Capable of Reinforcing Transient Performances for Networked Mobile Robots Over a Single Curve. *IEEE Trans. Instrum. Meas.* **2023**, *72*, 1–12. [CrossRef]

6. Shao, X.; Li, S.; Zhang, J.; Zhang, F.; Zhang, W.; Zhang, Q. GPS-free Collaborative Elliptical Circumnavigation Control for Multiple Non-holonomic Vehicles. *IEEE Trans. Intell. Veh.* **2023**, *8*, 3750–3761. [CrossRef]

7. Zhang, J.; Shao, X.; Zhang, W.; Zuo, Z. Multi-circular formation control with reinforced transient profiles for nonholonomic vehicles: A path-following framework. *Def. Technol.* **2023**. [CrossRef]

8. Bakhtadze, N.; Maximov, E.; Maximova, N. Digital Identification Algorithms for Primary Frequency Control in Unified Power System. *Mathematics* **2021**, *9*, 2875. [CrossRef]

9. Ranjan, M.; Shankar, R. A literature survey on load frequency control considering renewable energy integration in power system: Recent trends and future prospects. *J. Energy Storage* **2022**, *45*, 103717. [CrossRef]

10. Dreidy, M.; Mokhlis, H.; Mekhilef, S. Inertia response and frequency control techniques for renewable energy sources: A review. *Renew. Sustain. Energy Rev.* **2017**, *69*, 144–155. [CrossRef]

11. Shankar, R.; Pradhan, S.; Chatterjee, K.; Mandal, R. A comprehensive state of the art literature survey on LFC mechanism for power system. *Renew. Sustain. Energy Rev.* **2017**, *76*, 1185–1207. [CrossRef]

12. Khamies, M.; Magdy, G.; Ebeed, M.; Kamel, S. A robust PID controller based on linear quadratic gaussian approach for improving frequency stability of power systems considering renewables. *ISA Trans.* **2021**, *117*, 118–138. [CrossRef] [PubMed]

13. Fernández-Guillamón, A.; Gómez-Lázaro, E.; Muljadi, E.; Molina-García, Á. Power systems with high renewable energy sources: A review of inertia and frequency control strategies over time. *Renew. Sustain. Energy Rev.* **2019**, *115*, 109369. [CrossRef]

14. Pandey, S.K.; Mohanty, S.R.; Kishor, N. A literature survey on load–frequency control for conventional and distribution generation power systems. *Renew. Sustain. Energy Rev.* **2013**, *25*, 318–334. [CrossRef]

15. Lv, X.; Sun, Y.; Wang, Y.; Dinavahi, V. Adaptive Event-Triggered Load Frequency Control of Multi-Area Power Systems Under Networked Environment via Sliding Mode Control. *IEEE Access* **2020**, *8*, 86585–86594. [CrossRef]

16. Kavikumar, R.; Kwon, O.M.; Lee, S.H.; Sakthivel, R. Input-output finite-time IT2 fuzzy dynamic sliding mode control for fractional-order nonlinear systems. *Nonlinear Dyn.* **2022**, *108*, 3745–3760. [CrossRef]

17. Sakthivel, R.; Kavikumar, R.; Ma, Y.K.; Ren, Y.; Marshal Anthoni, S. Observer-Based H_∞ Repetitive Control for Fractional-Order Interval Type-2 TS Fuzzy Systems. *IEEE Access* **2018**, *6*, 49828–49837. [CrossRef]

18. Vrdoljak, K.; Perić, N.; Petrović, I. Sliding mode based load-frequency control in power systems. *Electr. Power Syst. Res.* **2010**, *80*, 514–527. [CrossRef]

19. Pan, C.; Liaw, C. An adaptive controller for power system load-frequency control. *IEEE Trans. Power Syst.* **1989**, *4*, 122–128. [CrossRef]

20. Rakhshani, E.; Rodriguez, P.; Cantarellas, A.M.; Remon, D. Analysis of derivative control based virtual inertia in multi-area high-voltage direct current interconnected power systems. *IET Gener. Transm. Distrib.* **2016**, *10*, 1458–1469. [CrossRef]

21. Kerdphol, T.; Rahman, F.S.; Mitani, Y.; Watanabe, M.; Kufeoglu, S. Robust Virtual Inertia Control of an Islanded Microgrid Considering High Penetration of Renewable Energy. *IEEE Access* **2018**, *6*, 625–636. [CrossRef]

22. Kocaarslan. I.; Çam, E. Fuzzy logic controller in interconnected electrical power systems for load-frequency control. *Int. J. Electr. Power Energy Syst.* **2005**, *27*, 542–549. [CrossRef]

23. Bu, X.; Yu, W.; Cui, L.; Hou, Z.; Chen, Z. Event-Triggered Data-Driven Load Frequency Control for Multiarea Power Systems. *IEEE Trans. Ind. Inform.* **2022**, *18*, 5982–5991. [CrossRef]

24. Latif, A.; Hussain, S.S.; Das, D.C.; Ustun, T.S.; Iqbal, A. A review on fractional order (FO) controllers' optimization for load frequency stabilization in power networks. *Energy Rep.* **2021**, *7*, 4009–4021. [CrossRef]

25. Morgan, E.F.; El-Sehiemy, R.A.; Awopone, A.K.; Megahed, T.F.; Abdelkader, S.M. Load Frequency Control of Interconnected Power System Using Artificial Hummingbird Optimization. In Proceedings of the 2022 23rd International Middle East Power Systems Conference (MEPCON), Cairo, Egypt, 13–15 December 2022; [CrossRef]

26. Elmelegi, A.; Mohamed, E.A.; Aly, M.; Ahmed, E.M. An Enhanced Slime Mould Algorithm Optimized LFC Scheme for Interconnected Power Systems. In Proceedings of the 2021 22nd International Middle East Power Systems Conference (MEPCON), Assiut, Egypt, 14–16 December 2021. [CrossRef]

27. Khokhar, B.; Dahiya, S.; Singh Parmar, K.P. Atom search optimization based study of frequency deviation response of a hybrid power system. In Proceedings of the 2020 IEEE 9th Power India International Conference (PIICON), Sonepat, India, 28 February–1 March 2020; pp. 1–5. [CrossRef]

28. Panda, S.; Mohanty, B.; Hota, P. Hybrid BFOA-PSO algorithm for automatic generation control of linear and nonlinear interconnected power systems. *Appl. Soft Comput.* **2013**, *13*, 4718–4730. [CrossRef]

29. Arora, K.; Kumar, A.; Kamboj, V.K.; Prashar, D.; Shrestha, B.; Joshi, G.P. Impact of Renewable Energy Sources into Multi Area Multi-Source Load Frequency Control of Interrelated Power System. *Mathematics* **2021**, *9*, 186. [CrossRef]

30. Gupta, D.K.; Soni, A.K.; Jha, A.V.; Mishra, S.K.; Appasani, B.; Srinivasulu, A.; Bizon, N.; Thounthong, P. Hybrid Gravitational–Firefly Algorithm-Based Load Frequency Control for Hydrothermal Two-Area System. *Mathematics* **2021**, *9*, 712. [CrossRef]

31. Yousri, D.; Babu, T.S.; Fathy, A. Recent methodology based Harris Hawks optimizer for designing load frequency control incorporated in multi-interconnected renewable energy plants. *Sustain. Energy Grids Netw.* **2020**, *22*, 100352. [CrossRef]
32. Salama, H.S.; Magdy, G.; Bakeer, A.; Vokony, I. Adaptive coordination control strategy of renewable energy sources, hydrogen production unit, and fuel cell for frequency regulation of a hybrid distributed power system. *Prot. Control. Mod. Power Syst.* **2022**, *7*, 34. [CrossRef]
33. Kumar, A.; Gupta, D.K.; Ghatak, S.R.; Appasani, B.; Bizon, N.; Thounthong, P. A Novel Improved GSA-BPSO Driven PID Controller for Load Frequency Control of Multi-Source Deregulated Power System. *Mathematics* **2022**, *10*, 3255. [CrossRef]
34. El Yakine Kouba, N.; Menaa, M.; Hasni, M.; Boudour, M. Optimal load frequency control based on artificial bee colony optimization applied to single, two and multi-area interconnected power systems. In Proceedings of the 2015 3rd International Conference on Control, Engineering & Information Technology (CEIT), Tlemcen, Algeria, 25–27 May 2015; pp. 1–6. [CrossRef]
35. Sharma, J.; Hote, Y.V.; Prasad, R. PID controller design for interval load frequency control system with communication time delay. *Control Eng. Pract.* **2019**, *89*, 154–168. [CrossRef]
36. Dey, P.; Saha, A.; Srimannarayana, P.; Bhattacharya, A.; Marungsri, B. A Realistic Approach Towards Solution of Load Frequency Control Problem in Interconnected Power Systems. *J. Electr. Eng. Technol.* **2021**, *17*, 759–788. [CrossRef]
37. Tripathi, S.; Singh, V.P.; Kishor, N.; Pandey, A. Load frequency control of power system considering electric Vehicles' aggregator with communication delay. *Int. J. Electr. Power Energy Syst.* **2023**, *145*, 108697. [CrossRef]
38. Fadheel, B.A.; Wahab, N.I.A.; Mahdi, A.J.; Premkumar, M.; Radzi, M.A.B.M.; Soh, A.B.C.; Veerasamy, V.; Irudayaraj, A.X.R. A Hybrid Grey Wolf Assisted-Sparrow Search Algorithm for Frequency Control of RE Integrated System. *Energies* **2023**, *16*, 1177. [CrossRef]
39. Said, S.M.; Mohamed, E.A.; Aly, M.; Ahmed, E.M. Enhancement of load frequency control in interconnected microgrids by SMES. In *Superconducting Magnetic Energy Storage in Power Grids*; Institution of Engineering and Technology: Stevenage, UK, 2022; pp. 111–148. [CrossRef]
40. Delassi, A.; Arif, S.; Mokrani, L. Load frequency control problem in interconnected power systems using robust fractional PI λ D controller. *Ain Shams Eng. J.* **2018**, *9*, 77–88. [CrossRef]
41. Fathy, A.; Alharbi, A.G. Recent Approach Based Movable Damped Wave Algorithm for Designing Fractional-Order PID Load Frequency Control Installed in Multi-Interconnected Plants With Renewable Energy. *IEEE Access* **2021**, *9*, 71072–71089. [CrossRef]
42. Ayas, M.S.; Sahin, E. FOPID controller with fractional filter for an automatic voltage regulator. *Comput. Electr. Eng.* **2021**, *90*, 106895. [CrossRef]
43. Sahu, R.K.; Panda, S.; Biswal, A.; Sekhar, G.C. Design and analysis of tilt integral derivative controller with filter for load frequency control of multi-area interconnected power systems. *ISA Trans.* **2016**, *61*, 251–264. [CrossRef]
44. Oshnoei, A.; Khezri, R.; Muyeen, S.M.; Oshnoei, S.; Blaabjerg, F. Automatic Generation Control Incorporating Electric Vehicles. *Electr. Power Compon. Syst.* **2019**, *47*, 720–732. [CrossRef]
45. Priyadarshani, S.; Subhashini, K.R.; Satapathy, J.K. Pathfinder algorithm optimized fractional order tilt-integral-derivative (FOTID) controller for automatic generation control of multi-source power system. *Microsyst. Technol.* **2020**, *27*, 23–35. [CrossRef]
46. Ahmed, E.M.; Mohamed, E.A.; Elmelegi, A.; Aly, M.; Elbaksawi, O. Optimum Modified Fractional Order Controller for Future Electric Vehicles and Renewable Energy-Based Interconnected Power Systems. *IEEE Access* **2021**, *9*, 29993–30010. [CrossRef]
47. Mohamed, E.A.; Aly, M.; Watanabe, M. New Tilt Fractional-Order Integral Derivative with Fractional Filter (TFOIDFF) Controller with Artificial Hummingbird Optimizer for LFC in Renewable Energy Power Grids. *Mathematics* **2022**, *10*, 3006. [CrossRef]
48. Sahu, R.; Panda, S.; Rout, U.K.; Sahoo, D. Teaching learning based optimization algorithm for automatic generation control of power system using 2-DOF PID controller. *Int. J. Electr. Power Energy Syst.* **2016**, *77*, 287–301. [CrossRef]
49. Hussain, I.; Das, D.C.; Latif, A.; Sinha, N.; Hussain, S.S.; Ustun, T.S. Active power control of autonomous hybrid power system using two degree of freedom PID controller. *Energy Rep.* **2022**, *8*, 973–981. [CrossRef]
50. Abdel-hamed, A.M.; Abdelaziz, A.Y.; El-Shahat, A. Design of a 2DOF-PID Control Scheme for Frequency/Power Regulation in a Two-Area Power System Using Dragonfly Algorithm with Integral-Based Weighted Goal Objective. *Energies* **2023**, *16*, 486. [CrossRef]
51. Dash, P.; Saikia, L.; Sinha, N. Automatic generation control of multi area thermal system using Bat algorithm optimized PD–PID cascade controller. *Int. J. Electr. Power Energy Syst.* **2015**, *68*, 364–372. [CrossRef]
52. Latif, A.; Paul, M.; Das, D.C.; Hussain, S.M.S.; Ustun, T.S. Price Based Demand Response for Optimal Frequency Stabilization in ORC Solar Thermal Based Isolated Hybrid Microgrid under Salp Swarm Technique. *Electronics* **2020**, *9*, 2209. [CrossRef]
53. Hossam-Eldin, A.; Mostafa, H.; Kotb, H.; AboRas, K.M.; Selim, A.; Kamel, S. Improving the Frequency Response of Hybrid Microgrid under Renewable Sources' Uncertainties Using a Robust LFC-Based African Vulture Optimization Algorithm. *Processes* **2022**, *10*, 2320. [CrossRef]
54. Sahu, P.R.; Simhadri, K.; Mohanty, B.; Hota, P.K.; Abdelaziz, A.Y.; Albalawi, F.; Ghoneim, S.S.M.; Elsisi, M. Effective Load Frequency Control of Power System with Two-Degree Freedom Tilt-Integral-Derivative Based on Whale Optimization Algorithm. *Sustainability* **2023**, *15*, 1515. [CrossRef]
55. Faramarzi, A.; Heidarinejad, M.; Mirjalili, S.; Gandomi, A.H. Marine Predators Algorithm: A nature-inspired metaheuristic. *Expert Syst. Appl.* **2020**, *152*, 113377. [CrossRef]
56. Houssein, E.H.; Ibrahim, I.E.; Kharrich, M.; Kamel, S. An improved marine predators algorithm for the optimal design of hybrid renewable energy systems. *Eng. Appl. Artif. Intell.* **2022**, *110*, 104722. [CrossRef]

57. Aly, M.; Ahmed, E.M.; Rezk, H.; Mohamed, E.A. Marine Predators Algorithm Optimized Reduced Sensor Fuzzy-Logic Based Maximum Power Point Tracking of Fuel Cell-Battery Standalone Applications. *IEEE Access* **2021**, *9*, 27987–28000. [CrossRef]
58. Soliman, M.A.; Hasanien, H.M.; Alkuhayli, A. Marine Predators Algorithm for Parameters Identification of Triple-Diode Photovoltaic Models. *IEEE Access* **2020**, *8*, 155832–155842. [CrossRef]
59. Yousri, D.; Babu, T.S.; Beshr, E.; Eteiba, M.B.; Allam, D. A Robust Strategy Based on Marine Predators Algorithm for Large Scale Photovoltaic Array Reconfiguration to Mitigate the Partial Shading Effect on the Performance of PV System. *IEEE Access* **2020**, *8*, 112407–112426. [CrossRef]
60. Sobhy, M.A.; Abdelaziz, A.Y.; Hasanien, H.M.; Ezzat, M. Marine predators algorithm for load frequency control of modern interconnected power systems including renewable energy sources and energy storage units. *Ain Shams Eng. J.* **2021**, *12*, 3843–3857. [CrossRef]
61. Ahmed, E.M.; Elmelegi, A.; Shawky, A.; Aly, M.; Alhosaini, W.; Mohamed, E.A. Frequency Regulation of Electric Vehicle-Penetrated Power System Using MPA-Tuned New Combined Fractional Order Controllers. *IEEE Access* **2021**, *9*, 107548–107565. [CrossRef]
62. Ahmed, E.M.; Selim, A.; Alnuman, H.; Alhosaini, W.; Aly, M.; Mohamed, E.A. Modified Frequency Regulator Based on TIλ-TDμFF Controller for Interconnected Microgrids with Incorporating Hybrid Renewable Energy Sources. *Mathematics* **2022**, *11*, 28. [CrossRef]
63. Hassan, A.; Aly, M.; Elmelegi, A.; Nasrat, L.; Watanabe, M.; Mohamed, E.A. Optimal Frequency Control of Multi-Area Hybrid Power System Using New Cascaded TID-PIλDμN Controller Incorporating Electric Vehicles. *Fractal Fract.* **2022**, *6*, 548. [CrossRef]
64. Morsali, J.; Zare, K.; Hagh, M.T. Comparative performance evaluation of fractional order controllers in LFC of two-area diverse-unit power system with considering GDB and GRC effects. *J. Electr. Syst. Inf. Technol.* **2018**, *5*, 708–722. [CrossRef]
65. Mohanty, B. TLBO optimized sliding mode controller for multi-area multi-source nonlinear interconnected AGC system. *Int. J. Electr. Power Energy Syst.* **2015**, *73*, 872–881. [CrossRef]
66. Elmelegi, A.; Mohamed, E.A.; Aly, M.; Ahmed, E.M.; Mohamed, A.A.A.; Elbaksawi, O. Optimized Tilt Fractional Order Cooperative Controllers for Preserving Frequency Stability in Renewable Energy-Based Power Systems. *IEEE Access* **2021**, *9*, 8261–8277. [CrossRef]
67. Mohamed, E.A.; Ahmed, E.M.; Elmelegi, A.; Aly, M.; Elbaksawi, O.; Mohamed, A.A.A. An Optimized Hybrid Fractional Order Controller for Frequency Regulation in Multi-Area Power Systems. *IEEE Access* **2020**, *8*, 213899–213915. [CrossRef]
68. Morsali, J.; Zare, K.; Hagh, M.T. Performance comparison of TCSC with TCPS and SSSC controllers in AGC of realistic interconnected multi-source power system. *Ain Shams Eng. J.* **2016**, *7*, 143–158. [CrossRef]

 *fractal and fractional*

Article

Design Optimization of Improved Fractional-Order Cascaded Frequency Controllers for Electric Vehicles and Electrical Power Grids Utilizing Renewable Energy Sources

Fayez F. M. El-Sousy [1,*], Mohammed H. Alqahtani [1], Ali S. Aljumah [1], Mokhtar Aly [2], Sulaiman Z. Almutairi [1] and Emad A. Mohamed [1,3]

[1] Department of Electrical Engineering, College of Engineering, Prince Sattam bin Abdulaziz University, Al Kharj 16278, Saudi Arabia; mh.alqahtani@psau.edu.sa (M.H.A.); as.aljumah@psau.edu.sa (A.S.A.); s.almutairi@psau.edu.sa (S.Z.A.); e.younis@psau.edu.sa or emad.younis@aswu.edu.eg (E.A.M.)

[2] Facultad de Ingeniería, Arquitectura y Diseño, Universidad San Sebastián, Bellavista 7, Santiago 8420000, Chile; mokhtar.aly@uss.cl

[3] Department of Electrical Engineering, Faculty of Engineering, Aswan University, Aswan 81542, Egypt

* Correspondence: f.elsousy@psau.edu.sa

Citation: El-Sousy, F.F.M.; Alqahtani, M.H.; Aljumah, A.S.; Aly, M.; Almutairi, S.Z.; Mohamed, E.A. Design Optimization of Improved Fractional-Order Cascaded Frequency Controllers for Electric Vehicles and Electrical Power Grids Utilizing Renewable Energy Sources. *Fractal Fract.* **2023**, *7*, 603. https://doi.org/10.3390/fractalfract7080603

Academic Editors: Behnam Mohammadi-Ivatloo and Arman Oshnoei

Received: 26 May 2023
Revised: 17 July 2023
Accepted: 18 July 2023
Published: 4 August 2023

Abstract: Recent developments in electrical power grids have witnessed high utilization levels of renewable energy sources (RESs) and increased trends that benefit the batteries of electric vehicles (EVs). However, modern electrical power grids cause increased concerns due to their continuously reduced inertia resulting from RES characteristics. Therefore, this paper proposes an improved fractional-order frequency controller with a design optimization methodology. The proposed controller is represented by two cascaded control loops using the one-plus-proportional derivative (1 + PD) in the outer loop and a fractional-order proportional integral derivative (FOPID) in the inner loop, which form the proposed improved 1 + PD/FOPID. The main superior performance characteristics of the proposed 1 + PD/FOPID fractional-order frequency controller over existing methods include a faster response time with minimized overshoot/undershoot peaks, an ability for mitigating both high- and low-frequency disturbances, and coordination of EV participation in regulating electrical power grid frequency. Moreover, simultaneous determination of the proposed fractional-order frequency controller parameters is proposed using the recent manta ray foraging optimization (MRFO) algorithm. Performance comparisons of the proposed 1 + PD/FOPID fractional-order frequency controller with existing PID, FOPID, and PD/FOPID controllers are presented in the paper. The results show an improved response, and the disturbance mitigation is also obtained using the proposed MRFO-based 1 + PD/FOPID control and design optimization methodology.

Keywords: electric vehicles (EVs); fractional-order control; frequency controller; manta ray foraging optimization; modern power grids

1. Introduction

Climate change and its serious effects on different environmental conditions has motivated the urgent transition to new renewable and clean sources in energy-related sectors [1,2]. In the energy generation sector, renewable energy sources (RESs) have dominated the recent installation shares. Specifically, photovoltaic (PV) and wind energy sources are the most commonly used in modern electrical power grids [3]. In addition, the increased use of electric vehicles (EVs) in the transportation sector helps to combat climate change and utilize clean energy sources. In [4], a study on the benefits and impacts of adding more functionalities of EVs into electrical grids with RESs was presented. The study proved the ability of EVs to actively share the required tasks of RES inverters, and hence there was no need for high-capacity energy storage devices. Therefore, cleaner and sustainable energy systems characterize modern electrical power grid systems [5].

Connecting RESs with electrical power grids requires power electronic-based power converters for grid integration of different RESs and for maximizing the extracted power from RESs. This makes RESs different from traditional synchronous-generator-based power generation; therefore, RESs impact electrical power grids with their lowered inertia [6]. The reduced inertial characteristics of modern electrical power grids with high participation rates of RESs make them highly susceptible to voltage and frequency stability problems. Therefore, proper frequency regulation is mandatory to mitigate the reduced inertia of RESs in modern electrical power grids. The load frequency control (LFC) has been shown to be a more suitable method for regulating frequency in modern electrical power grids with RESs [7].

The literature includes several research works on the topic of developing proper frequency regulation controllers [8]. Data-driven-based neural network (NN) controllers were recently presented for several electrical power grid case studies. However, they require huge computational burdens for data processing and designing NNs. From another side, model-based predictive controllers (MPCs) were introduced in the literature with different MPC structures. However, they are sensitive to the modelling of the process and system parameters. Additional LFC proposals have been proposed in the literature, such as the sliding mode LFC, machine and deep learning-based LFC, linear matrix inequalities (LQR), etc. [9]. Additionally, type-2 fuzzy modelling and control methods have been shown to have better performance in the literature. For instance, the type-2 fuzzy modelling has provided improved modelling for non-linear systems in [10,11]. Interval type-2 fuzzy modelling with fractional-order (FO)-based LFC methods has been shown in [12,13]. These methods merge the benefits of fuzzy type-2 with FO control methods.

From another perspective, integer-order (IO) frequency regulators and fractional-order (FO) frequency regulators have been widely discussed in the literature [14]. Different combinations have been presented using basic terms such as proportional (P), derivative (D), integral (I), tilt (T), filtering (F), fractional filtering (FF), and FO operators. Furthermore, various cascaded solutions of IO and FO have been developed in the literature to provide a better disturbance rejection capability. Metaheuristic-based optimized determination methods for control parameters have been developed in the literature to ensure the best control parameters without the need for complex control modelling and design tools [15].

An I-based controller was designed for LFC in interconnected electrical power grids in the literature [16]. Another adaptive I controller with an optimized design using Jaya balloon optimizers (JBO) was discussed in [17]. A PI controller was introduced in [18] with a binary-moth flame optimizer (MFO), whereas the hybrid gravitation searching with the firefly optimizer algorithm (hGFA) was introduced in [19]. Moreover, a PI controller with Harris–Hawks optimization (HHO) was provided in [20], and the grey wolf optimizers (GWO) were provided in [21]. PI-based LFC methods have succeeded in improving system dynamics; however, they fail to deal with system non-linearities and parameter uncertainties. Additionally, a PID controller with the imperialist competitive optimization algorithm (ICA) was proposed in [22]. Moreover, a PID virtual inertia controller was proposed in [23] in order to improve the inertial response for a real electrical power grid case study. IO-based frequency controllers showed simple design requirements, easy implementation, and lower costs for implementation. However, IO-based frequency regulators cannot fully mitigate uncertainty and fluctuations resulting from the electrical power grid parameters. Moreover, they showed a poor response to any uncertainties in the power system's parameters and at low-inertia operation.

Different control structures have been provided in the literature using cascaded control loops [24]. In cascaded LFC methods, two loops are used to construct the frequency regulation controller. The area control error (ACE) is used as an input for the outer loop, and the frequency deviation signal is employed for the inner loop [25]. This leads to a higher degree of freedom and better rejection of existing disturbances. In some cascaded controllers, the tie-line power is also included in the inner loop. The ACE signal represents the slow dynamics loop, and the frequency deviation represents a faster loop of system frequency disturbances.

The cascaded IO-based frequency regulator using the PD-PI scheme was introduced in [26], applying the enhanced slime mould optimization algorithm (ESMOA) for design optimization. The PI-PDF controller was provided in [27] with the driver training-based optimizer (DTBO). The results showed improved disturbance mitigation of the power grids. Another fuzzy logic control (FLC) cascaded PI-PDF controller with scaled factors and the modified Dragonfly optimizer algorithm was proposed in [28]. Additional IO-based frequency regulators have been proposed in the literature, such as PIDF [29], 2DoF-PID [24], PD-PID [30], PI-(1 + DD) [25], PID2D [31], IPD-(1 + I) [32], FLC-PID [33], and the neuro-fuzzy LFC [34]. The associated optimizers include the slap-swarm-based algorithm (SSA) [35], and the flower pollination-based algorithm (FPA) [36]. Some local modelling LFC was provided in [37,38]. The cascaded IO-based frequency regulators have shown better mitigation for existing electrical power grid disturbances. However, the use of IO control methods has a lower number of tunable parameters compared with FO-based control methods.

In the literature, FO-based frequency controllers have also found widespread use in regulating frequency in electrical power grids with different structures. Some metaheuristic optimizers have been presented for the design optimization of FO-based frequency regulators, such as the sine cosine-based algorithm (SCA) [39], genetics algorithm (GA) [22], and movable-damped-wave-based algorithm (MDWA) [40]. In [41], a review of the possible cascaded and multiple input-based LFC methods has been presented. Several two- and three-input schemes exist in the literature. A cascaded FOPID FO-based frequency regulator was proposed in [42] for stand-alone electrical power grids, whereas the FOPIDF was provided in [43]. A higher degree-of-freedom cascaded 3DOF TID-FOPID was provided in [44] to enhance electrical power grid stability. Other cascaded FO-based frequency controllers have been provided based on the FOID-FOPIDF in [45], FO-IDF in [46], and PI-TDF in [47]. Examples of associated optimizers include the pathfinder-based algorithm (PFA) in [48], the artificial-bee-colony-based algorithm (ABC) in [49], differential evolution-based algorithm (DE) in [50], and the SSA optimization in [47]. A cascaded FOPD-PI controller, considering plug-in EVs (PEVs), was presented in [51]. Another ID-T cascaded controller was proposed in [52] with an Archimedes optimizer algorithm (AOA). The inclusion of FO operators in cascaded LFC methods increase the flexibility and the number of tunable parameters. This can be reflected as a better optimization of the system's response to disturbances. A common difficulty in applying FO control systems is their implementation complexity. For instance, using Oustaloup's recursive approximation (ORA) representation of the fifth-order leads to eleven-order IO equivalent representations. However, this difficulty can be solved with the recent powerful microcontroller systems.

Paper Contribution

It has become obvious that several types and structures exist of IO- and FO-based frequency regulators for LFC in interconnected electrical power grids. Moreover, several techniques of metaheuristic optimization algorithms have been associated with the presented controllers for design optimization and reaching the best parameters. Proper and optimum design and selection of LFC and optimization method are crucial when facing expected reduced inertia with high participation levels of RESs. Additionally, probable local minimum settling represents another issue for several metaheuristic optimizers. Therefore, this paper presents an improved 1 + PD/FOPID FO-based frequency regulator, using the recent manta ray foraging optimization (MRFO) for design optimization.

Based on authors' knowledge, this is the first time the (1 + PD) has been combined with the FOPID controller in a cascaded way to provide a hybrid high-flexibility frequency regulation controller. Additionally, the integration of the MRFO algorithm with the proposed 1 + PD/FOPID controller leads to providing joint optimum behaviour and a better parameter determination process. This confirms the aforementioned findings that the power grid frequency regulation performance is determined jointly by the used control methodology and the applied optimization algorithm. Table 1 provides a summary of the existing controllers and the proposed controller. The main contributions in this paper are as follows:

1. An improved controller and design optimization method is proposed for frequency regulation in interconnected electrical power grids with high participation levels of RESs in addition to active participation of EVs in regulating frequency. The proposed controller and design methodology can effectively lead to mitigating various existing frequency fluctuations in electrical power grids. The proposed method can be generalized and applied to various electrical power grid systems and components.
2. The proposed frequency regulation control methodology is formed using a cascaded 2DoF 1 + PD/FOPID control method, which utilizes two input signals (namely the frequency deviation in each area, and control error in each area (ACE)). The utilization of two different signals is beneficial for the mitigation process of low- and high-frequency existing disturbances.
3. The proposed frequency regulation methodology using a 2DoF 1 + PD/FOPID controller provides better frequency regulation responses compared with the widely utilized PID, FOPID, and PD/FOPID LFCs, providing better disturbances rejections capabilities. The proposed 2DoF 1 + PD/FOPID structure is capable of mitigating various deviations in area frequency and electrical power grid tie-line power as a direct result of employing two cascaded loops with frequency and ACE signals.
4. Benefiting from EVs' batteries in the effective participation in frequency regulation is coordinated through the proposed 2DoF 1 + PD/FOPID structure. Therefore, the proposed 2DoF 1 + PD/FOPID structure reduces the frequency regulation complexities due to employing the centralized frequency regulation structure that coordinates the connected EVs' batteries and LFC regulator.
5. An improved design optimization methodology using the recent manta ray foraging optimization (MRFO) to determine the best parameters for the proposed 2DoF 1 + PD/FOPID frequency regulation. The optimized values of LFCs in different electrical power grids are simultaneously searched using the MRFO optimizer, thus minimizing the desired objectives.

The remainder of the paper is divided as follows: Section 2 provides the mathematical model representations for the interconnected electrical power grids. The proposed 2Dof 1 + PD/FOPID frequency regulation is detailed in Section 3. The proposed design optimization of the 1 + PD/FOPID controller is described in Section 4. Section 5 provides the obtained simulation results of the interconnected two-area electrical power grids with EVs participations and RESs. Finally, the paper's conclusions are presented in Section 6.

Table 1. Summarized comparison of existing LFC methods and the paper's contribution.

Ref.	Category	Control Schemes	Characteristics
[16–20,22]	Conventional IO LFC (single input)	I, PI, PID, PIDF	• Simple structure and implementation; • Low ability to mitigate disturbances; • Lower robustness against parameters uncertainty.
[22,39,40,42,43]	Conventional FO LFC (single input)	FOPI, FOPID, FOPIDF	• Increased flexibility and number of parameters compared to IO LFC methods; • Limited rejection capability of existing disturbances; • Lower mitigation of high-frequency deviations.

Table 1. *Cont.*

Ref.	Category	Control Schemes	Characteristics
[24,25,27–30,32,33]	Cascaded IO LFC (multiple inputs)	PD-PI, PI-PDF, 2DoF-PID, PD-PID, PI-(1 + DD), IPD-(1 + I), FLC-PID	• Easy to be implemented; • Improved ability to mitigate disturbances; • limited number of tunable parameters (only gains can be tuned).
[44–49,51,52]	Cascaded FO LFC (multiple inputs)	Cascaded FOPID, 3DOF TID-FOPID, FOID-FOPIDF, FO-IDF, PI-TDF, ID-T, FOPD-PI	• Higher number of tunable parameters compared to IO LFC methods; • Increased design flexibility compared to other IO-based LFC methods; • Enhanced disturbance rejection capability compared to single-input FO LFC methods; • Improved mitigation ability of high-frequency deviations.
Proposed	Proposed cascaded FO LFC (multiple inputs)	Proposed cascaded 2DoF 1 + PD/FOPID LFC method	• New hybrid (1 + PD) cascaded with FOPID controller for LFC application; • Merging the characteristics of IO-based control with the added flexibility of FO control; • Simultaneous design optimization process of all tunable parameters, considering connected devices using a powerful MRFO algorithm; • Active contribution of EVs in regulating frequency, and this functionality is considered during the design process; • High disturbance rejection capability; • Better mitigation of high-frequency fluctuations due to using two loops and employing the frequency deviation signal.

2. Modelling of Interconnected Electrical Power Grids

2.1. Electrical Power Grid Description

The interconnected electrical power grid using two areas is selected to verify the proposed 1 + PD/FOPID and design optimization method. Figure 1 shows two electrical power grid areas (area *a* and area *b*) connected using an AC bus as tie-line between the areas. The electrical power grid area *a* includes a thermal power plant, local loads, EV batteries, and a wind RES, whereas electrical power grid area *b* includes a hydraulic power plant, local loads, EV batteries, and a PV RES. It is assumed in the analysis that EVs are equally shared by the two electrical power grids. Each electrical power grid has a frequency regulation controller that controls the power injection of each power/storage devices in the area. The parameters for the studied electrical power grids in this work are taken from [14] and listed in Table 2. Figure 2 presents the complete implemented model of the studied electrical power grid elements with the EV batteries and RESs.

The controlled system consists of the aforementioned two-area interconnected power grids with the connected elements in each area. The input error to be controlled is the ACE signal of each area, and it has to be maintained at a zero reference value. The output of the control method adjusts the contribution of the connected generation and/or energy storage devices to mitigate the frequency changes. When the system has unbalanced generation and loading, this is reflected as an increase/decrease in the system frequency. Therefore, the LFC method mitigates the frequency fluctuations and preserves the frequency deviations at a zero value.

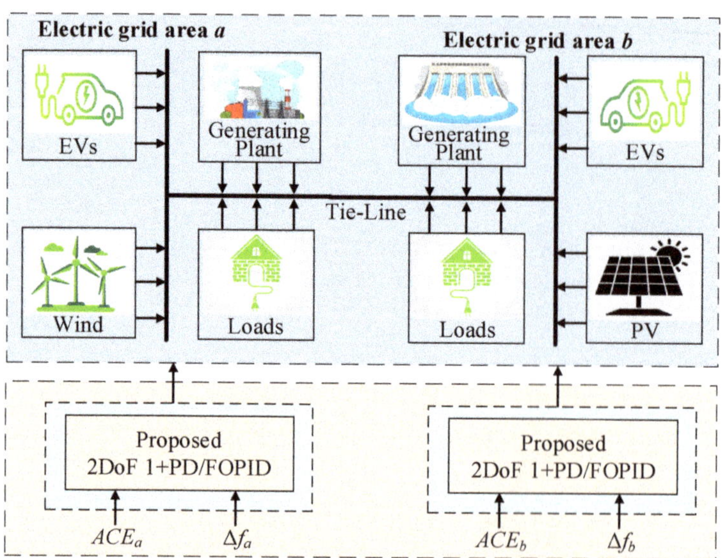

Figure 1. Structure of studied electrical power grids and the proposed 1 + PD/FOPID control.

2.2. RES Behaviour Models

In the studied electrical power grids, participation of PV and wind is considered with their intermittency proprieties. They rely on environmental conditions, such as wind speed, solar irradiance levels, and temperatures. For these reasons, continuous tracking is mandatory for the maximum power operating point (MPPT controller), in which, power electronics conversion blocks play a vital role. In addition, they are responsible for grid integration and synchronization of RESs. The outputted powers from RESs are continuously fluctuating due to variations in the weather conditions.

For PV generation, PV outputted power is unpredictable due to its intermittency. This, in turn, results in high-frequency power fluctuations that can lead to severe stability problems in the electrical power grids. In this work, the PV model from [53] is employed for PV power as follows:

$$P_{PV} = \eta \Phi_{solar} A_{PV} [1 - 0.005(T_a + 25)] \tag{1}$$

where η stands for PV panel's conversion efficiency (in %), Φ_{solar} stands for solar insolation (W/m^2), A_{PV} stands for the area of the PV unit (m^2), and T_a stands for the ambient temperature (°C). The implemented PV configuration model represents a realistic PV power generation model such as in the presented model in [54].

From another side, the outputted mechanical wind power from the turbine possesses high fluctuations resulting from its intermittency characteristics. The wind speed continuously varies, and hence the outputted power differs from instant to another. The calculation of the mechanical power is as follows [55]:

$$P_{wind} = \frac{1}{2} \rho A_r C_p V_w^3 \tag{2}$$

where ρ stands for the density of air (in kg/m^3), A_r stands for the swept area (in m^2), C_p stands for the power coefficient, and V_w stands for the speed of wind (in m/s). The implemented configuration of the wind power model is based on the realistic representation using the presented modelling in [54].

2.3. EV Behaviour Model

In this study, participation of the installed PV batteries is also considered for the frequency regulation control of the electrical power grids. This, in turn, leads to eliminating the need for additional energy storage capacities in the electrical power grids. The connected EV batteries are charged/discharges based on control signals coming from the electrical power grids. Accordingly, a better performance of the electrical power grid's reliability, stability, efficiency, and transient response is achieved. In frequency regulation in the electrical power grids, the EV batteries regulate frequencies against the fluctuations resulting from the RESs and connected load proprieties. The used EV battery model is included in Figure 2 for frequency regulation studies in electrical power grids as in [56], in which, Nernst's equation-based EV battery model determines the dependency of the open circuit battery voltage (V_{oc}) on their state of charge (SOC) as follows:

$$V_{oc}(SOC) = V_{nom} + S_p \frac{RT}{F} ln \left(\frac{SOC}{C_{nom} - SOC} \right) \tag{3}$$

where $V_{oc}(SOC)$ stands for the V_{oc} dependency on SOC, V_{nom} stands for the nominal battery voltage, and C_{nom} stands for the nominal battery capacities of the EVs (in Ah). Furthermore, S_p denotes the sensitivity parameter of the V_{oc} and SOC of the batteries. R and F are the gas and Faraday constants, respectively, and T is the temperature.

2.4. System State Space Model

The studied system in Figure 2 is represented mathematically in Appendix A. The state space-based linear representation is employed for the proposed analysis. The models in Appendix A are collected in the state space model. The general representation of the state space model is modelled as:

$$\dot{x} = Ax + B_1 \omega + B_2 u \tag{4}$$

$$y = Cx \tag{5}$$

where x stands for a vector including state variables, y stands for a vector including output states, ω stands for a vector including the disturbance of a system, and u stands for a vector including the control output. The generation system is modelled using Laplace transform, and this model is employed to define the system state variables vector x. Whereas the load and RESs are considered as disturbances in this model for defining the vector ω. In (4), vectors x and ω are expressed as:

$$x = \begin{bmatrix} \Delta f_a & \Delta P_{ga} & \Delta P_{ga1} & \Delta P_{WT} & \Delta f_b & \Delta P_{gb} & \Delta P_{gb1} & \Delta P_{gb2} & \Delta P_{PV} & \Delta P_{tie} \end{bmatrix}^T \tag{6}$$

$$\omega = \begin{bmatrix} \Delta P_{la} & P_{WT} & \Delta P_{lb} & P_{PV} \end{bmatrix}^T \tag{7}$$

where ΔP_{ga} and ΔP_{ga1} are the governor and turbine outputs of the thermal unit in area a, respectively. Whereas ΔP_{gb}, ΔP_{gb1}, and ΔP_{gb2} are the governor, droop compensation, and penstock turbines outputs of the hydraulic generation in area b, respectively. In (4), vector u includes frequency regulation signals of each electrical power grid $ACEo_a$, and $ACEo_b$ in addition to the participated power by EVs (ΔP_{EVa} and ΔP_{EVb}) as follows:

$$u = \begin{bmatrix} ACEo_a & \Delta P_{EV_a} & ACEo_b & \Delta P_{EV_b} \end{bmatrix}^T \tag{8}$$

The representative matrices in state space modelling in (4), (A, B_1, B_2, and C) are obtained from the electrical power grid model in Figure 2 and its parameters. They are represented as:

$$A = \begin{bmatrix} -\frac{D_a}{2H_a} & \frac{1}{2H_a} & 0 & \frac{1}{2H_a} & 0 & 0 & 0 & 0 & 0 & -\frac{1}{2H_a} \\ 0 & -\frac{1}{T_t} & \frac{1}{T_t} & 0 & 0 & 0 & 0 & 0 & 0 & 0 \\ -\frac{1}{R_aT_g} & 0 & -\frac{1}{T_g} & 0 & 0 & 0 & 0 & 0 & 0 & 0 \\ 0 & 0 & 0 & -\frac{1}{T_{WT}} & 0 & 0 & 0 & 0 & 0 & 0 \\ 0 & 0 & 0 & 0 & -\frac{D_b}{2H_b} & \frac{1}{2H_b} & 0 & 0 & \frac{1}{2H_b} & \frac{1}{2H_b} \\ 0 & 0 & 0 & 0 & \frac{2T_R}{R_bT_1T_2} & -\frac{2}{T_w} & \frac{2T_2+2T_w}{T_2T_w} & \frac{2T_R-2T_1}{T_1T_2} & 0 & 0 \\ 0 & 0 & 0 & 0 & -\frac{T_R}{R_bT_1T_2} & 0 & -\frac{1}{T_2} & \frac{T_1-T_R}{T_1T_2} & 0 & 0 \\ 0 & 0 & 0 & 0 & -\frac{1}{R_bT_1} & 0 & 0 & -\frac{1}{T_1} & 0 & 0 \\ 0 & 0 & 0 & 0 & 0 & 0 & 0 & 0 & -\frac{1}{T_{PV}} & 0 \\ 2\pi T_{tie,eq} & 0 & 0 & 0 & -2\pi T_{tie,eq} & 0 & 0 & 0 & 0 & 0 \end{bmatrix} \tag{9}$$

$$B_1 = \begin{bmatrix} -\frac{1}{2H_a} & 0 & 0 & 0 \\ 0 & 0 & 0 & 0 \\ 0 & 0 & 0 & 0 \\ 0 & \frac{K_{WT}}{T_{WT}} & 0 & 0 \\ 0 & 0 & -\frac{1}{2H_b} & 0 \\ 0 & 0 & 0 & 0 \\ 0 & 0 & 0 & 0 \\ 0 & 0 & 0 & 0 \\ 0 & 0 & 0 & \frac{K_{PV}}{T_{PV}} \\ 0 & 0 & 0 & 0 \end{bmatrix} \text{, and} \quad B_2 = \begin{bmatrix} 0 & -\frac{1}{2H_a} & 0 & 0 \\ 0 & 0 & 0 & 0 \\ \frac{1}{T_g} & 0 & 0 & 0 \\ 0 & 0 & 0 & 0 \\ 0 & 0 & 0 & -\frac{1}{2H_b} \\ 0 & 0 & \frac{2T_R}{T_1T_2} & 0 \\ 0 & 0 & \frac{T_R}{T_1T_2} & 0 \\ 0 & 0 & \frac{1}{T_1} & 0 \\ 0 & 0 & 0 & 0 \\ 0 & 0 & 0 & 0 \end{bmatrix} \tag{10}$$

$$C = \begin{bmatrix} 1 & 0 & 0 & 0 & 0 & 0 & 0 & 0 & 0 & 0 \\ B_a & 0 & 0 & 0 & 0 & 0 & 0 & 0 & 0 & 1 \\ 0 & 0 & 0 & 0 & 1 & 0 & 0 & 0 & 0 & 0 \\ 0 & 0 & 0 & 0 & B_b & 0 & 0 & 0 & 0 & -1 \end{bmatrix} \tag{11}$$

Table 2. The electrical power grid parameters for the modelled case study (with $x \in \{a,b\}$), [14].

Symbols	Value	
	Area a	Area b
P_{rx} (MW)	1200	1200
R_x (Hz/MW)	2.4	2.4
B_x (MW/Hz)	0.4249	0.4249
Valve min. limit V_{vlx} (p.u.MW)	−0.5	−0.5
Valve max. limit V_{vux} (p.u.MW)	0.5	0.5
T_g (s)	0.08	-
T_t (s)	0.3	-
T_1 (s)	-	41.6
T_2 (s)	-	0.513
T_R (s)	-	5
T_w (s)	-	1
H_x (p.u.s)	0.0833	0.0833
D_x (p.u./Hz)	0.00833	0.00833
T_{PV} (s)	-	1.3
K_{PV} (s)	-	1
T_{WT} (s)	1.5	-
K_{WT} (s)	1	-

Table 2. *Cont.*

Symbols	Value	
	Area *a*	Area *b*
EVs Modelling		
Penetration level	5–10%	5–10%
V_{nom} (V)	364.8	364.8
C_{nom} (Ah)	66.2	66.2
R_s (ohms)	0.074	0.074
R_t (ohms)	0.047	0.047
C_t (farad)	703.6	703.6
RT/F	0.02612	0.02612
Minimum EVs SOC %	10	10
Maximum EVs SOC %	95	95
Minimum capacity EVs limit (p.u.MW)	−0.1	−0.1
Maximum capacity EVs limit (p.u.MW)	+0.1	+0.1
C_{batt} (kWh)	24.15	24.15

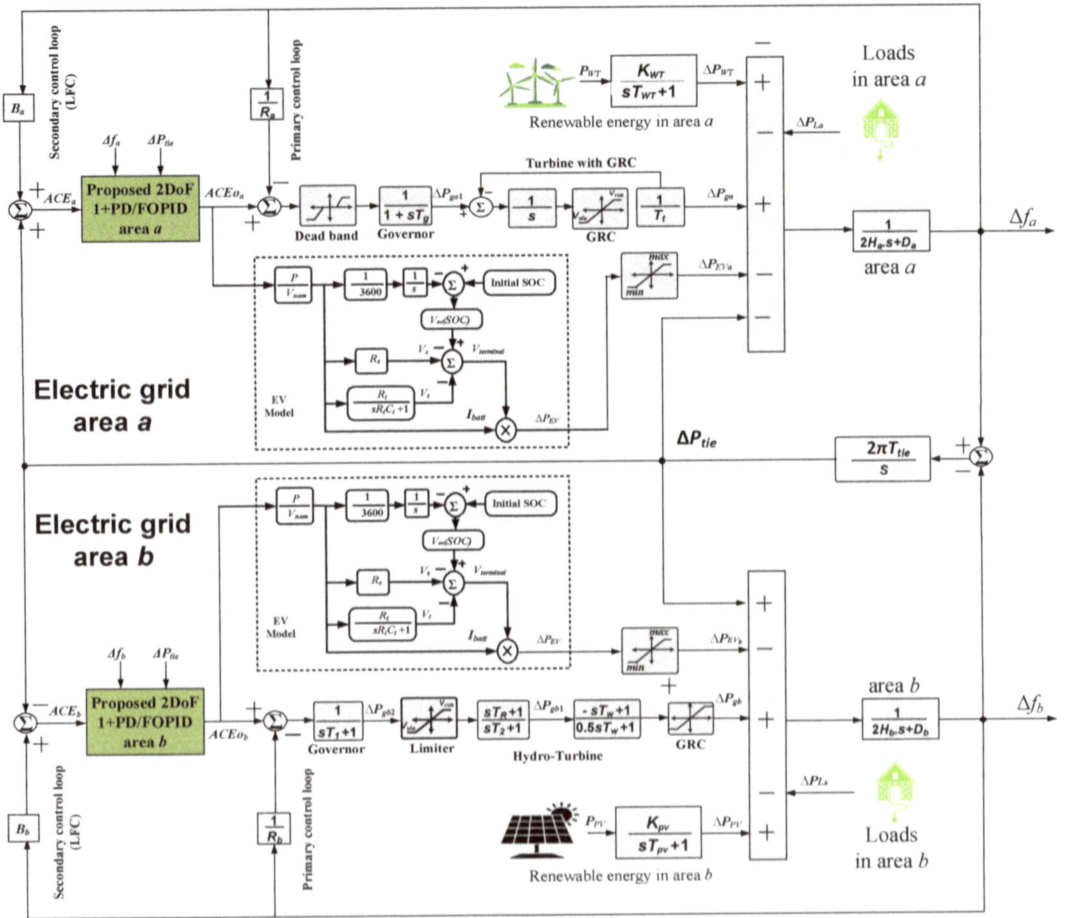

Figure 2. Modelling of the various components of the studied electrical power grids.

3. The Proposed 2Dof 1 + PD/FOPID Frequency Regulation

3.1. FO-Based Frequency Regulator Representation

The FO-based frequency regulator systems are based on FO calculus and representations of non-integer systems. There are several schemes provided in the literature, such as Grunwald–Letnikov, Riemann–Liouville, and Caputo [57]. The α_{th} FO-based derivative for function f between a and t is defined using the Grunwald–Letnikov method:

$$D^{\alpha}|_a^t = \lim_{h \to 0} \frac{1}{h^{\alpha}} \sum_{r=0}^{\frac{t-a}{h}} (-1)^r \binom{n}{r} f(t - rh) \tag{12}$$

where h is the employed step time, and $[\cdot]$ represents the used operator based on integer terms for the Grunwald–Letnikov method. Whereas n should satisfy the condition $(n - 1 < \alpha < n)$. The binomial-based coefficients are represented using:

$$\binom{n}{r} = \frac{\Gamma(n+1)}{\Gamma(r+1)\Gamma(n-r+1)'} \tag{13}$$

where the used gamma function within (13) is usually defined using:

$$\Gamma(n+1) = \int_0^{\infty} t^{x-1} e^{-t} \, dt \tag{14}$$

From another side, the Riemann–Liouville method defines FO-based derivatives by avoiding the utilization of the sum and limits, while it uses the integer derivative in addition to the integral representations as follows:

$$D^{\alpha}|_a^t = \frac{1}{\Gamma(n-\alpha)} \left(\frac{d}{dt}\right)^n \int_a^t \frac{f(\tau)}{(t-\tau)^{\alpha-n+1}} \, d\tau \tag{15}$$

The Caputo representation for the FO derivative is as follows:

$$D^{\alpha}|_a^t = \frac{1}{\Gamma(n-\alpha)} \int_a^t \frac{f^{(n)}(\tau)}{(t-\tau)^{\alpha-n+1}} \, d\tau \tag{16}$$

Generally, the FO non-integer operator $D^{\alpha}|_a^t$ can be denoted as follows:

$$D^{\alpha}|_a^t = \begin{cases} \alpha > 0 \;\to\; \frac{d^{\alpha}}{dt^{\alpha}} & \text{FO derivative} \\ \alpha < 0 \;\to\; \int_{t_0}^{t_f} dt^{\alpha} & \text{FO integral} \\ \alpha = 0 \;\to\; 1 \end{cases} \tag{17}$$

Implementing FO-based frequency regulators using Oustaloup's recursive approximation (ORA) has been shown as a suitable digital representation scheme in real-time implementations [57]. The ORA is selected in this paper for FO-based frequency regulator implementations. In which the αth derivatives (s^{α}) are represented as [57]:

$$s^{\alpha} \approx \omega_h^{\alpha} \prod_{k=-N}^{N} \frac{s + \omega_k^z}{s + \omega_k^p} \tag{18}$$

where ω_k^p and ω_k^z stand for the pole/zero locations within the ω_h sequence, and their calculations are as follows:

$$\omega_k^z = \omega_b \left(\frac{\omega_h}{\omega_b}\right)^{\frac{k+N+\frac{1-\alpha}{2}}{2N+1}} \tag{19}$$

$$\omega_k^p = \omega_b \left(\frac{\omega_h}{\omega_b}\right)^{\frac{k+N+\frac{1+\alpha}{2}}{2N+1}} \tag{20}$$

$$\omega_h^\alpha = \left(\frac{\omega_h}{\omega_b}\right)^{\frac{-\alpha}{2}} \prod_{k=-N}^{N} \frac{\omega_k^p}{\omega_k^z} \tag{21}$$

where the approximated FO-based frequency regulator operator function possesses a $(2N+1)$ pole/zero number. Therefore, N is related to ORA's order of representation and equals $(2N+1)$. In this work, ORA is employed for FO-based frequency regulator representation with $(M=5)$ within the frequency range $(\omega \in [\omega_b, \omega_h])$ between $[0.001, 1000]$ rad/s.

3.2. Controllers from the Literature

The literature has several proposals of frequency regulation control schemes, including IO- and FO-based frequency controllers. Some existing IO-based frequency regulators include the I, PI, PID, and PIDF LFCs, and their transfer functions (TFs) $C(s)$ are as follows:

$$C_I(s) = \frac{Y(s)}{E(s)} = \frac{K_i}{s}$$

$$C_{PI}(s) = \frac{Y(s)}{E(s)} = K_p + \frac{K_i}{s}$$

$$C_{PID}(s) = \frac{Y(s)}{E(s)} = K_p + \frac{K_i}{s} + K_d\, s \tag{22}$$

$$C_{PIDF}(s) = \frac{Y(s)}{E(s)} = K_p + \frac{K_i}{s} + K_d\, s\, \frac{N_f}{s + N_f}$$

Whereas FO-based frequency regulators in the literature include the following:

$$C_{FOI}(s) = \frac{Y(s)}{E(s)} = \frac{K_i}{s^\lambda}$$

$$C_{FOPI}(s) = \frac{Y(s)}{E(s)} = K_p + \frac{K_i}{s^\lambda}$$

$$C_{FOPID}(s) = \frac{Y(s)}{E(s)} = K_p + \frac{K_i}{s^\lambda} + K_d\, s^\mu$$

$$C_{FOPIDF}(s) = \frac{Y(s)}{E(s)} = K_p + \frac{K_i}{s^\lambda} + K_d\, s^\mu\, \frac{N_f}{s + N_f} \tag{23}$$

$$C_{TID}(s) = \frac{Y(s)}{E(s)} = K_t\, s^{-\left(\frac{1}{n}\right)} + \frac{K_i}{s} + K_d\, s$$

$$C_{FOTID}(s) = \frac{Y(s)}{E(s)} = K_t\, s^{-\left(\frac{1}{n}\right)} + \frac{K_i}{s^\lambda} + K_d\, s^\mu$$

$$C_{PFOTID}(s) = \frac{Y(s)}{E(s)} = K_p + K_t\, s^{-\left(\frac{1}{n}\right)} + \frac{K_i}{s^\lambda} + K_d\, s^\mu$$

It has become evident that each frequency regulator includes a number of tunable parameters to optimize its performance. The number of tunable parameters relies on the employed control scheme.

3.3. Proposed 1 + PD/FOPID Controllers

Figure 3 shows the structure of the proposed 1 + PD/FOPID controller for interconnected electrical power grids. The proposed 1 + PD/FOPID controller uses two cascaded loops using the ACE signal of each electrical power grid in the area in the outer loop and the frequency deviation signal of the area in the inner loop. The 1 + PD is employed for the outer loop and FOPID is employed for the inner loop. Therefore, the proposed 1 + PD/FOPID controller can benefit from both the characteristics of 1 + PDF in the outer loop and the FOPID FO controller in the inner loop. Moreover, the proposed 1 + PD/FOPID controller uses two different inputs signals with different characteristics. That is, the ACE can mitigate low-frequency-related disturbances, while frequency deviation can mitigate high-frequency-related disturbances. Therefore, the proposed 1 + PD/FOPID achieves a fast and robust response, low values of overshoot/undershoot peaks, and a high rate for rejecting disturbances with various frequency scales. The modelling of inputs into the ACE loop ((ACE_a) and (ACE_b)) and into the controller inputs ($E_{a1}(s)$ and $E_{b1}(s)$) is provided as:

$$E_{a1}(s) = ACE_a = \Delta P_{tie} + B_a \, \Delta f_a$$
$$E_{b1}(s) = ACE_b = A_{ab} \, \Delta P_{tie} + B_b \, \Delta f_b \tag{24}$$

where (A_{ab}) represents the ratio among the capacities of the electrical power grids' areas *a* and *b*, whereas the outputted signals from $Y_{a1}(s)$ and $Y_{b1}(s)$ of the 1 + PD controller's loop are expressed as:

$$Y_{a1}(s) = [1 + K_{p1} + K_{d1} \, s] \cdot E_{a1}(s)$$
$$Y_{b1}(s) = [1 + K_{p2} + K_{d2} \, s] \cdot E_{b1}(s) \tag{25}$$

From (25), each electrical power grid has two parameters for tuning in the outer loop. The electrical power grid in area *a* has two tunable parameters (K_{p1} and K_{d1}), whereas the electrical power grid in area *b* has K_{p2} and K_{d2} for the tuning process. The output of the outer loop is fed into the inner FOPID loop. The representations of the error signals $E_{a2}(s)$ and $E_{b2}(s)$ are as follows:

$$E_{a2}(s) = Y_{a1}(s) - \Delta P_{tie} - \Delta f_a$$
$$E_{b2}(s) = Y_{b1}(s) - \Delta P_{tie} - \Delta f_b \tag{26}$$

The representations of the FOPID loops are as follows:

$$Y_{a2}(s) = [K_{p3} + \frac{K_{i1}}{s^{\lambda_1}} + K_{d3} \, s^{\mu_1}] \cdot E_{a2}(s)$$
$$Y_{b2}(s) = [K_{p4} + \frac{K_{i2}}{s^{\lambda_2}} + K_{d4} \, s^{\mu_2}] \cdot E_{b2}(s) \tag{27}$$

From (27), the electric power grid in area *a* has five tunable control parameters (K_{p3}, K_{i1}, K_{d3}, λ_1 and μ_1), and the electrical power grid in area *b* has K_{p4}, K_{i2}, K_{d4}, λ_2 and μ_2 as tunable parameters.

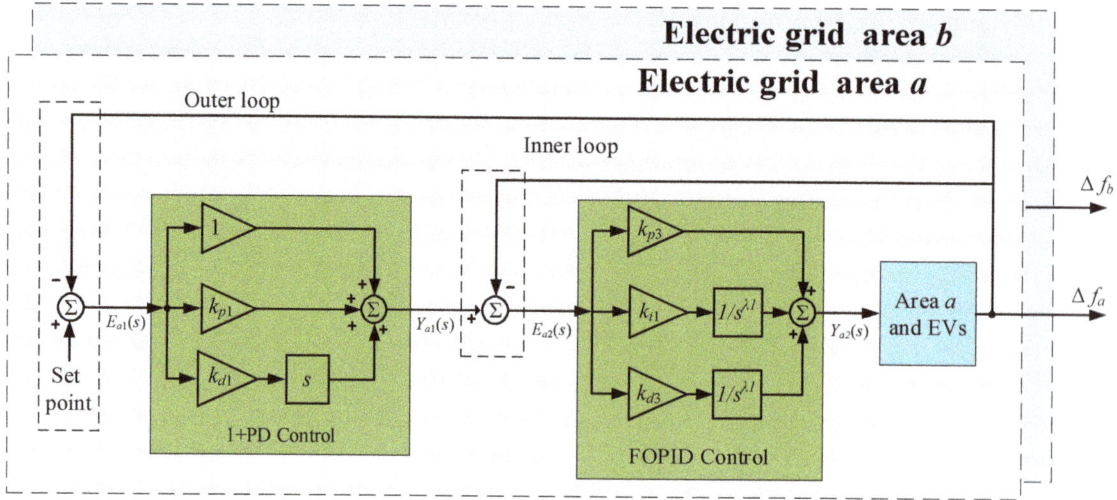

Figure 3. Proposed 1 + PD/FOPID controller.

4. The Proposed Design Optimization

4.1. MRFO Optimizer

The MRFO is a metaheuristic optimizer that belong to foraging strategies followed by manta rays during the catching process of their prey [58]. It is mainly composed of three foraging processes (chain, cyclone and somersault foraging). During chain foraging, manta rays consider highly concentrated plankton, which represent the desired optimization objectives and tracker. This, in turn, makes them align with the foraging chain. In which, everyone is directed towards food by the manta rays within its front. Then, an updated process of individuals is obtained from the best solution in each iterations. The chain process of foraging is mathematically expressed as [58]:

$$
x_i^{t+1} = \begin{cases} x_i^t + r.(x_{best}^t - x_i^t) + \omega_1(x_{best}^t - x_i^t), i = 1 \\ x_i^t + r.(x_{i-1}^t - x_i^t) + \omega_1(x_{best}^t - x_i^t), i = 2 : N \end{cases}
\tag{28}
$$

where x_i^t stands for the ith position for the current iteration t, r represents a random vector, x_{best}^t stands for best solution in the tth iteration, N is the number of manta rays, and ω_1 is the weighting coefficient, expressed as:

$$
\omega_1 = 2 \times r \times \sqrt{|\log(r)|}
\tag{29}
$$

From (28), the individuals' positions are calculated for all individuals except the first $(i-1)^{th}$ individual and the best individual x_{best}^t. After plankton patch position determinations using manta rays, a chain is formed by their combination and they swim in a spiral shape towards their prey. Additionally, individuals swim towards the front-sided manta ray. The cyclone foraging process is expressed as:

$$
\begin{aligned}
x_i^{t+1} &= x_{best} + r.(x_{i-1}^t - x_i^t) + e^{b\omega}.\cos(2\pi\omega).(x_{best} - x_i^t) \\
y_i^{t+1} &= y_{best} + r.(y_{i-1}^t - y_i^t) + e^{b\omega}.\cos(2\pi\omega).(y_{best} - y_i^t)
\end{aligned}
\tag{30}
$$

where ω stands for a random number. Then, the cyclone process foraging is as follows:

$$
x_i^{t+1} = \begin{cases} x_{best} + r.(x_{best}^t - x_i^t) + \omega_2(x_{best}^t - x_i^t), i = 1 \\ x_{best} + r.(x_{i-1}^t - x_i^t) + \omega_2(x_{best}^t - x_i^t), i = 2 : N \end{cases}
\tag{31}
$$

where ω_2 represents the weighting factor of cyclone foraging and is determined as:

$$
\omega_2 = 2e^{r_1}\left(\frac{T-t+1}{T}\right).\sin(2\pi r_1)
\tag{32}
$$

where t stands for the current iteration, T stands for the maximum iterations number, and r_1 stands for a random number. The improved exploitation is obtained through the cyclone foraging process to determine the best solution region. This is because all existing manta rays contribute to the food search processes based on their reference positions. Additionally, the exploitation process is also enhanced by forcing the individuals to search for new positions located away from the current best position. The random position in the search space is determined as:

$$
x_{rand} = Lb + r.(Ub - Lb)
\tag{33}
$$

$$
x_i^{t+1} = \begin{cases} x_{rand} + r.(x_{rand}^t - x_i^t) + \beta(x_{rand}^t - x_i^t), i = 1 \\ x_{rand} + r.(x_{i-1}^t - x_i^t) + \beta(x_{rand}^t - x_i^t), i = 2 : N \end{cases}
\tag{34}
$$

where Ub and Lb are the upper/lower limits, respectively, of the desired variables, x_{rand} is an assigned random position in the search space. In somersault foraging, the food is recognized in this stage as a hinge. Wherein, each manta ray swims backwards and forwards around the food hinge, and then they tumble to the new position. The somersault foraging process is performed as follows:

$$x_i^{t+1} = x_i^t + S.(r_2.x_{best}^t - r_3.x_i^t), i = 1, 2, ..., N \tag{35}$$

where S is the somersault factor employed to determine the somersault range for the manta rays, and r_2 and r_3 are random numbers. A flowchart representation of the MRFO stages is shown in Figure 4.

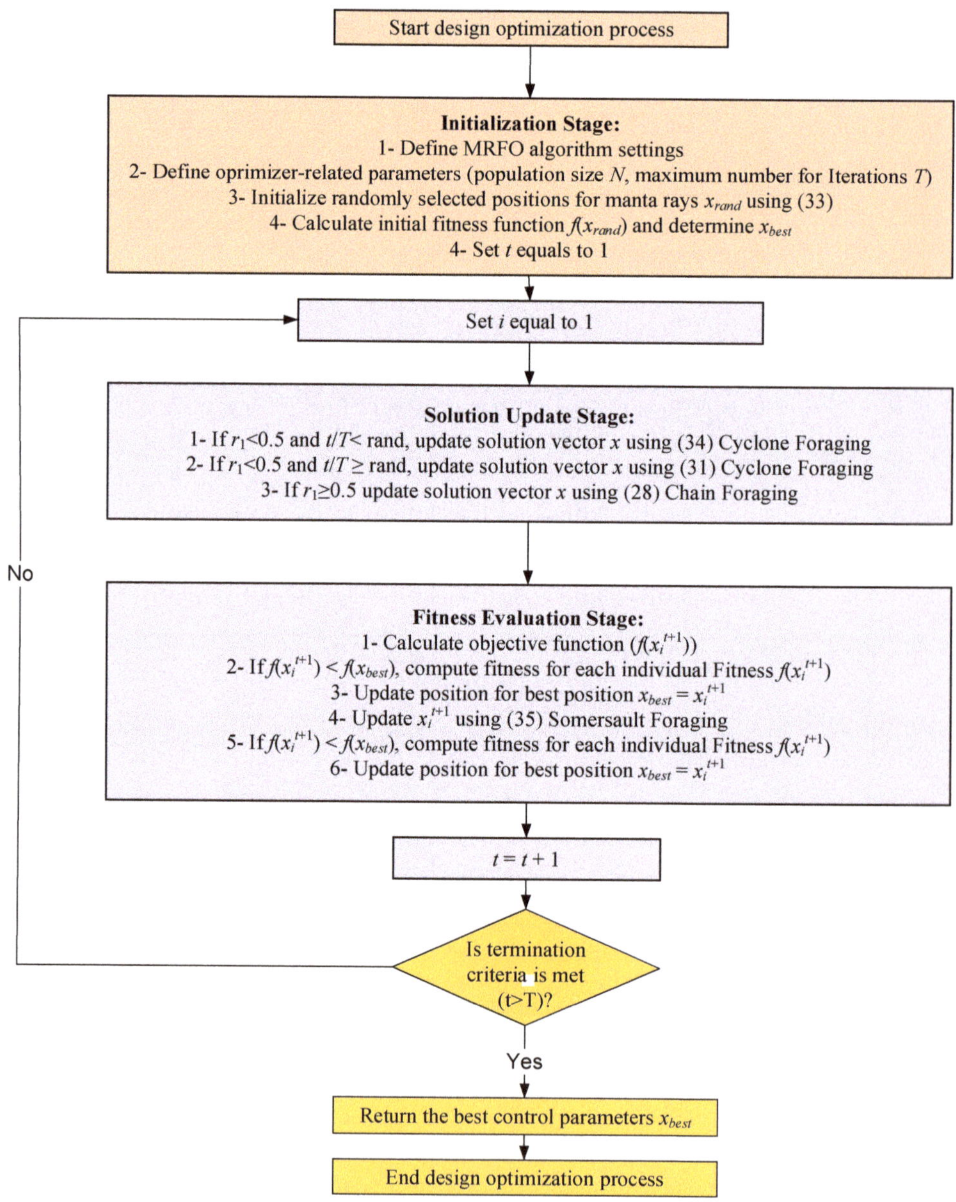

Figure 4. Flowchart representation of the MRFO stages.

4.2. Design Optimization

The MRFO is proposed in this paper to determine the best parameters for the proposed 1 + PD/FOPID for both electrical power grids. The main driving objectives for the optimum parameters include minimizing the existing fluctuations in the frequency in both areas as well as the tie-line power among the electrical power grids. A measure of frequency deviation (Δf_a of area a and Δf_b of area b) and tie-line power (ΔP_{tie} between electrical power grids) is performed and fed into the optimization process. Then, they are combined in a single-objective function to drive the optimization process for simulation time t_s while taking problem constraints into consideration in the process. The employed objective functions can be expressed as follows:

$$ISE = \text{integral squared-error} = \int_0^{t_s} ((\Delta f_a)^2 + (\Delta f_b)^2 + (\Delta P_{tie})^2)\, dt \tag{36}$$

$$ITSE = \text{integral time-squared-error} = \int_0^{t_s} t.((\Delta f_a)^2 + (\Delta f_b)^2 + (\Delta P_{tie})^2)\, dt \tag{37}$$

$$IAE = \text{integral absolute-error} = \int_0^{t_s} (abs(\Delta f_a) + abs(\Delta f_b) + abs(\Delta P_{tie}))\, dt \tag{38}$$

$$ITAE = \text{integral time-absolute-error} = \int_0^{t_s} t.(abs(\Delta f_a) + abs(\Delta f_b) + abs(\Delta P_{tie}))\, dt \tag{39}$$

The ISE and IAE are based on using the integration of the square and absolute error values, respectively, within the simulation time t_s. The ISE provides better consideration of the large error values due to the square operation (when errors are more than 1). In addition, IAE provides equal consideration for large and low error values. Whereas ITSE and ITAE consider the time during the integration compared to the ISE and IAE objectives, which leads to lower/zero steady state error compared to ISE and IAE. The four objectives are considered in this paper to provide a comprehensive comparison of the studied controllers.

The proposed design optimization process based on MRFO is summarized in Figure 5. The employed optimization constraints in our proposed design process are:

$$\begin{aligned}
k_p^{min} &\leq k_{p1}, k_{p2}, k_{p3}, k_{p4} \leq k_p^{max} \\
k_i^{min} &\leq k_{i1}, k_{i2} \leq k_i^{max} \\
k_d^{min} &\leq k_{d1}, k_{d2}, k_{d3}, k_{d4} \leq k_d^{max} \\
\lambda^{min} &\leq \lambda_1, \lambda_2 \leq \lambda^{max} \\
\mu^{min} &\leq \mu_1, \mu_2 \leq \mu^{max}
\end{aligned} \tag{40}$$

where the lower/upper constraints are represented by *min*, and *max*, respectively, for the proposed 1 + PD/FOPID LFC. The used minimum constraints for k_p^{min}, k_i^{min}, and k_d^{min} were set at zero, and k_p^{max}, k_i^{max}, and k_d^{max} were set at five during the proposed optimization stages. The set minimum values for λ^{min} and μ^{min} were zero and for λ^{max} and μ^{max} were 1.

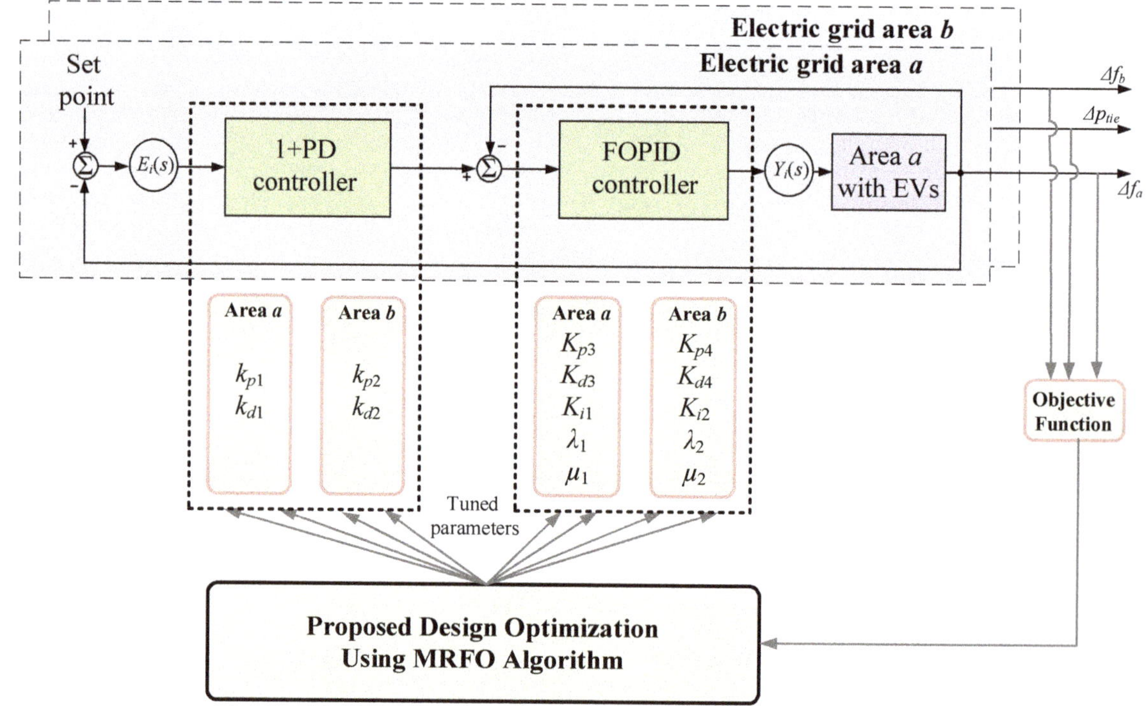

Figure 5. Proposed design optimization of the 1 + PD/FOPID controller.

5. Simulation Results and Performance Verification

The proposed system and proposed design optimization were implemented using the MATLAB R2021a joint m-file and Simulink platforms. The objective function and optimizers were programmed using m-file and linked with Simulink platform. The proposed design optimization process is based on using 20 populations with a maximum of 100 iterations for all the studied optimizers. The same process was used for the design of all the compared controllers for a fair comparison. The two-area electrical power grid system was tested, and the performance of the proposed 1 + PD/FOPID controller was evaluated and compared with the PID, FOPID, and PD/FOPID controllers. Moreover, the convergence performance of the MRFO optimizer was compared with some metaheuristic optimization algorithms from the literature.

The considered optimizers are the GA, PSO, and MPA. The design optimization was made using a personal computer with an Intel Core i7, CPU of 2.9 GHz, and a 64-bit version. Figure 6 compares the ISE and IAE convergence curves of the studied optimizers, whereas Figure 7 shows the ITSE and ITAE comparisons. It has become evident that the MRFO-based design optimization possesses the best convergence with the lowest objective function for all the studied ISE, IAE, ITSE, and ITAE objective functions. In addition, MRFO achieves a very fast conversion with better determination of the control parameters compared with the other studied optimization algorithms. Table 3 summarizes the obtained controllers' parameters using the proposed design optimization process. The considered test scenarios are organized as follows:

- **Scenario (1):** Impacts of the stepped load perturbations (SLP);
- **Scenario (2):** Impacts of multiple SLPs on the two interconnected electrical power grids;
- **Scenario (3):** Impacts of multiple connection/disconnection of RESs;
- **Scenario (4):** Impacts of randomly varying loads;

- **Scenario (5):** Joint impacts of RES fluctuations with various load-type variations.

Table 3. The optimum controllers' parameters using MRFO design optimization.

Controller	Area	Parameters						
		k_{p1}	k_{p2}	k_{i1}	k_{d1}	k_{d2}	λ_1	μ_1
PID	Area *a*	1.9062	—	1.8547	1.8637	—	—	—
	Area *b*	0.8808	—	0.2823	0.4233	—	—	—
FOPID	Area *a*	1.8184	—	1.567	0.9969	—	0.83	0.56
	Area *b*	1.9809	—	1.189	1.9497	—	0.89	0.73
PD/FOPID	Area *a*	4.3749	4.9837	1.9231	3.1152	1.6403	0.91	0.76
	Area *b*	2.5839	4.7702	0.9544	0.7011	3.3158	0.62	0.93
1 + PD/FOPID	Area *a*	4.5281	3.2751	3.4007	4.2212	4.9497	0.97	0.82
	Area *b*	3.7113	0.6361	1.6341	4.3158	2.9466	0.77	0.91

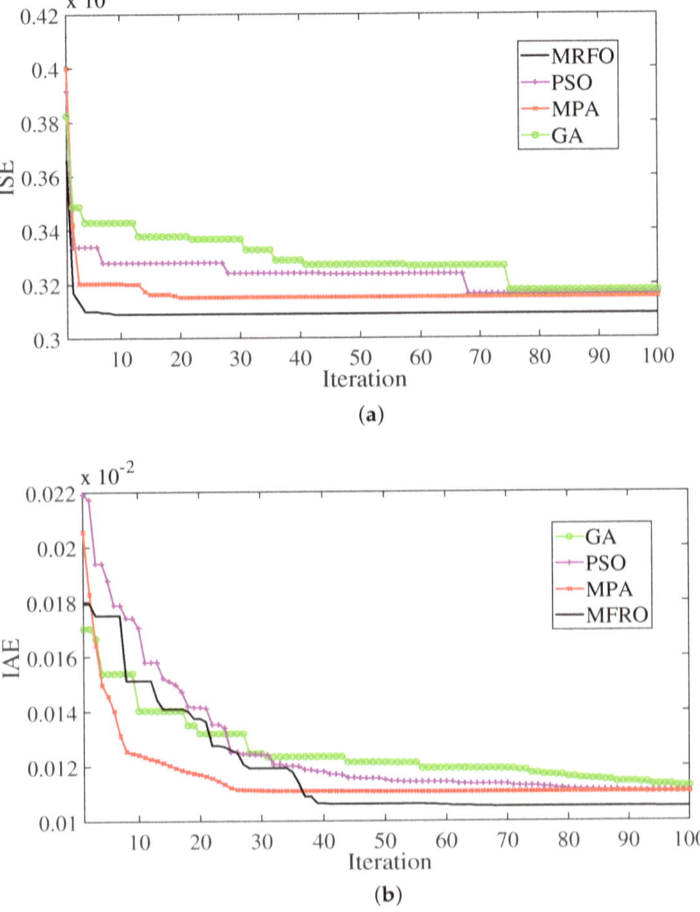

Figure 6. Convergence curves for the proposed design optimization; (**a**) ISE; (**b**) IAE.

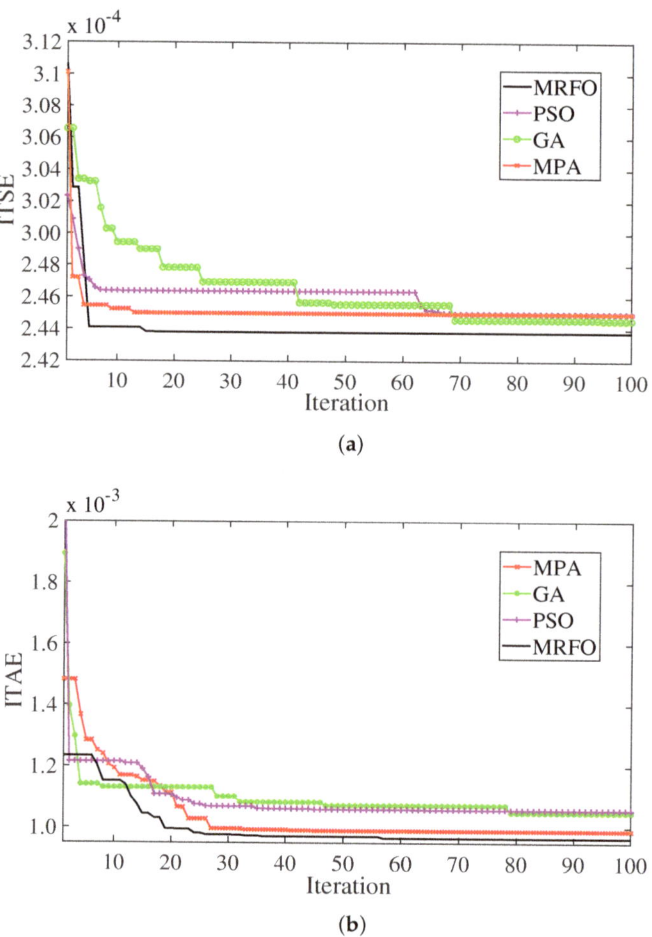

Figure 7. Convergence curves for the proposed design optimization; (**a**) ITSE; (**b**) ITAE.

5.1. Results of Scenario (1)

Figure 8 shows the obtained results during Scenario (1) with an SLP of 2%. The proposed 1 + PD/FOPID LFC achieves the best transient response compared with the studied controllers. The proposed 1 + PD/FOPID LFC has a maximum undershoot (MU) in Δf_a of 0.0018 compared with 0.0101, 0.0061, and 0.0044 under the PID, FOPID, and PD/FOPID, respectively. Moreover, the MU in Δf_b was 0.0002 under the proposed 1 + PD/FOPID controller compared with 0.0071, 0.0023, and 0.0016 under the PID, FOPID, and PD/FOPID, respectively. Furthermore, the proposed 1 + PD/FOPID had the lowest settling time (ST) compared with the studied controllers. In addition, the effect of the proposed controller on the thermal, hydraulic, and EV performances is depicted on Figure 9. It can be noted from this figure that the output powers from the thermal power unit and EVs in area *a* do not exceed their maximum bounds based on different control signals from the frequency variations or area control error. This is reflected as an improvement in the stability behaviour and response of the studied electrical power grids with various expected sudden load changes. For this, the proposed controller succeeded at preserving a better response with subjected load changes compared with the studied controllers.

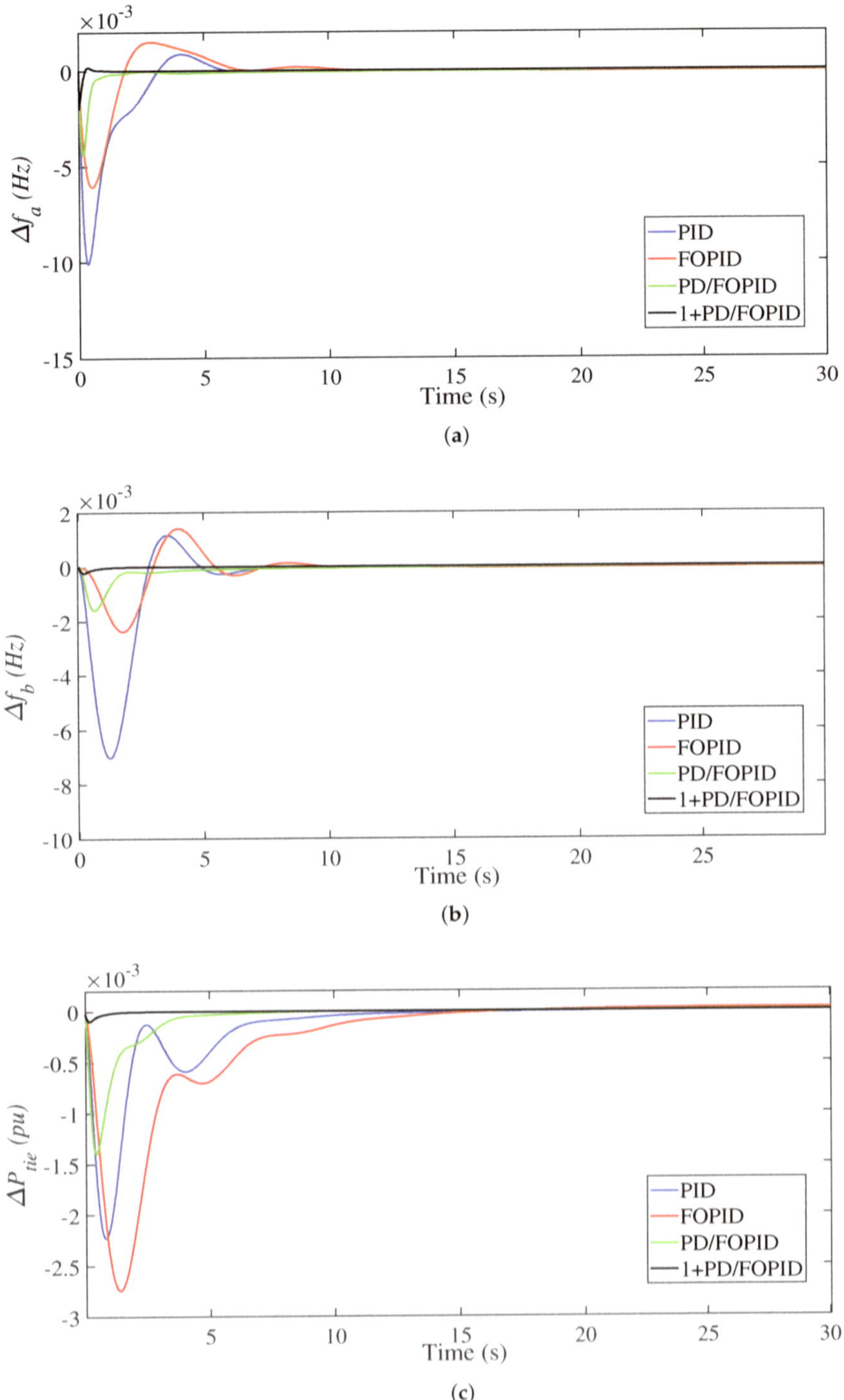

Figure 8. Performance evaluations with an SLP of 2%. Scenario (1): (**a**) Δf_a; (**b**) Δf_b; (**c**) ΔP_{tie}.

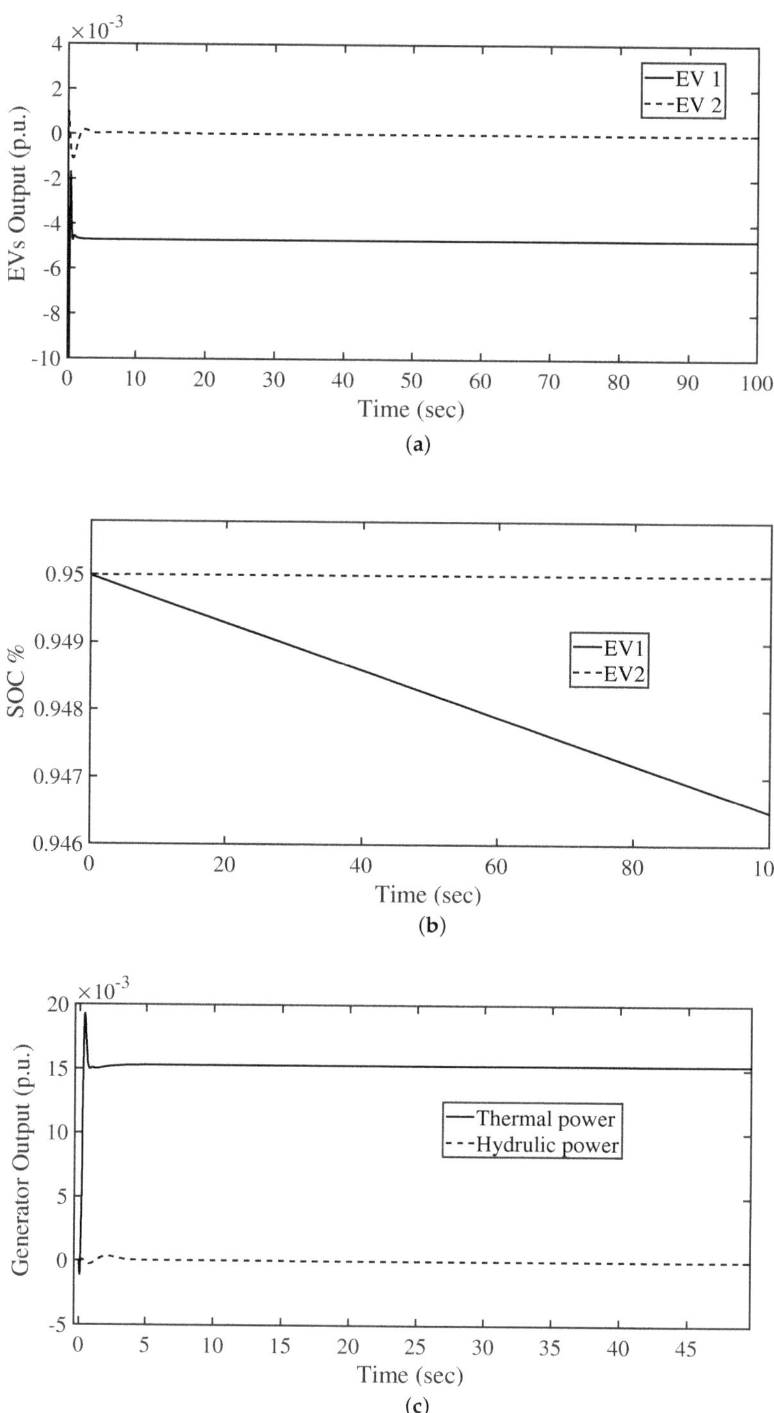

Figure 9. EV participation in Scenario (1): (**a**) EV output power; (**b**) EV battery SOC; (**c**) generator output power.

5.2. Results of Scenario (2)

From another side, Scenario (2) was made using multiple SLPs in different areas. Figure 10 shows the applied load powers in different electrical power grids as a test scenario. Figure 11a–c shows the obtained performance response in this scenario. The response of the frequency and tie-line power transients shows the proposed controller with better transients in all the tested SLP changes in this scenario. From the measured response, the proposed 1 + PD/FOPID has the best frequency deviation response in areas a and b. The proposed 1 + PD/FOPID has an MU in area a of 0.0005 and in area b of 0.0011. Whereas the PID has values of 0.0078 and 0.0103 in areas a and b, respectively; FOPID has values of 0.0051 and 0.0081 in areas a and b, respectively; and PD/FOPID has values of 0.0037 and 0.00581 in areas a and b, respectively. Therefore, the proposed 1 + PD/FOPID achieves the lowest peaks during this scenario. Furthermore, the superior impact of the proposed controller on the performance of the hydraulic generation unit and EVs outputs and its static of charge can be seen in Figure 12, which shows that both of them can regulate the system frequency without exceeding their maximum limits. In addition, it can be observed that there is a cross-coupling between the areas during the transient state, and hence each area produces its own power at a steady state.

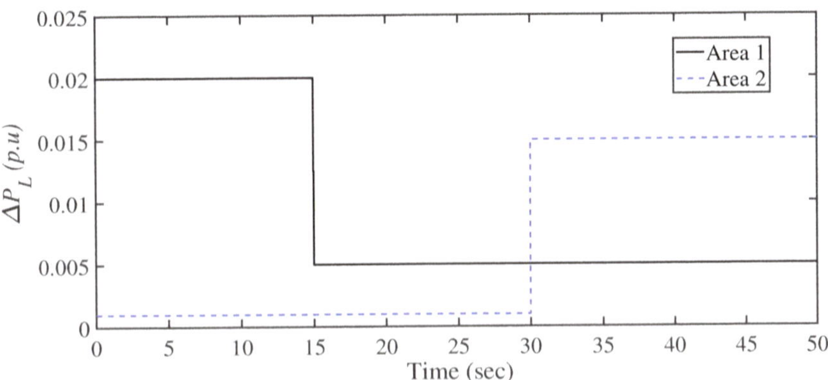

Figure 10. Load profiles at multiple SLPs in Scenario (2).

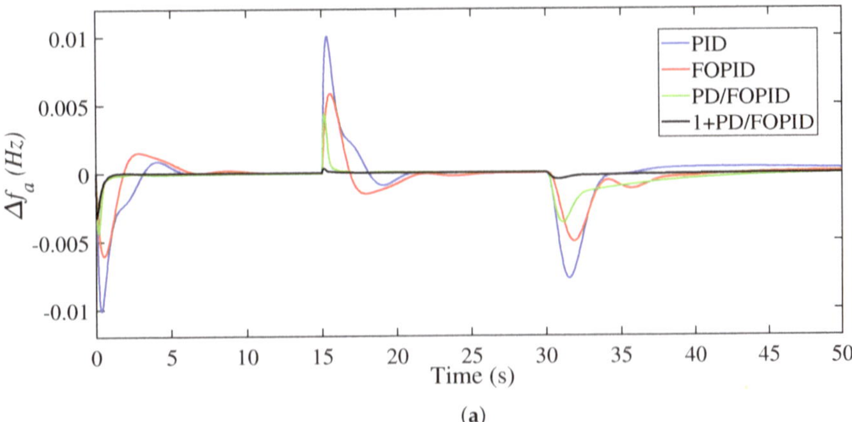

(a)

Figure 11. *Cont.*

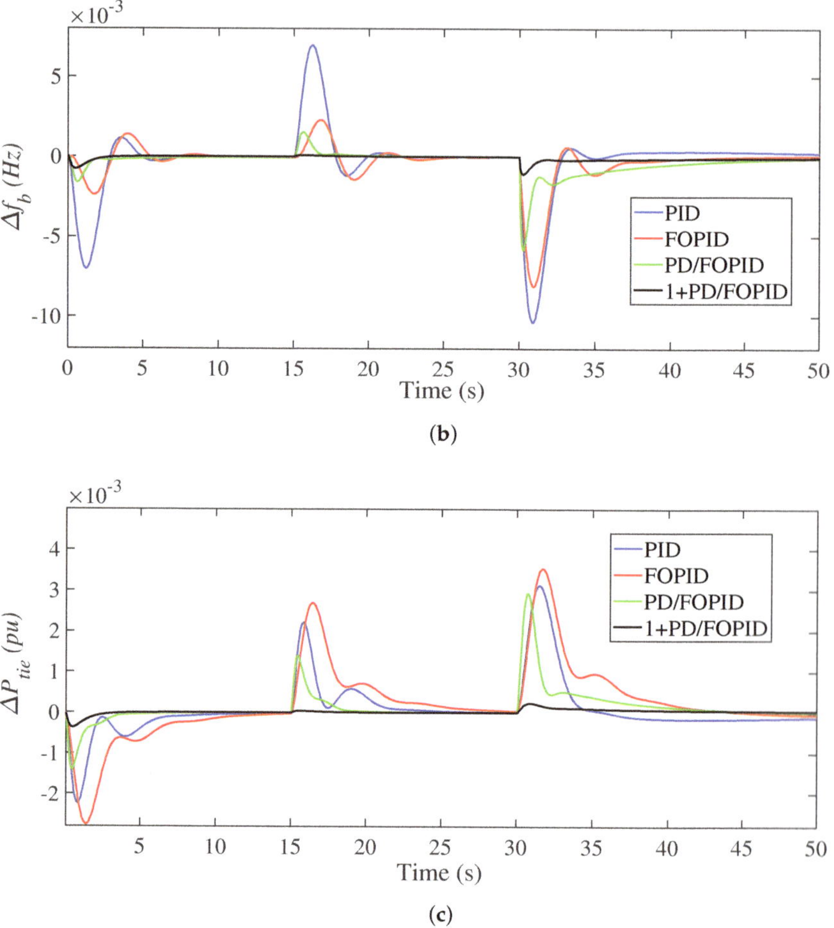

(b)

(c)

Figure 11. Performance evaluations at multiple SLPs Scenario (2): (**a**) Δf_a; (**b**) Δf_b; (**c**) ΔP_{tie}.

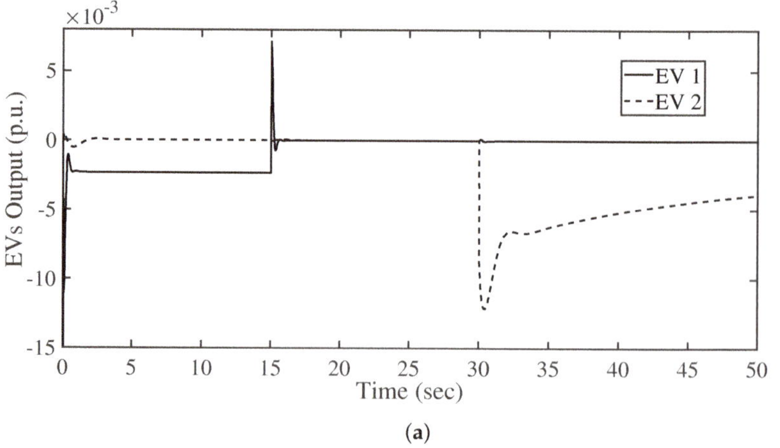

(a)

Figure 12. *Cont.*

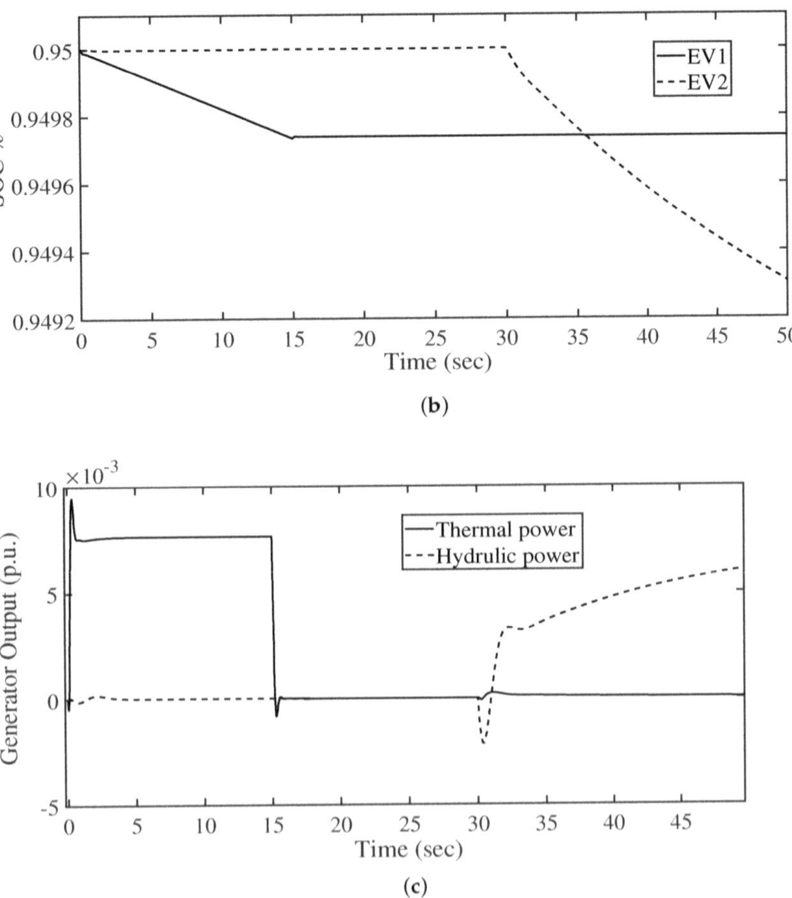

Figure 12. EV participation in Scenario (2): (**a**) EV output power; (**b**) EV battery SOC; (**c**) generator output power.

5.3. Results at Scenario (3)

In this scenario, multiple RES connections/disconnections have been made to test the proposed controller. Figure 13 shows the PV and wind powers in this scenario. The wind is connected at time 30 s and disconnected at time 80 s. Whereas the PV power is connected at time 50 s. In addition, an SLP is made at the start of the scenario. The obtained results are shown in Figure 14a–c for this scenario. It can be seen that connecting/disconnecting wind/PV affects the response of the system due to its participation level. The proposed 1 + PD/FOPID has the best performance metrics in this scenario, and the PIF has the worst response. For instance, the obtained MO values in area a at 30 s are 0.1202, 0.0596, 0.0456, and 0.0085 for the PID, FOPID, PPD/FOPID, and proposed 1 + PD/FOPID LFC, respectively. The measured performance metrics for all the studied scenarios are shown in Table 4. In which, a performance enhancement is observed when using the proposed control and design optimization method.

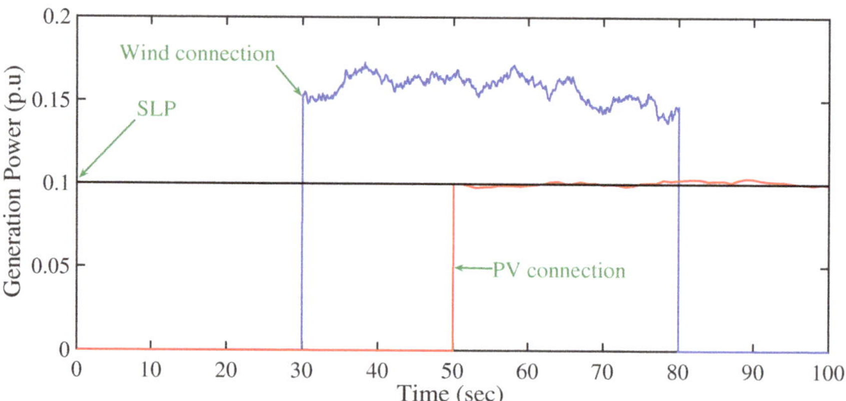

Figure 13. Load and RES generation profiles in Scenario (3).

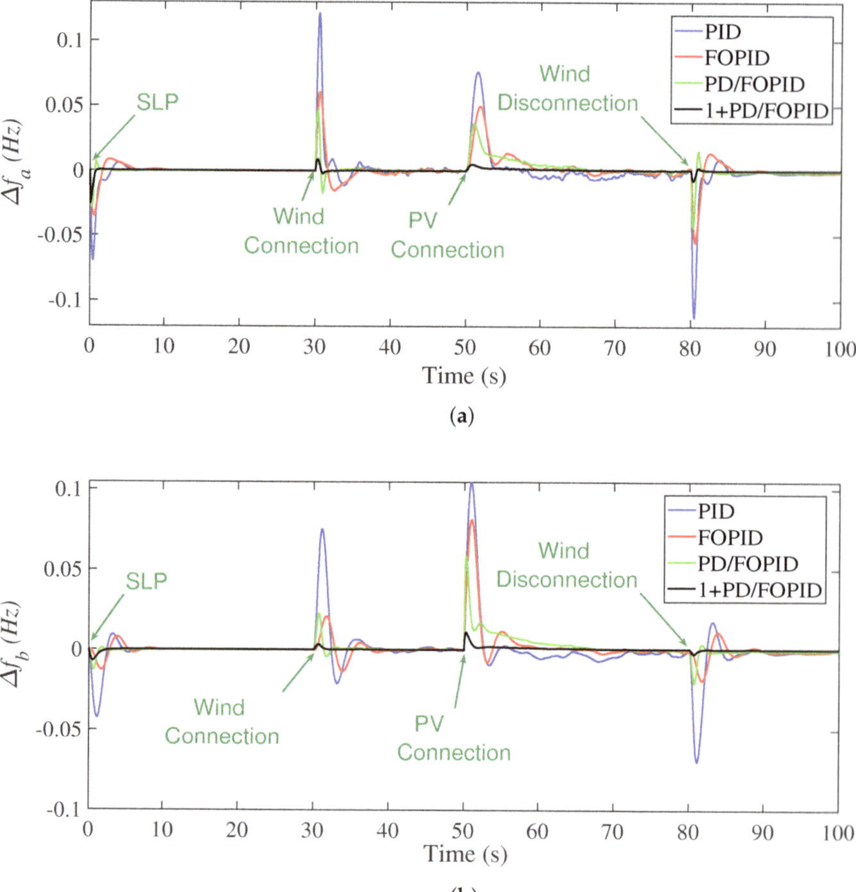

Figure 14. *Cont.*

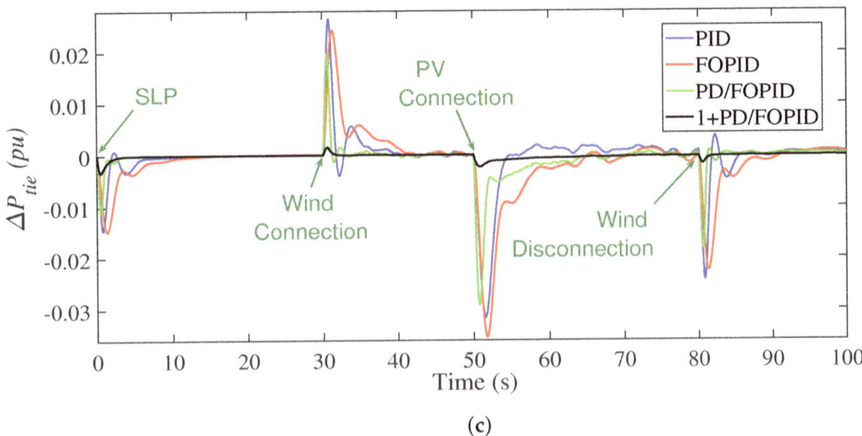

(c)

Figure 14. Performance evaluations for RES changes in Scenario (3): (**a**) Δf_a; (**b**) Δf_b; (**c**) ΔP_{tie}.

5.4. Results at Scenario (4)

An important factor to be considered is the characteristics of the connected electrical load in the electrical power grids. The load varies all day, and hence so too do the expected different demands of power in each moment. These variations are often reflected as fluctuations in the operating frequency of the electrical power grids. Therefore, in this tested scenario, a randomly changing electrical loading is considered as shown in Figure 15. The associated results for this scenario are shown in Figure 16. It can be seen that the proposed 1 + PD/FOPID has the lowest peak fluctuations with varying load profiles. Whereas the PID-based LFC has the highest level of fluctuations during this level. The PD/FOPID comes in second place in terms of improving the electrical power grid response; this followed by the response of the FOPID control.

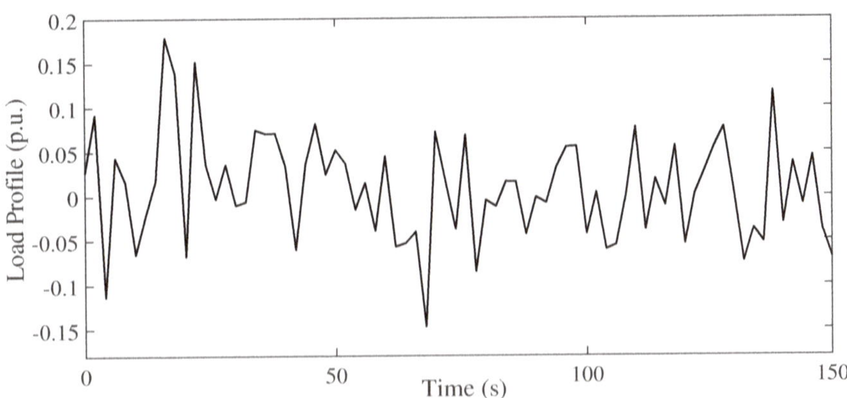

Figure 15. Random load change profiles of Scenario (4).

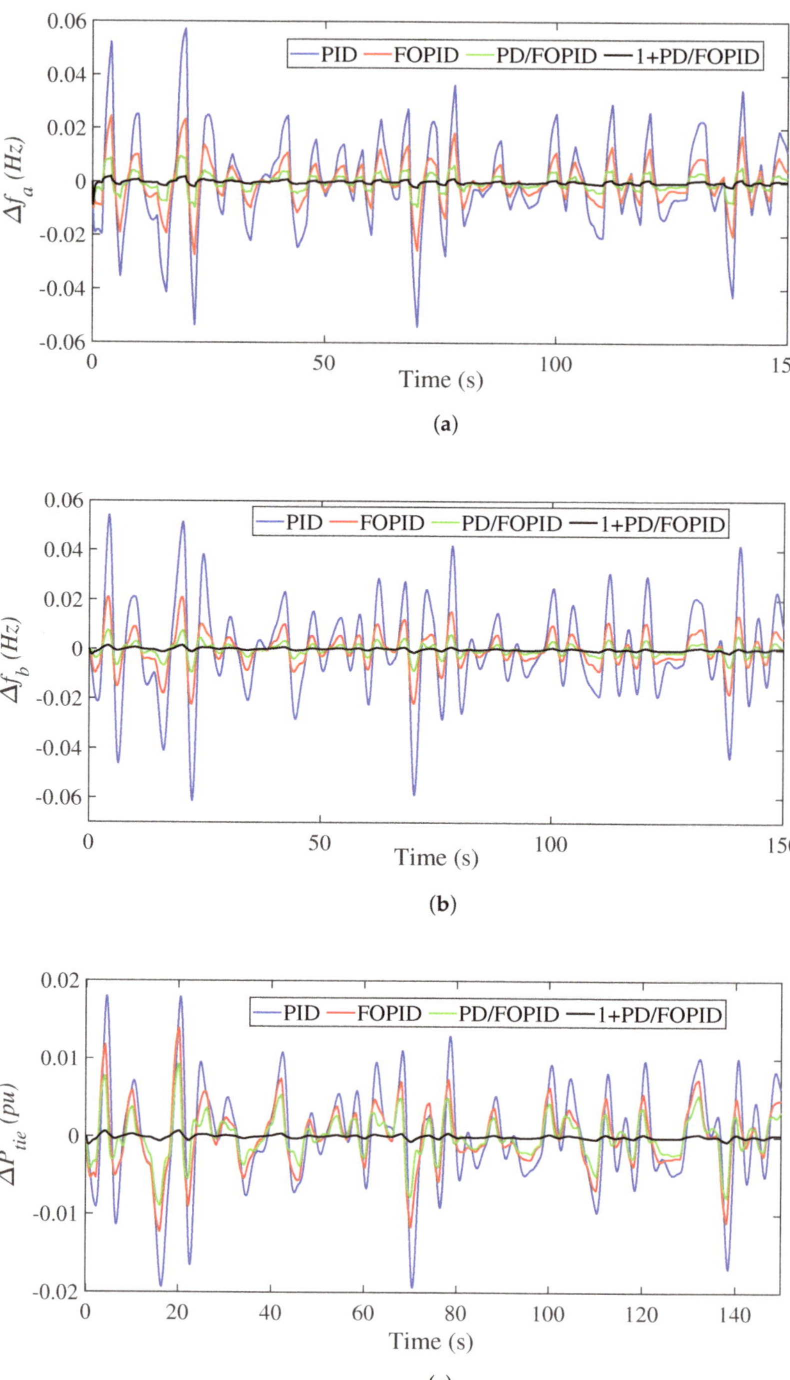

Figure 16. Performance evaluations at random load changes in Scenario (4): (**a**) Δf_a; (**b**) Δf_b; (**c**) ΔP_{tie}.

5.5. Results at Scenario (5)

The interconnected electrical power grids are subjected to joint intermittency of the connected RESs with the connected electrical loading. In this scenario, the domestic and industrial load profiles are considered and studied as shown in Figure 17. Additionally, the fluctuated RESs are included in this scenario with their participation levels and connection/disconnection events in the scenario, as shown in Figure 18. At time 0 s, all the renewable sources and loads are connected, which represents the worst-case scenario to test all the studied controllers and the proposed design optimization method. Figure 19 presents the obtained results in this scenario. The PID at the start of scenario with all RESs and loading step has 0.0927 MO in Δf_a of area a and 0.1463 MO in Δf_b of area b. Furthermore, regarding the deviations in ΔP_{tie}, the PID has an MU of 0.0461 in this scenario. Therefore, it has the lowest performance of the four studied controllers. From another side, the proposed controller has an MU of 0.0334 and 0.0879 in Δf_a, and Δf_b, respectively. In addition, the proposed control has deviations in ΔP_{tie} of 0.0026 in this scenario. Therefore, the best performance is achieved through the proposed 1 + PD/FOPID LFC method. A full measure of system performance during the five scenarios through various metrics is detailed in Table 4.

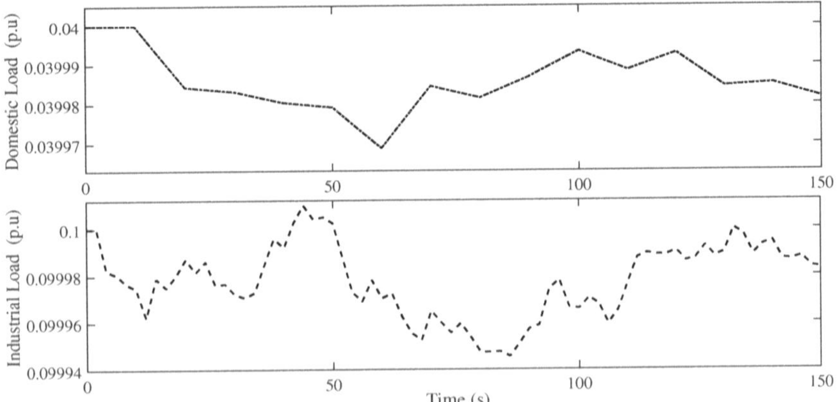

Figure 17. Load profiles at high RES participation in Scenario (5).

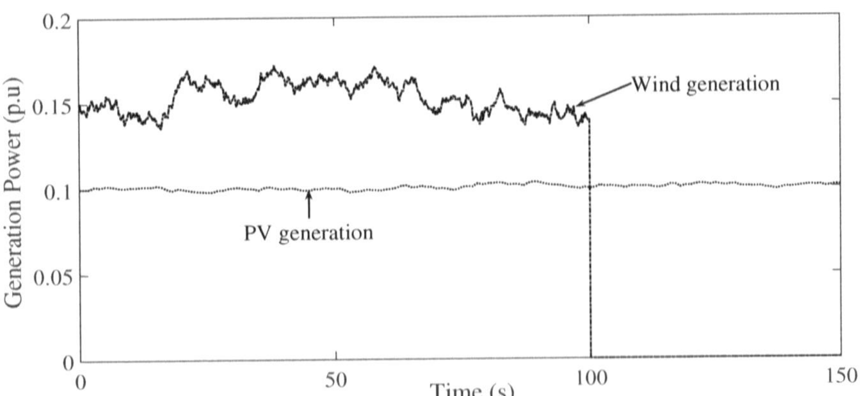

Figure 18. PV and wind generation profiles at high RES participation in Scenario (5).

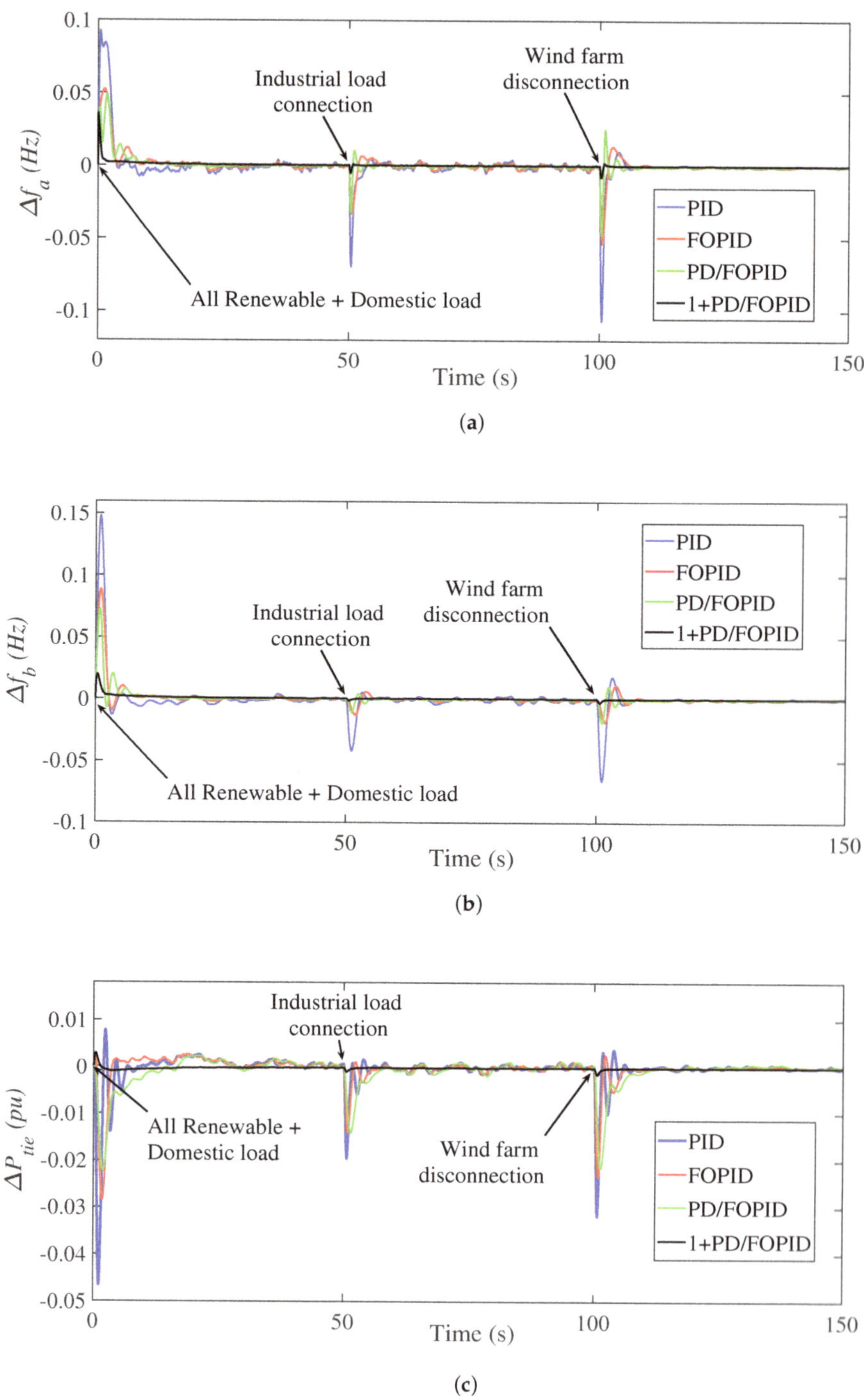

Figure 19. Performance evaluations at high RES participation in Scenario (5): (**a**) Δf_a; (**b**) Δf_b; (**c**) ΔP_{tie}.

Table 4. Measurements of the settling time (ST), peak undershoot (PU), and peak overshoot (PO) for the studied scenarios (where FU stands for a fluctuated condition.

Scenario	Controller	Δf_1			Δf_2			ΔP_{tie}		
		PO	PU	ST (s)	PO	PU	ST (s)	PO	PU	ST (s)
No. 1 at 0 s	PID	0.0008	0.0101	13	0.0011	0.0071	9	0.0006	0.0027	19
	FOPID	0.0015	0.0061	11	0.0014	0.0023	11	0.0001	0.0023	16
	PD/FOPID	0.0002	0.0044	8	-	0.0016	10	0.0004	0.0014	10
	1 + PD/FOPID	0.0001	0.0018	4	-	0.0002	3	-	9.3×10^{-5}	3
No. 2 at 30 s	PID	0.0003	0.0078	>20 s	0.0005	0.0103	>20 s	0.0035	0.0001	>20 s
	FOPID	0.0006	0.0051	19	0.0006	0.0081	>20 s	0.0031	0.0009	>20 s
	PD/FOPID	-	0.0037	22	0.0012	0.0058	19	0.0029	0.0005	20
	1 + PD/FOPID	-	0.0005	7	-	0.0011	5	0.0003	-	6
No. 3 at 30 s	PID	0.1202	0.0111	FU	0.0739	0.0209	FU	0.0264	0.0039	FU
	FOPID	0.0596	0.0155	FU	0.0236	0.0132	FU	0.0243	0.0058	FU
	PD/FOPID	0.0456	0.0174	13	0.0205	0.0043	11	0.0194	0.0012	FU
	1 + PD/FOPID	0.0085	0.0021	7	0.0031	-	5	0.0015	-	5
No. 4	PID	0.0569	0.0534	FU	0.0537	0.0608	FU	0.0179	0.0192	FU
	FOPID	0.0239	0.0272	FU	0.0219	0.0209	FU	0.0139	0.0124	FU
	PD/FOPID	0.0094	0.0088	FU	0.0076	0.0092	FU	0.0094	0.0089	FU
	1 + PD/FOPID	0.0017	0.0015	FU	0.0012	0.0013	FU	0.0007	0.0006	FU
No. 5 at 0 s	PID	0.0927	0.0073	FU	0.1463	0.0127	FU	0.0077	0.0461	FU
	FOPID	0.0527	0.0004	FU	0.0879	0.0092	FU	0.0019	0.0285	FU
	PD/FOPID	0.0471	0.0141	FU	0.0711	0.0065	FU	0.0022	0.0221	FU
	1 + PD/FOPID	0.0334	-	15	0.0191	-	19	-	0.0026	23

6. Conclusions

An improved fractional-order controller based on a cascaded 1 + PD/FOPID control was proposed in this paper with MRFO-based design optimization to regulate the frequency in interconnected electrical grids. The proposed controller is advantageous at mitigating disturbances at a wide range of frequencies due to employing two cascaded control loops. Additionally, the employment of the MRFO and design optimization process leads to simultaneous design and determination of the best control parameters. The proposed 1 + PD/FOPID and MRFO-based design optimization were implemented and simulated in MATLAB. Various scenarios of load power changes and renewable power connection/interconnection scenarios were considered in the performance investigation. In addition, the PID, FOPID and PD/FOPID LFCs were compared with the proposed controller. Results verify the reduced peak overshoot and settling times under the proposed 1 + PD/FOPID LFC compared with the studied controllers. Future research includes the use of more practical modelling of electric power systems (integer and fractional-order modelling), stability analysis with system non-linearities, and comprehensive comparisons of existing LFC methods.

Author Contributions: Conceptualization, F.F.M.E.-S., M.H.A., E.A.M. and M.A.; data curation, A.S.A., S.Z.A., E.A.M. and M.H.A.; formal analysis, F.F.M.E.-S., S.Z.A., A.S.A. and M.A.; funding acquisition, M.H.A., A.S.A., F.F.M.E.-S. and S.Z.A.; investigation, F.F.M.E.-S., M.H.A., E.A.M. and M.A.; methodology, F.F.M.E.-S., M.H.A., E.A.M., A.S.A. and S.Z.A.; project administration, S.Z.A., E.A.M., F.F.M.E.-S. and M.H.A.; resources, F.F.M.E.-S., A.S.A., E.A.M. and M.A.; software, F.F.M.E.-S., M.H.A., A.S.A. and M.A.; supervision, A.S.A., S.Z.A., E.A.M. and M.A.; validation, F.F.M.E.-S., M.H.A.,

E.A.M., M.A. and S.Z.A.; visualization, F.F.M.E.-S., E.A.M., M.H.A. and M.A.; writing—original draft, F.F.M.E.-S., M.H.A., A.S.A., E.A.M. and M.A.; writing—review and editing, F.F.M.E.-S., M.H.A. and M.A. All authors have read and agreed to the published version of the manuscript.

Funding: The authors extend their appreciation to the Deputyship for Research and Innovation, Ministry of Education in Saudi Arabia for funding this research work through the Project number (IF2/PSAU/2022/01/23085).

Data Availability Statement: Not applicable.

Conflicts of Interest: The authors declare no conflict of interest.

Appendix A

Appendix A.1

For power system representation in areas a and b, the representation of Δf_a and Δf_b based on the system representation in Figure 2 are represented as:

$$\Delta f_a = (\Delta P_{ga} + \Delta P_{WT} - \Delta P_{EV_a} - \Delta P_{la} - \Delta P_{tie}) \frac{1}{2H_a s + D_a} \tag{A1}$$

$$\Delta f_b = (\Delta P_{gb} + \Delta P_{PV} - \Delta P_{EV_b} - \Delta P_{lb} + \Delta P_{tie}) \frac{1}{2H_b s + D_b} \tag{A2}$$

By multiplying both parts in (A1) by $2H_a s + D_a$ and also both parts in (A2) by $2H_b s + D_b$, we obtain:

$$2H_a s \Delta f_a + D_a \Delta f_a = \Delta P_{ga} + \Delta P_{WT} - \Delta P_{EV_a} - \Delta P_{la} - \Delta P_{tie} \tag{A3}$$

$$2H_b s \Delta f_b + D_b \Delta f_b = \Delta P_{gb} + \Delta P_{PV} - \Delta P_{EV_b} - \Delta P_{lb} + \Delta P_{tie} \tag{A4}$$

where the term $s\Delta f_a$ represents $\Delta \dot{f}_a$ and $s\Delta f_b$ represents $\Delta \dot{f}_b$. Thus, (A3) and (A4) become:

$$\Delta \dot{f}_a = \frac{-D_a}{2H_a} \Delta f_a + \frac{1}{2H_a} \Delta P_{ga} + \frac{1}{2H_a} \Delta P_{WT} - \frac{1}{2H_a} \Delta P_{EV_a} - \frac{1}{2H_a} \Delta P_{la} - \frac{1}{2H_a} \Delta P_{tie} \tag{A5}$$

$$\Delta \dot{f}_b = \frac{-D_b}{2H_b} \Delta f_b + \frac{1}{2H_b} \Delta P_{gb} + \frac{1}{2H_b} \Delta P_{PV} - \frac{1}{2H_b} \Delta P_{EV_b} - \frac{1}{2H_b} \Delta P_{lb} + \frac{1}{2H_b} \Delta P_{tie} \tag{A6}$$

Appendix A.2

For thermal generation, it is represented by:

$$\Delta P_{ga} = \frac{1}{T_t s + 1} \Delta P_{ga1} \tag{A7}$$

$$\Delta P_{ga1} = \frac{1}{T_g s + 1} \left(ACEo_a - \frac{1}{R_a} \Delta f_a \right) \tag{A8}$$

By simplifying (A7) and (A8), and $s\Delta P_{ga}$ is replaced with $\Delta \dot{P}_{ga}$ and $s\Delta P_{ga1}$ is replaced with $\Delta \dot{P}_{ga1}$, we obtain:

$$\Delta \dot{P}_{ga} = \frac{-1}{T_t} \Delta P_{ga} + \frac{1}{T_t} \Delta P_{ga1} \tag{A9}$$

$$\Delta \dot{P}_{ga1} = \frac{-1}{T_g} \Delta P_{ga1} + \frac{1}{T_g} ACEo_a - \frac{1}{T_g R_a} \Delta f_a \tag{A10}$$

For hydraulic generation, it is represented by:

$$\Delta P_{gb} = \frac{-T_w s + 1}{0.5 T_w s + 1} \Delta P_{gb1} \tag{A11}$$

$$\Delta P_{gb1} = \frac{T_R s + 1}{T_2 s + 1} \Delta P_{gb2} \tag{A12}$$

$$\Delta P_{gb2} = \frac{1}{T_1 s + 1}\left(ACEo_b - \frac{1}{R_b} \Delta f_b \right) \tag{A13}$$

By simplifying (A11)–(A13), and $s\Delta P_{gb}$ is replaced with $\Delta \dot{P}_{gb}$, $s\Delta P_{gb1}$ is replaced with $\Delta \dot{P}_{gb1}$ and $s\Delta P_{gb2}$ is replaced with $\Delta \dot{P}_{gb2}$, we obtain:

$$\Delta \dot{P}_{gb} = \frac{-2}{T_w} \Delta P_{gb} + \frac{2T_2 + 2T_w}{T_2 T_w} \Delta P_{gb1} + \frac{2T_R - 2T_1}{T_1 T_2} \Delta P_{gb2} + \frac{2T_R}{T_1 T_2 R_b} \Delta f_b + \frac{2T_R}{T_1 T_2} ACEo_b \tag{A14}$$

$$\Delta \dot{P}_{gb1} = \frac{-1}{T_2} \Delta P_{gb1} + \frac{T_1 - T_R}{T_1 T_2} \Delta P_{gb2} - \frac{T_R}{T_1 T_2 R_b} \Delta f_b - \frac{T_R}{T_1 T_2} ACEo_b \tag{A15}$$

$$\Delta \dot{P}_{gb2} = \frac{-1}{T_1} \Delta P_{gb2} + \frac{1}{T_1} ACEo_b - \frac{1}{T_1 R_b} \Delta f_b \tag{A16}$$

Appendix A.3

For wind generation, it is represented by:

$$\Delta P_{WT} = \frac{K_{WT}}{T_{WT} s + 1} P_{WT} \tag{A17}$$

By simplifying (A17), and $s\Delta P_{WT}$ is replaced with $\Delta \dot{P}_{WT}$, we obtain:

$$\Delta \dot{P}_{WT} = \frac{-1}{T_{WT}} \Delta P_{WT} + \frac{K_{WT}}{T_{WT}} P_{WT} \tag{A18}$$

For PV generation, it is represented by:

$$\Delta P_{PV} = \frac{K_{PV}}{T_{PV} s + 1} P_{WT} \tag{A19}$$

By simplifying (A19), and $s\Delta P_{PV}$ is replaced with $\Delta \dot{P}_{PV}$, we obtain:

$$\Delta \dot{P}_{PV} = \frac{-1}{T_{PV}} \Delta P_{PV} + \frac{K_{PV}}{T_{PV}} P_{PV} \tag{A20}$$

References

1. Blaabjerg, F.; Yang, Y.; Kim, K.A.; Rodriguez, J. Power Electronics Technology for Large-Scale Renewable Energy Generation. *Proc. IEEE* **2023**, *111*, 335–355. [CrossRef]
2. Biswas, S.; Mahata, S.; Roy, P.K.; Chatterjee, K. Application of Empirical Bode Analysis for Delay-Margin Evaluation of Fractional-Order PI Controller in a Renewable Distributed Hybrid System. *Fractal Fract.* **2023**, *7*, 119. [CrossRef]
3. Sami, I.; Ullah, S.; Khan, L.; Al-Durra, A.; Ro, J.S. Integer and Fractional-Order Sliding Mode Control Schemes in Wind Energy Conversion Systems: Comprehensive Review, Comparison, and Technical Insight. *Fractal Fract.* **2022**, *6*, 447. [CrossRef]
4. Salama, H.S.; Said, S.M.; Aly, M.; Vokony, I.; Hartmann, B. Studying Impacts of Electric Vehicle Functionalities in Wind Energy-Powered Utility Grids With Energy Storage Device. *IEEE Access* **2021**, *9*, 45754–45769. [CrossRef]
5. Alilou, M.; Azami, H.; Oshnoei, A.; Mohammadi-Ivatloo, B.; Teodorescu, R. Fractional-Order Control Techniques for Renewable Energy and Energy-Storage-Integrated Power Systems: A Review. *Fractal Fract.* **2023**, *7*, 391. [CrossRef]
6. Daraz, A.; Malik, S.A.; Basit, A.; Aslam, S.; Zhang, G. Modified FOPID Controller for Frequency Regulation of a Hybrid Interconnected System of Conventional and Renewable Energy Sources. *Fractal Fract.* **2023**, *7*, 89. [CrossRef]
7. Fayek, H.H. 5G Poor and Rich Novel Control Scheme Based Load Frequency Regulation of a Two-Area System with 100% Renewables in Africa. *Fractal Fract.* **2020**, *5*, 2. [CrossRef]

8. Ahmed, E.M.; Mohamed, E.A.; Selim, A.; Aly, M.; Alsadi, A.; Alhosaini, W.; Alnuman, H.; Ramadan, H.A. Improving load frequency control performance in interconnected power systems with a new optimal high degree of freedom cascaded FOTPID-TIDF controller. *Ain Shams Eng. J.* **2023**, *14*, 102207. [CrossRef]

9. Mohamed, E.A.; Aly, M.; Elmelegi, A.; Ahmed, E.M.; Watanabe, M.; Said, S.M. Enhancement the Frequency Stability and Protection of Interconnected Microgrid Systems Using Advanced Hybrid Fractional Order Controller. *IEEE Access* **2022**, *10*, 111936–111961. [CrossRef]

10. Kavikumar, R.; Kwon, O.M.; Lee, S.H.; Sakthivel, R. Input-output finite-time IT2 fuzzy dynamic sliding mode control for fractional-order nonlinear systems. *Nonlinear Dyn.* **2022**, *108*, 3745–3760. [CrossRef]

11. Kavikumar, R.; Sakthivel, R.; Kwon, O.M.; Selvaraj, P. Robust tracking control design for fractional-order interval type-2 fuzzy systems. *Nonlinear Dyn.* **2022**, *107*, 3611–3628. [CrossRef]

12. Gheisarnejad, M.; Khooban, M.H.; Dragicevic, T. The Future 5G Network-Based Secondary Load Frequency Control in Shipboard Microgrids. *IEEE J. Emerg. Sel. Top. Power Electron.* **2020**, *8*, 836–844. [CrossRef]

13. Gheisarnejad, M.; Karimaghaie, P.; Boudjadar, J.; Khooban, M.H. Real-Time Cellular Wireless Sensor Testbed for Frequency Regulation in Smart Grids. *IEEE Sens. J.* **2019**, *19*, 11656–11665. [CrossRef]

14. Ahmed, E.M.; Selim, A.; Alnuman, H.; Alhosaini, W.; Aly, M.; Mohamed, E.A. Modified Frequency Regulator Based on TIλ-TDμFF Controller for Interconnected Microgrids with Incorporating Hybrid Renewable Energy Sources. *Mathematics* **2022**, *11*, 28. [CrossRef]

15. Aly, M.; Mohamed, E.A.; Noman, A.M.; Ahmed, E.M.; El-Sousy, F.F.M.; Watanabe, M. Optimized Non-Integer Load Frequency Control Scheme for Interconnected Microgrids in Remote Areas with High Renewable Energy and Electric Vehicle Penetrations. *Mathematics* **2023**, *11*, 2080. [CrossRef]

16. Dahab, Y.A.; Abubakr, H.; Mohamed, T.H. Adaptive Load Frequency Control of Power Systems Using Electro-Search Optimization Supported by the Balloon Effect. *IEEE Access* **2020**, *8*, 7408–7422. [CrossRef]

17. Ewais, A.M.; Elnoby, A.M.; Mohamed, T.H.; Mahmoud, M.M.; Qudaih, Y.; Hassan, A.M. Adaptive frequency control in smart microgrid using controlled loads supported by real-time implementation. *PLoS ONE* **2023**, *18*, e0283561. [CrossRef]

18. Arora, K.; Kumar, A.; Kamboj, V.K.; Prashar, D.; Shrestha, B.; Joshi, G.P. Impact of Renewable Energy Sources into Multi Area Multi-Source Load Frequency Control of Interrelated Power System. *Mathematics* **2021**, *9*, 186. [CrossRef]

19. Gupta, D.K.; Soni, A.K.; Jha, A.V.; Mishra, S.K.; Appasani, B.; Srinivasulu, A.; Bizon, N.; Thounthong, P. Hybrid Gravitational–Firefly Algorithm-Based Load Frequency Control for Hydrothermal Two-Area System. *Mathematics* **2021**, *9*, 712. [CrossRef]

20. Yousri, D.; Babu, T.S.; Fathy, A. Recent methodology based Harris Hawks optimizer for designing load frequency control incorporated in multi-interconnected renewable energy plants. *Sustain. Energy Grids Netw.* **2020**, *22*, 100352. [CrossRef]

21. Salama, H.S.; Magdy, G.; Bakeer, A.; Vokony, I. Adaptive coordination control strategy of renewable energy sources, hydrogen production unit, and fuel cell for frequency regulation of a hybrid distributed power system. *Prot. Control Mod. Power Syst.* **2022**, *7*, 34. [CrossRef]

22. Shabani, H.; Vahidi, B.; Ebrahimpour, M. A robust PID controller based on imperialist competitive algorithm for load-frequency control of power systems. *ISA Trans.* **2013**, *52*, 88–95. [CrossRef] [PubMed]

23. Youssef, A.R.; Mallah, M.; Ali, A.; Shaaban, M.F.; Mohamed, E.E.M. Enhancement of Microgrid Frequency Stability Based on the Combined Power-to-Hydrogen-to-Power Technology under High Penetration Renewable Units. *Energies* **2023**, *16*, 3377. [CrossRef]

24. Sahu, R.K.; Panda, S.; Rout, U.K.; Sahoo, D.K. Teaching learning based optimization algorithm for automatic generation control of power system using 2-DOF PID controller. *Int. J. Electr. Power Energy Syst.* **2016**, *77*, 287–301. [CrossRef]

25. Kouba, N.E.Y.; Menaa, M.; Hasni, M.; Boudour, M. Optimal load frequency control based on artificial bee colony optimization applied to single, two and multi-area interconnected power systems. In Proceedings of the 2015 3rd International Conference on Control, Engineering Information Technology (CEIT), Tlemcen, Algeria, 25–27 May 2015. [CrossRef]

26. Abid, S.; El-Rifaie, A.M.; Elshahed, M.; Ginidi, A.R.; Shaheen, A.M.; Moustafa, G.; Tolba, M.A. Development of Slime Mold Optimizer with Application for Tuning Cascaded PD-PI Controller to Enhance Frequency Stability in Power Systems. *Mathematics* **2023**, *11*, 1796. [CrossRef]

27. Zhang, G.; Daraz, A.; Khan, I.A.; Basit, A.; Khan, M.I.; Ullah, M. Driver Training Based Optimized Fractional Order PI-PDF Controller for Frequency Stabilization of Diverse Hybrid Power System. *Fractal Fract.* **2023**, *7*, 315. [CrossRef]

28. Singh, B.; Slowik, A.; Bishnoi, S.K.; Sharma, M. Frequency Regulation Strategy of Two-Area Microgrid System with Electric Vehicle Support Using Novel Fuzzy-Based Dual-Stage Controller and Modified Dragonfly Algorithm. *Energies* **2023**, *16*, 3407. [CrossRef]

29. Mohanty, B.; Panda, S.; Hota, P. Controller parameters tuning of differential evolution algorithm and its application to load frequency control of multi-source power system. *Int. J. Electr. Power Energy Syst.* **2014**, *54*, 77–85. [CrossRef]

30. Dash, P.; Saikia, L.C.; Sinha, N. Automatic generation control of multi area thermal system using Bat algorithm optimized PD–PID cascade controller. *Int. J. Electr. Power Energy Syst.* **2015**, *68*, 364–372. [CrossRef]

31. Raju, M.; Saikia, L.C.; Sinha, N. Automatic generation control of a multi-area system using ant lion optimizer algorithm based PID plus second order derivative controller. *Int. J. Electr. Power Energy Syst.* **2016**, *80*, 52–63. [CrossRef]

32. Latif, A.; Hussain, S.M.S.; Das, D.C.; Ustun, T.S. Optimization of Two-Stage IPD-(1I) Controllers for Frequency Regulation of Sustainable Energy Based Hybrid Microgrid Network. *Electronics* **2021**, *10*, 919. [CrossRef]

33. Gheisarnejad, M. An effective hybrid harmony search and cuckoo optimization algorithm based fuzzy PID controller for load frequency control. *Appl. Soft Comput.* **2018**, *65*, 121–138. [CrossRef]
34. Prakash, S.; Sinha, S. Simulation based neuro-fuzzy hybrid intelligent PI control approach in four-area load frequency control of interconnected power system. *Appl. Soft Comput.* **2014**, *23*, 152–164. [CrossRef]
35. Latif, A.; Paul, M.; Das, D.C.; Hussain, S.M.S.; Ustun, T.S. Price Based Demand Response for Optimal Frequency Stabilization in ORC Solar Thermal Based Isolated Hybrid Microgrid under Salp Swarm Technique. *Electronics* **2020**, *9*, 2209. [CrossRef]
36. Hussain, I.; Das, D.C.; Latif, A.; Sinha, N.; Hussain, S.S.; Ustun, T.S. Active power control of autonomous hybrid power system using two degree of freedom PID controller. *Energy Rep.* **2022**, *8*, 973–981. [CrossRef]
37. Bakeer, A.; Magdy, G.; Chub, A.; Jurado, F.; Rihan, M. Optimal Ultra-Local Model Control Integrated with Load Frequency Control of Renewable Energy Sources Based Microgrids. *Energies* **2022**, *15*, 9177. [CrossRef]
38. Yakout, A.H.; AboRas, K.M.; Kotb, H.; Alharbi, M.; Shouran, M.; Samad, B.A. A Novel Ultra Local Based-Fuzzy PIDF Controller for Frequency Regulation of a Hybrid Microgrid System with High Renewable Energy Penetration and Storage Devices. *Processes* **2023**, *11*, 1093. [CrossRef]
39. Ayas, M.S.; Sahin, E. FOPID controller with fractional filter for an automatic voltage regulator. *Comput. Electr. Eng.* **2021**, *90*, 106895. [CrossRef]
40. Fathy, A.; Alharbi, A.G. Recent Approach Based Movable Damped Wave Algorithm for Designing Fractional-Order PID Load Frequency Control Installed in Multi-Interconnected Plants With Renewable Energy. *IEEE Access* **2021**, *9*, 71072–71089. [CrossRef]
41. Latif, A.; Hussain, S.S.; Das, D.C.; Ustun, T.S.; Iqbal, A. A review on fractional order (FO) controllers' optimization for load frequency stabilization in power networks. *Energy Rep.* **2021**, *7*, 4009–4021. [CrossRef]
42. Singh, K.; Amir, M.; Ahmad, F.; Refaat, S.S. Enhancement of Frequency Control for Stand-Alone Multi-Microgrids. *IEEE Access* **2021**, *9*, 79128–79142. [CrossRef]
43. Zaheeruddin; Singh, K. Load frequency regulation by de-loaded tidal turbine power plant units using fractional fuzzy based PID droop controller. *Appl. Soft Comput.* **2020**, *92*, 106338. [CrossRef]
44. Oshnoei, S.; Aghamohammadi, M.; Oshnoei, S.; Oshnoei, A.; Mohammadi-Ivatloo, B. Provision of Frequency Stability of an Islanded Microgrid Using a Novel Virtual Inertia Control and a Fractional Order Cascade Controller. *Energies* **2021**, *14*, 4152. [CrossRef]
45. Peddakapu, K.; Srinivasarao, P.; Mohamed, M.; Arya, Y.; Kishore, D.K. Stabilization of frequency in Multi-Microgrid system using barnacle mating Optimizer-based cascade controllers. *Sustain. Energy Technol. Assess.* **2022**, *54*, 102823. [CrossRef]
46. Arya, Y.; Kumar, N.; Dahiya, P.; Sharma, G.; Çelik, E.; Dhundhara, S.; Sharma, M. Cascade-I λDμN controller design for AGC of thermal and hydro-thermal power systems integrated with renewable energy sources. *IET Renew. Power Gener.* **2021**, *15*, 504–520. [CrossRef]
47. Malik, S.; Suhag, S. A Novel SSA Tuned PI-TDF Control Scheme for Mitigation of Frequency Excursions in Hybrid Power System. *Smart Sci.* **2020**, *8*, 202–218. [CrossRef]
48. Priyadarshani, S.; Subhashini, K.R.; Satapathy, J.K. Pathfinder algorithm optimized fractional order tilt-integral-derivative (FOTID) controller for automatic generation control of multi-source power system. *Microsyst. Technol.* **2020**, *27*, 23–35. [CrossRef]
49. Oshnoei, A.; Khezri, R.; Muyeen, S.M.; Oshnoei, S.; Blaabjerg, F. Automatic Generation Control Incorporating Electric Vehicles. *Electr. Power Compon. Syst.* **2019**, *47*, 720–732. [CrossRef]
50. Sahu, R.K.; Panda, S.; Biswal, A.; Sekhar, G.C. Design and analysis of tilt integral derivative controller with filter for load frequency control of multi-area interconnected power systems. *ISA Trans.* **2016**, *61*, 251–264. [CrossRef]
51. Elkasem, A.H.; Khamies, M.; Hassan, M.H.; Nasrat, L.; Kamel, S. Utilizing controlled plug-in electric vehicles to improve hybrid power grid frequency regulation considering high renewable energy penetration. *Int. J. Electr. Power Energy Syst.* **2023**, *152*, 109251. [CrossRef]
52. Ahmed, M.; Magdy, G.; Khamies, M.; Kamel, S. Modified TID controller for load frequency control of a two-area interconnected diverse-unit power system. *Int. J. Electr. Power Energy Syst.* **2022**, *135*, 107528. [CrossRef]
53. Khokhar, B.; Dahiya, S.; Parmar, K.P.S. A Robust Cascade Controller for Load Frequency Control of a Standalone Microgrid Incorporating Electric Vehicles. *Electr. Power Compon. Syst.* **2020**, *48*, 711–726. [CrossRef]
54. Das, D.C.; Roy, A.; Sinha, N. GA based frequency controller for solar thermal–diesel–wind hybrid energy generation/energy storage system. *Int. J. Electr. Power Energy Syst.* **2012**, *43*, 262–279. [CrossRef]
55. Ray, P.K.; Mohanty, S.R.; Kishor, N. Proportional–integral controller based small-signal analysis of hybrid distributed generation systems. *Energy Convers. Manag.* **2011**, *52*, 1943–1954. [CrossRef]
56. Ahmed, E.M.; Mohamed, E.A.; Elmelegi, A.; Aly, M.; Elbaksawi, O. Optimum Modified Fractional Order Controller for Future Electric Vehicles and Renewable Energy-Based Interconnected Power Systems. *IEEE Access* **2021**, *9*, 29993–30010. [CrossRef]
57. Micev, M.; Ćalasan, M.; Oliva, D. Fractional Order PID Controller Design for an AVR System Using Chaotic Yellow Saddle Goatfish Algorithm. *Mathematics* **2020**, *8*, 1182. [CrossRef]
58. Zhao, W.; Zhang, Z.; Wang, L. Manta ray foraging optimization: An effective bio-inspired optimizer for engineering applications. *Eng. Appl. Artif. Intell.* **2020**, *87*, 103300. [CrossRef]

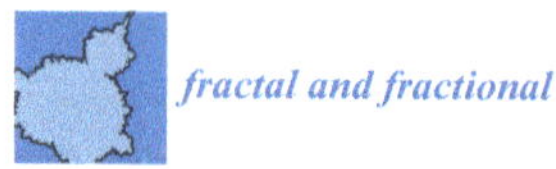
fractal and fractional

Article

A New Intelligent Fractional-Order Load Frequency Control for Interconnected Modern Power Systems with Virtual Inertia Control

Sherif A. Zaid [1,2,*], Abualkasim Bakeer [3], Gaber Magdy [4,5], Hani Albalawi [1,2], Ahmed M. Kassem [6], Mohmed E. El-Shimy [1], Hossam AbdelMeguid [7] and Bassel Manqarah [1]

[1] Electrical Engineering Department, Faculty of Engineering, University of Tabuk, Tabuk 47913, Saudi Arabia
[2] Renewable Energy & Energy Efficiency Centre (REEEC), University of Tabuk, Tabuk 47913, Saudi Arabia
[3] Department of Electrical Engineering, Faculty of Engineering, Aswan University, Aswan 81542, Egypt
[4] Faculty of Engineering, King Salman International University, El-Tor 46511, Egypt
[5] Department of Electrical Engineering, Faculty of Energy Engineering, Aswan University, Aswan 81528, Egypt
[6] Electrical Engineering Department, Faculty of Engineering, Sohag University, Sohag 82524, Egypt
[7] Mechanical Engineering Department, Faculty of Engineering, University of Tabuk, Tabuk 47913, Saudi Arabia
* Correspondence: shfaraj@ut.edu.sa

Citation: Zaid, S.A.; Bakeer, A.; Magdy, G.; Albalawi, H.; Kassem, A.M.; El-Shimy, M.E.; AbdelMeguid, H.; Manqarah, B. A New Intelligent Fractional-Order Load Frequency Control for Interconnected Modern Power Systems with Virtual Inertia Control. *Fractal Fract.* **2023**, *7*, 62. https://doi.org/10.3390/fractalfract7010062

Academic Editors: Behnam Mohammadi-Ivatloo and Arman Oshnoei

Received: 13 December 2022
Revised: 28 December 2022
Accepted: 2 January 2023
Published: 5 January 2023

Abstract: Since modern power systems are susceptible to undesirable frequency oscillations caused by uncertainties in renewable energy sources (RESs) and loads, load frequency control (LFC) has a crucial role to get these systems' frequency stability back. However, existing LFC techniques may not be sufficient to confront the key challenge arising from the low-inertia issue, which is due to the integration of high-penetration RESs. Therefore, to address this issue, this study proposes an optimized intelligent fractional-order integral (iFOI) controller for the LFC of a two-area interconnected modern power system with the implementation of virtual inertia control (VIC). Here, the proposed iFOI controller is optimally designed using an efficient metaheuristic optimization technique, called the gray wolf optimization (GWO) algorithm, which provides minimum values for system frequency deviations and tie-line power deviation. Moreover, the effectiveness of the proposed optimal iFOI controller is confirmed by contrasting its performance with other control techniques utilized in the literature, such as the integral controller and FOI controller, which are also designed in this study, under load/RES fluctuations. Compared to these control techniques from the literature for several scenarios, the simulation results produced by the MATLAB software have demonstrated the efficacy and resilience of the proposed optimal iFOI controller based on the GWO. Additionally, the effectiveness of the proposed controller design in regulating the frequency of interconnected modern power systems with the application of VIC is confirmed.

Keywords: load frequency control; intelligent fractional-order integral (iFOI) control; grey wolf optimization (GWO); renewable energy sources (RESs); interconnected modern power system; virtual inertia control

1. Introduction

Nowadays, renewable energy sources (RESs) are increasingly being integrated into utility power systems. RESs have the advantages of being clean, infinite, and inexpensive energy sources [1,2]. However, RESs can cause some issues and challenges for utility grids. One of these challenges is the reduction in the inertia of interconnected power systems caused by the incorporation of RESs into modern interconnected power systems. Usually, RESs are interfaced with the utility grid via power converters. These converters have been proven to cause an inertia drop in the power system. Consequently, the frequency and voltage stability of the power system are altered. Therefore, the main obstacle to RESs penetration in modern interconnected power systems is the drop in power system inertia [3,4].

To counteract the reduction of system inertia, creating more inertia virtually is one solution to stabilize the frequency and voltage of modern power systems. This phenomenon of virtual inertia is denominated as a virtual inertia control (VIC) or virtual synchronous generator [5]. The concept of VIC in modern power systems can be established using the inverter-based energy storage system (ESS) and the suitable control technique. Therefore, several applications related to the VIC concept have been presented in the Refs. [6–9]. The derivative control approach represents one of the most efficient VIC techniques to enhance the frequency stability of modern power systems by establishing virtual inertia [9]. Additionally, in the framework of the VIC system, the derivative control technique is integrated into the proportional–integral (PI) controller [5], and the H_infinity technique [6,8] has been provided for the frequency stability improvement of modern power systems with high RES penetration.

On the other hand, a considerable drop in system frequency occurs in conjunction with a change in load demand when the power system, in particular, a microgrid operates independently, thus endangering the integrity of such systems. It is possible to restore the power system frequency to the desired value by ensuring a match between generation and demand, which is greatly facilitated by implementing load frequency control (LFC). Therefore, many control techniques have been used for the LFC of modern power systems, such as conventional proportional-integral-derivative (PID) controllers [10], fuzzy logic control [11], model predictive control [12], sliding mode control [13], and others. Recently, control systems engineering has paid great attention to fractional calculus approaches that have excellent facilities for control applications. One of these approaches is the fractional-order (FO) approach. The controller used in that approach is called the FO controller (FOC). A new version of the FOC, called intelligent FOC (iFOC), has just been introduced [14]. The multiresolution analysis introduced robust tuning and an excellent response of the closed-loop systems, which are the main advantages of the iFOC [15]. Tuning the iFOC control parameters is the original challenge of its application. Many tuning techniques have been introduced for tuning iFOC, such as auto-tuning, analytic tuning, numerical tuning, and rule-based tuning [16]. However, the optimal response of the iFOC could be achieved using mathematical optimization algorithms [17].

Many optimization algorithms are proposed to tune the controller parameters [18–20]. Those algorithms have produced good responses. However, their convergence speed, objective functions, and complexity differed. Recently, a metaheuristic optimization technique called Gray Wolf Optimization (GWO) has been introduced [21]. Due to its simplicity, the GWO algorithm has too many applications in power system control and mechatronics [22]. The GWO has great merit over other optimization techniques. It is considered flexible, uses few variables, has a comprehensiveness feature, and has simple programming and implementation. Furthermore, an adaptive fuzzy PID controller based on a modified GWO algorithm has been proposed for the frequency regulation of hybrid power systems [23], where the optimal designed controller has been performed using different optimization techniques to demonstrate the superiority of the GWO algorithm.

Motivated by these aforementioned observations, this study proposes an intelligent fractional-order integral (iFOI) controller for the LFC of a two-area interconnected modern power system with the implementation of VIC, where ultralocal model (ULM) control is incorporated into the LFC based on an FOI controller to enhance the frequency stability of interconnected modern power systems. Moreover, the design of the proposed controller considers high penetration levels of RESs and system nonlinearities. In contrast, previous LFC techniques may not be sufficient to confront the key challenge arising from the low-inertia issue, which is due to the integration of high-penetration RESs. Therefore, this paper proposes an optimized iFOI controller in conjunction with VIC technology to strongly enhance the frequency stability of a two-area interconnected power system considering the high penetration of RESs. Moreover, the optimal proposed controller is fine-tuned using an efficient metaheuristic optimization technique, that is, the GWO algorithm. Furthermore, the effectiveness of the proposed iFOI controller is confirmed by contrasting its performance with other control techniques utilized in the literature, such as the integral (I) controller [1] and the FOI controller, which were designed beside the proposed controller in this study, under load/RESs fluctuations.

The rest of this paper is presented as follows. The modelling and description of the proposed two-area test system are described in Section 2. In Section 3, the proposed iFOI controller-based LFC is explained. Section 3 discusses the optimization approach based on the GWO. Section 4 provides the simulation results. Finally, the research conclusions are presented in Section 5.

2. System Description and Modeling

Figure 1 presents the structure of the investigated two-area interconnected modern power system integrated with the VIC strategy. The two areas have been interconnected through a power transmission line called the tie-line. Each area has a thermal power plant, load, and VIC based on ESS. In addition, photovoltaic (PV) solar and wind farms have been included in areas 1, and 2, respectively. The first area has a solar farm of 8 MW, however, the second area has a wind farm of 6 MW. The system controller deals with the frequency and tie-line power of the two areas. Then, it generates the control signals for each area.

To model the studied system, the low-order models of the individual generation units may be considered suitable for controlling interconnected power systems [1]. The linear model of the studied two-area interconnected power system with the VIC technology is shown in Figure 2. Additionally, the model parameters are presented in Table 1. Each area has a domestic load that is treated as a disturbance in the model. Wind and solar farms are also treated as disturbances. The deviation in the frequency of the proposed interconnected power system of the *j-th* area considering the impact of droop control, LFC, and VIC can be represented as follows:

$$\Delta f_j = \frac{1}{2H_j s + D_j}\left(\Delta P_{mj} + \Delta P_{WTj} + \Delta P_{PVj} + \Delta P_{Inertiaj} - \Delta P_{Lj} - \Delta P_{Tie,jk}\right) \tag{1}$$

where:

$$\Delta P_{mj} = \frac{1}{1 + sT_{tj}}\left(\Delta P_{gj}\right) \tag{2}$$

$$\Delta P_{gj} = \frac{1}{1 + sT_{gj}}\left(\Delta P_{ACE,j} - \frac{1}{R_j}\Delta f_j\right) \tag{3}$$

$$\Delta P_{WTj} = \frac{1}{1 + sT_{WTj}}\left(\Delta P_{Wind,j}\right) \tag{4}$$

$$\Delta P_{PVj} = \frac{1}{1 + sT_{PVj}}\left(\Delta P_{Solar,j}\right) \tag{5}$$

where ΔP_{mj} and ΔP_{gj} denote the power produced in area j by the relevant thermal generation system and turbine system, respectively. The area control error (ACE) is determined as the linear combination of the weighted frequency variation and the tie-line power flow, as follows [1,9].

$$\Delta P_{ACE,j} = \frac{ACE_j}{s} = \frac{K_j}{s}\left(\beta_j f_j\right) + \frac{1}{s}\Delta P_{Tie,j} \tag{6}$$

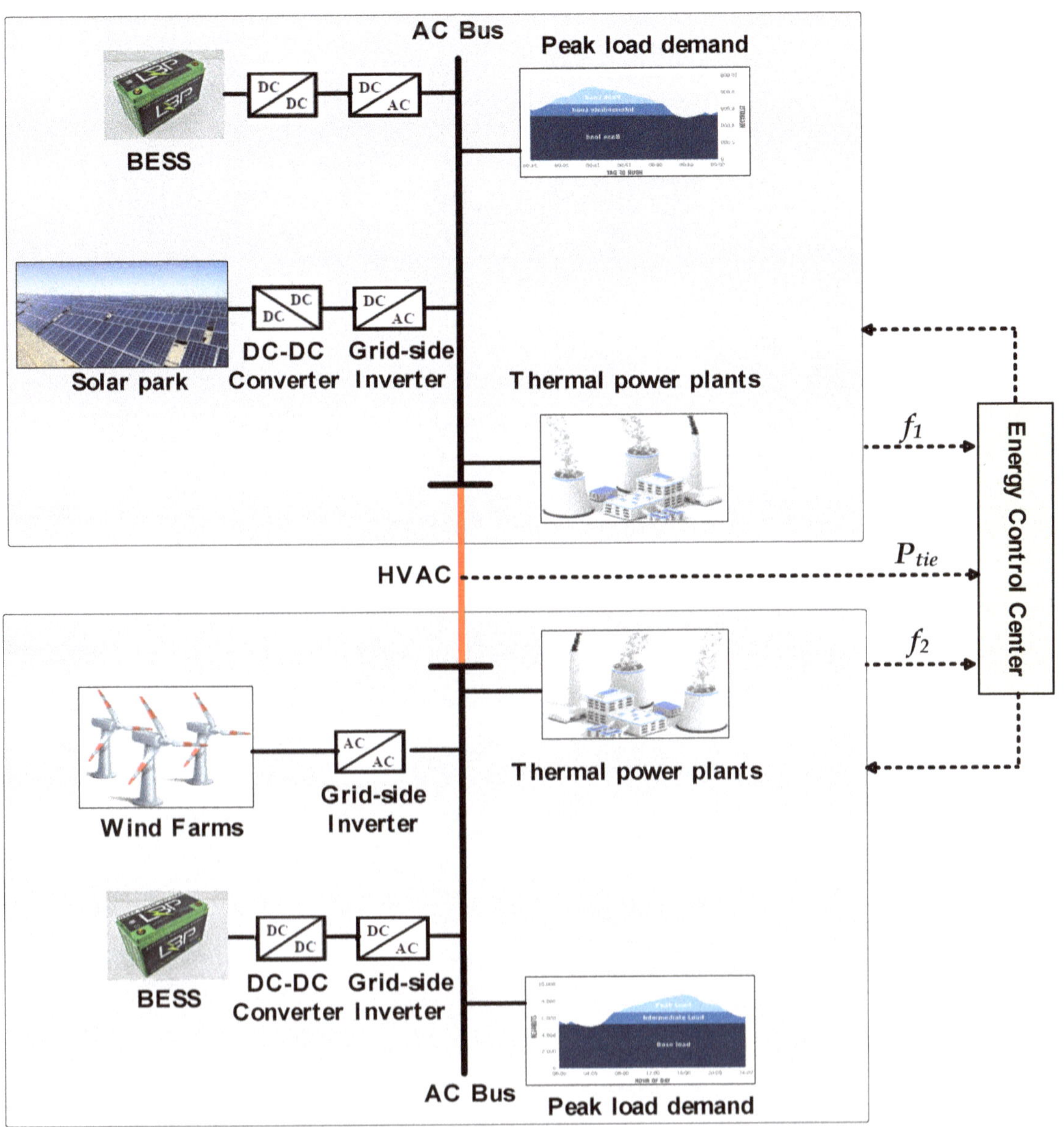

Figure 1. The basic structure of the proposed two-area interconnected power system.

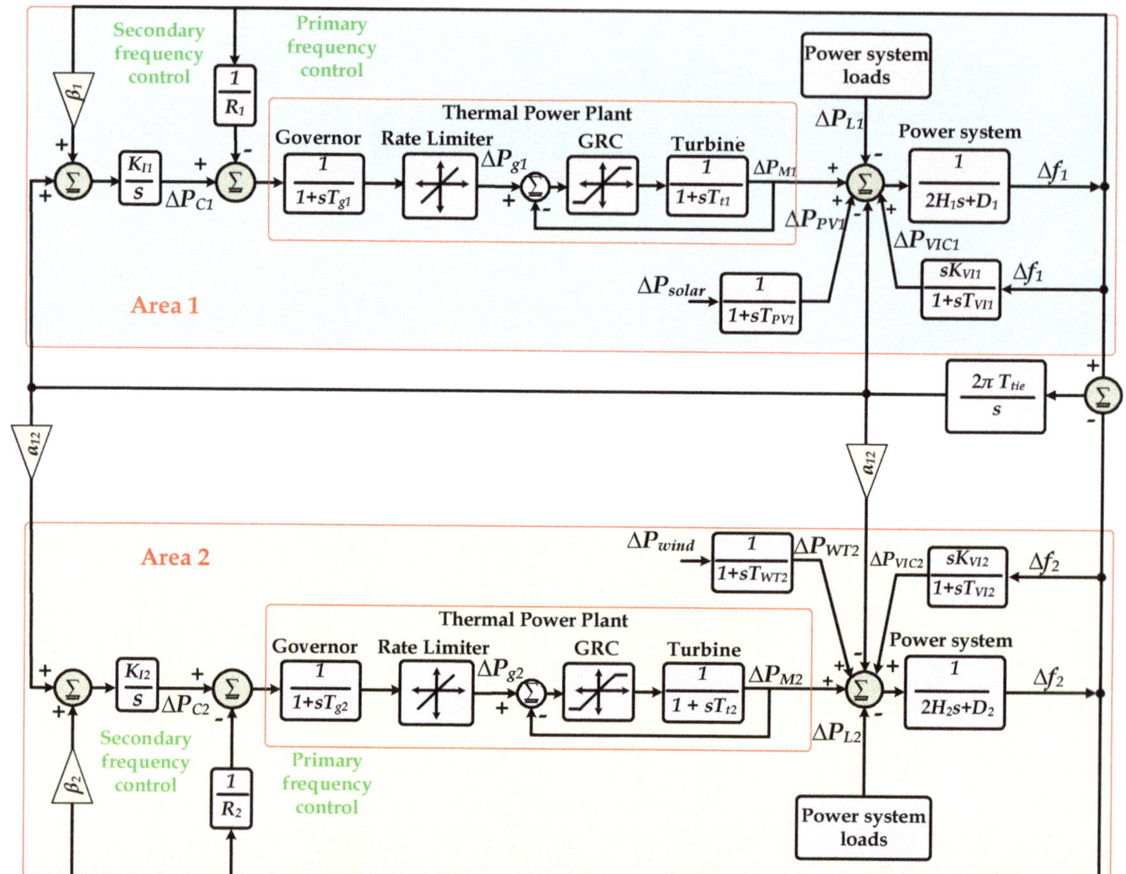

Figure 2. The dynamic model of the studied interconnected two-area power system considering a VIC-based BESS.

Table 1. Parameters of the two-area interconnected system.

Technical Parameter	Value	
	Area1	Area1
System damping coefficient, D (pu)	0.015	0.016
System inertia, H (pu)	0.083	0.101
The time constant of the governor, T_g (s)	0.080	0.060
The time constant of the turbine, T_t (s)	0.400	0.440
Droop constant, R (pu)	3.000	2.730
Integral control variable gain, K_I	0.300	0.200
Frequency bias factor, β (pu MW/Hz)	0.3483	0.3827
The time constant of the PV system, T_{PV} (s)	1.300	-
The time constant of the WT system, T_{WT} (s)	-	1.500
Virtual inertia control gain, K_{VI} (s)	1.540	1.750
Virtual inertia time constant, T_{VI} (s)	10.000	10.000
Nominal system frequency, f (Hz)	50.000	50.000
Synchronizing coefficient between two areas, T_{tie}	0.080	0.080
The capacity ratio between two areas, α_{12}	−0.600	−0.600

The tie-line power deviation between the j-area and the remaining areas can be calculated to create the interconnections among m areas in an interconnected power system,

$$\Delta P_{Tie,j} = \sum_{\substack{k=1 \\ k \neq j}}^{m} \Delta P_{Tie,jk} = \frac{2\pi}{s}\left(\sum_{\substack{k=1 \\ k \neq j}}^{m} T_{jk}\Delta f_j - \sum_{\substack{k=1 \\ k \neq j}}^{m} T_{jk}\Delta f_k \right) \tag{7}$$

where T_{jk} is the coefficient of synchronization between areas and $\Delta P_{Tie,jk}$ denotes the area-j and area-k tie-line power exchange.

The linearized state-space model of the proposed power system may be derived using state variables from (1) to (7):

$$\dot{X} = AX + BU + EW \tag{8}$$

$$Y = CX + DU + FW \tag{9}$$

where:

$$X^T = \begin{bmatrix} \Delta f_1 & \Delta P_{g1} & \Delta p_{m1} & \Delta P_{PV} & \Delta P_{Inertia,1} & \Delta P_{Tie,12} & \Delta f_2 & \Delta P_{g2} & \Delta p_{m2} & \Delta p_{WT} & \Delta P_{Inertia,2} \end{bmatrix}$$

$$U^T = \begin{bmatrix} \Delta P_{ACE,1} & \Delta P_{ACE,2} \end{bmatrix}$$

$$W^T = \begin{bmatrix} \Delta P_{Solar} & \Delta P_{Wind} & \Delta P_{L1} & \Delta P_{L2} \end{bmatrix}$$

$$Y^T = \begin{bmatrix} \Delta f_1 & \Delta f_2 \end{bmatrix}$$

Thus, U is the ACE signal for the considered power system, W is the input perturbations (i.e., loads and RESs) vector, Y is the output, and X is the state vector of the considered system (i.e., frequency deviation for each area). While E stands for the disturbance inputs, B is for the control output signal, D is for the zero vector of the same size as the controlled signal, F is for the zero vector of the same size as the disturbance input vector, and C is for the output of the system. As a result, it is possible to obtain the full matrices of the state space representation of the proposed two-area interconnected power system considering RESs as below:

$$A = \begin{bmatrix}
-\frac{D_1}{2H_1} & 0 & \frac{1}{2H_1} & \frac{1}{2H_1} & \frac{1}{2H_1} & -\frac{1}{2H_1} & 0 & 0 & 0 & 0 & 0 \\
-\frac{1}{R_1 T_{g1}} & -\frac{1}{T_{g1}} & 0 & 0 & 0 & 0 & 0 & 0 & 0 & 0 & 0 \\
0 & \frac{1}{T_{t1}} & -\frac{1}{T_{t1}} & 0 & 0 & 0 & 0 & 0 & 0 & 0 & 0 \\
0 & 0 & 0 & -\frac{1}{T_{PV}} & 0 & 0 & 0 & 0 & 0 & 0 & 0 \\
-\frac{D_1 K_{VI1}}{2H_1 T_{VI1}} & 0 & \frac{K_{VI1}}{2H_1 T_{VI1}} & \frac{K_{VI1}}{2H_1 T_{VI1}} & \left(\frac{K_{VI1}}{2H_1 T_{VI1}} - \frac{1}{T_{VI1}}\right) & -\frac{K_{VI1}}{2H_1 T_{VI1}} & 0 & 0 & 0 & 0 & 0 \\
2\pi T_{12} & 0 & 0 & 0 & 0 & 0 & -2\pi T_{12} & 0 & 0 & 0 & 0 \\
0 & 0 & 0 & 0 & 0 & -\frac{\alpha_{12}}{2H_1} & -\frac{D_2}{2H_2} & 0 & \frac{1}{2H_2} & \frac{1}{2H_2} & \frac{1}{2H_2} \\
0 & 0 & 0 & 0 & 0 & 0 & -\frac{1}{R_2 T_{g2}} & -\frac{1}{T_{g2}} & 0 & 0 & 0 \\
0 & 0 & 0 & 0 & 0 & 0 & 0 & \frac{1}{T_{t2}} & -\frac{1}{T_{t2}} & 0 & 0 \\
0 & 0 & 0 & 0 & 0 & 0 & 0 & 0 & 0 & -\frac{1}{T_{WT}} & 0 \\
0 & 0 & 0 & 0 & 0 & -\frac{\alpha_{12} K_{VI2}}{2H_2 T_{VI2}} & -\frac{D_2 K_{VI2}}{2H_2 T_{VI2}} & 0 & \frac{K_{VI2}}{2H_2 T_{VI2}} & \frac{K_{VI2}}{2H_2 T_{VI2}} & \left(\frac{K_{VI2}}{2H_2 T_{VI2}} - \frac{1}{T_{VI2}}\right)
\end{bmatrix}$$

$$B = \begin{bmatrix}
0 & 0 \\
\frac{1}{T_{g1}} & 0 \\
0 & 0 \\
0 & 0 \\
0 & 0 \\
0 & 0 \\
0 & 0 \\
0 & \frac{1}{T_{g2}} \\
0 & 0 \\
0 & 0 \\
0 & 0
\end{bmatrix}
\qquad
E = \begin{bmatrix}
0 & 0 & -\frac{1}{2H_1} & 0 \\
0 & 0 & 0 & 0 \\
0 & 0 & 0 & 0 \\
\frac{1}{T_{PV}} & 0 & 0 & 0 \\
0 & 0 & -\frac{K_{VI1}}{2H_1 T_{VI1}} & 0 \\
0 & 0 & 0 & 0 \\
0 & 0 & 0 & -\frac{1}{2H_2} \\
0 & 0 & 0 & 0 \\
0 & 0 & 0 & 0 \\
0 & \frac{1}{T_{WT}} & 0 & 0 \\
0 & 0 & 0 & -\frac{K_{VI1}}{2H_1 T_{VI1}}
\end{bmatrix}$$

$$C = \begin{bmatrix} 1 & 0 & 0 & 0 & 0 & 0 & 1 & 0 & 0 & 0 & 0 \end{bmatrix}$$

3. Proposed Intelligent Fractional-Order Integral Controller for LFC

3.1. Fractional-Order Controller

With the use of fractional operators in the controller, any real number can be represented as a general differential or integral notation [24]. The basic mathematical relation of the FO differentiator may be seen as a common form of differential or integral operators, and this can be expressed as follows:

$$D_{l,u}^{q} = \begin{cases} \frac{d^d}{dt^q} & q > 0 \\ 1 & q = 0 \\ \int_{l}^{u} (d\tau)^{-q} & q < 0 \end{cases} \tag{10}$$

where q is the FO operator, and u and l denote the upper and lower bands, respectively. Two theories can be used to define the FO principle, and one of them is Riemann–Liouville (RL), which is the first method used to determine the order derivative of a function $f(t)$, as given below [25,26]:

$$D_{l,u}^{q} f(t) = \frac{1}{\Gamma(n-q)} \left(\frac{d}{dt} \right)^{n} \int_{l}^{u} \frac{f^n(\tau)}{(t-\tau)^{q-n+1}} d\tau \tag{11}$$

where $\Gamma(z) = \int_{0}^{\infty} t^{z-1} e^{-t} dt$, $\Re(z) > 0$ is the function of Gamma and the variable q is limited as $n - 1 < q < n$.

After a fractional derivative of RL is found in (11), it can be transformed by the Laplace method as shown in (12) [27]. Additionally, there is the definition of Caputo as a second definition associated with the concept of FO by which we can express the time-domain representation of the a order of the function $f(t)$ as in (13).

$$\mathcal{L}\{D_0^a f(t)\} = s^a F(s) - \sum_{y=0}^{n-1} s^y \left(D_0^{a-y-1} f(t) \right) \Big|_{t=0} \tag{12}$$

$$D_{t_0}^{t} f(t) = \begin{cases} \frac{1}{\Gamma(n-a)} \left(\int_{t_0}^{t} \frac{f^n(\tau)}{(t-\tau)^{1-(n-a)}} d\tau \right) & n - 1 < a < n \\ \left(\frac{d}{dt} \right)^{n} f(t) & a = n \end{cases} \tag{13}$$

A physical meaning for the integral order can be understood by looking at (14), which is a Laplace transformation of (13). Therefore, the integral order has an initial condition that indicates its physical meaning.

$$\mathcal{L}\{D_0^a f(t)\} = s^a F(s) - \sum_{k=0}^{n-1} s^{a-k-1} f^{(k)}(0) \tag{14}$$

A FOPID controller consists of five parameters, including two additional parameters of integral and fractional order, as shown in Figure 3, which is given in terms of the integral and fractional order of the parameters. The complete transfer function of the FOPID is given in (15). It has been proven that these parameters can increase the controller's transient time, stability, and steady-state error compared to a traditional PID. Additionally, it provides the controller with more flexibility, allowing it to cope with a wide range of system disturbances:

$$G_c(s) = k_p + k_i \left(\frac{1}{s} \right)^{\lambda} + k_d s^{\mu} \tag{15}$$

where λ and μ are frequently in the [0, 1] range.

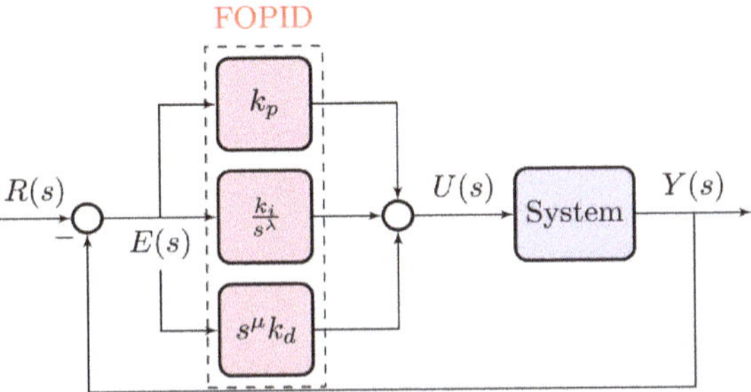

Figure 3. The basic structure of the FOPID controller.

3.2. Ultra-Local Model Control

The changes in the single-input single-output (SISO) system may be approximated with the help of a differential equation with finite dimensionality as in Refs. [28,29]:

$$E\left(t, y, y^{(1)}, \ldots, y^{(n)}, u, u^{(1)}, \ldots, u^{(m)}\right) = 0 \tag{16}$$

where E represents a non-linear function, u indicates the system's input, y indicates the system's output, and (n, m) specify the control orders of the system's output and input, respectively. Figure 4 shows the basic structure of the ULM in general, where F is the ULM total uncertainties and disturbance of the system. Measurement of the system output and prior control input determines the unknown function F. In addition, it is possible to simplify (16) using the ULM principle as follows:

$$y^{(n)} = F + \alpha u \tag{17}$$

where $y^{(n)}$ denotes the nth derivative of y (i.e., the practitioner typically selects n equal to 1 or 2, with 1 being the case in all real-world situations [29]), and $\alpha \in R$ represents a non-physical parameter.

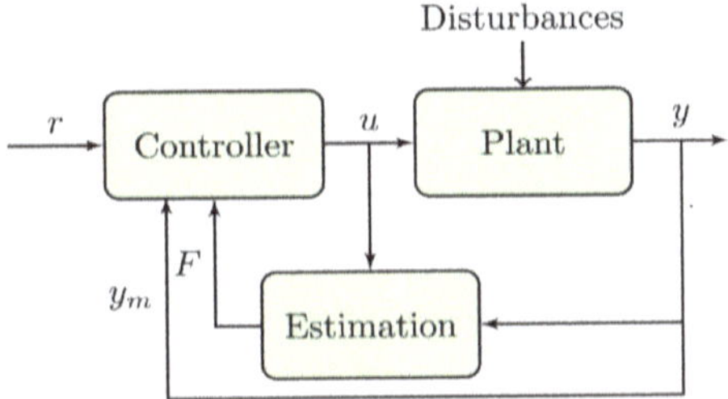

Figure 4. The structure of the ultra-local model.

Since the value of F is only an estimate and lacks precision, it can be substituted with the letter $\hat{F}$ to represent the estimated value when using identification techniques [30]:

$$\hat{F}(t) = -\frac{3!}{L^3} \int_{t-L}^{t} ((L - 2\sigma)y(\sigma) + \alpha\sigma(L - \sigma)u(\sigma))d\sigma \tag{18}$$

where L has a small value based on the noise intensity and the sampling period T_s. Finally, the Heun method can be used to find the value of $\hat{F}$ as in (12), where N_f is the window length.

$$\hat{F} = -\frac{3}{N_f{}^3 T_s} \sum_{i=1}^{N_f} (A + B) \tag{19}$$

where

$$A = \left(N_f - 2(i-1)\right)y(k-1) + \left(N_f - 2i\right)y(k)$$

$$B = \left(\alpha(i-1)T_s\left(N_f - (i-1)\right)\right)u(k-1) + \alpha i T_s\left(N_f - i\right)u(k)$$

3.3. Proposed iFOI for LFC

The secondary controller's goal is to keep the system's frequency at its nominal value within a few minutes after a disturbance. As a result of the great stability and robustness of the FO-based I controller, it has been widely applied for power system stability, whereas most studies have focused on controlling the LFC of power systems [31]. In the current work, Figure 5 indicates the ULM part is incorporated into the FOI controller to reduce the frequency deviation Δf. Hence, the overall structure of the controller may be named an intelligent integral (i.e., iFOI) because it avoids the need for a model to be applied to the controller. This control strategy makes it possible to get rid of the parameter mismatches that cause the power system to malfunction. The incremental power of each area ΔP_{ci}, which represents the control input of the proposed controller, is formulated as in (20) where s is the Laplace operator.

$$\Delta P_{ci} = \left(\frac{\hat{F}_i}{\alpha_i} + k_{ii}\left(\frac{1}{s}\right)^{\lambda_i}\right)\Delta f_i \tag{20}$$

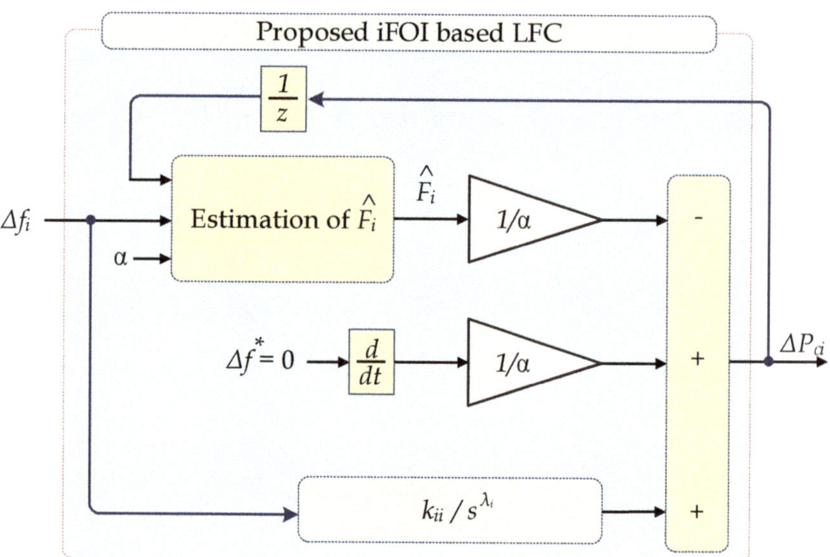

Figure 5. Structure of the proposed iFOI for the secondary controller for each area.

The trial-and-error methodology can be used to define the iFOI parameter, making the tuning process challenging and practitioner-dependent. It is also challenging to find the ideal parameter value for the proposed iFOI, which enhances system performance and ensures the system's stability against disturbances using these trial-and-error methods, and thus the robustness of the system goes down. Therefore, in this paper, to find the best value for the parameter of the iFOI controller of each area, a metaheuristic optimization approach based on GWO is employed, and the complete tuning process of the entire system is shown in Figure 6. The main flowchart of the GWO approach is sketched in Figure 7. The maximum iteration of the GWO is set at 100 and the number of grey wolves is 50.

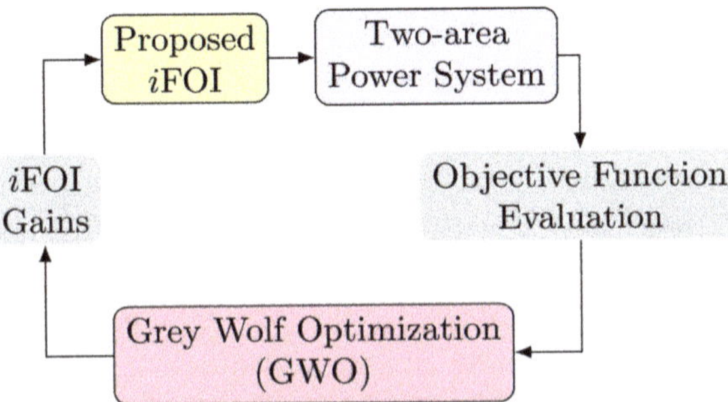

Figure 6. Tuning process of the proposed iFOI.

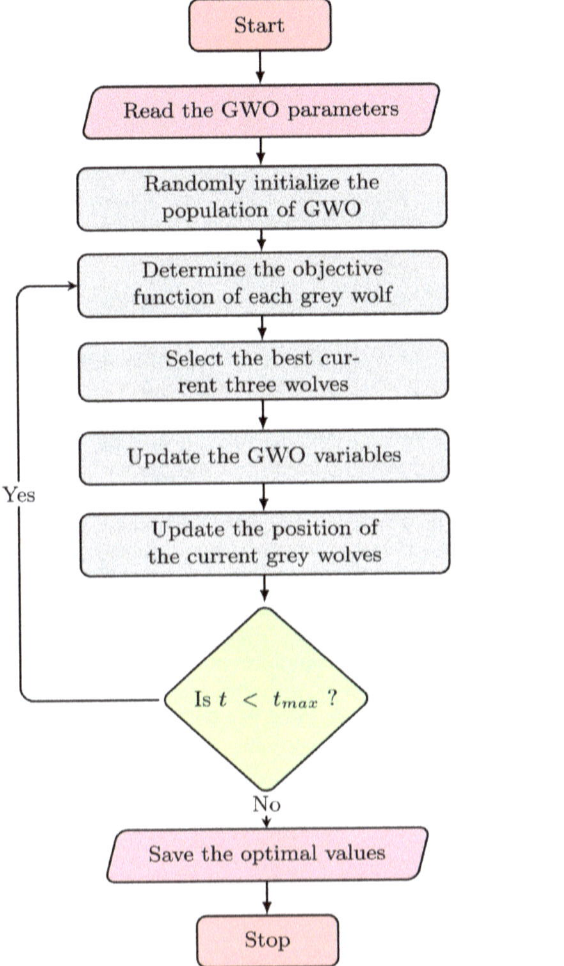

Figure 7. Flowchart of the metaheuristic GWO algorithm.

4. Simulation Results

The proposed two-area interconnected power system with the VIC technique and the proposed iFOI controller, presented in Figure 2, is simulated using the Matlab/Simulink platform with the technical parameters shown in Table 1. However, the optimal gains of the FOI and the proposed iFOI for each area are summarized in Table 2. The results have been extracted into four scenarios or cases. The system has a step-change in load with a uniform RESs profile in the first scenario. The second scenario considers a step load change with a random RES profile. The system has a random load with a uniform RES profile in the third case. The fourth case provides that the system has a random load change with a random RESs profile. Three controllers are compared for each scenario. The compared three controllers are the proposed iFOI, the FOI which is designed beside the proposed controller in this study, and the conventional I controller [1]. The detailed results and discussion are found in the next few paragraphs.

Table 2. Summary of the optimal values of the tested controllers based on GWO.

Controller	Parameter	Value	
		Area1	**Area1**
FOI	λ	0.808	0.805
	k_i	-1.367	-1.739
Proposed iFOI	λ	0.663	0.734
	k_i	-0.609	-0.574
	α	4.020	4.870

4.1. Case 1: Fixed Load Step Change and Uniform RESs Profile

The loads' profiles for the two areas in this scenario are shown in Figure 8a. Each load has a 0.5 pu step change, but it happens at different times. Figure 8b,c shows the generation profiles of the PV and wind energy sources, respectively. They have uniform step variations at different times. The peak power output from the solar panels in Area 1 is 0.125 pu, while the peak power output from the wind turbines in area 2 is 0.1 pu. Figure 8d,e shows the frequency deviation in the two areas under different controllers. The conventional integral controller has the worst response with high overshoots and slow responses. However, the proposed iFOI has the best responses in these two areas. Additionally, the frequency deviations remained within the recommended ranges when the load change is rather minimal. Additionally, the iFOI has the best performance over the other controllers. The change in BESS power is presented in Figure 8f,g, which is controlled based on the virtual inertia control. The variation in the tie-line power exchange between the two areas is shown in Figure 8h. The control output of the LFC controller of each area, which represents the incremental power to the governor, is shown in Figure 8i,j. It is clear that ΔP_{Ci} with the proposed iFOI follows the dynamic of the load and RES changes with minimum oscillations. The overall response of the interconnected system performs best with the proposed iFOI.

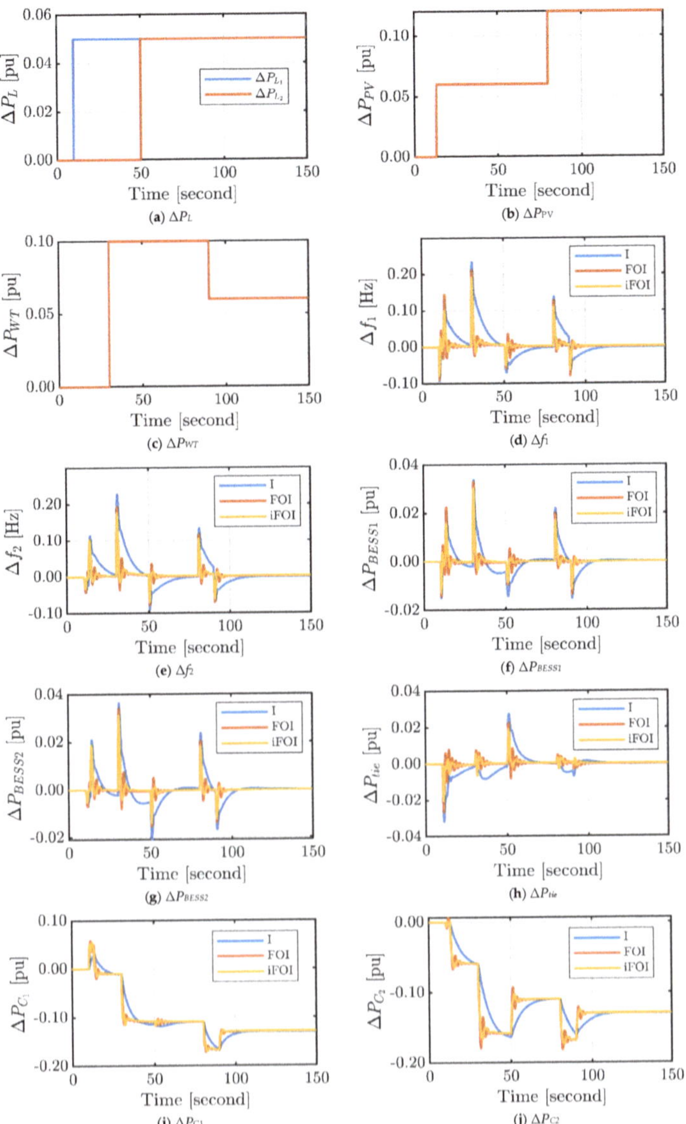

Figure 8. Dynamic response of the two-area power system in case 1.

4.2. Case 2: Fixed Load Step Change and Random RESs Profile

In this case, the loads' profiles, for the two areas, are shown in Figure 9a. The loads have 0.1 pu and 0.05 pu steps, respectively, but at different times. The considered loads are uniform over time. Additionally, the generation profiles of the PV and wind energy sources are presented in Figure 9b,c, respectively. The PV power has a random variation limited to $0.02 \text{ pu} < \Delta P_{PV} < 0.12 \text{ pu}$. In the other area, the wind power has a random variation limited to $0.03 \text{ pu} < \Delta P_{WT} < 0.2 \text{ pu}$. Figure 8d,e shows the frequency deviation in the two areas under different controllers. The conventional integral controller has the worst response of high overshoots and slow response. Nevertheless, the iFOI has the best responses in the two areas where it provides the minimum variations in the Δf of each area compared to other controllers. Additionally, the iFOI has the best performance over the other controllers. The change in the power of the BESS is presented in Figure 9f,g. The variation in the tie-line power exchange between the two areas is shown in Figure 9h. Moreover, the control output of the LFC controller of each area is shown in Figure 9i,j. With both FO controllers, ΔP_{Ci} has the same dynamics of changes similar to the RES changes. The overall response of the interconnected system performs best with the proposed iFOI.

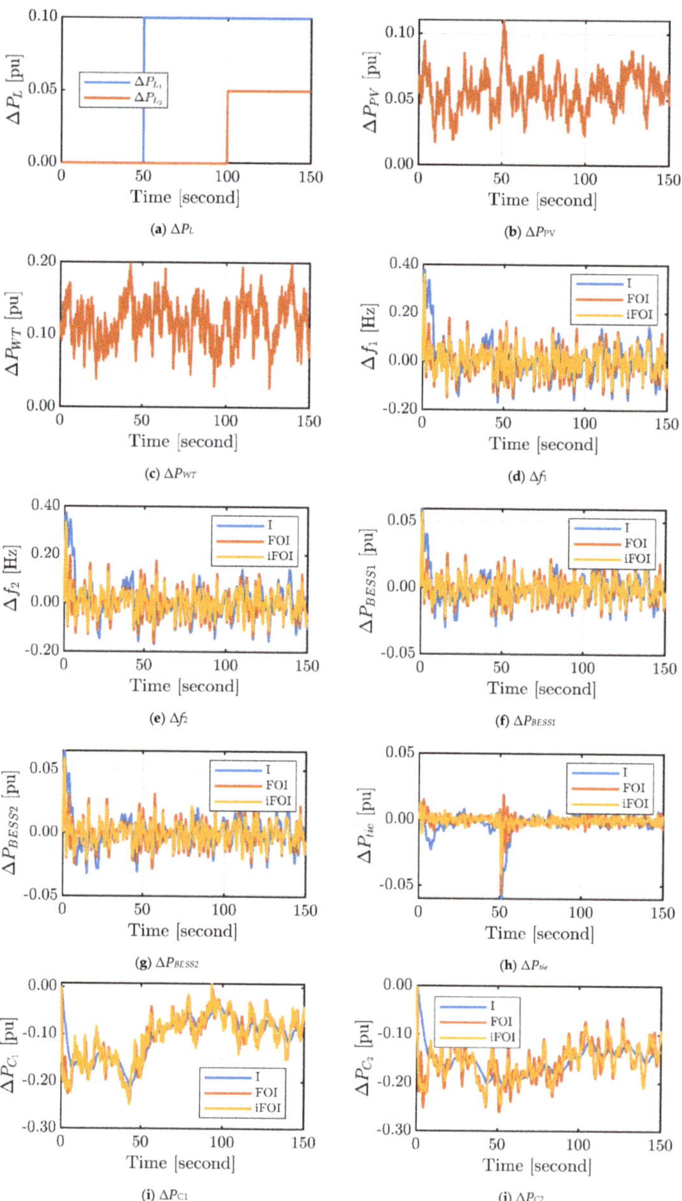

Figure 9. Dynamic response of the two-area power system in case 2.

4.3. Case 3: Random Load and Uniform RESs

In this case, the load profiles for the two areas are random, as shown in Figure 10a. The first area load variations have been limited to 0.18 pu $< \Delta P_{L1} < 0.23$ pu. However, the second area load variations have been limited to 0.05 pu $< \Delta P_{L2} < 0.13$ pu. Additionally, the generation profiles of the PV and wind energy sources are present in Figure 10b,c, respectively. They have step variations at different times. Figure 10d,e shows the frequency deviation in the two areas under different controllers. The conventional integral controller has the worst response of high overshoots and a slow response. It is clear that there is a frequency deviation at a time instant of zero as the load of area 1 and area 2 starts with changes of 0.21 pu and 0.11 pu, respectively. Nevertheless, the iFOI has the best responses in the two areas. Additionally, the iFOI has the best performance over the other controllers. The change in the power of the BESS is presented in Figure 10f,g. The variation in the tie-line power exchange between the two areas is shown in Figure 10h. The LFC controller of each area produces output signals, Figure 10i,j, that indicate the amount of power change that the area is supplying to maintain the system frequency at the nominal value. The overall response of the interconnected system performs best with the proposed iFOI.

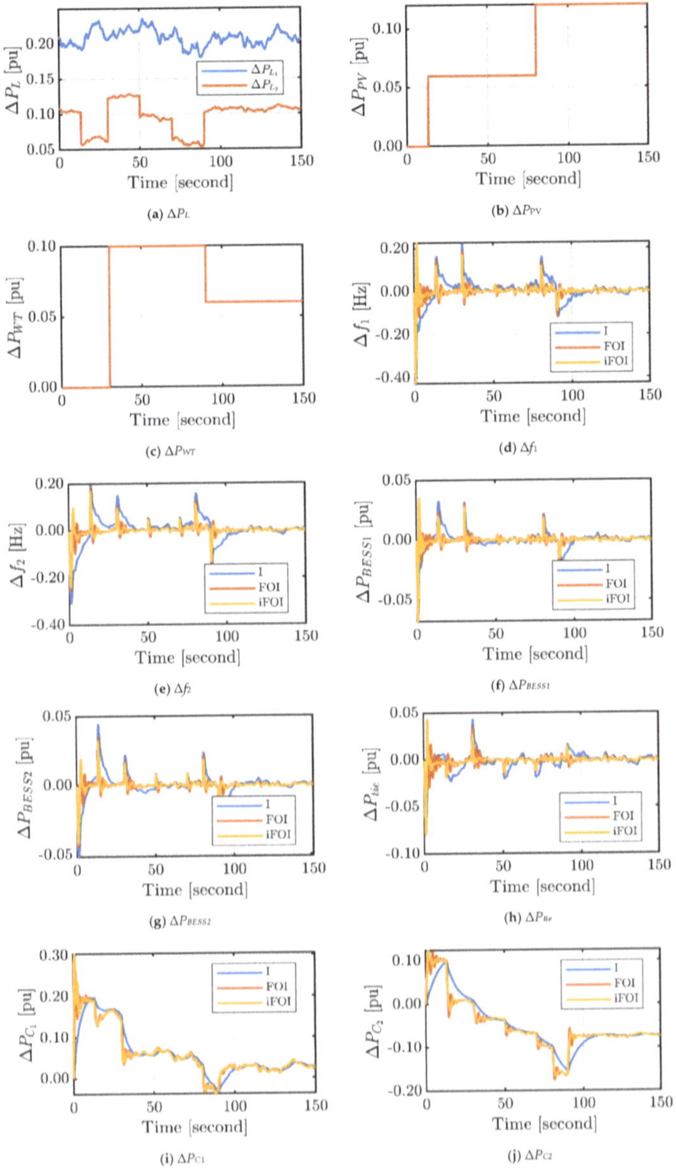

Figure 10. Dynamic response of the two-area power system in case 3.

4.4. Case 4: Random Load and Random RESs

In this case, the load profiles for the two areas are random, as shown in Figure 11a. The first area load variations have been limited to 0.18 pu $< \Delta P_{L1} <$ 0.23 pu. However, the second area load variations have been limited to 0.05 pu $< \Delta P_{L2} <$ 0.13 pu. Additionally, the generation profiles of the PV and wind energy sources are present in Figure 11b,c. The PV power has a random variation limited to 0.02 pu $< \Delta P_{PV} <$ 0.12 pu. In the other area, wind power has a random variation limited to 0.03 pu $< \Delta P_{WT} <$ 0.2 pu. Figure 11d,e shows the frequency deviation in the two areas under different controllers. The conventional integral controller has the worst response of high overshoots and slow response. Nevertheless, the iFOI has the best responses in the two areas. Additionally, the frequency deviations were kept within the recommended standards. Additionally, the iFOI has the best performance over the other controllers. The change in BESS power is presented in Figure 11f,g. The variation in the tie-line power exchange between the two areas is shown in Figure 10h. Furthermore, the control output of the LFC controller for each area is shown in Figure 11i,j. The power change of the controller's control output is different because each area has different loading and RES penetration levels. The overall response of the interconnected system performs best with the proposed iFOI.

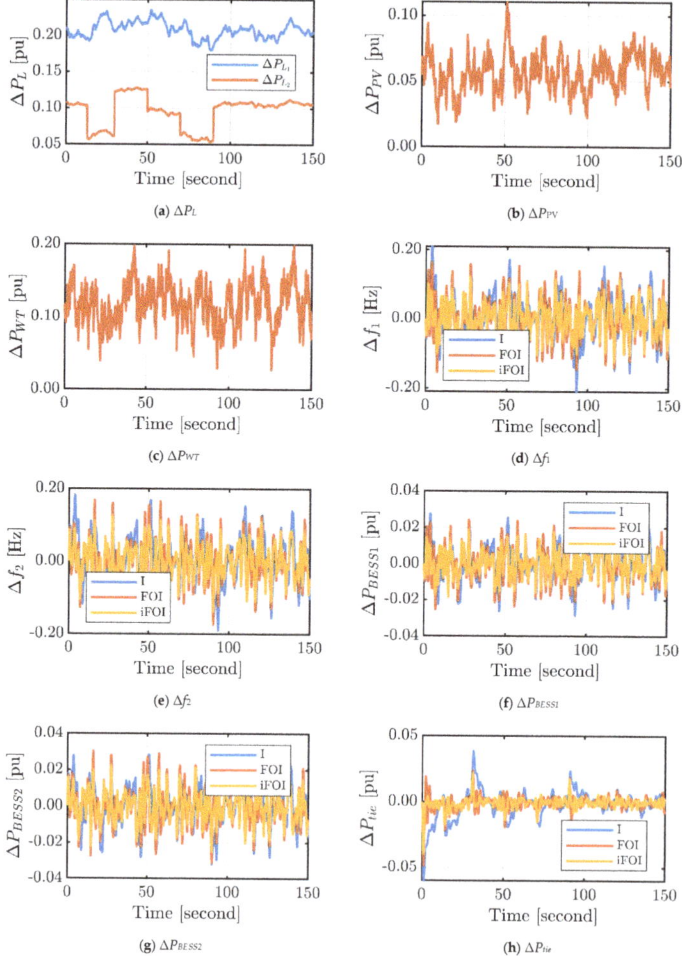

(a) ΔP_L

(b) ΔP_{PV}

(c) ΔP_{WT}

(d) Δf_1

(e) Δf_2

(f) ΔP_{BESS1}

(g) ΔP_{BESS2}

(h) ΔP_{tie}

Figure 11. *Cont.*

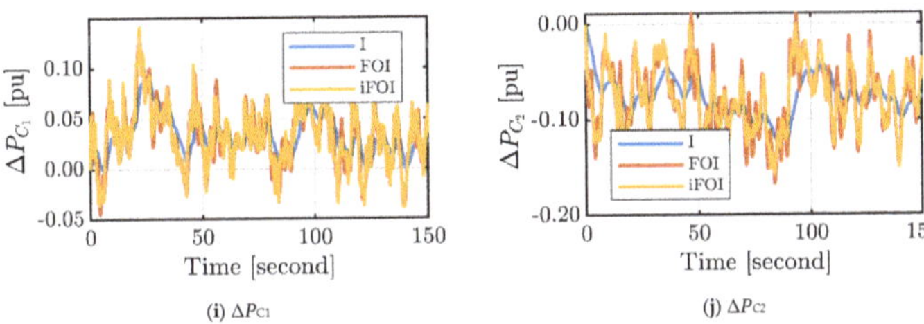

Figure 11. Dynamic response of the two-area power system in case 4.

4.5. Objective Function Analysis for the Previous Four Cases

The objective function based on the integral square error (ISE) of each controller in each studied case is depicted in Figure 12. It is obvious that the proposed iFOI has the minimum value of the objective function compared to the other two controllers in each case. Additionally, case 2 has the highest record of ISE for the three controllers as it has a random RES profile beside the uniform loading in each area. In this case, the proposed iFOI reduces the ISE objective function by 33.67% and 59.18% compared to the LFC according to the FOI controller and the integral controller, respectively.

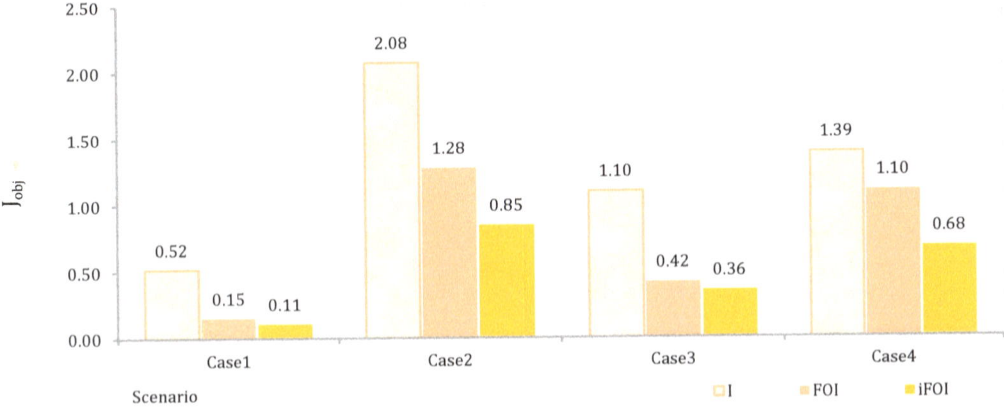

Figure 12. Evaluation of the objective function of each controller for each studied case.

5. Conclusions

This study proposed an optimized iFOI controller for the LFC of a two-area interconnected modern power system with the implementation of virtual inertia control. Here, the proposed iFOI controller is optimally designed using the gray wolf optimization, which is known as an efficient metaheuristic optimization technique and provides minimum frequency deviations and tie-line power deviation. The effectiveness of the proposed optimal iFOI controller is confirmed by contrasting its performance with other control techniques utilized in the literature, such as the integral controller and the FOI controller. The performance of the two-area power system has been tested under load/RESs fluctuations. Compared to these control techniques from the literature for several scenarios, the simulation results produced by the MATLAB software demonstrated the efficacy and resilience of the proposed optimal iFOI controller based on the GWO. Nevertheless, the iFOI had the best responses in the two areas. Additionally, the frequency deviations were kept within the recommended standards. The results from all scenarios show that the overall response of the interconnected system performs best with the proposed iFOI, compared to the integral controller and the FOI controller. On the other hand, the objective functions for the controllers used with the studied scenarios have been analyzed based on the basis of the integral square error. It is noticed that the proposed iFOI has the minimum value of the objective function compared to the other two controllers in each scenario.

Author Contributions: A.B., G.M. and S.A.Z. conceived and designed the system, derived the model, and analyzed the results. H.A. (Hossam AbdelMeguid), M.E.E.-S., B.M. and A.M.K. helped with collecting the funding, resources validation, and visualization, and H.A. (Hani Albalawi) supported the manuscript proofreading and writing. All authors have read and agreed to the published version of the manuscript.

Funding: This research was funded by the University of Tabuk, Grant Number S-1443-0018 at https://www.ut.edu.sa/web/deanship-of-scientific-research/home (accessed on 21 August 2022).

Institutional Review Board Statement: Not applicable.

Informed Consent Statement: Informed consent was obtained from all individual participants included in the study.

Data Availability Statement: Data are available from the authors upon reasonable request.

Acknowledgments: The authors extend their appreciation to the Deanship of Scientific Research at the University of Tabuk for funding this work through research no. S-1443-0018.

Conflicts of Interest: The authors declare no conflict of interest.

References

1. Kerdphol, T.; Rahman, F.S.; Mitani, Y. Virtual Inertia Control Application to Enhance Frequency Stability of Interconnected Power Systems with High Renewable Energy Penetration. *Energies* **2018**, *11*, 981. [CrossRef]
2. El-Hendawi, M.; Gabbar, H.; El-Saady, G.; Ibrahim, E.-N. Control and EMS of a Grid-Connected Microgrid with Economical Analysis. *Energies* **2018**, *11*, 129. [CrossRef]
3. Deepak, M.; Abraham, R.J.; Gonzalez-Longatt, F.M.; Greenwood, D.M.; Rajamani, H.S. A novel approach to frequency support in a wind integrated power system. *Renew. Energy* **2017**, *108*, 194–206. [CrossRef]
4. Magdy, G.; Bakeer, A.; Nour, M.; Petlenkov, E. A New Virtual Synchronous Generator Design Based on the SMES System for Frequency Stability of Low-Inertia Power Grids. *Energies* **2020**, *13*, 5641. [CrossRef]
5. Magdy, G.; Shabib, G.; Elbaset, A.A.; Mitani, Y. Renewable power systems dynamic security using a new coordination of frequency control strategy based on virtual synchronous generator and digital frequency protection. *Int. J. Electr. Power Energy Syst.* **2019**, *109*, 351–368. [CrossRef]
6. Kerdphol, T.; Rahman, F.S.; Mitani, Y.; Watanabe, M.; Küfeoğlu, S.K. Robust Virtual Inertia Control of an Islanded Microgrid Considering High Penetration of Renewable Energy. *IEEE Access* **2018**, *6*, 625–636. [CrossRef]
7. Magdy, G.; Bakeer, A.; Alhasheem, M. Superconducting energy storage technology-based synthetic inertia system control to enhance frequency dynamic performance in microgrids with high renewable penetration. *Prot. Control Mod. Power Syst.* **2021**, *6*, 36. [CrossRef]
8. Fathi, A.; Shafiee, Q.; Bevrani, H. Robust Frequency Control of Microgrids Using an Extended Virtual Synchronous Generator. *IEEE Trans. Power Syst.* **2018**, *33*, 6289–6297. [CrossRef]
9. Magdy, G.; Ali, H.; Xu, D. A new synthetic inertia system based on electric vehicles to support the frequency stability of low-inertia modern power grids. *J. Clean. Prod.* **2021**, *297*, 126595. [CrossRef]
10. Nour, M.; Magdy, G.; Chaves-Ávila, J.P.; Sánchez-Miralles, Á.; Petlenkov, E. Automatic Generation Control of a Future Multisource Power System Considering High Renewables Penetration and Electric Vehicles: Egyptian Power System in 2035. *IEEE Access* **2022**, *10*, 51662–51681. [CrossRef]
11. Khooban, M.H.; Gheisarnejad, M. A Novel Deep Reinforcement Learning Controller Based Type-II Fuzzy System: Frequency Regulation in Microgrids. *IEEE Trans. Emerg. Top. Comput. Intell.* **2021**, *5*, 689–699. [CrossRef]
12. Kumar, A.; Kumari, N.; Shankar, G.; Elavarasan, R.M.; Kumar, S.; Srivastava, A.K.; Khan, B. Load Frequency Control of Distributed Generators Assisted Hybrid Power System Using QOHSA Tuned Model Predictive Control. *IEEE Access* **2022**, *10*, 109311–109325. [CrossRef]
13. Khooban, M.-H. Secondary Load Frequency Control of Time-Delay Stand-Alone Microgrids With Electric Vehicles. *IEEE Trans. Ind. Electron.* **2018**, *65*, 7416–7422. [CrossRef]
14. Paul, R.; Sengupta, A. Fractional Order Intelligent Controller for Single Tank Liquid Level System. In Proceedings of the 2021 IEEE Second International Conference on Control, Measurement and Instrumentation (CMI), Kolkata, India, 8–10 January 2021; pp. 24–29. [CrossRef]
15. Birs, I.; Muresan, C.; Nascu, I.; Ionescu, C. A Survey of Recent Advances in Fractional Order Control for Time Delay Systems. *IEEE Access* **2019**, *7*, 30951–30965. [CrossRef]
16. Shah, P.; Agashe, S. Review of fractional PID controller. *Mechatronics* **2016**, *38*, 29–41. [CrossRef]
17. Cao, J.Y.; Liang, J.; Cao, B.G. Optimization of Fractional Order PID controllers based on genetic algorithms. In Proceedings of the International Conference Machine and Learning Cybernetics ICMLC, Guangzhou, China, 18–21 August 2005; Springer: Berlin/Heidelberg, Germany, 2005; pp. 5686–5689.
18. Mosaad, M.I. Direct power control of SRG-based WECSs using optimised fractional-order PI controller. *IET Electr. Power Appl.* **2020**, *14*, 409–417. [CrossRef]

19. Ate¸s, A.; Yeroglu, C. Optimal fractional order PID design via Tabu Search based algorithm. *ISA Trans.* **2016**, *60*, 109–118. [CrossRef]
20. Kumar, M.R.; Deepak, V.; Ghosh, S. Fractional-order controller design in frequency domain using an improved nonlinear adaptive seeker optimization algorithm. *Turkish J. Electr. Eng. Comput. Sci.* **2017**, *25*, 4299–4310. [CrossRef]
21. Mirjalili, S.; Mirjalili, S.M.; Lewis, A. Grey Wolf Optimizer. *Adv. Eng. Softw.* **2014**, *69*, 46–61. [CrossRef]
22. Qais, M.H.; Hasanien, H.M.; Alghuwainem, S. A Grey Wolf optimizer for optimum parameters of multiple PI controllers of a grid connected PMSG driven by variable speed wind turbine. *IEEE Access* **2018**, *6*, 44120–44128. [CrossRef]
23. Padhy, S.; Panda, S. Application of a simplified Grey Wolf optimization technique for adaptive fuzzy PID controller design for frequency regulation of a distributed power generation system. *Prot. Control Mod. Power Syst.* **2021**, *6*, 2. [CrossRef]
24. Mohamed, E.A.; Ahmed, E.M.; Elmelegi, A.; Aly, M.; Elbaksawi, O.; Mohamed, A.-A.A. An Optimized Hybrid Fractional Order Controller for Frequency Regulation in Multi-Area Power Systems. *IEEE Access* **2020**, *8*, 213899–213915. [CrossRef]
25. Podlubny, I. *Fractional Differential Equations: An Introduction to Fractional Derivatives, Fractional Differential Equations, to Methods of Their Solution and Some of Their Applications*; Elsevier: Amsterdam, The Netherlands, 1999.
26. Morsali, J.; Zare, K.; Tarafdar Hagh, M. Applying fractional order PID to design TCSC-based damping controller in coordination with automatic generation control of interconnected multi-source power system. *Eng. Sci. Technol. Int. J.* **2017**, *20*, 1–17. [CrossRef]
27. Moschos, I.; Parisses, C. A novel optimal PIλDND2N2 controller using coyote optimization algorithm for an AVR system. *Eng. Sci. Technol. Int. J.* **2022**, *26*, 100991. [CrossRef]
28. Fliess, M.; Join, C. Model-free control, and intelligent pid controllers: Towards a possible trivialization of nonlinear control? *IFAC Proc. Vol.* **2009**, *42*, 1531–1550. [CrossRef]
29. Michel, L.; Join, C.; Fliess, M.; Sicard, P.; Cheriti, A. Model-free control of dc/dc converters. In Proceedings of the IEEE Workshop on Control and Modeling for Power Electronics (COMPEL), Boulder, CO, USA, 28–30 June 2010. [CrossRef]
30. Thabet, H.; Ayadi, M.; Rotella, F. Experimental comparison of new adaptive PI controllers based on the ultra-local model parameter identification. *Int. J. Control Autom. Syst.* **2016**, *14*, 1520–1527. [CrossRef]
31. Saxena, S. Load frequency control strategy via fractional-order controller and reduced-order modeling. *Int. J. Electr. Power Energy Syst.* **2019**, *104*, 603–614. [CrossRef]

fractal and fractional

Article

Driver Training Based Optimized Fractional Order PI-PDF Controller for Frequency Stabilization of Diverse Hybrid Power System

Guoqiang Zhang [1], Amil Daraz [1,2,*], Irfan Ahmed Khan [3], Abdul Basit [1,2], Muhammad Irshad Khan [4] and Mirzat Ullah [5]

1. School of Information Science and Engineering, NingboTech University, Ningbo 315100, China
2. College of Information Science and Electronic Engineering, Zhejiang University, Hangzhou 310027, China
3. Department of Electrical Engineering, Faculty of Engineering, Universiti Malaya, Federal Territory of Kuala Lumpur 50603, Malaysia
4. College of Electronics and Information Engineering, Nanjing University of Aeronautics and Astronautics (NUAA), Nanjing 210000, China
5. Graduate School of Economics and Management, Ural Federal University, Yekaterinburg 620002, Russia
* Correspondence: amil.daraz@nbt.edu.cn

Abstract: This work provides an enhanced novel cascaded controller-based frequency stabilization of a two-region interconnected power system incorporating electric vehicles. The proposed controller combines a cascade structure comprising a fractional-order proportional integrator and a proportional derivative with a filter term to handle the frequency regulation challenges of a hybrid power system integrated with renewable energy sources. Driver training-based optimization, an advanced stochastic meta-heuristic method based on human learning, is employed to optimize the gains of the proposed cascaded controller. The performance of the proposed novel controller was compared to that of other control methods. In addition, the results of driver training-based optimization are compared to those of other recent meta-heuristic algorithms, such as the imperialist competitive algorithm and jellyfish swarm optimization. The suggested controller and design technique have been evaluated and validated under a variety of loading circumstances and scenarios, as well as their resistance to power system parameter uncertainties. The results indicate the new controller's steady operation and frequency regulation capability with an optimal controller coefficient and without the prerequisite for a complex layout procedure.

Keywords: renewable energy resources; optimization techniques; fractional order controller; power system; load frequency control; heuristic techniques; driver training-based optimization

Citation: Zhang, G.; Daraz, A.; Khan, I.A.; Basit, A.; Khan, M.I.; Ullah, M. Driver Training Based Optimized Fractional Order PI-PDF Controller for Frequency Stabilization of Diverse Hybrid Power System. *Fractal Fract.* **2023**, *7*, 315. https://doi.org/10.3390/fractalfract7040315

Academic Editors: António Lopes, Behnam Mohammadi-Ivatloo and Arman Oshnoei

Received: 3 March 2023
Revised: 28 March 2023
Accepted: 2 April 2023
Published: 6 April 2023

1. Introduction

Electrical power has played a significant role in technological development for many years. The demand for electricity has greatly increased because of population growth and related technological advancements. Conventional, non-renewable energies led to energy sector installations in the past. However, because of their dearth and unfavorable effects on the environment, concerns are shifting away from these sources and toward the installation of renewable energy-based sources (RESs) [1]. To replace non-renewable supplies with RESs, such as wind energy, photovoltaic (PV) generation, biodiesel, etc., it is necessary to put more emphasis on sustainable development. Additionally, the use of energy storage devices to improve green energy-based power grids and the collaborative management of installed electric cars have drawn significant interest from researchers, businesses, and governmental incentives and regulations. They may contribute to maintaining the robustness and dependability of electricity grids [2]. Furthermore, by using modern single/multi-constraint optimization methods, such as stochastic optimization [3] and resilient optimization approaches [4], the performance of the power sector can be improved. Renewable-based power grids must overcome several obstacles, including intermittency, decreased inertia, irregular loading patterns, etc. The connectivity of grids powered by renewable energy is advantageous in several ways. However, renewable energies bring about unstable electricity grids that respond poorly to disturbances [5]. When compared to typical grids that are non-renewable-based, the poor inertial response is the main reason for power grid instability. The inability of photovoltaic and wind generation to sustain a significant inertial response results from their interaction with power interface converters, which restricts their ability to balance power demands [6]. Low inertial responses cause severely unbalanced power grids and lower flexibility of harmonic distortion in renewable-based power grids when renewable penetration level increases [7].

The literature contains several study recommendations for incorporating electrical vehicles (EVs) into the power system [8,9]. Green transportation has become a challenging issue with the current load equilibrium techniques, however, due to the complexity of managing a networked, multi-area system. The literature has suggested several integrated orders, predictive models, fuzzy logic controllers, neural networks, fractional orders, and advanced control systems as the best controllers for load frequency control (LFC) [10–12]. The tilt, derivative, proportional, integrator, and filter derivative have all been extensively linked in the literature to create several LFC systems. The PI regulator was introduced for EVs in [13]. However, stability issues with this controller exist, specifically when the time delay (TD) is taken into account. The filter-based tilt integral derivative controller for hybrid power networks has been optimized using the differential evolution algorithm, which was presented in [13]. The PI, TD, and filter controller parameters were combined to analyze the power networks in [14]. A hybrid approach using an updated form of particle swarm optimization (PSO) and the genetic algorithm was reported in [15] for establishing the controller employed to stabilize the frequency of power networks. An imperialist competitive search (ICA) method with a fractional order controller has been suggested in [16] for multi-generational networks. The stated controller can successfully enhance the performance of the power technique when there are several step variations in the production and/or loading. In two-area power networks, the FOPID and FLC are cascaded to accomplish frequency regulation [17]. Additionally, it has been suggested to use the grey wolf optimization algorithm to develop the load frequency controller multi-generation power networks [18].

The FOPID with FO filter was suggested by the authors in [19], and the SCA technique was utilized to successfully improve the controller parameters. The authors of [20] utilized an algorithm known as Harris hawk's optimization to design the P-I based LFC parameters in the best possible way. With the addition of capacitive energy storage, Daraz et al. exploited FOTIDN for multisource IPS while taking into account various non-linearities [21]. By using a hybrid of SCA and fitness-dependent algorithms, the parameters of the suggested method are changed. The authors in [22] used control EVs with TID controllers and optimized bee colony heuristics to change the settings of the suggested controller. The virtual inertia monitoring approach reported in [23] was expanded using PSO. In [24], an ultra-capacitor energy storage device has been developed to address AGC issues in connected PS. An improved design for the FOTID controller has also been provided using the path finder optimization technique [25]. Amil et al. recommended fine-tuned MFOPID/FOPID controllers for a hybrid system in [26], utilizing the jellyfish search algorithm. The authors in [27] proposed a different method of using the imperialist competitor optimizer to find the ideal settings of the second-order proposed controller for frequency stabilization systems. A modified tilt derivative with a filter controller based on fractional order is presented by Mohamed et al. in [28] and has been tuned using the artificial hummingbird optimizer technique. The salp swarm algorithm was introduced in [29] to tune the gains of PID controllers considering two area networks. Additionally, the dual-stage controller was developed in [30] using the butterfly optimization approach. A unique cascaded FO-ID with filter controller is suggested for AGC systems in PS with wind/solar/fuel systems in the study mentioned in [31].

It is now clear that the literature has a variety of LFC concepts that employ various optimization methods. The combination of the LFC-type and the selected optimizer greatly affects how well the power grid performs during transients. To lessen the projected loading impacts of RESs in future low-inertial grids, however, enhanced LFC method performance and design approaches are needed. This paper first introduces a cascaded structure, FOI, and PD with filter regulators in order to develop a revolutionary modified FO LFC method. From a different angle, their parameters need a lot of work to be adjusted. Several meta-heuristic optimization techniques lack reliability because of their greater inclination to settle at local minimums [32]. Correct tuning is also required for a variety of parameters, especially for FO-based LFC methods. The decision to optimize the parameters is therefore fraught with difficulty [33]. Extended delay times, exhaustion, sensitivity, and selectivity to parameter changes are other issues that certain optimizers face. Another issue with some optimizers is their lengthy processing periods, which require numerous iterations to ensure solution convergence. This study introduces driver training-based optimization (DTBO), a new stochastic optimization technique that imitates the human activity of driving training. The DTBO design was primarily influenced by how people learn to drive in driving schools and by instructor-training programs. Three stages of the proposed algorithm are mathematically modeled: (1) instruction from the driving coach, (2) modeling of student behavior after instructor techniques, and (3) practice. The effectiveness of DTBO is assessed using 23 common objective functions, including unimodal, multimodal, and IEEE CEC(2017) test function types [34]. The suggested algorithm has a number of benefits for difficult optimization challenges as well as its anticipated versatility in handling many types of optimization problems, given that many problems require more flexibility than DTBO can provide. Due to its mathematical foundation, this algorithm can be used to address a variety of engineering optimization problems, especially those with high dimensionality. Based on the inspiration given by the current gap in LFCs and their layout techniques, the study's main contributions are summarized below:

- For the connected PS taking into account electrical vehicles, a novel cascade structure of the proportional integral (PI)-proportional derivative with filter (PDF) is adopted.
- The proposed cascaded control structure is compared to a number of other control approaches, such as PIDF, PID, and PI controllers.
- The performance of the suggested LFC technique is enhanced using driver–teacher-based optimization (DTBO), which optimally selects the parameters of the suggested controller. The outcomes of DTBO are contrasted with those of other contemporary meta-heuristic algorithms, including the ICA and JSO.
- To ensure the viability of the system, a variety of non-linearities, such as time delay (TD), governor dead zone (GDZ), boiler dynamic (BD), and generation rate limitations (GRL), have been examined for the proposed hybrid power system.
- A synchronized participation of EVs with current-generating power units is offered using the proposed FOPI-PDF central controller.
- Finally, utilizing load changes of ±25% and ±50% and system parameters within a ±40% tolerance, the suggested cascaded controller's robustness is verified.

2. Power System Investigation

The suggested FOPI-PDF controller's design is shown in Figure 1, employing the two area-connected PS with the selected EVs and RESs. The RESs are placed in all of the areas, with solar energy in region 1 and wind energy in area 2. Area 1 comprises a reheat thermal plant, whereas area 2 holds the hydro generation unit. Furthermore, it is presumed that both regions have an equal distribution of EVs. The scheme is built in Matlab/Simulink using the PS information from [35], which is presented in Appendix A. Additionally, the physical limitations of PS, including GRL and GDZ, are taken into consideration by using the GRL rate (0.003 and 0.0017 pu/s), allowing for non-linearity and a more precise thermal unit analysis. Likewise, hydro power plants have a maximum production rate of 0.045 pu/s for increasing rates and 0.06 p.u. for declining rates [36–38].

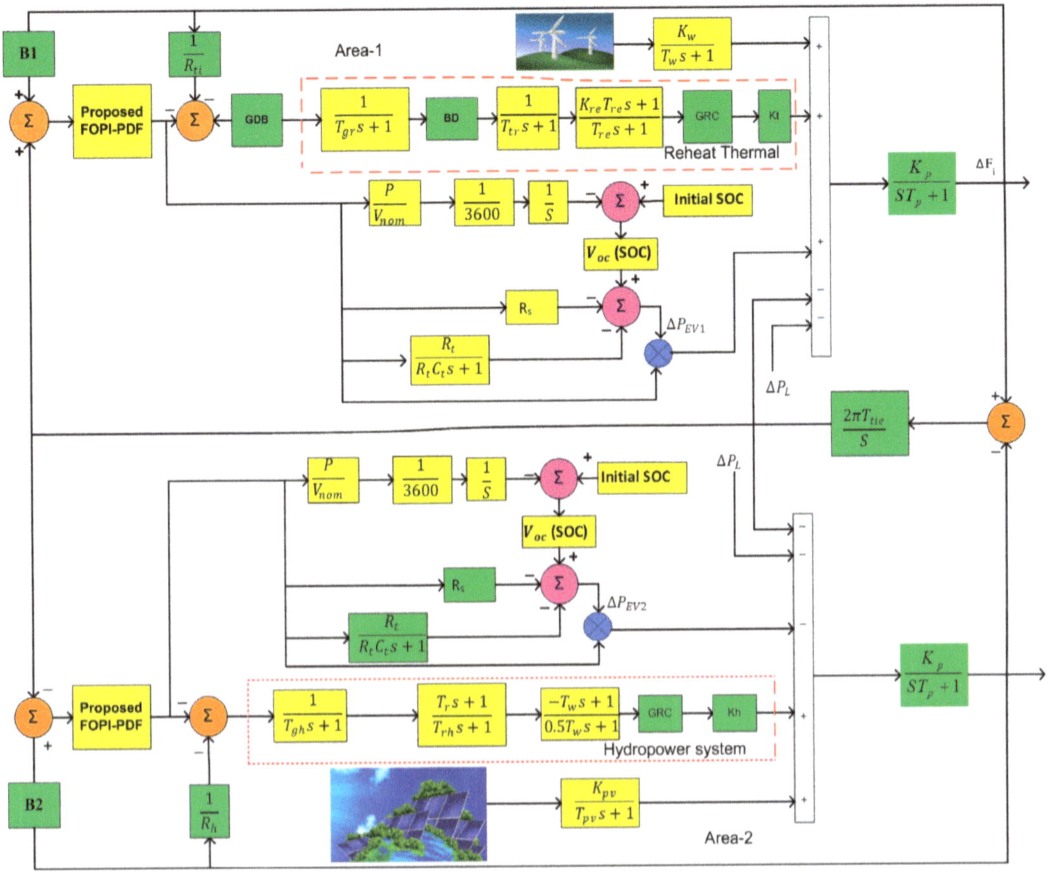

Figure 1. Transfer function model of hybrid power system.

The transfer function (TF) given in Equation (1) represents the governor dead zone (GDZ) with a margin of 0.50% [39].

$$\frac{\text{GDZ}}{\text{GDB}} = \frac{N_1 + N_2 s}{T_{sg}s + 1} \tag{1}$$

where $N_1 = 0.8$ and,

$$N_2 = \frac{-0.2}{\pi} \tag{2}$$

Time delay (TD) can influence controller implementation, which can amplify oscillations in the system. Consequently, this work contains a dynamic simulation that considers TD in the controller error field as well as various operational nonlinearities. Figure 2 denotes the transfer function typical for the BD. This paradigm can be used to assess both inefficiently managed gas/oil-fired power units as well as efficiently managed coal-fired power units. When the boiler regulator senses a change in pressure/steam flow rate, the pertinent controls are instantly initiated. This is how traditional steam power plants change their production. Equation (3) is an illustration of the TF boiler dynamics concept [39,40].

$$T_{cpu}(s) = \frac{K_{1b}(1 + T_{1b}s)(1 + T_{rb}s)}{(1 + 0.1T_{rb}s)s} \tag{3}$$

$$T_f(s) = \frac{e^{-t_d(s)}}{Ts + 1} \tag{4}$$

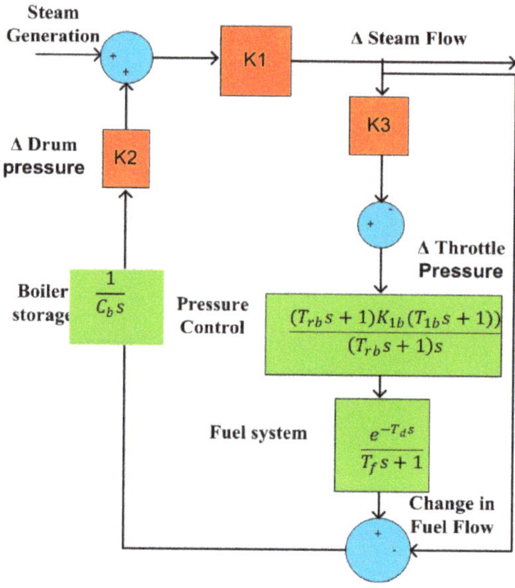

Figure 2. Drum type structure of boiler dynamics.

2.1. Modeling of Conventional Power Systems

The general TF model for the thermal reheat unit ($GT(s)$), which is represented by Equations (5)–(8) correspondingly, includes the reheat ($G_{T1}(s)$), turbine ($G_{T2}(s)$), and governor ($G_{T3}(s)$).

$$G_{T1}(s) = \frac{1 + T_{re}K_{re}s}{(1 + T_{re}s)} \tag{5}$$

$$G_{T2}(s) = \frac{1}{(1 + T_{tr}s)} \tag{6}$$

$$G_{T3}(s) = \frac{1}{(1 + T_{gr}s)} \tag{7}$$

$$G_T(s) = \frac{1 + T_{re}K_{re}s}{(1 + T_{gr}s)(1 + T_{re}s)(1 + T_{tr}s)} \tag{8}$$

Likewise, Equations (9)–(12), respectively, reflect the total TF of the hydropower system ($GH(s)$) in addition to the TF of the droop compensation ($G_{H1}(s)$), TF of the hydro governor ($G_{H2}(s)$), and TF of the penstock with turbine ($G_{H3}(s)$).

$$G_{H1}(s) = \frac{(1 - T_w s)}{(1 + 0.5T_w s)} \tag{9}$$

$$G_{H2}(s) = \frac{(1 + T_{rs}s)}{(1 + T_{rh}s)} \tag{10}$$

$$G_{H3}(s) = \frac{1}{\left(1 + T_{gh}s\right)} \tag{11}$$

$$G_H(s) = \frac{(1 - T_w s)(1 + T_{rs}s)}{\left(1 + T_{gh}s\right)(1 + 0.5T_w s)(1 + T_{rh}s)} \tag{12}$$

2.2. Renewable Energy Resources (RES,s) Modelling

The following models are used to express the $G_{PV}(s)$ of a solar energy system and $G_w(s)$ of a wind energy system [41]:

$$G_{PV}(s) = \frac{K_{PV}}{T_{PV}s + 1} \tag{13}$$

$$G_w(s) = \frac{K_T}{T_T s + 1} \tag{14}$$

where K_{pv} and T_{pv} stand for the PV plant's gain and time constant, respectively. Similarly, K_T and T_T stand for the wind farm's gain and time constant, respectively.

2.3. Modeling of EV Systems

The batteries of today's EVs may successfully regulate the PS performance. In response to electrical system management demands, they can be activated or deactivated. They might also increase the power system's reliability, efficiency, and dynamic response, among other things. Due to the fluctuating pattern of RESs and the associated electrical demands, one significant task of their use is the role of an EV in preserving the system stability of a PS. Figure 3 [42] displays the EV dynamical model that was used for the frequency response analysis in this paper.

The Nernst equation [42] is used in the model to illustrate the relationship between the linked EVs' open circuit voltage (Voc) and state of charge (SOC):

$$V_{oc}(SOC) = S\frac{RT}{F} \ln\left(\frac{SOC}{C_{nom} - SOC}\right) + V_{nom} \tag{15}$$

where C_{nom} and V_{nom} are the nominal capacities and voltages of the EV batteries, respectively. R stands for the gasoline constant, F for the Faraday constant, and T for temperature. S stands for the sensitivity parameter.

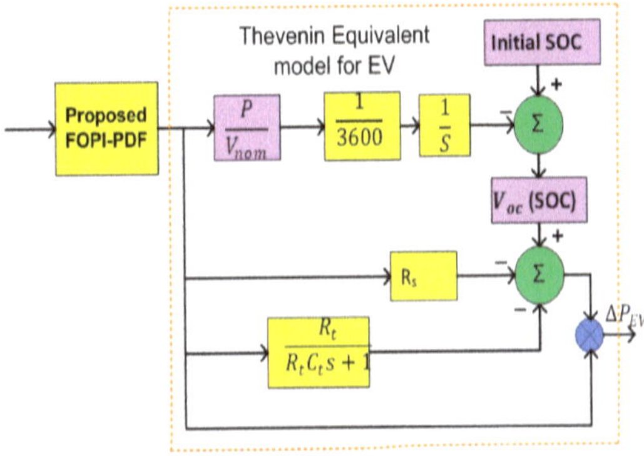

Figure 3. Dynamic model of EV system.

3. Driving Training Based Optimization (DTBO)

DTBO is a new stochastic optimization technique recently proposed in [34] that emulates the human action of driving guidance. The DTBO design was primarily influenced by how people learn to drive in driving schools and by instructor-training programs. Three stages of DTBO are mathematically modeled: (1) instruction from the driving coach, (2) modeling of student behavior after instructor techniques, and (3) practice. The effectiveness of DTBO is assessed using 23 common objective functions, including unimodal, multimodal, and IEEE CEC(2017) test function forms. The suggested DBOA has a number of benefits for difficult optimization challenges as well as its anticipated versatility in handling many types of optimization problems, given that many problems require more flexibility than DTBO can provide. Due to its mathematical foundation, DTBO can be used to address a variety of engineering optimization problems, especially those with high dimensionality. The detail of DTBO algorithm comprises of the subsequent steps:

3.1. Mathematical Representations of DTBO

Driving instructors and students make up the members of the population-based metaheuristic known as DTBO. Members of the DTBO are potential answers to the specified problem, which is depicted using a population matrix in Equation (16). Equation (17) is used to initialize these member positions at random at the beginning of implementation [34].

$$
X = \begin{bmatrix} x_{11} & \cdots & x_{ij} & \cdots & x_{im} \\ \vdots & \ddots & \vdots & \ddots & \vdots \\ x_{i1} & \cdots & x_{ij} & \cdots & x_{im} \\ \vdots & \ddots & \vdots & \ddots & \vdots \\ x_{N1} & \cdots & x_{Nj} & \cdots & x_{Nm} \end{bmatrix}_{N \times M} = \begin{bmatrix} X_1 \\ \vdots \\ X_i \\ \vdots \\ X_N \end{bmatrix}_{N \times M} \tag{16}
$$

$$
x_{i,j} = lb_j + \left(ub_j - lb_j\right) \times \mathrm{r}, \quad i = 1, 2, 3\ldots\ldots\ldots N, \quad J = 1, 2,\ldots, m \tag{17}
$$

where N is the population dimension, m denotes the problem of variables, r belongs to a random number between [0, 1], and ub_j and lb_j are the upper and lower bounds, respectively. X is the inhabitants of DTBO, x_i is the ith applicant solution, and $x_{i,j}$ is the value of the jth mutable represented by the ith applicant solution. The objective function's standards are modeled by the vector in Equation (18).

$$
F = \begin{bmatrix} F_1 \\ \vdots \\ F_i \\ \vdots \\ F_N \end{bmatrix}_{N \times 1} = \begin{bmatrix} F(X_1) \\ \vdots \\ F(X_i) \\ \vdots \\ F(X_N) \end{bmatrix}_{N \times 1} \tag{18}
$$

where F_i is the cost function provided by the ith applicant solution and F denotes the vector of the objective functions. Applicant solutions in DTBO are restructured during the following three steps: (i) beginner driver training by a driving tutor; (ii) beginner driver modeling using tutor skills; and (iii) learner driver rehearsal.

3.2. Phase 1: (Learner Driver Training by a Driving Instructor)

The trainee driver selects the driving instructor in the first phase of the DTBO update, and the instructor then instructs the learner driver in driving. The best members of the DTBO community are divided into trainee drivers and a limited group of driving instructors. Members of the population will go to various locations in the search space after selecting the driving teacher and mastering their techniques. This will strengthen the DTBO's investigation capabilities in the broad quest for and detection of the perfect region. As a result, this stage of the DTBO update illustrates the exploratory capabilities of this algorithm. The N memberships of the DTBO are chosen as driving tutors for an individual rehearsal based on an evaluation of the values of the cost function, as given in Equation (19).

$$
DI = \begin{bmatrix} DI_1 \\ \vdots \\ DI_i \\ \vdots \\ DI_{N_{DI}} \end{bmatrix}_{N_{DI} \times m} = \begin{bmatrix} DI_{11} & \cdots & DI_{1i} & \cdots & DI_{1m} \\ \vdots & \ddots & \vdots & \ddots & \vdots \\ DI_{i1} & \cdots & DI_{ij} & \cdots & DI_{im} \\ \vdots & \ddots & \vdots & \ddots & \vdots \\ DI_{N_{DI}1} & \cdots & DI_{N_{DI}j} & \cdots & DI_{N_{DI}m} \end{bmatrix}_{N_{DI} \times m} \tag{19}
$$

where $N_{DI} = \left[0.1 \cdot N \cdot \left(1 - \frac{t}{T}\right)\right]$ is the number of driving tutors, DI is the driving instructor matrix, DI_i is the ith driving teacher, $DI_{i,j}$ is the jth dimension, and T is the maximum number of iterations. The new location for each element in this DTBO phase is first determined using Equation (20) according to the mathematical modeling of this phase. Then, if the new position increases the value of the function, it replaces the old one in accordance with Equation (21).

$$
x_{i,j}^{PI} = \begin{cases} x_{i,j} + r \cdot \left(DI_{k_i,j} - I \cdot x_{i,j}\right), & FDI_{k_i,j} < F_i; \\ x_{i,j} + r \cdot \left(I \cdot x_{i,j} - DI_{k_i,j}\right), & Otherwise \end{cases} \tag{20}
$$

$$
X_i = \begin{cases} X_i^{PI}, & F_i^{PI} < F_i; \\ X_i, & Otherwise \end{cases} \tag{21}
$$

where I and r are random numbers chosen from the range $[0, 1]$ and $[1, 2]$, respectively. DI_{k_i} is arbitrarily selected from the range $[1, 2, ..., N_{DI}]$, that represents a driving instructor, $x_{i,j}^{PI}$ is its jth dimension, F is its objective function value, and X_i^{PI} is the new intended location for the ith applicant solution based on the first stage.

3.3. Phase-2 (Modeling of Student Behavior after Instructor Techniques)

The trainee driver imitates the instructor in this stage by trying to mimic all of the instructor's gestures and driving techniques. This method shifts DTBO participants to several locations within the quest space, boosting the DTBO's exploration capacity. A novel location is created based on the weighted sum of each participant with the teacher in accordance with Equation (22) to mathematically mimic this idea. According to Equation (23), the updated location will replace the prior one if it increases the objective function rate.

$$
x_{i,j}^{P2} = P \cdot x_{i,j} + r \cdot (I - P) \cdot DI_{k_i,j} \tag{22}
$$

$$
X_i = \begin{cases} X_i^{P2}, & F_i^{P2} < F_i; \\ X_i, & Otherwise \end{cases} \tag{23}
$$

where F_i^{P2} represents the objective function value, X_i^{P2} sis the updated position for i^{th} candidates, $x_{i,j}^{P2}$ represents its jth dimension while the pattern index (P) is denoted by below equation.

$$
P = 0.01 + 0.09(I - t/T) \tag{24}
$$

3.4. Phase 3 (Practice)

The third stage of the DTBO upgrade is based on each trainee driver's individual practice to strengthen and improve their driving abilities. In this stage, each novice driver aims to get a little bit closer to his best abilities. This phase is set up so that each participant can find a more advantageous position by conducting a local search near where they are currently located. The ability of DTBO to leverage confined pursuit is demonstrated in this step. This DTBO phase is precisely described so that, in accordance with Equation (25), a random position is initially created close to each population member. If this location increases the value of the goal function, Equation (26) states that it should take the place of the prior position.

$$x_{i,j}^{P3} = x_{i,j} + R \cdot (1 - 2r)\left(1 - \frac{t}{T}\right) \cdot x_{i,j} \tag{25}$$

$$X_i = \begin{cases} X_i^{P3}, & F_i^{P3} < F_i; \\ X_i, & Otherwise \end{cases} \tag{26}$$

where R is a constant with a value of 0.05. A DTBO iteration is finished after modifying the sample population in accordance with the first through third phases. The algorithm entered the following DTBO iteration with the modified population. Through the maximum number of repetitions, the update procedure is repeated during the mentioned phases and according to Equations (20)–(26). After DTBO has been applied to the provided problem, the best possible choice solution that was noted during execution is presented as the solution. Figure 4 shows the flowchart for the suggested DTBO approach.

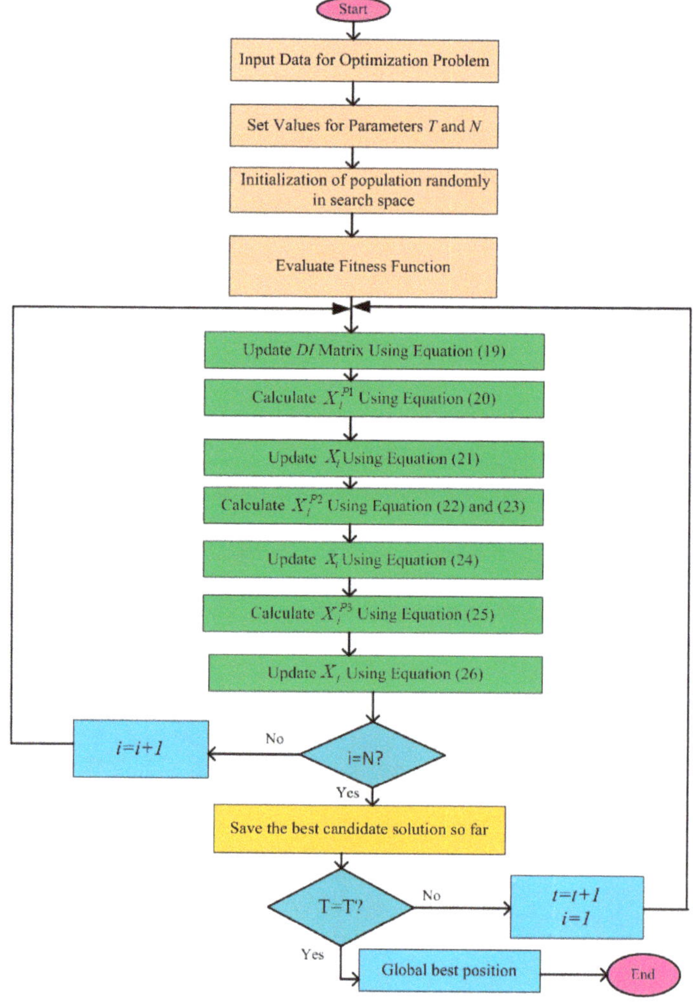

Figure 4. The flowchart for the suggested DTBO approach.

4. Proposed Control Structure and Fitness Function

Traditional PID control can improve controller stability and response time. However, because of the derivative mode, excessive control inputs are injected into the plant. The primary culprit in this problem is the noise that is already present in the control indicators. By including a filtering portion in the derivative part, the inserted noise is removed. The chattering noise can be reduced by fine-tuning the pole [43,44]. As a result, the FOPI-PDF is used in the proposed cascaded controller to improve the effectiveness of the control methodology by combining fractional order integer with proportional and the derivative filter. The transfer function of FOPI, PDF, and FOPIDF is depicted below:

$$C_1(s) \;=\; \frac{Y(s)}{R(s)} \;=\; K_p + \frac{K_i}{s^\lambda} \tag{27}$$

$$C_2(s) \;=\; \frac{Y(s)}{R(s)} \;=\; K_p + K_d\left[\frac{N_d s}{s + N_d}\right] \tag{28}$$

$$FOPIDF \;=\; \frac{Y(s)}{R(s)} \;=\; K_p + \frac{K_i}{s^\lambda} + K_d s^\mu \left[\frac{N_d s}{s + N_d}\right] \tag{29}$$

The schematic diagrams of the *FOPID, FOPI-PDF,* and combined controller structures are shown in Figure 5, Figure 6, and Figure 7, respectively. The proposed configuration has the capability to reduce the influence of turbulence on the control system's performance. Equation (30) could also be used to express the primary loop transfer function.

$$Y(s) \;=\; G(s)U(s) + d(s) \tag{30}$$

where $G(s)$ represents the execution and $U(s)$ represents the input pulse. Equation (31) can be used to calculate $U(s)$.

$$U(s) \;=\; C_1(s) \cdot C_2(s) \tag{31}$$

The cascaded (*FOPI-PDF*) controller gains will be ascertained by minimizing the cost function (*CF*) using the DTBO algorithm. The integral of time weighted by the squared error (*ITSE*) [4,26] is chosen as the CF because it can reduce time settling and overwhelm high oscillations quickly [30]:

$$ITSE \;=\; J \;=\; \int_0^t t\left[\Delta F_1^2 + \Delta F_2^2 + \Delta P_{tie}^2\right] dt \tag{32}$$

The following restrictions apply to the proposed FOI-PDN controller gains.

$$K_p^{Min} \le K_p \le K_p^{Max};\; K_d^{Min} \le K_d \le K_d^{Max};\; K_i^{Min} \le K_i \le K_i^{Max};\; \lambda^{Min} \le \lambda \le \lambda^{Max};\; N_d^{Min} \le N_d \le N_d^{Max};\; \mu^{Min} \le \mu \le \mu^{Max} \tag{33}$$

Several studies have shown that the Oustaloup recursive approximation (ORA) of FO derivatives can be implemented in real-time digitally [45]. It has become more familiar to the ORA with regard to the tuning processes involved with FO controllers. Since it is widely used in the literature in order to model the integrals and derivatives of FO, the ORA method has been used in this paper. In mathematical terms, the αth FO derivative (s^α) can be expressed as follows [45]:

$$s^\alpha \approx w_h^\alpha \prod_{K=-N}^{N} \frac{s + w_k^z}{s + w_k^p} \tag{34}$$

where w_k^z denotes the zeros and w_k^p denotes the poles, which can be represented by the below equations, respectively.

$$w_k^z \;=\; w_b\left(\frac{w_h}{w_b}\right)^{\frac{k+N+\frac{1-\alpha}{2}}{2N+1}} \tag{35}$$

$$\omega_h^{\alpha} = \left(\frac{\omega_h}{\omega_b}\right)^{\frac{-\alpha}{2}} \prod_{k=-N}^{N} \frac{\omega_k^p}{\omega_k^z} \tag{36}$$

The approximate FO operator's function has $(2N + 1)$ zeroes/ poles. ORA filter order is determined by the number N (order $= (2N + 1)$). This paper uses the ORA with ($M = 5$) and a frequency range ($\omega \in [\omega_h, \omega_b]$) of $[10^3, 10^{-3}]$ rad/s.

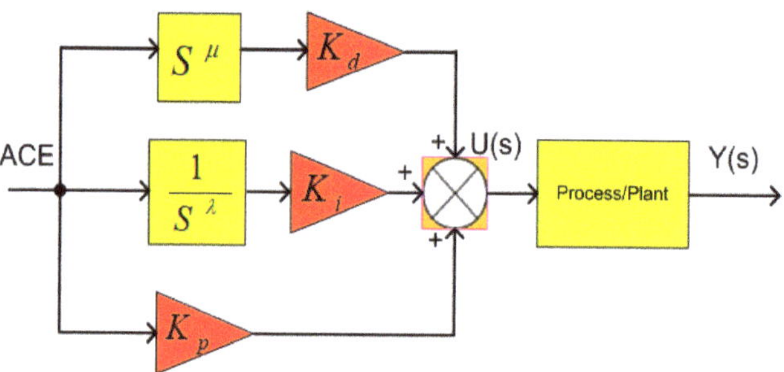

Figure 5. Design of FOPID controller.

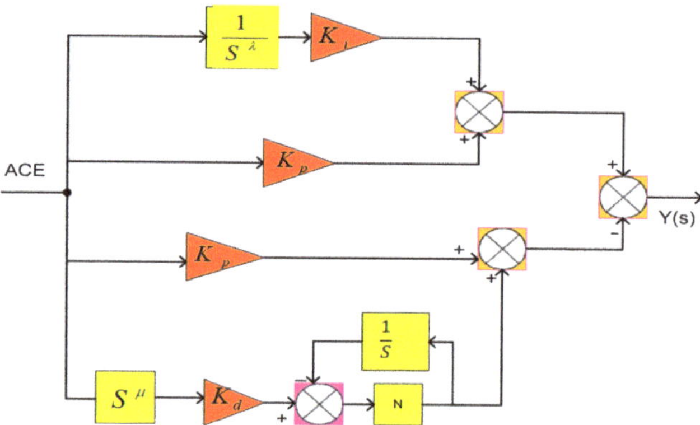

Figure 6. Design of FOPI-PDF controller.

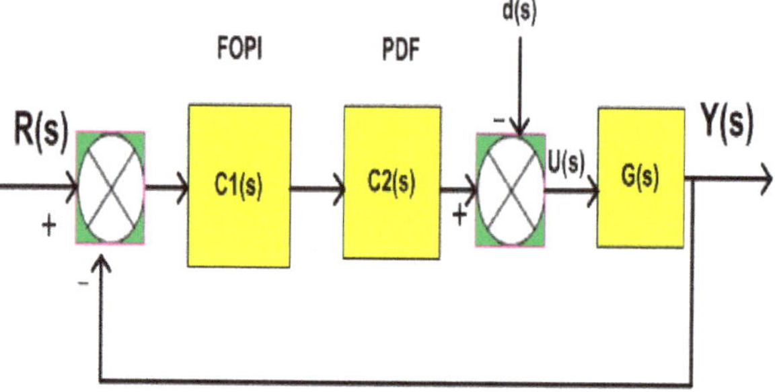

Figure 7. Cascaded form of controller.

5. Implementation, Results and Discussion

This part investigates the efficacy and validity of the unique FOPI-PDF controller implementation, depicted in Figure 1, in conjunction with EVs for enhancing IPS with the LFC problem. To ensure fairness, a newly suggested DTBO method was employed to tune the various control parameters of the FOPI-PDF and other controllers such as the FOPIDF, PI, and PID. The DTBO technique was constructed using the MATLAB program m-file code and linked up with the simulink mechanism of the researched interconnected PS to reach the LFC objective function. Table 1 shows the DTBO-based controller parameters for the given case study after running the optimization algorithms 15 times using the data from Appendix B. The robustness of the proposed FOPI-PDF controller is tested by comparing it to traditional and advanced controllers such as PID, PI, and FOPIDF, using the same alignment as the EV system that uses the DTBO approach. The per unit load change in each case is set at (5%) =0.05 p.u. The following case studies critically evaluate the results obtained from the analyzed multi-area IPS.

Table 1. Optimal values obtained for the proposed techniques.

Parameters	Case-1					Case-2	
	DTBO	JSO	ICA	FOPI-PDF	FOPIDF	PID	PI
K_{p1}	1.998	1.877	1.900	1.098	1.950	1.405	1.893
K_{i1}	1.678	1.458	0.400	1.878	1.340	1.012	1.032
K_{d1}	1.998	1.877	1.200	1.998	0.902	1.405	-
K_{p2}	0.345	0.123	1.145	1.889	-	-	-
λ_1	0.710	0.556	1.620	0.710	0.620	-	-
μ_1	0.671	0.601	1.863	0.671	0.823	-	-
N_1	8.678	3.234	9.972	8.678	9.972	-	9.899
K_{p3}	1.678	1.234	2.000	1.678	2.000	1.232	1.767
K_{d2}	1.998	1.877	1.405	1.998	1.989	1.405	-
K_{p4}	0.644	1.990	1.235	1.009	-	-	-
μ_2	0.710	0.456	0.620	0.710	0.620	-	-
λ_2	0.878	0.972	0.678	0.878	0.678	-	-
N_2	9.900	9.897	7.893	9.900	7.894	-	-

5.1. Case-1

In this case, the effectiveness of the DTBO approach was contrasted with the performances of the JSO, hDE-PS, ICA, and FPA algorithms. As shown in Figure 8a–c, the dynamic response for each optimization algorithm technique has been evaluated for the interconnected tie line (ΔP_{tie}), area 2 (ΔF2), and area 1 (ΔF1). Table 2 shows the overall comparison for (ΔF1), (ΔF2), and (ΔPtie) in terms of maximum overshoot (MO), minimum undershoot (MU), and settling time (ST). Figure 8a–c, demonstrates that the FOPI-PDF controller tuned with the DTBO approaches has improved STs for (ΔPtie) and (ΔF2) of 29.11% and 35.08%, respectively, but almost the same peak overshoot as the FOPI-PDF adjusted with the ICA approaches. Table 2 demonstrates that the DTBO method outperforms the JSO strategies for (ΔF1), (ΔF2), and (ΔPtie) in terms of ST (46.63%, 30.32%, and 14.11%) and MU (79.12%, 73.99%, and 90.00%). When compared to an JSO approach, the DTBO algorithm reduced peak overshoot by 70.11%, 78.12%, and 69.01% when taking into account (ΔF1), (ΔF2), and (ΔPtie), respectively. For the interconnected tie line (ΔPtie), area 2 (ΔF2), and area 1 (ΔF1), it is evident from Table 2 that our suggested DTBO algorithm outperforms JSO, ICA, hDE-PS [42], hTLBO with PS [10], and FPA [25] techniques.

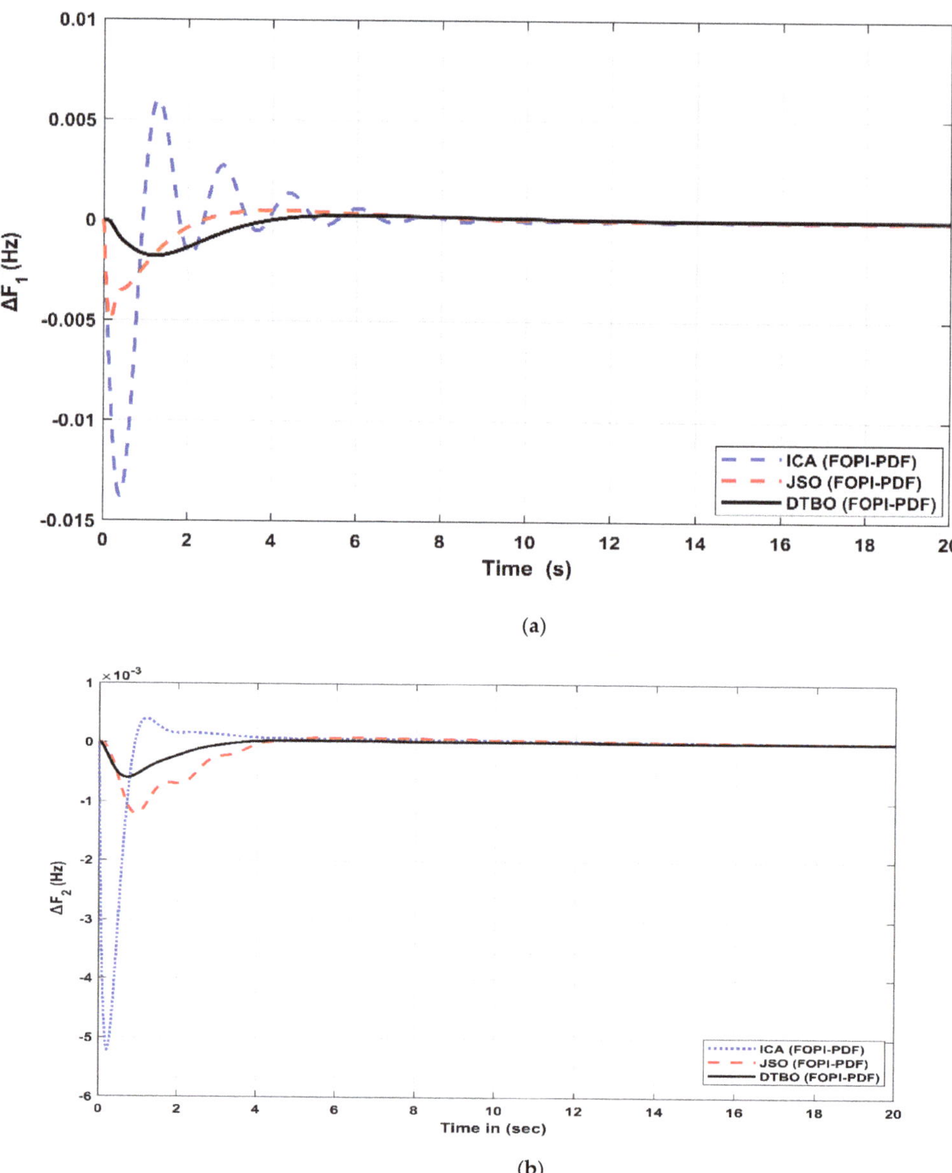

(a)

(b)

Figure 8. *Cont.*

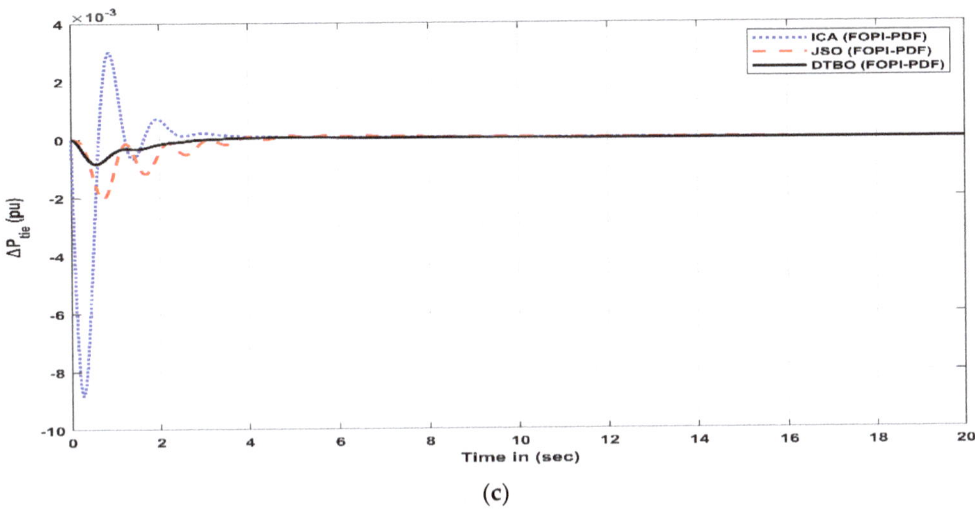

Figure 8. Dynamic response of the power system for case-1 (**a**) ΔF_1 (**b**) ΔF_2 (**c**) ΔP_{tie}.

Table 2. Transient results for PS considering Case-1.

Techniques	ST (Settling Time)			MO (Maximum Overshoot)			MU (Minimum Undershoot)		
	Area 1	Area 2	(ΔP_{tie})	Area 1	Area 2	(ΔP_{tie})	Area 1	Area 2	(ΔP_{tie})
DTBO: FOPI-PDF	8.23	3.93	2.96	0.000041	0.000272	0.000000	−0.00059	−0.00178	−0.00084
JSO: FOPI-PDF	8.09	9.13	7.83	0.000090	0.000509	0.000127	−0.00121	−0.00500	−0.00202
ICA: FOPI-PDF	10.4	6.44	4.68	0.000402	0.006035	0.003012	−0.00521	−0.01376	−0.00883
[10] hTLBO-PS	13.7	9.53	10.36	0.070400	0.007222	0.003500	−0.24010	−0.18888	−0.06330
[42] hDE-PS	19.0	18.09	12.69	0.00080	0.001700	0.000600	−0.00100	−0.01500	−0.00800
[25] FPA	25.5	23.2	18.77	0.00680	0.01170	0.00260	−0.02450	−0.02288	−0.00440

5.2. Case-2

In this case, the effectiveness of a FOPI-PDF controller using the DTBO technique was compared to the performances of FOPIDF, FOPID, PID, FOTID, and PI controllers. As shown in Figure 9a–c, the dynamic response for each controller has been evaluated for the interconnected tie line (ΔPtie), area 2 (ΔF2), and area 1 (ΔF1). Table 3 shows the overall comparison for various controllers in terms of transient contents, including MO, MU, and ST for (ΔF1), (ΔF2), and (ΔPtie). It is noticeable from Table 3 and Figure 9c that our suggested FOPI-PDF controller (MO = 0.000129, MU = −0.00065) has the least undershoot and overshoot as compared to FOPIDF (MO = 0.000218, MU = −0.00119), PID (MO = 0.000437, MU = −0.00627), PI (MO = 0.001045, MU = −0.00722), MID (MO = 0.000600, MU =−0.00800), and FOTID controller (MO = 0.00260, MU = −0.00440) for interconnected tie-line. It can also be seen from Table 3 and Figure 9c that FOPIDF controllers optimized with DTBO have the lowest settling time for area 1 (ST = 4.420), followed by PID controllers (ST = 5.020), PI controllers (6.533), FOPI-PDF controllers (ST = 8.434), MID controllers (ST = 19.01), and FOTID controllers (ST = 25.5). In a tie-line, the FOPI-PDF controller (ST = 5.98) is very excellent in terms of other controllers, including FOPIDF (ST = 12.60), PID (8.83), PI (ST = 6.82), MID (ST = 12.69), and FOTID (ST = 18.77). Therefore, it is evident from Figure 9c that the current described approach outperforms FOPIDF, PID, PI, and FOTID controllers in terms of ST, MO, and MU for interconnected tie-lines. From Figure 9b, it can also be observed that the PID controller tuned with the DTBO algorithm has superior performance (ST = 6.23) as compared to the FOPIDF controller with (ST = 8.61), the PI controller with (ST = 9.93), the FOPI-PIDF controller with (ST = 10.9), the MID controller with (ST = 18.09), and the FOTID controller with (ST = 23.2).

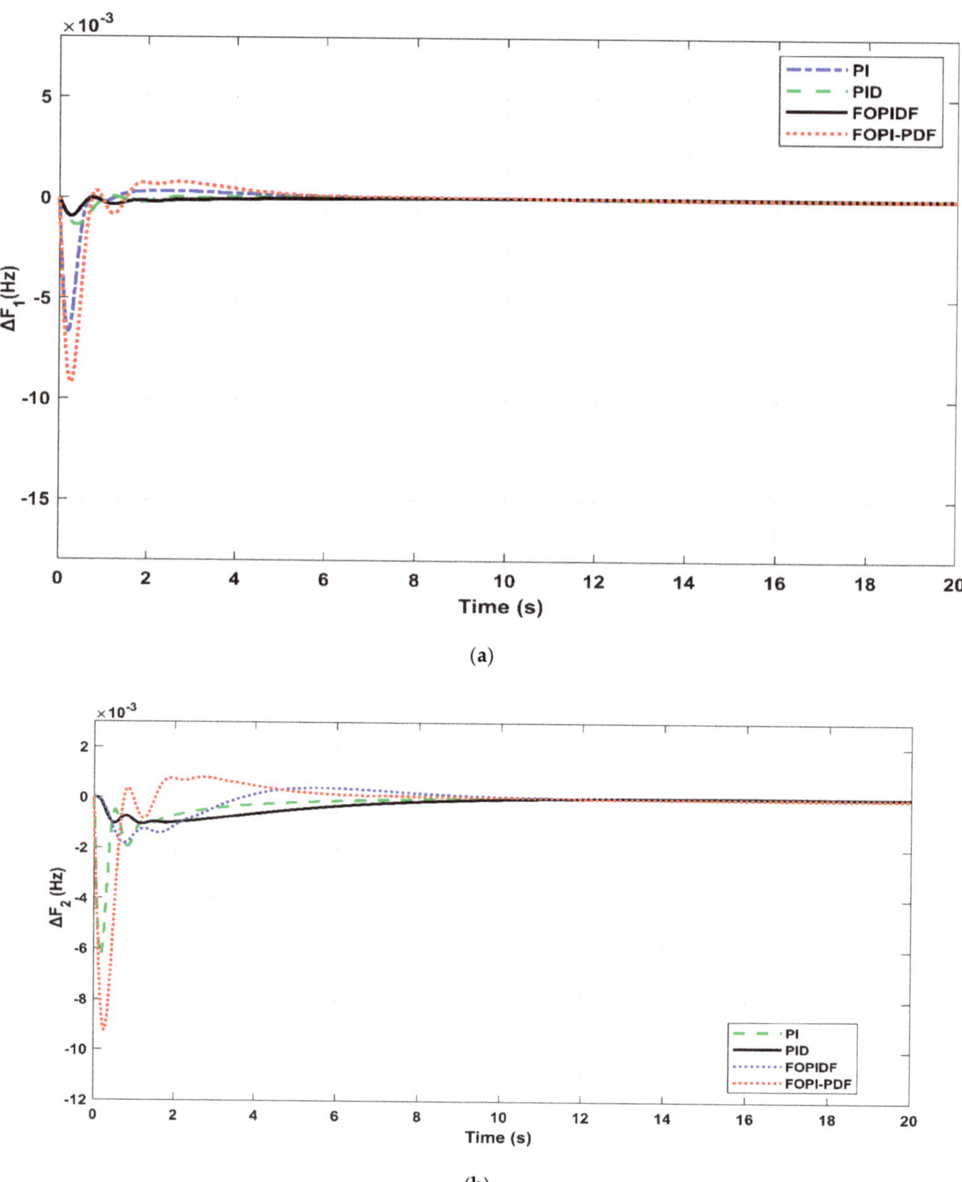

Figure 9. *Cont.*

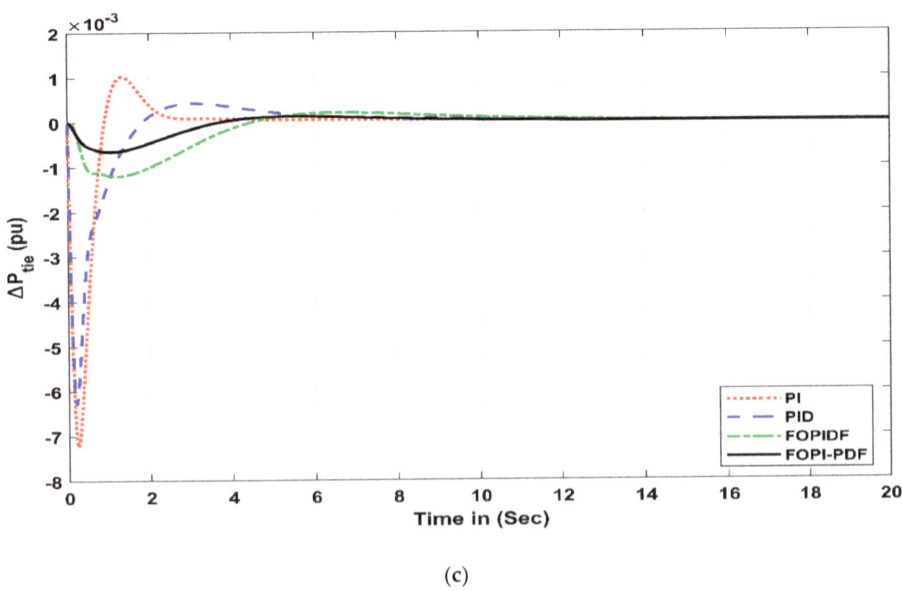

(c)

Figure 9. Dynamic response of the PS for Case-2 (**a**) ΔF_1 (**b**) ΔF_2 (**c**) ΔP_{tie}.

Table 3. Transient results for hybrid PS considering Case-2.

Controllers	ST (Settling Time)			MO (Maximum Overshoot)			MU (Minimum Undershoot)		
	Area 1	Area 2	(ΔP_{tie})	Area 1	Area 2	(ΔP_{tie})	Area 1	Area 2	(ΔP_{tie})
FOPI-PDF: DTBO	8.434	10.9	5.98	0.000813	0.000813	0.000129	−0.00922	−0.00922	−0.00065
FOPIDF: DTBO	4.420	8.61	12.6	0.000082	0.000406	0.000218	−0.00135	−0.00179	−0.00119
PID: DTBO	5.020	6.23	8.83	0.000363	0.000048	0.000437	−0.00664	−0.00628	−0.00627
PI:DTBO	6.533	9.93	6.82	0.000017	0.000041	0.001045	−0.00094	−0.00104	−0.00722
[42] MID: hDE-PS	19.01	18.09	12.69	0.00080	0.001700	0.000600	−0.00100	−0.01500	−0.00800
[25] FOTID:FPA	25.5	23.2	18.77	0.00680	0.01170	0.00260	−0.02450	−0.0228	−0.00440

5.3. Case-3

As shown in Figure 10a–c, the convergence curves of various algorithms, including DTBO, ICA, and JSO, have been assessed for hybrid interconnected PS in this case. Using the ITSE assessments as a cost function, the suggested FOPI-PDF controller parameters are fine-tuned. The DTBO parameters listed in Appendix A were selected to yield the best possible controller improvements. There are 30 simulated runs with 80 iterations, and the rest of the parameters are detailed in Appendix B. Each optimization method uses 20 populations. As can be seen in Figure 10a–c, the suggested DTBO optimization procedure outperforms the investigated JSO and ICA optimizers in terms of conversion characteristics for ITSE objective functions. Figure 10a–c demonstrates that, in comparison to JSO and ICA, whose ITSE values are 8.27×10^{-4} and 5.92×3, respectively, the DTBO method converges quickly under ITSE situations and obtains a value of (ITSE = 6.83×10^{-4}).

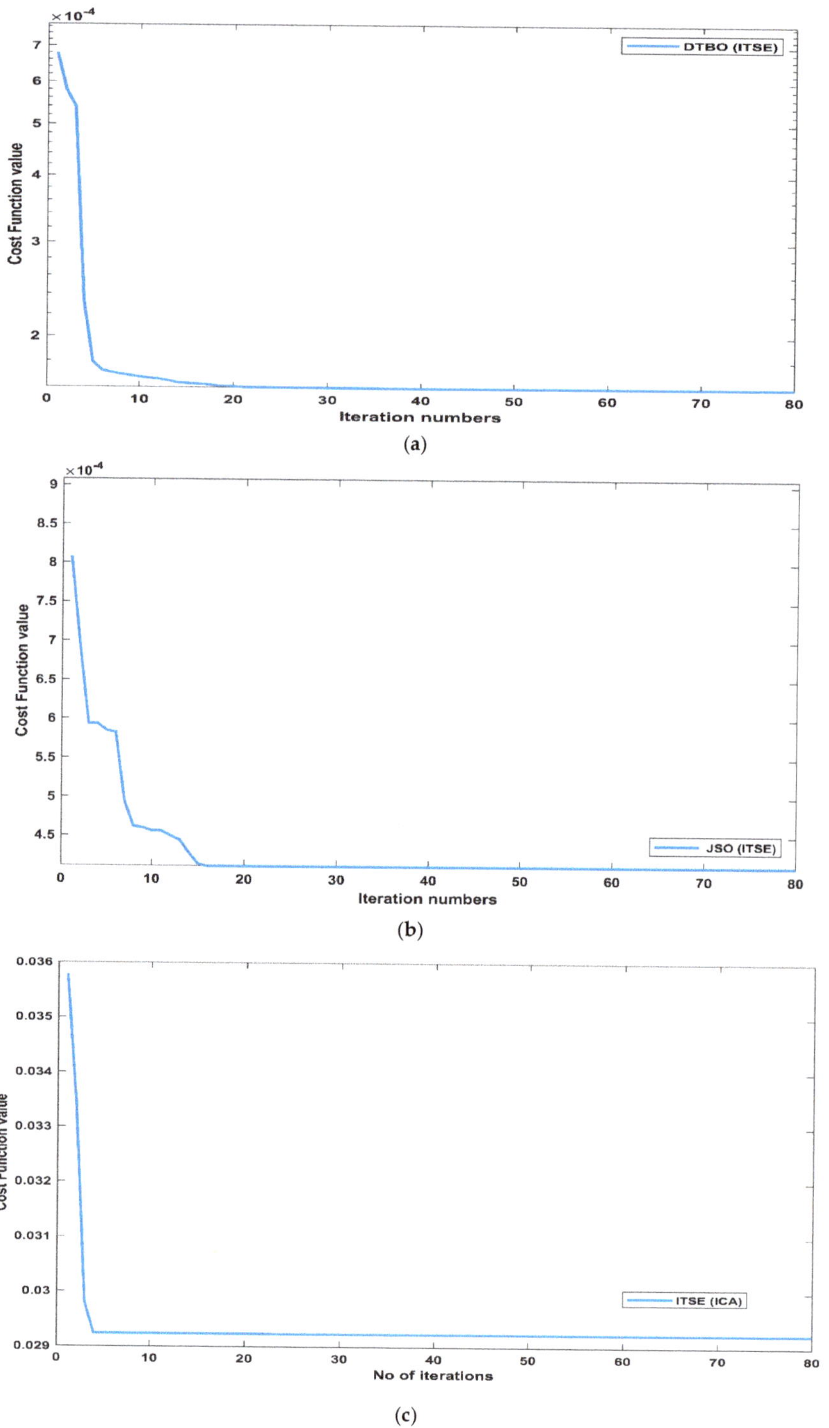

Figure 10. Convergence characteristics curve for algorithms (**a**) DTBO (**b**) JSO (**c**) ICA.

5.4. Sensitivity Analysis/Rubustness

Although system models can be described mathematically in a variety of ways, and because system parameters and configuration might vary over time as a result of the deterioration of system components, the given controller must be robust in the face of parameter uncertainties. Parametric uncertainties in the system can occasionally disrupt stability when the proposed control structure is unable to account for them. Parameters such as Kw, R, Kre, and Tgr are all varied by roughly ±40% from their nominal values and compared to their minimal responses in order to verify the robustness of the proposed controller. Figure 11a–c displays validation of the DTBO: FOPI-PDF controller performance under varying load disturbances up to 25% and ±50%, representing real-world circumstances. Results obtained with varying system parameters are shown in Figure 12 and Table 4, proving the proposed controller's robustness in the face of parameter uncertainty. Furthermore, the load characteristics of a real-world power system are highly unpredictable and varied. The mechanism of control needs to be flexible enough to handle unpredictable changes in load. Consequently, the proposed controller is resilient under a wide range of loads. As can be seen in Table 4, the actual system response is quite close to the nominal values for several parameters. The results show that the proposed DTBO-based FOPI-PDF controller consistently executes within a ± 40% tolerance band for the PS parameters. Furthermore, for a large variety of parameters at the rated value, the suggested controller's optimal values do not necessitate retuning.

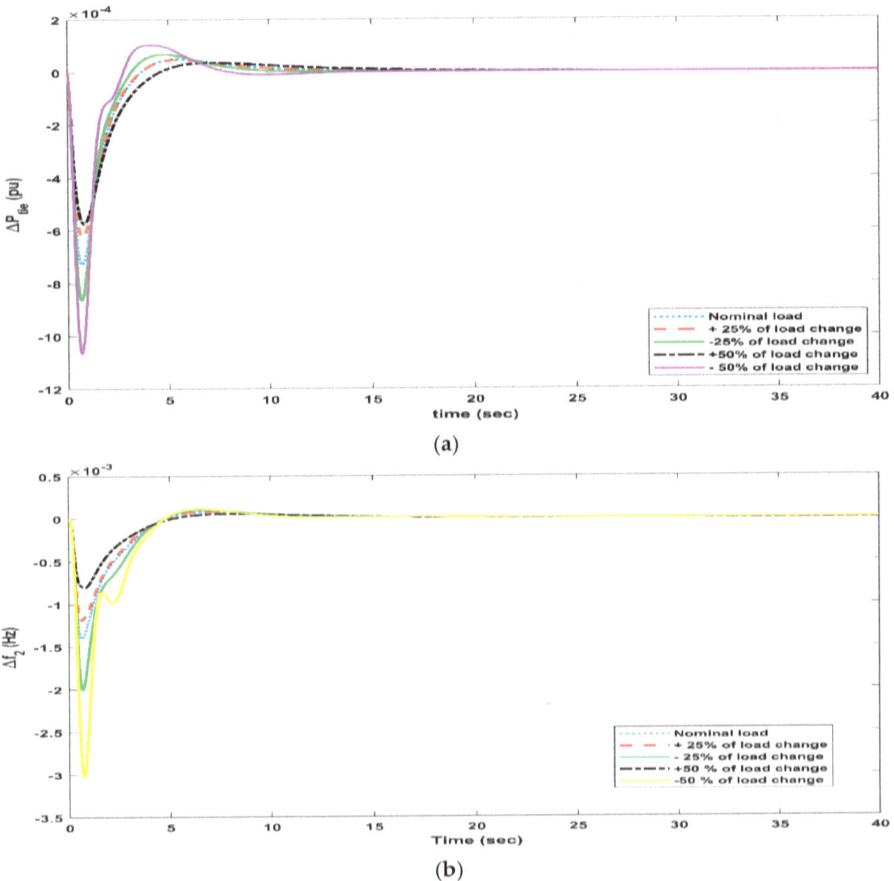

Figure 11. *Cont.*

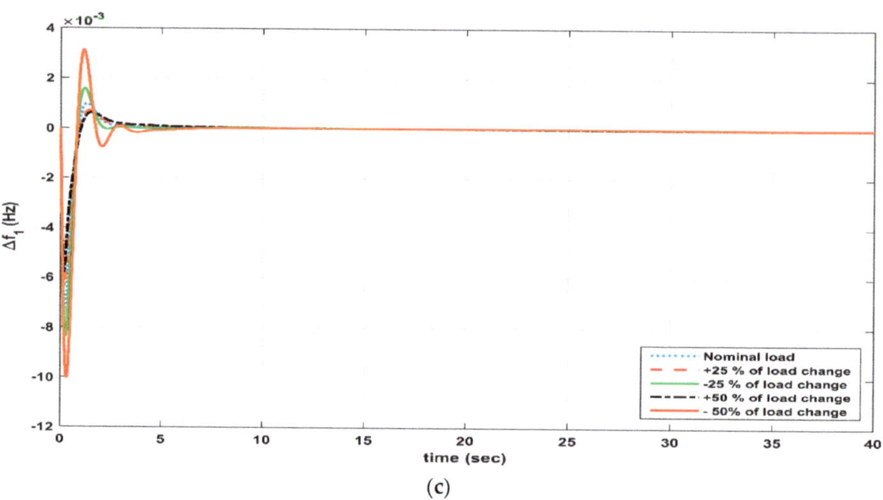

(c)

Figure 11. Different load change for the system considering (**a**) ΔPtie (**b**) ΔF2 (**c**) ΔF3.

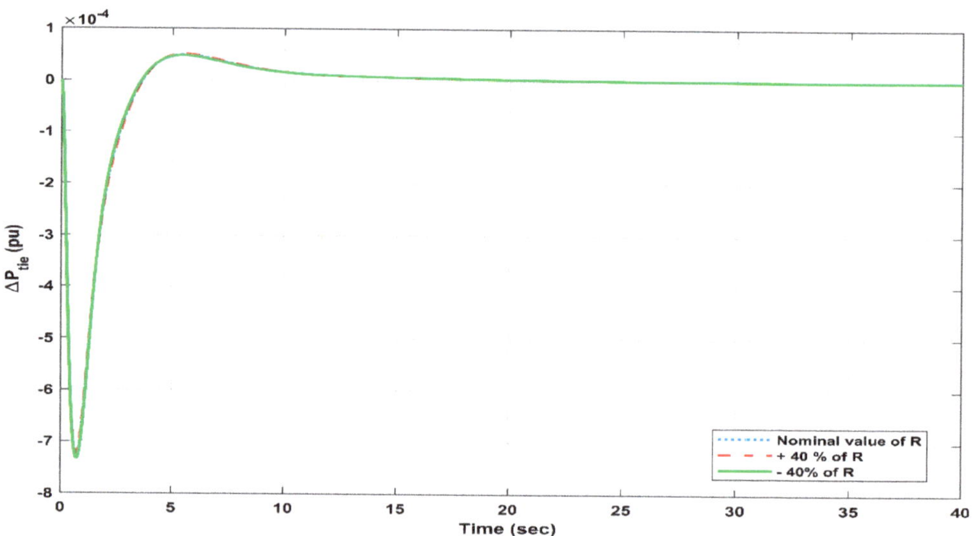

Figure 12. Sensitivity analysis for the system parameters.

Table 4. Transient response computation for change in parameters of the power system.

Parameter	% Change	ST			MO			MU		
		Area 1	Area 1	ΔP_{tie}	Area 1	Area 1	(ΔP_{tie})	Area 1	Area 1	(ΔP_{tie})
K_w	+40	6.09	13.23	14.89	0.00031	0.00032	0.00063	-0.00251	-0.00830	-0.00623
	-40	7.82	13.23	14.90	0.00031	0.00031	0.00061	-0.00257	-0.00840	-0.00618
K_{re}	+40	6.38	13.45	14.21	0.0002	0.00037	0.00094	-0.00489	-0.00713	-0.00693
	-40	8.03	13.46	14.23	0.0002	0.00030	0.00098	-0.00482	-0.00913	-0.00678
R	+40	6.10	12.79	14.60	0.0003	0.00014	0.00083	-0.00361	-0.00780	-0.00731
	-40	7.80	12.80	14.61	0.0003	0.00017	0.00075	-0.00361	-0.00740	-0.00725
T_{gr}	+40	3.47	12.72	14.09	0.0003	0.00068	0.00064	-0.00315	-0.00240	-0.00610
	-40	3.51	12.73	14.10	0.0002	0.00047	0.00054	-0.00313	-0.00236	-0.00600

6. Conclusions

The proposed FOPI-PDN controller for the LFC of two regions, hybrid renewable energies and conventional power sources, with the incorporation of numerous nonlinearities including GDZ, GRL, TD, and BD, was investigated in this research work. The Driver Training Based Optimization (DTBO), an advanced stochastic meta-heuristic algorithm, was used to optimize the settings of the recommended controller. The simulation results show that the DTBO-based tuned FOPI-PDF controller successfully decreases peak overshoot by 89.12%, 83.11%, and 78.10% for area-2, area-1, and link power variation, respectively, while delivering a minimum undershoot of 79.12%, 73.99%, and 90.00% for both areas and link power. Similarly, as compared to the conventional controller, the DTBO-based FOPI-PDF controllers improve the ST by 46.63%, 30.32%, and 14.11% for the load frequencies (ΔF1), (ΔF2), and (ΔPtie), respectively. Finally, the FOPI-PDF controller resilience is tested by deviating from the minimal values for the system parameters. The results show that when the system coefficients or load conditions change, the suggested controller gains are not reset. The efficiency of the DTBO-based FOPI-PDF controller shows that it can successfully manage LFC difficulties in hybrid power systems with protracted oscillations. In the future, the proposed control scheme could be extended to include three or more areas as well as regulation of the combined effect of frequency and voltage for multigeneration interconnected renewable/non-renewable power systems.

Author Contributions: Conceptualization, G.Z. and A.B.; data curation, A.D. and M.I.K.; formal analysis, A.B.; investigation, I.A.K. and M.I.K.; methodology, M.U.; project administration, G.Z.; resources, I.A.K. and M.U.; software, A.D. and A.B.; supervision, G.Z.; validation, G.Z.; writing—original draft, A.D.; writing—review & editing, A.D., I.A.K., M.I.K. and M.U. All authors have read and agreed to the published version of the manuscript.

Funding: This work is supported by the "Young Talent Sub-project of Ningbo Yongjiang Talent Introduction Programme under grant no. 20100859001."

Data Availability Statement: No new data were analyzed or generated in this study. Data sharing is not appropriate to this manuscript.

Conflicts of Interest: The authors declare no conflict of interest.

Appendix A

Table A1. Hybrid PS and Their Parametric Values [27,42,44].

LFC model			
Parameter	Value	Parameter	Value
T_{ps1}	11.49	Kps1	68.97
R_H	2.4	Kps2	68.97
T_{ps2}	11.49	$\beta 2$	0.4312
R_T	2.4	B1	2.4
Reheat Thermal PS			
K_t	0.54367	T_{tr}	0.3
T_{re}	10	K_{re}	0.3
T_{gr}	0.08		
Parameters and their values for Electric Vehicles			
V_{nom}	364.8	C_{nom}	66.2
R_s	0.074	R_t	0.047
C_t	703.6	RT/F	0.02612
Minimum SOC (in Percentage)	10	Maximum SOC (in Percentage)	95
C_{Batt}	24.15		

Table A1. *Cont.*

Hydro Power System			
T_w	1	T_{rh}	28.749
Kh	0.32586	Tr	5
T_{gh}	0.2		
Renewable energy resources			
Ks	0.5	K_T	1
Ts	1	T_T	0.3
K_{WTG}	1	T_{WTG}	1.5
Boiler Dynamic			
Cb	200	K3	0.92
Trb	0.545	Tf	0.23
Tr	1.4	T_{rh}	28.75
K1	0.85	K2	0.095
T_{1b}	0.545	K1b	0.950

Appendix B

Table A2. DTBO Coefficient and Their Values.

Coefficient	Values	Coefficient	Values	Coefficient	Values	Coefficient	Values
No of Iteration	80	Lower limit (Lb)	−2	No of dimension	7	Coefficient	2
No of Population (Np)	30	Constant (R)	0.05	Random Number (r)	[0, 1]	Coefficient	

References

1. Jan, M.U.; Xin, A.; Abdelbaky, M.A.; Rehman, H.U.; Iqbal, S. Adaptive and Fuzzy PI Controllers Design for Frequency Regulation of Isolated Microgrid Integrated With Electric Vehicles. *IEEE Access* **2020**, *8*, 87621–87632. [CrossRef]
2. Hassan, A.; Aly, M.; Elmelegi, A.; Nasrat, L.; Watanabe, M.; Mohamed, E.A. Optimal Frequency Control of Multi-Area Hybrid Power System Using New Cascaded TID-PI$^\lambda$D$^\mu$N Controller Incorporating Electric Vehicles. *Fractal Fract.* **2022**, *6*, 548. [CrossRef]
3. Xiao, D.; Chen, H.; Wei, C.; Bai, X. Statistical Measure for Risk-Seeking Stochastic Wind Power Offering Strategies in Electricity Markets. *J. Mod. Power Syst. Clean Energy* **2021**, *10*, 1437–1442. [CrossRef]
4. Gulzar, M.M.; Murawwat, S.; Sibtain, D.; Shahid, K.; Javed, I.; Gui, Y. Modified Cascaded Controller Design Constructed on Fractional Operator 'β' to Mitigate Frequency Fluctuations for Sustainable Operation of Power Systems. *Energies* **2022**, *15*, 7814. [CrossRef]
5. Zhang, P.; Daraz, A.; Malik, S.A.; Sun, C.; Basit, A.; Zhang, G. Multi-resolution based PID controller for frequency regulation of a hybrid power system with multiple interconnected systems. *Front. Energy Res.* **2023**, *10*, 1109063. [CrossRef]
6. Arya, Y. Effect of electric vehicles on load frequency control in interconnected thermal and hydrothermal power systems utilizing CFFOIDF controller. *IET Gener. Transm. Distrib.* **2020**, *14*, 2666–2675. [CrossRef]
7. Zaid, S.A.; Bakeer, A.; Magdy, G.; Albalawi, H.; Kassem, A.M.; El-Shimy, M.E.; AbdelMeguid, H.; Manqarah, B. A New Intelligent Fractional-Order Load Frequency Control for Interconnected Modern Power Systems with Virtual Inertia Control. *Fractal Fract.* **2023**, *7*, 62. [CrossRef]
8. Jia, H.; Li, X.; Mu, Y.; Xu, C.; Jiang, Y.; Yu, X.; Wu, J.; Dong, C. Coordinated control for EV aggregators and power plants in frequency regulation considering time-varying delays. *Appl. Energy* **2018**, *210*, 1363–1376. [CrossRef]
9. Arias, N.B.; Hashemi, S.; Andersen, P.B.; Træholt, C.; Romero, R. Assessment of economic bene_ts for EV owners participating in the primary frequency regulation markets. *Int. J. Electr. Power Energy Syst.* **2020**, *120*, 105985. [CrossRef]
10. Khamari, D.; Sahu, R.K.; Gorripotu, T.S.; Panda, S. Automatic generation control of power system in deregulated environment using hybrid TLBO and pattern search technique. *Ain Shams Eng. J.* **2019**, *11*, 553–573. [CrossRef]
11. Vrdoljak, K.; Peri´c, N.; Petrovi´c, I. Sliding modelbased load-frequency control in power systems. *Electr. Power Syst. Res.* **2010**, *80*, 514–527. [CrossRef]
12. Pan, C.; Liaw, C. An adaptive controller for power system load-frequency control. *IEEE Trans. Power Syst.* **1989**, *4*, 122–128. [CrossRef]

13. Arya, Y.; Kumar, N. BFOA-scaled fractional order fuzzy PID controller applied to AGC of multi-area multi-source electric power generating systems. *Swarm Evol. Comput.* **2017**, *32*, 202–218. [CrossRef]
14. Sahu, R.K.; Panda, S.; Biswal, A.; Sekhar, G.C. Design and analysis of tilt integral derivative controller with filter for load frequency control of multi-area interconnected power systems. *ISA Trans.* **2016**, *61*, 251–264. [CrossRef]
15. Malik, S.; Suhag, S. A Novel SSA Tuned PI-TDF Control Scheme for Mitigation of Frequency Excursions in Hybrid Power System. *Smart Sci.* **2020**, *8*, 202–218. [CrossRef]
16. Elmelegi, A.; Mohamed, E.A.; Aly, M.; Ahmed, E.M.; Mohamed, A.A.A.; Elbaksawi, O. Optimized Tilt Fractional OrderCooperative Controllers for Preserving Frequency Stability in Renewable Energy-Based Power Systems. *IEEE Access* **2021**, *9*, 8261–8277. [CrossRef]
17. Arya, Y. A new optimized fuzzy FOPI-FOPD controller for automatic generation control of electric power systems. *J. Frankl. Inst.* **2019**, *356*, 5611–5629. [CrossRef]
18. Paliwal, N.; Srivastava, L.; Pandit, M. Application of grey wolf optimization algorithm for load frequency control in multi-source single area power system. *Evol. Intell.* **2020**, *15*, 563–584. [CrossRef]
19. Ayas, M.S.; Sahin, E. FOPID controller with fractional filter for an automatic voltage regulator. *Comput. Electr. Eng.* **2021**, *90*, 106895. [CrossRef]
20. Yousri, D.; Babu, T.S.; Fathy, A. Recent methodology-based Harris Hawks optimizer for designing load frequency control incorporated in multi-interconnected renewable energy plants. *Sustain. Energy Grids Netw.* **2020**, *22*, 100352. [CrossRef]
21. Daraz, A.; Malik, S.A.; Azar, A.T.; Aslam, S.; Alkhalifah, T.; Alturise, F. Optimized Fractional Order Integral-Tilt Derivative Controller for Frequency Regulation of Interconnected Diverse Renewable Energy Resources. *IEEE Access* **2022**, *10*, 43514–43527. [CrossRef]
22. Oshnoei, A.; Khezri, R.; Muyeen, S.M.; Oshnoei, S.; Blaabjerg, F. Automatic Generation Control Incorporating Electric Vehicles. *Electr. Power Components Syst.* **2019**, *47*, 720–732. [CrossRef]
23. Magdy, G.; Bakeer, A.; Nour, M.; Petlenkov, E. A new virtual synchronous generator design based on the SMES system for frequency stability of low-inertia power grids. *Energies* **2020**, *13*, 5641. [CrossRef]
24. Arya, Y. Impact of ultra-capacitor on automatic generation control of electric energy systems using an optimal FFOID controller. *Int. J. Energy Res.* **2019**, *43*, 8765–8778. [CrossRef]
25. Priyadarshani, S.; Subhashini, K.R.; Satapathy, J.K. Path finder algorithm optimized fractional order tilt-integral-derivative (FOTID) controller for automatic generation control of multi-source power system. *Microsyst. Technol.* **2020**, *27*, 23–35. [CrossRef]
26. Daraz, A.; Malik, S.A.; Basit, A.; Aslam, S.; Zhang, G. Modified FOPID Controller for Frequency Regulation of a Hybrid Interconnected System of Conventional and Renewable Energy Sources. *Fractal Fract.* **2023**, *7*, 89. [CrossRef]
27. Arya, Y. Impact of hydrogen Aqua electrolyzer-fuel cell units on automaticgeneration control of power systems with a new optimal fuzzy TIDFII controller. *Renew. Energy* **2019**, *139*, 468–482. [CrossRef]
28. Mohamed, E.A.; Aly, M.; Watanab, M. New Tilt Fractional-Order Integral Derivative with Fractional Filter (TFOIDFF)Controller with Artificial Hummingbird Optimizer for LFC in Renewable Energy Power Grids. *Mathematics* **2022**, *10*, 3006. [CrossRef]
29. Hasanien, H.M.; El-Fergany, A.A. Salp swarm algorithm-based optimal load frequency control of hybrid renewable power systems with communication delay and excitation cross-coupling effect. *Electr. Power Syst. Res.* **2019**, *176*, 105938. [CrossRef]
30. Latif, A.; Hussain, S.M.S.; Das, D.C.; Ustun, T.S. Optimum synthesis of a BOA optimized novel dual-stage PI-(1 C ID) controller for frequency response of a microgrid. *Energies* **2020**, *13*, 3446. [CrossRef]
31. Arya, Y.; Kumar, N.; Dahiya, P.; Sharma, G.; Çelik, E.; Dhundhara, S.; Sharma, M. Cascade-IDN controller design for AGC of thermal and hydro-thermal power systems integrated with renewable energy sources. *IET Renew. Power Gener.* **2020**, *15*, 504–520. [CrossRef]
32. Khamies, M.; Magdy, G.; Selim, A.; Kamel, S. An improved Rao algorithm for frequency stability enhancement of nonlinearpower system interconnected by AC/DC links with high renewables penetration. *Neural Comput. Appl.* **2021**, *34*, 2883–2911. [CrossRef]
33. Ali, H.H.; Fathy, A.; Kassem, A.M. Optimal model predictive control for LFC of multi-interconnected plants comprising renewable energy sources based on recent sooty terns approach. *Sustain. Energy Technol. Assess.* **2020**, *42*, 100844. [CrossRef]
34. Dehghani, M.; Trojovská, E.; Trojovský, P. A new human-based metaheuristic algorithm for solving optimization problems on the base of simulation of driving training process. *Sci. Rep.* **2022**, *12*, 9924. [CrossRef]
35. Gulzar, M.M.; Iqbal, A.; Sibtain, D.; Khalid, M. An Innovative Converterless Solar PV Control Strategy for a Grid Connected Hybrid PV/Wind/Fuel-Cell System Coupled with Battery Energy Storage. *IEEE Access* **2023**, *11*, 23245–23259. [CrossRef]
36. Yakout, A.H.; Kotb, H.; Hasanien, H.M.; Aboras, K.M. Optimal Fuzzy PIDF Load Frequency Controller for Hybrid Microgrid System Using Marine Predator Algorithm. *IEEE Access* **2021**, *9*, 54220–54232. [CrossRef]
37. Yousri, D.; Babu, T.S.; Beshr, E.; Eteiba, M.B.; Allam, D. A robust strategy based on marine predators algorithm for large scale photovoltaic array reconguration to mitigate the partial shading effect onthe performance of PV system. *IEEE Access* **2020**, *8*, 112407–112426.
38. Elkasem, A.H.A.; Kamel, S.; Hassan, M.H.; Khamies, M.; Ahmed, E.M. An Eagle Strategy Arithmetic Optimization Algorithm for Frequency Stability Enhancement Considering High Renewable Power Penetration and Time-Varying Load. *Mathematics* **2022**, *10*, 854. [CrossRef]
39. Ahmed, E.M.; Mohamed, E.A.; Elmelegi, A.; Aly, M.; Elbaksawi, O. Optimum Modified Fractional Order Controller for FutureElectric Vehicles and Renewable Energy-Based Interconnected Power Systems. *IEEE Access* **2021**, *9*, 29993–30010. [CrossRef]

40. Ali, T.; Malik, S.A.; Daraz, A.; Aslam, S.; Alkhalifah, T. Dandelion Optimizer-Based Combined Automatic Voltage Regulation and Load Frequency Control in a Multi-Area, Multi-Source Interconnected Power System with Nonlinearities. *Energies* **2022**, *15*, 8499. [CrossRef]
41. Mohamed, E.A.; Ahmed, E.M.; Elmelegi, A.; Aly, M.; Elbaksawi, O.; Mohamed, A.A.A. An Optimized Hybrid Fractional Order Controller for Frequency Regulation in Multiarea Power Systems. *IEEE Access* **2020**, *8*, 213899–213915. [CrossRef]
42. Sahu, R.; Gorripotu, T.; Panda, S. A hybrid DE-PS algorithm for load frequency control under deregulated power system withUPFC and RFB. *Ain Shams Eng. J.* **2015**, *6*, 893–911. [CrossRef]
43. Tasnin, W.; Saikia, L.C.; Raju, M. Deregulated AGC of multi-area system incorporating dish-Stirling solar thermal and geothermalpower plants using fractional order cascade controller. *Int. J. Electr. Power Energy Syst.* **2018**, *101*, 60–74. [CrossRef]
44. Ahmed, E.M.; Selim, A.; Alnuman, H.; Alhosaini, W.; Aly, M.; Mohamed, E.A. Modified Frequency Regulator Based on TI$^\lambda$-TD$^\mu$FF Controller for Interconnected Microgrids with Incorporating Hybrid Renewable Energy Sources. *Mathematics* **2023**, *11*, 28. [CrossRef]
45. Micev, M.; Calasan, M.; Oliva, D. Fractional Order PID Controller Design for an AVR System Using Chaotic Yellow Saddle Goatfish Algorithm. *Mathematics* **2020**, *8*, 1182. [CrossRef]

MDPI

St. Alban-Anlage 66

4052 Basel

Switzerland

www.mdpi.com

Fractal and Fractional Editorial Office

E-mail: fractalfract@mdpi.com

www.mdpi.com/journal/fractalfract